Amino Acids, Peptides, and Proteins

Volume 25

A Specialist Periodical Report

Amino Acids, Peptides, and Proteins

Volume 25

A Review of the Literature Published during 1992

Senior Reporter
J.S. Davies, *University College of Swansea*

Reporters
G.C. Barrett, *Oxford Brookes University*
C.M. Bladon, *University of Kent*
D.T. Elmore, *University of Oxford*
R.W. Hay, *University of St Andrews*
J.A. Littlechild, *University of Exeter*
K.B. Nolan, *The Royal College of Surgeons in Ireland*
A.A. Soudi, *The Royal College of Surgeons in Ireland*

ROYAL
SOCIETY OF
CHEMISTRY

ISBN 0-85186-234-9
ISSN 0269-7521

Published by the Royal Society of Chemistry
Thomas Graham House, Science Park, Cambridge CB4 4WF

Typeset by Computape (Pickering) Ltd, Pickering, North Yorkshire
Printed and bound in England by Athenaeum Press Ltd,
Newcastle upon Tyne

Preface

This series of Specialist Reports,originally named Amino Acids Peptides and Proteins has reached its Silver Anniversary Edition. Many aspects of this broad field of endeavour have undergone great changes over the past 25 years, and the volume of publications has continually expanded. By the time Volume 17 was reached the Senior Reporter (Dr.John Jones) stated that 'an exhaustive review of a year's output could no longer be contained in a single volume' and hence it was decided to contract the scope, and shift the emphasis of the title to Amino Acids and Peptides only. Even after this major surgery the literature continued to expand, but many yearned for coverage of the exciting developments which were also happening in the Protein field. Hence the great satisfaction this year in securing the assistance of Dr. Jennifer Littlechild to provide an overview in Chapter 5 of the very interesting current developments in Protein Research. The title of the series,in its silver anniversary edition, reverts to its inaugural form, and the sincere hope is that it will continue with this breadth of coverage.

Re-visiting the covers of Volume 1 in the series confirms the phenomenal changes which have occurred over the 25 years. In 1969 there were only 213 citations in the peptide synthesis section, with only about a dozen involving 'synthesis on a polymeric support'. In recent volumes, citations in this area usually top the 600 mark. In Volume 1 the Senior Reporter (Dr.Geoffrey Young) listed as the important advance of the year, the 'chemical synthesis of material having the enzymic activity of ribonuclease' which on reflection seems a good choice of words for those pre-hplc times. There are no citations in Volume 1 for, multi-dimensional nmr techniques, molecular dynamics, FAB-mass spectrometry, multiple peptide syntheses, custom-designed peptide surrogates, and the development of peptidomimetic drugs based on protease inhibition. These are now very much part of everyday usage in this field. International Peptide/Protein Symposia in Europe, America and Japan have no problems in attracting a thousand participants. Within this expansion the discipline has also undergone a change of emphasis from its chemical origins into more inter-disciplinary fields. The wide-ranging and demanding problems being approached by the peptide and protein fraternity really demand the multi-disciplinary approach.

What new horizons the next 25 years will reveal are difficult to predict. We are already experiencing a new thrust into glyco-peptides and -proteins, a greater understanding of the mechanism of immune response, and peptido-receptor ligand interactions are slowly being rationalised. Peptide-based vaccines could be around the corner, and further useful peptido-mimetics will be designed from more selective screening programmes based on natural sources or possibly from synthetically created peptide libraries.

With this series continuing into Volume 26, it is hoped that the fruits of the considerable labour of our Reporters in providing an authoritative collection of the annual research output will assist in guiding workers into the vast literature in the field. For a significant number of the last 25 volumes Graham Barrett (AminoAcids) and Donald Elmore (Peptide Synthesis) have been unstinting in their efforts in compiling their annual chapters, and I wish to thank them again for their continued support. Christine Bladon has again taken care of the peptido-mimetic area and this year we benefit from the biennial coverage of Metal Complexes of Amino Acids and Proteins by Kevin Nolan and Hay, assisted by Dr. Soudi. To everyone, a most sincere thanks for your great help.

John S. Davies
University of Wales, Swansea

Contents

Abbreviations

The abbreviations for amino acids and their use in the formulation of derivatives follow in general the 1983 Recommendations of the IUB-IUPAC Joint Commission on Biochemical Nomenclature, which were reprinted as an Appendix in Volume 16 of this series. Chapter authors have been encouraged to include new abbreviations in their texts.

The synthetic peptide chemist has over the years been a strong supporter of the three-letter amino-acid abbreviation. However, with the increasing sizes of the peptides being produced the single-letter code has found increasing use. A complete listing of the code appeared in the Abbreviations section of Volume 24 of this Series of Reports. Structures and abbreviations for the closely related members of the BOP family of coupling agents also appeared in the prelims of Volume 24.

1
Amino Acids

By G.C.BARRETT

1 Introduction

The chemistry and biochemistry of the amino acids as represented in the 1992 literature, is covered in this Chapter. The usual policy for this Specialist Periodical Report has been continued, with almost exclusive attention in this Chapter, to the literature covering the natural occurrence, chemistry, and analysis methodology for the amino acids. Routine literature covering the natural distribution of well-known amino acids is excluded.

The discussion offered is brief for most of the papers cited, so that adequate commentary can be offered for papers describing significant advances in synthetic methodology and mechanistically-interesting chemistry. Patent literature is almost wholly excluded but this is easily reached through Section 34 of *Chemical Abstracts*. It is worth noting that the relative number of patents carried in Section 34 of *Chemical Abstracts* is increasing (e.g. Section 34 of *Chem.Abs.*, 1992, Vol.116, Issue No.11 contains 45 patent abstracts, 77 abstracts of papers and reviews), reflecting the perception that amino acids and peptides are capable of returning rich commercial rewards due to their important physiological roles and consequent pharmaceutical status. However, there is no slowing of the flow of journal papers and secondary literature, as far as the amino acids are concerned. The coverage in this Chapter is arranged into sections as used in all previous Volumes of this Specialist Periodical Report, and major Journals and *Chemical Abstracts* (to Volume 118, issue 11) have been scanned to provide the material surveyed here.

2 Textbooks and Reviews

Reviews cover the asymmetric synthesis of unusual amino acids starting from serine or cysteine[1] and the uses of amino acids as chiral synthons in synthesis.[2] A review of selenocysteine, an amino acid that has leapt to prominence as a new addition (codon UGA) to the universal genetic code, has appeared.[3] A plea for consistent representation on paper of chiral formulae, and details of a new method for doing so, has been published.[4] Several other reviews have appeared, and are listed at the start of appropriate Sections in this Chapter.

3 Naturally Occurring Amino Acids

3.1 *Isolation of Amino Acids from Natural Sources.*

This Section continues to hold a position early in this Chapter even though it would be thought of as a routine aspect of the literature. However, the validity of reports on the presence of amino acids in natural locations is only reliable if it is certain that artefacts are not introduced through extraction procedures. Although the extreme sensitivity of current analytical methods for amino acids enhances confidence in the results obtained for samples, at the same time it enhances the possibility that erroneous conclusions may be reached on the indigeneity of amino acids in natural sources.

Extraction of tyrosine from aqueous solutions using n-butanol is 87% complete after two partitions if the aqueous solution is saturated with an alkali-metal salt.[5] Similar partition leading to separation of amino acid mixtures and enrichment of amino acids in liquid surfactant membranes using tri-n-octylmethylammonium chloride, has been described.[6] After extraction of samples with water, acidic amino acids (aspartic and glutamic acids) may be adsorbed preferentially from the solutions on to acid-treated alumina.[7] A similar principle has been applied to isolate glutamic acid, glycine and lysine sequentially from aqueous solutions using silica-magnesia mixtures, and histidine and lysine using silica-titania mixtures.[8] Isolation of tryptophan from aqueous solutions using the MK-40 cation exchange membrane[9] and the adsorption of this amino acid onto the gel matrix during gel chromatography of amino acid hydrolysates on Bio-Gel P-2 has been described.[10] Tryptophan is the most readily adsorbed from a mixture of amino acids [Gly < Ala < Cys < Val < Met < Pro < Ile < Leu < Tyr < Phe < Trp] from aqueous solutions by benzylsilylated silica gel when the amino acids in the mixture are derivatized as N-acetylamino acid N'-methylamides.[11]

3.2 *Occurrence of Known Amino Acids.*

The unusual amino acids present in mushrooms[12] and the distribution in plant gall tumours and the chemistry of N-(carboxyalkyl)amino acids,[13] have been reviewed. The occurrence and identification of N-carbamyl-β-amino acids in urine has relevance in the monitoring of metabolic processes.[14]

Substantial studies over the years, of cross-linking amino acids in proteins, continue unabated, a representative review this year being the identification of cross-links in cattle hide after maturation.[15] These cross-links are predominantly between histidine and hydroxylysinonorleucine, but the discovery of proteoglycans as a source of bridging sites, is notable.

Other studies identifying the presence of known, but unusual, amino acids in hydrolysates, include alloisoleucine, allothreonine, N-methylphenylalanine, and p-methoxyphenylglycine from the antibiotic xanthostatin,[16] and O-methylserine and αβ-dehydrotryptophan from the cyclic peptide keramamide F from the marine sponge *Theonella*.[17] Herbaceamide (1), from the marine sponge *Dysidea herbacea*, is an N-acylated (2S,4S)-5,5,5-trichloroleucine, adding to the lengthening list of halogenated amino acids found in such organisms.[18] Careful studies have established the presence of D-enantiomers of alanine, serine, and proline in the mouse kidney.[19] Similarly careful studies are obligatory in assessing the common amino acids present in meteorites, and the problem of ensuring the indigeneity of protein amino acids in such sources was recently considered solved, since in test cases, the stable isotope distribution of organic constituents differed from the terrestrial distribution (see Vol. 24, p.2). A salutary warning arises from independent studies, that hydrolysis and derivatization reactions performed on fossils and meteorites can be accompanied by kinetic fractionation, influencing the stable isotope signatures.[20,21] In a particular context,[21] as a dipeptide is progressively hydrolysed the residual unhydrolysed dipeptide is increasingly enriched in ^{13}C and ^{15}N.

3.3 *New Naturally Occurring Amino Acids.*

Close relatives of protein amino acids that have been newly-discovered, are the novel immunomodulator metacyclofilin (2), from *Metarhizium* sp.TA2759 (the structure assigned lacks details of absolute configuration),[22] and the potent insecticide ulosantoin (3) from the marine sponge *Ulosa ruetzleri*.[23] New α-amino acids with heterocyclic side-chains are two analogues (4) and (5) of acromelic acid, from *Clitocybe acromelalga*.[24] Full details have been published[25] of the characteristics of the new amino acids from *Clitocybe acromelalga* L-3-(2-pyrrolyl)alanine (see Vol.24, p.3) and L-3-(2-oxo-5-pyridyl)alanine, as well as their biosynthetic precursor, the already-known stizolobic acid (6). Also from the poisonous mushroom *Clitocybe acromelalga*, another neuroexcitatory α-amino acid L-3-(6-carboxy-2-oxo-4-pyridyl)alanine (7) has been identified.[26] The isoxazolinone (8) from *Streptomyces platensis* A-136 shows antifungal activity against *Candida albicans* and low toxicity against mice.[27]

Greater separation between the amino and carboxy functions is shown in the herbicidal γ-amino acid cis-2-amino-1-hydroxycyclobutane-1-acetic acid (9)[28] from *Streptomyces rochei*, and the new spermine

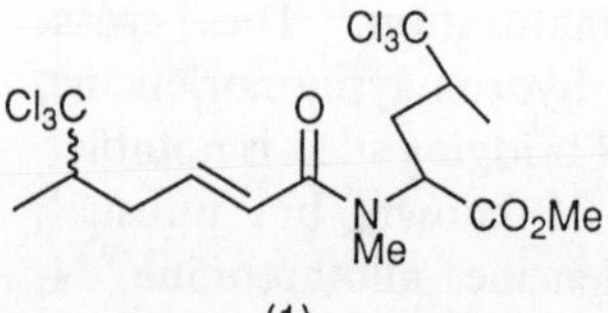

(1) 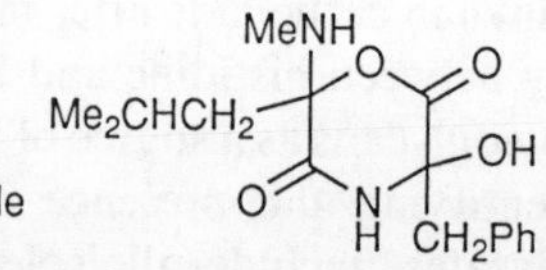 (2) 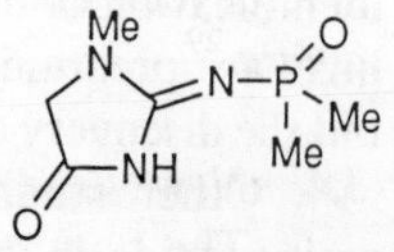 (3)

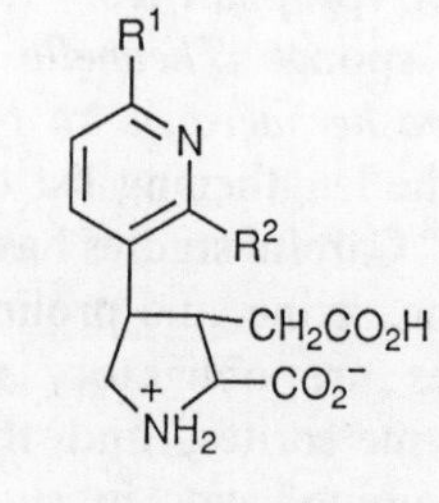

(4; $R^1 = CO_2H$, $R^2 = H$)
(5; $R^1 = H$, $R^2 = CO_2H$)

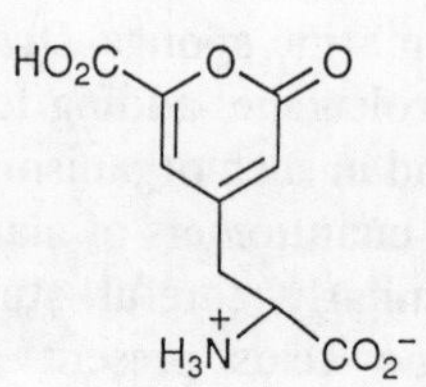

(6) 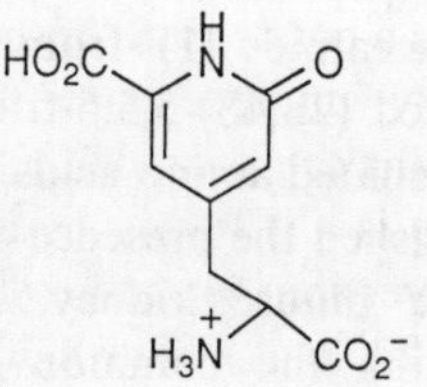 (7)

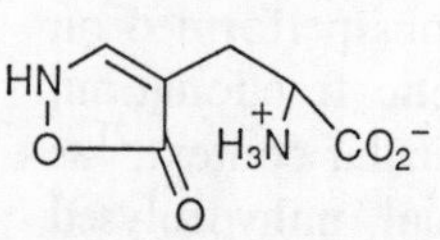

(8) 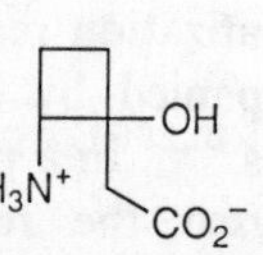 (9) 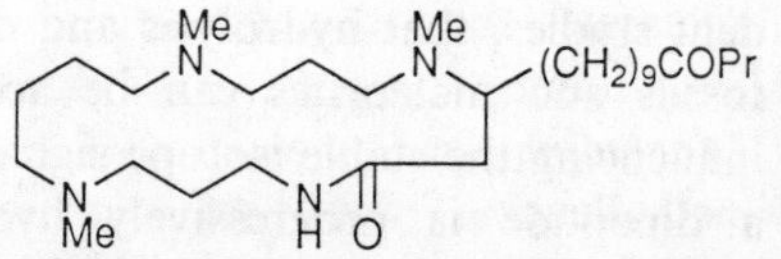 (10)

Three-dimensional features at chiral centres of structures depicted in this chapter follow the convention: –

(a) horizontally-ranged atoms, and their bonds, and atoms in rings, are understood to be in the plane of the paper;
(b) atoms and groups attached to these atoms in (a) are ABOVE the page if ranged LEFTWARDS and BELOW the page if ranged RIGHTWARDS;

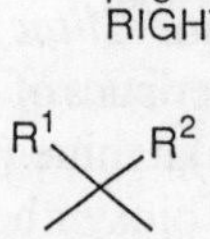

means 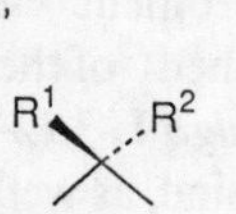 ; 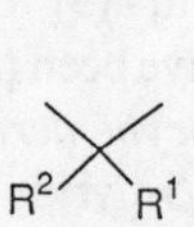 means

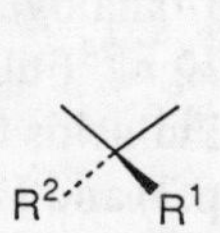

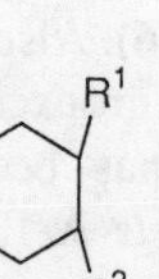

means

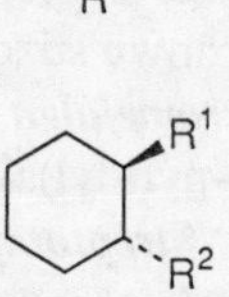

macrocyclic alkaloid budmunchiamine (10) present in seeds of *Albizia amara*.[29]

3.4 *New Amino Acids from Hydrolysates.*

The flow continues unabated, of new discoveries of peptides and related derivatives comprising new amino acids that can be formally considered to be obtained from them by hydrolysis. In the protein cross-linking category, "cyclopentensine" has joined other elastin cross-links and is a condensate of three allysine residues[30] to generate a cyclopentene moiety, unprecedented in crosslinking amino acids. Lysine residues in proteins can provide the amino group required for the Maillard reaction. There is a growing realization that crosslinking may result from an *in vivo* version of this reaction between proteins and carbohydrates (see Vol. 22, p.49), and a model reaction between α-N-acetyl-L-lysine and glucose has been shown to lead to fluorescent compounds [11; R = $-(CH_2)_4CH(NHAc)CO_2H$] that have properties similar to age- and diabetes-related cross-linking moieties in proteins.[31] Simpler analogues (11; R = n-pentyl) were also made in this study, and, like the lysine analogues, the configuration at one chiral centre is still to be determined.

Immunosuppressive lipopeptides, microcolins A and B, from *Lyngba majuscula*, contain N-methylvaline, O-acetyl-D-threonine, N-methylleucine, proline (or hydroxyproline, in the case of microcolin B), and 2-methylpyrrolidin-5-one condensed in that order from the N-terminus. The microcolins carry an N-terminal 2,4-dimethyloctanoyl moiety.[32] Botanical interest predominates in the case of BZR-cotoxin II, the cause of Leaf Spot disease in corn and produced by Bipolaris zeicola race 3 (the factor is a cyclic nonapeptide containing N-methyl δ-hydroxyleucine, 1-aminocyclopropane-1-carboxylic acid, and of γ-methylproline, together with some common amino acids),[33] and in the case of the antimitotic tetrapeptide ustiloxin (12) produced by *Ustilaginoidea vireus* growing on rice plant panicles.[34] The latter compound is related to Phomopsin A, but contains a novel di-amino di-acid carrying a sulphinyl function.

4 Chemical Synthesis and Resolution of Amino Acids

4.1 *General Methods of Synthesis of α-Amino Acids.*

Named preparative methods are as widely used as ever, almost literally "as ever", since they were established many decades ago. Many of the methods described in the next, and later, Sections are also general methods. The Ploechl-Erlenmeyer process based on $Ph_2C{=}NMe$ for alkylation of 2-phenyloxazol-5(4H)-one has been illustrated,[35] and an

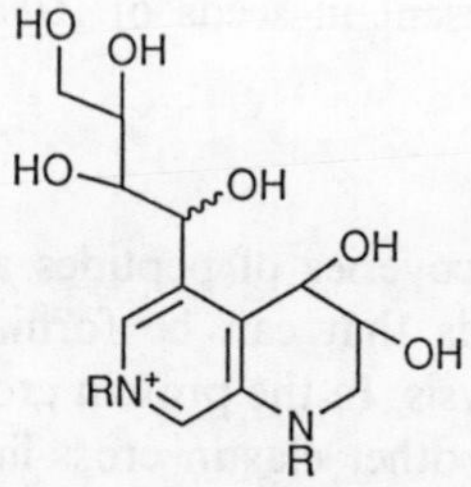

(11)

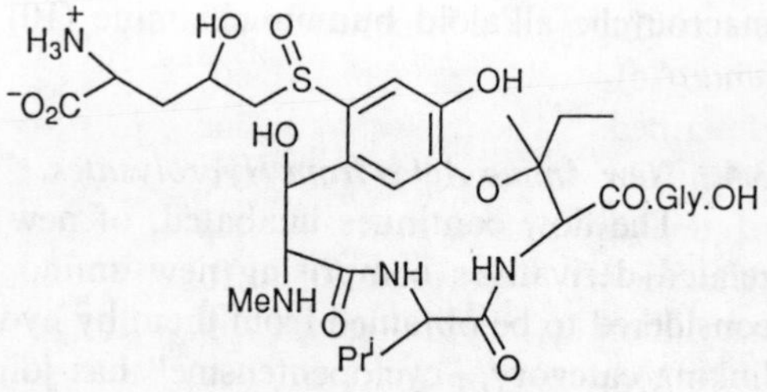

(12)

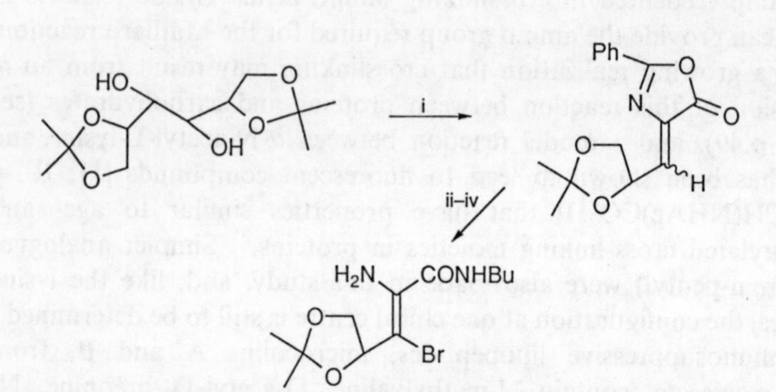

Reagents: i, $Pb(OAc)_4$, Ac_2O, hippuric acid, THF, reflux; ii, $BuNH_2$; iii, Br_2/CH_2Cl_2; iv, DABCO→ more stable *Z*-isomer

Scheme 1

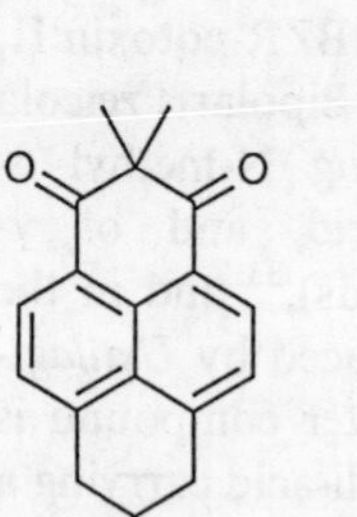

(13)

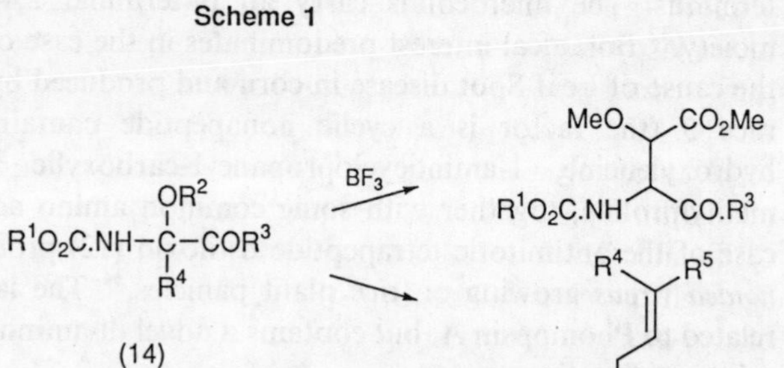

(14)

Ref.53: $R^1 = Me$; $R^2 = Me$; $R^3 = NMe$; $R^4 = H, Me$
Ref.54: $R^1 = H$; $R^2 = SiMe_3$; $R^3 = NMe$; $R^4 = H, Me$

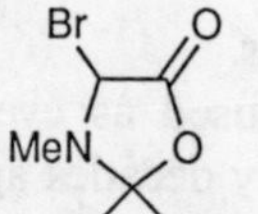

(15)

interesting *in situ* generation of a triose aldehyde for the process using $Pb(OAc)_4$ to cleave a protected hexose derivative (Scheme 1) has been described.[36] The customary concluding step for the process is reduction of the C = C bond but the latter example is designed to lead to β-bromo-αβ-dehydro-amino acids. A related procedure employing 2,5-di-oxopiperazine has been exemplified for a synthesis of phenylalanine, through alkylation with benzaldehyde followed by Zn-acid reduction.[37]

More fundamental processes are at work in carbonylation reactions illustrated for perfluoroalkyl α-amino acids (Scheme 2).[38] A yield of 64% obtained in this study for the preparation of 3,3,3-trifluoro-alanine, compares well with a figure (12%) through the best previous route for this compound. The Strecker synthesis applied to N-alkyl α-amino acids uses a primary amine, HCN, and an aldehyde.[39] The similarly-oriented hydantoin synthesis but using $(NH_4)_2CO_3$, has been used to provide a series of phenylalanine analogues with two substituents in the phenyl moiety.[40]

Alkylation of glycine derivatives has expanded in scope from the time-honoured acetamidomalonate synthesis[159,204] to the more recent processes based on the alkylation of glycine Schiff bases, e.g. $Ph_2C=NCH_2CO_2Me \rightarrow Ph_2C=NCHRCO_2Me$. Schiff base alkylation can be achieved by various strategies after carbanion generation, illustrated by Michael addition to $ArCH=CRCO_2Et$ giving 3-arylglutamates,[41] and by the use of nitro-aldols formed from RCH_2NO_2 + $HCHO \rightarrow RCH(NO_2)CH_2OH$ (R = D-xylopyranosyl).[42] The purpose of the last-mentioned study, synthesis of C-glycosyl serines, was achieved after reductive de-nitration of the alkylated Schiff base with Bu_3SnH/AIBN. More conventional alkylation protocols, using alkyl halides, are represented in the synthesis of phenylalanine analogues[43] including α-dialkylation after NaHDMS de-protonation, leading to the remarkable αα-dialkylated glycines (13; such crowded structures are difficult to prepare through the classical Bucherer-Bergs method).[44] An improved synthesis of α-amino phosphonic acids follows this strategy with $Ph_2C=NCH_2P(O)(Bu^iO)_2$.[45] αα-Disubstituted amino acid amides can be prepared by phase-transfer alkylation of benzylidene-amino acid amides $PhCH=NCHRCONH_2$.[46]

The most obvious short cut for this alkylation approach for some applications, is to use an N-acylated or N-carbamylated amino acid, i.e. not an ester; thus, Boc-glycine gives the tri-anion with LDA/THF which gives 40-80% yields on alkylation without di-alkylation side-products.[47] Contrary to earlier reports,[48] only two equivalents of strong base (rather than one, which results in N-alkylation) are needed for C-alkylation of methyl hippurate, though it should be noted that the choice of additive

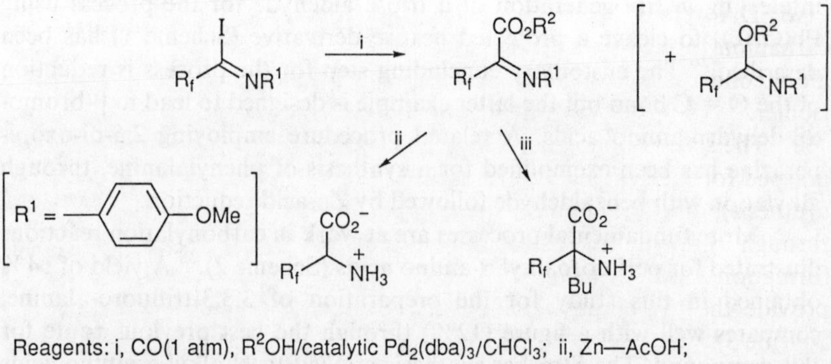

I; R_f; NR^1 —i→ CO_2R^2; R_f; NR^1 [+ OR^2; R_f; NR^1]

ii; iii

[R^1 = —C6H4—OMe]

CO_2^-; R_f; $\overset{+}{N}H_3$

CO_2^-; R_f; $\overset{+}{N}H_3$; Bu

Reagents: i, CO(1 atm), R^2OH/catalytic $Pd_2(dba)_3/CHCl_3$; ii, Zn—AcOH; iii, BuLi then H_2/Pd—C

Scheme 2

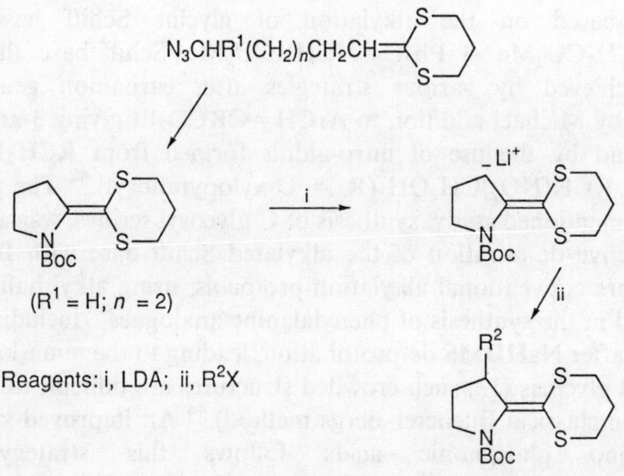

$N_3CHR^1(CH_2)_nCH_2CH=$ S S

S S N Boc —i→ $^-$ Li$^+$ S S N Boc

(R^1 = H; n = 2)

ii

R^2 S S N Boc

Reagents: i, LDA; ii, R^2X

Scheme 3

(TMEDA versus HMPA) influences the product composition.[49] The carbanion formed with LDA from Boc-sarcosine ethyl ester, undergoes aldol condensation with acrolein to give predominantly (85%) the anti-isomer.[50]

The special opportunities offered by proline as a synthon are often grasped for the synthesis of substituted prolines (and indeed, in alkaloid synthesis), and a circuitous approach giving 3-substituted prolines through γ-alkylation of 2-aminoketene S,S-acetals after carbanion formation LDA (Scheme 3).[51] Intramolecular azide cycloaddition can provide the pyrrolidinone S,S-acetals and higher homologues (also in Scheme 3).[52]

The alternative approach in which α-hetero-atom-substituted glycines act as cation equivalents continues to be usefully explored. α-Methoxyglycines of various types (14) undergo BF_3-catalysed addition to ester enolates so as to give ββ-disubstituted aspartates,[53] and α-(trimethylsilyl)oxy alanines[54] can be alkylated by allylsilanes with TMSOTf to give γδ-unsaturated α-methyl-α-amino acids. Some dehydro-alanine is formed as side-product in the latter study, and further clarification of reaction conditions will be needed if this is to be avoided. A new electrophilic sarcosine synthon (15) has been advocated, and employed in Michaelis-Arbuzov-type synthesis of sarcosines carrying phosphorus-containing side-chains.[55]

αβ-Dehydroamino acids are becoming more attractive synthons for use in α-amino acid synthesis, though of course, as is the case for a number of outcomes of other alkylation methods, they yield racemic products when used for Lewis-acid catalysed α-acylamidoalkylation of furans and anisole.[56] A most promising route, in which N-acetyl dehydro-alanine methyl ester complexed with $Fe(CO)_3$ [*alias* (methyl 2-acetamido acrylate)tricarbonyl iron(0)], treated with MeLi and a tertiary alkyl halide, gives t-leucine and new amino acids with β-branched alkyl side-chains (2-amino-3,3-dimethyl-pentanoic and -hexanoic acids).[57]

Amination of a range of substrates has long been a favoured approach to α-amino acids, and is exemplified in its classical form using phase-transfer catalysed amination of α-halogeno-esters by Cl_3CONH_2[58] or by $F_3C(CF_2)_nCH_2CH_2NH_2$.[59] Reductive amination of α-keto-acids is also represented in the recent literature.[245] Synthesis of N,N-bis(carbamylated) amino acids can be accomplished using potassium iminodicarboxylates (readers will recognize the formula $Boc_2N^-K^+$ more readily!) with 2-bromo-alkanoates or through a Mitsunobu reaction involving ethyl lactate (cf. also Ref. 78).[60] New-style Hofmann rearrangement leading to amination employs $Pb(OAc)_4$ and Bu^tOH, and is illustrated (Scheme 4) in a synthesis of sterically-constrained surrogates

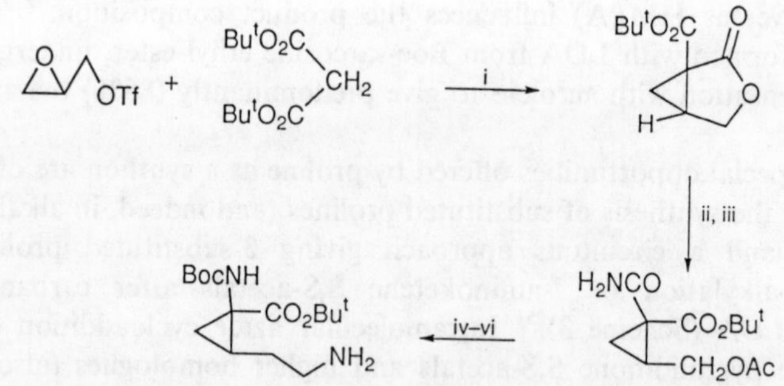

(16)

Reagents: i, NaH; ii, NH_3; iii, Ac_2O; iv, $Pb(OAc)_4$ /Bu^tOH; v, Boc_2O; vi, routine elaboration → ornithine and arginine side-chains

Scheme 4

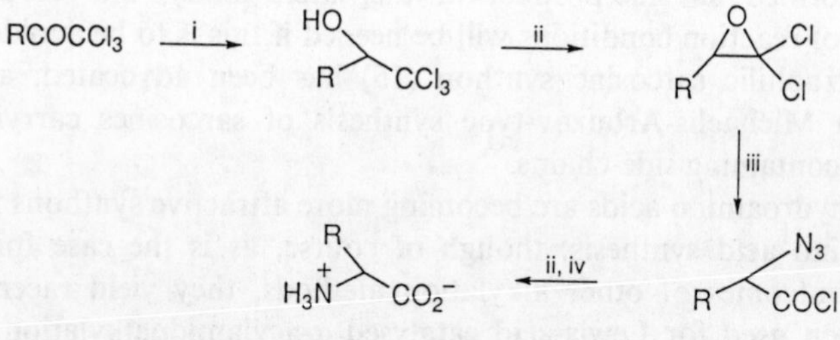

Reagents: i, (1,2–$C_6H_4O_2$)BH, (*S*)–Oxazaborolidine catalyst; ii, OH^-; iii, N_3^-; iv, H_2/Pd–C

Scheme 5

(17)

(18)

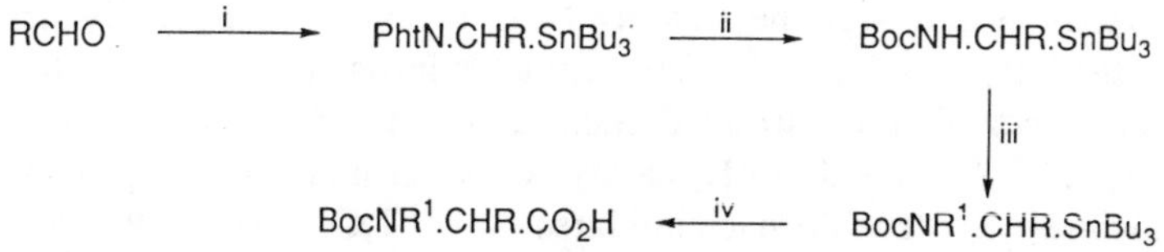

$$RCHO \xrightarrow{i} PhtN.CHR.SnBu_3 \xrightarrow{ii} BocNH.CHR.SnBu_3 \xrightarrow{iii} BocNR^1.CHR.SnBu_3 \xrightarrow{iv} BocNR^1.CHR.CO_2H$$

Reagents: i, Bu_3SnLi/ phthalimide /PPh_3 /DEAD; ii, routine deprotection–re-protection; iii, R^1X; iv, BuLi, then CO_2

Scheme 6

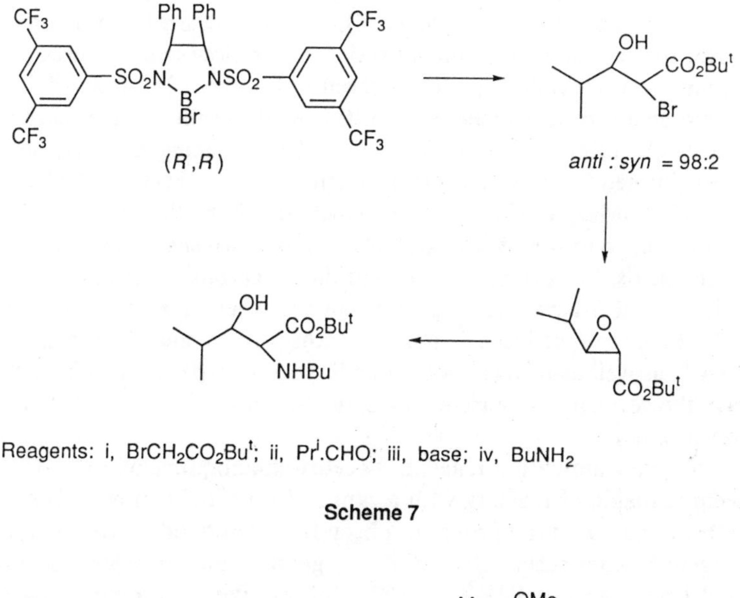

CF3, Ph Ph, SO2N, NSO2, B, Br, CF3, (*R*,*R*)

OH, CO_2Bu^t, Br

anti : *syn* = 98:2

O, CO_2Bu^t

OH, CO_2Bu^t, NHBu

Reagents: i, $BrCH_2CO_2Bu^t$; ii, $Pr^i.CHO$; iii, base; iv, $BuNH_2$

Scheme 7

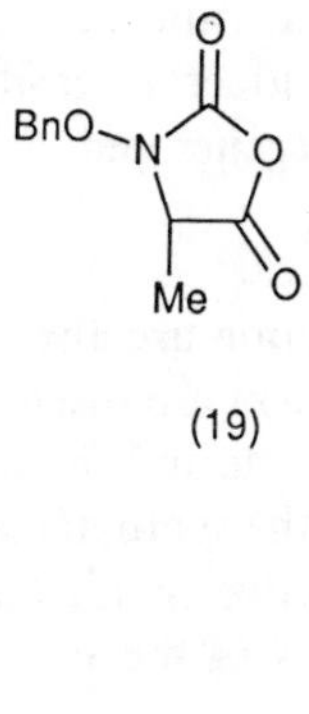

BnO, N, O, O, O, Me

(19)

Me, OMe, NH_2, Me, OH

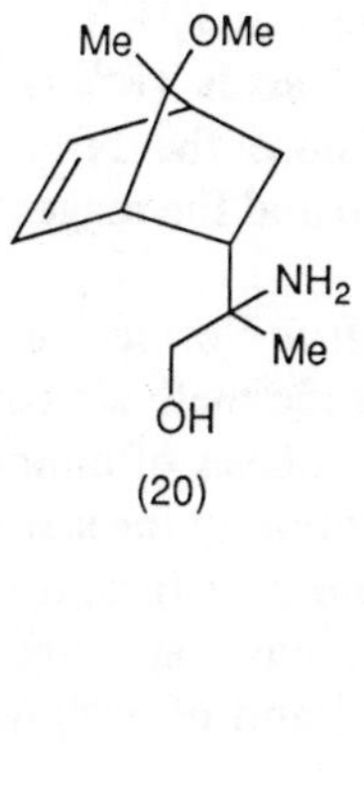

(20)

for ornithine and arginine starting from a homochiral glycidyl triflate and di-t-butyl malonate.[61] The same approach leads to all four stereoisomers of a constrained methionine [16; CH_2SMe in place of $(CH_2)_nNH_2$].[62] Palladium(II)-catalysed Overman rearrangement of homochiral trichloroacetimidates $Cl_3CC(=NH)OCHRCH=CHCH_2OTBDPS$, yields (E)-βγ-unsaturated α-amino acids via optically-pure mono-protected allyldiols $Cl_3CCONHCH(OTBDPS)CH=CHR$ in this particular case.[63] Amination of trichloromethyl carbinols, formed from trichloromethyl ketones, provides the basis of a new enantioselective synthesis of α-amino acids,[64] illustrated (Scheme 5) for t-butylglycine (*alias* t-leucine). A convenient synthesis of trichloromethyl ketones needed for this purpose, has been published.[65] Phthalimides yield Boc-α-amino-organostannanes (Scheme 6), formed by Sn-Li exchange with BuLi, that are configurationally stable at very low temperatures (−95°/10 minutes) but suffer significant racemization at −78° and −55°.[66] An amino acid synthesis emerges from this work on the basis of carboxylation (CO_2) after carbanion formation. Ring-opening of aziridines, e.g. those formed from electron-rich alkenes (such as $MeCH=CHCO_2Me$) and $HN(OMe)_2$/TMSOTf, from which the N-methoxy group can be reductively removed with Na/NH_3,[67] give correspondingly-substituted amino acids. For example, 3-arylaziridine-2-carboxylic esters (17) give 3-chloro- and 3-benzenethio-phenylalanines when treated with HCl and with PhSH respectively,[68] and the trans-3-hexyl analogue reacts similarly,[69] as well as being shown to undergo ring-expansion with MeCN to give the corresponding cis-imidazoline-2-carboxylic acid (essentially a 3-aza-proline).

A new amination reagent, 3-acetoxyaminoquinazolinone (18), that seems capable of reacting with a range of carbonyl compounds so as to introduce an α-amino group, has been fully described.[70] Addition of this reagent to enol ethers $CH_2=CHOEt$ giving α-amino aldehydes, and to silyl ketene acetals $R^1R^2C=C(R^3)OSiMe_3$ giving α-amino ketones (R^3 = alkyl) and acids (R^3 = OMe), offers a valuable route to these compounds, since the N-N bond in the adduct is easily reductively cleaved (SmI_2) and the reagent is then easily regenerated.

4.2 *Asymmetric Synthesis of α-Amino Acids.*

Most of the methods covered in this section are also viable general methods of synthesis of amino acids, but papers are discussed here with exclusive attention to the stereochemical outcome and the way that this is governed by the structural features sought in the synthetic target.

Reviews have appeared of stereoselective syntheses of unusual amino acids,[71] and of arylglycines.[72] A review of the synthesis of chiral cyanohydrins for use as starting materials in synthesis is relevant to the content of a Chapter on Amino Acids.[73]

An enantioselective version of a standard general method of α-amino acid synthesis is found in a Strecker synthesis of N-[(R)-2-hydroxy-1-phenylethyl]aminonitriles, $HOCH_2CHPhNHCHRCN$, readily elaborated into L-amino acids (R = aryl, Me, Pr^i, Bu^t) and owing their homochirality to the use of (R)-phenylglycinol as a reactant,[74] and transimination-hydrocyanation of O-methoxyisopropyl-(R)-mandelonitrile giving (2R,3R)-β-hydroxy-α-aminonitrile derivatives is a further illustration.[75] Amination of homochiral alkyl α-halogenoalkanoates illustrates an enantiospecific version of a standard method, and there are some unusual variants of this; α-bromoalkanoyl bromides on treatment with a tertiary base, give bromoketenes from which, by addition of a chiral alkanol, the homochiral α-bromoester is obtained. Azide displacement with inversion and routine stages leads to the homochiral α-amino acid, and (R)-pantolactone was used as the chiral alkanol in this study.[76] t-Butyl bromoacetate can participate in diastereoselective aldol reactions (Scheme 7) giving mainly the anti-isomer of a t-butyl α-bromo-β-hydroxyalkanoate.[77] Cyclisation to the chiral oxirane and ring opening with an amine or with azide ion gives the corresponding α-amino-β-hydroxy acid, and the method has been illustrated for a synthesis of (2R,3S)-hydroxyleucine that can be operated on a large scale.

A synthesis of N-benzyloxy-D-alanine anhydride (19) by amination of L-(+)-lactic acid represents relatively routine methodology (cf. also Ref.60),[78] but an unusual amination procedure is based on a new retro-aza-ene reaction; a keto-acid esterified with the (S)-β-aminoethanol (20; abbreviated in Scheme 8) can undergo intramolecular Schiff base formation by cyclization and the ester is liable to thermolysis and acid hydrolysis to give the (R)-α-amino acid.[79] This may not become an established general method because of the complexities but particularly because the chiral auxiliary is not recovered in the process.

A version of a standard method is seen in phase-transfer alkylation of (S)-aldimine Schiff bases $RCH{=}NCHMeCO_2Me$.[80]

The use of chiral auxiliaries is more explicit in a number of other enantioselective amino acid synthetic methods that have become standard practice over recent years. Thus, the Schöllkopf bis-lactim ether method continues to gain adherents, as shown in reports of syntheses of (S)-cis- and -trans-crotylglycines using $MeC{\equiv}CCH_2Br$ as alkylating agent with the lactim ether (21),[81] and analogous applications leading to allocoronamic acid (22) via alkylation with (E)-$CH_2CH{=}CHCH_2Cl$/LDA,[82] and 1-benzenesulphonyl-6-methoxy-D-tryptophan methyl ester

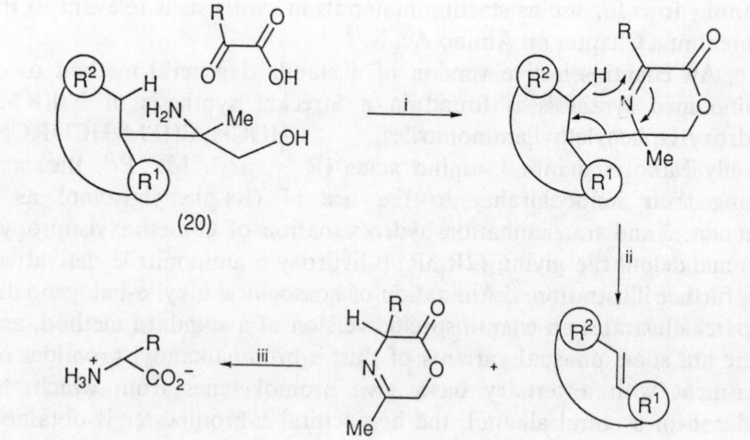

Reagents: i, Schiff base formation and esterification; ii, >200 °C; iii, H_3O^+

Scheme 8

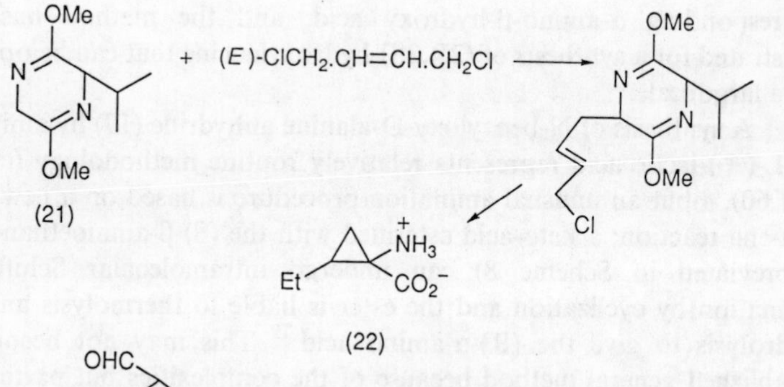

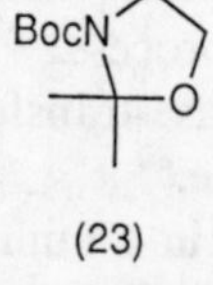

(starting from 6-methoxyindole and generating the relevant alkylating agent from it.[83] In a synthesis of a phenylalanine analogue carrying an N-methyl benzomorpholine moiety in place of phenyl, a claim that the Schöllkopf procedure must be performed in a modified manner to achieve this target in view of the "highly electron-rich side-chain" holds no credibility in view of the lack of comparative data.[84] A thorough assessment of the use of the Schöllkopf procedure in the straightforward (but expensive) way (reactions in MeO^2H and ^2H_2O) for the synthesis of (R)- and (S)-[2-^{2}H]-labelled α-amino acids has appeared.[85]

A use has been explored for the Schöllkopf procedure independently in two laboratories - the synthesis of L,L-di-aminopimelic acid [*alias* (2S,6S)-2,6-di-aminoheptanedioic acid][86] and its meso-diastereoisomer.[87] To attain the synthetic target requires that one of the two eventual chiral centres is located in an alkylating species and that the stereochemistry of the chiral centre that is generated by bis-lactim ether alkylation is controlled by the bis-lactim ether chiral centre. Two-carbon homologation of L-glutamic acid semialdehyde, or the Garner oxazolidine derived from L- or D-serine (23), [-CHO → -CH=CHCHO → -$CH_2CH_2CH_2Br$ or -CH_2CH_2CHO] was followed by routine use of the Schöllkopf procedure. Analogues carrying a 3,4-double bond, or a 4-fluoro- or 3-chloro-substituent, were prepared in one of these studies, in which an improved synthesis of the Garner synthon (23) was reported. The chirality at the 3-chloro-substituent, developed from the hydroxy substituent arising through the use of the aldehyde as alkylating agent, was defined by the enantio- and diastereoselective nature of the aldol addition to the Schöllkopf titanium enolate.

Novel variations of the Schöllkopf procedure include a demonstrated synthesis of L-alanine in 98% diastereoisomeric excess employing cyclo(Gly-L-Ala) carrying an (S)-phenylethyl substituent on each amide nitrogen[88] (perhaps an example of excessive provision of chiral director groups?). The titanated bis-lactim ether (21) undergoes ready Michael addition to nitro-alkenes to give diastereoisomerically pure α-(γ-nitroalkyl) α-amino acids.[89]

Numerous related strategies by which enantioselection can be imposed on an amino acid synthesis by a chiral auxiliary have been proposed and continue to be exemplified. Among the simplest is the 8-phenylmenthyl ester of glycine, which has been shown to undergo free radical bromination (NBS) and (amazingly) with high diastereoselectivity. The products are convertible into L-amino acids by bromide displacement with Grignard reagents.[90]

The Evans amination methodology, in which a chiral oxazolidin-2-one (24) is acylated by an alkanoic acid that is to undergo α-amination

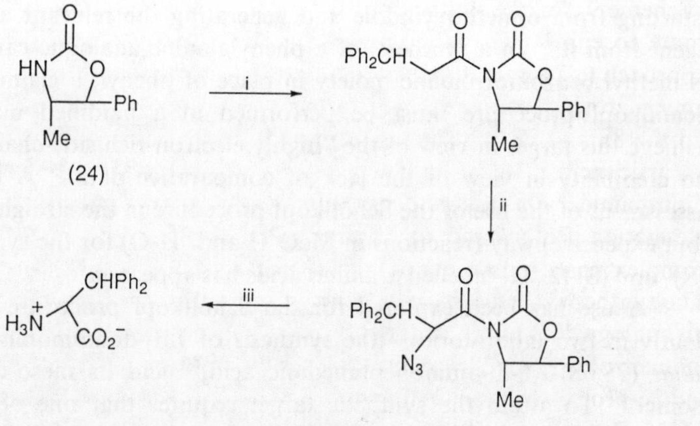

HN O O Ph Me (24) —i→ Ph_2CH O O N O Ph Me

↓ ii

$H_3\overset{+}{N}$ $CHPh_2$ CO_2^- ←iii— Ph_2CH O O N O N_3 Ph Me

Reagents: i, $Ph_2CH.CH_2.CO_2H/Bu^tCOCl$ + NaH; ii, KHDMS then trisyl azide; iii, $LiOH/H_2O_2$ then H_2/Pd–C

Scheme 9

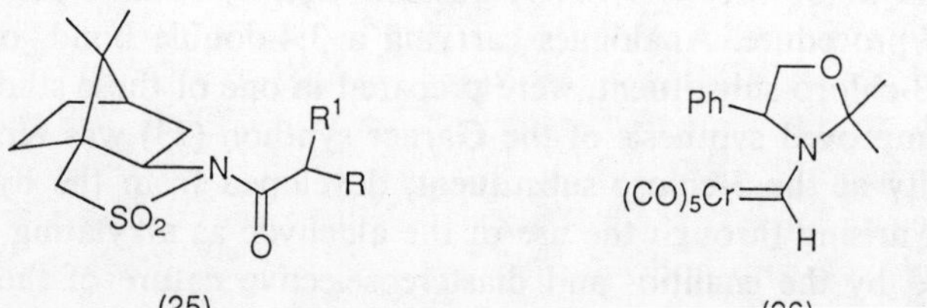

N SO_2 O R^1 R (25)

Ph O N $(CO)_5Cr$ H (26)

Ph Ph HN O O (27) —RCHO, $(MeO_2CCH{=})_2$→ [Ph Ph O N^+ R O] —→ Ph Ph O N R O MeO_2C CO_2Me

↓ NBS P(OMe)$_3$

Ph Ph ZN O $(MeO)_2P(O)$ —NaH, HCHO→ Ph Ph ZN O O —$PhCH_2N(CH_2OMe)CH_2SiMe_3$ ↓ $PhCH_2\overset{+}{N}{=}CH_2$ CH_2^-→ Ph Ph ZN O O $PhCH_2N$ —→ $H_3\overset{+}{N}$ CO_2^- N H

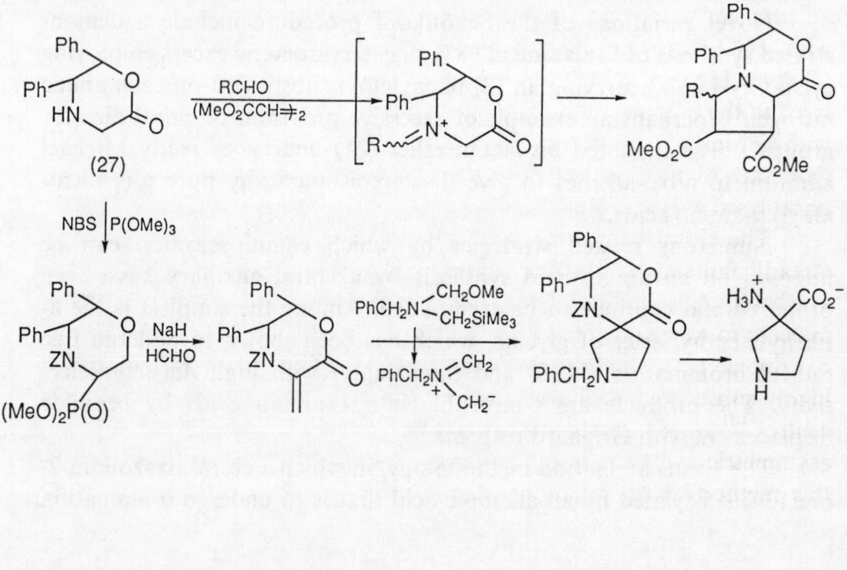

[(Scheme 9) for the synthesis of β-phenyl-(R)-phenylalanine][91] and for a route to either (R)- or (S)-β-methylphenylalanine[92] relies on a similar approach (α-bromination, substitution by azide ion and reduction to the amino group). An oxazolidin-2-one is used as chiral template for setting up the 8S and 9S centres of the unusual amino acid "Adda", using D-aspartic acid.[93] A spectacular use is described, for a synthesis of diphthamide. This amino acid is a residue in EF-2, the protein synthesis elongation factor, and diphthamide has been described as the most complex post-translationally-modified amino acid known in proteins.[94] The synthesis of diphthamide from D-and L-glutamic acids proceeds via an analogue of (24; Ph or $PhCH_2$ in place of Me; H in place of Ph) in the manner of Scheme 9. Further examples have been published, leading to indole-protected β-methyltryptophans from 3-indole-acrylic acid and MeMgBr/CuBr-Me_2S for alkylation of the resulting N-acyloxazolinone,[95] and to (3S)-and (3R)-piperazic acids by α-hydrazination of N-(5-bromovaleroyl)oxazolinones followed by cyclization.[96]

The principle extends to amination with 1-chloro-1-nitrosocyclohexane, of enolates of N-crotonylsultams (25) formed with $NaN(SiMe_3)_2$. The synthon has been christened a "NH_2^+-equivalent" even though it ws used in this study to give an optically-pure N-hydroxy amino acid.[97] A particularly interesting area of study in which the N-substituent on the eventual amino acid is also homochiral, concerns the oxazolidine carbene Cr complexes (26), which can be regarded as Cr-bound ketenes. These, on photolysis in the presence of ^2H_2O, give (S)/(R)-[2-^{2}H]glycine with good diastereoselectivity.[98] Comparison is made in this study, with the stereoselectivity observed in quenching equivalent ketenes and ester enolates [i.e. R-CH=C=O and R-CH=C(OMe)O^-] with ^2H_2O. The same substrate (but with Me in place of H at the carbene C) gives L-alanyl dipeptides with high diastereoselectivity when photolysed in the presence of an L-amino acid ester.[99] The same substrate (but with $O^-Me_4N^+$ in place of H at the carbene C) gives arylglycines with moderate diastereoselectivity via aryl substituted oxazinones when photolysed in (1R,2S)-(-)-2-amino-1,2-diphenylethanol.[100]

The chiral oxazinones mentioned in the preceding paragraph are already familiar as starting materials for enantioselective amino acid synthesis. A route involving a [1,3]-dipolar cycloaddition of diethyl maleate to an azomethine ylide, formed from the saturated 2,3-diphenyl oxazinone (27) and an aldehyde, is highly endoselective, leading to highly-substituted prolines containing up to 4 continuous stereogenic centres.[101] A synthesis of (S)-(-)-cucurbitine, for which no efficient asymmetric synthesis is otherwise available, illustrates the principle of this method using the diphenyloxazinone in the way originally introduced, for a specified synthetic target.[102] The diphenyloxazinone has also been used in an asymmetric synthesis of 2,6-di-aminopimelic acid enantiomers by the same group,[103] and to L-2,7-diaminoheptanoic acid and L-2,8-diaminooctanoic acid (alkylation of the diphenyloxazinone by $I(CH_2)_nI$ and substitution by azide of the ω-iodo atom).[104] (S,S)-2,6-Diamino-2-hydroxymethylheptanedioic acid has been synthesized from the same oxazinone (27).[105] A related six-membered heterocyclic compound has expanded the horizons of asymmetric synthesis in the β-amino acid field, and is described in the later Section 4.15.

The other homochiral heterocyclic synthons are of the five-membered family, and have also become widely used, particularly the parent oxazolidinone of (28), for the asymmetric synthesis of α-amino acids. Nitronates, e.g. $Me_2C^-NO_2\ K^+$, undergo Michael addition to the methylidene derivative of (28) to give an 84:16-mixture of diastereoisomers.[106] The method offers a convenient entry to amino acids with β-substituted-alkyl side-chains, based upon the easy reductive cleavage of the nitro group (see also preceding Section 4.1: General Methods of Synthesis). The corresponding imidazolone formed from L-methionine and converted into the vinylglycine analogue (by $NaIO_4$ then pyrolysis) has been used to prepare avicin analogues by 1,3-dipolar cycloaddition to nitrile oxides.[107]

The chiral imidazolidinone (29) has been used in a further example of the "alkylation of chiral heterocycle" approach to asymmetric synthesis of α-amino acids, this time of enantiomers of the NMDA antagonist 2-amino-5-(phosphonomethyl)[1,1′-biphenyl]-3-propanoic acid (*alias* 3′-phosphonomethyl-5′-phenylphenylalanine).[108] This paper includes a preparative procedure for the 2-isopropyl analogue of (29), with glycine, methylamine, isobutyraldehyde and Boc_2O as essential ingredients.

Chiral oxazolines (30 in Scheme 10) are intermediates in a distant relative of the Strecker synthesis, and provide D-amino acid esters from aldehydes via corresponding α-aminonitriles.[109]

Routes to prolines and pipecolic acids based on ene-iminium ion cyclizations that undergo a cationic aza-Cope rearrangement, proceed with complete stereoselectivity, and have been used in syntheses of (R,R)-4-(1-hydroxy-1-methylethyl)proline (31) for example.[110] Further details have been published relating to applications of aza-Diels-Alder reactions in asymmetric synthesis of pipecolic acids.[111]

Schiff bases of glycine esters, in which the Schiff base moiety alone is the chiral director, continue to provide a template for alkylation, in studies of links between structure and diastereoselectivity. A long-standing interest in nickel(II) (S)-2-[N-(benzylprolyl)amino] benzophenone

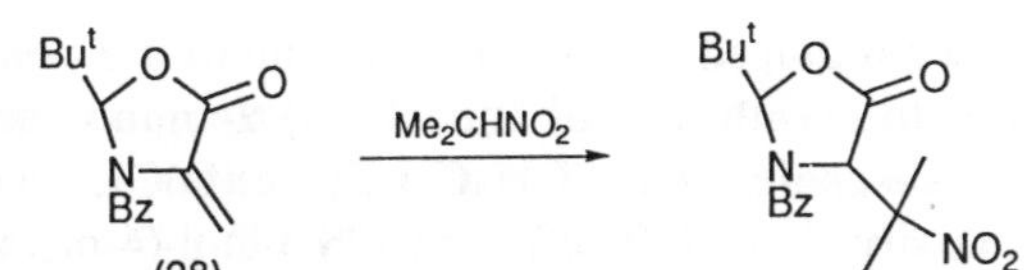

(28)

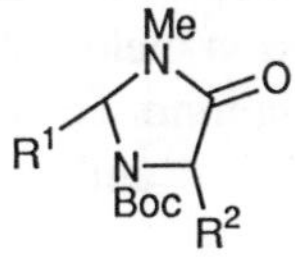

(29)

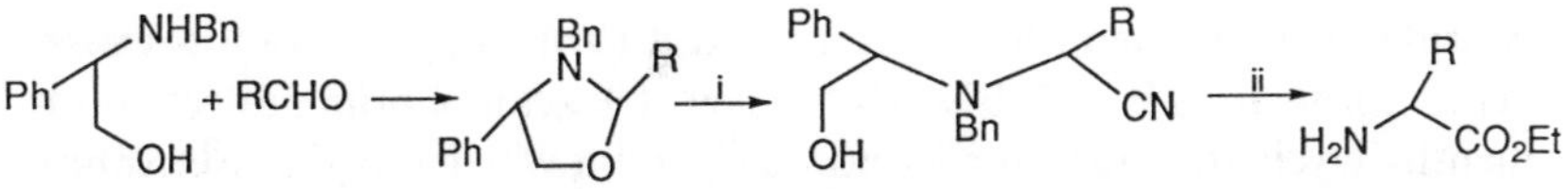

Reagents: i, Et_2AlCN; ii, EtOH; iii, H_2/Pd–C

Scheme 10

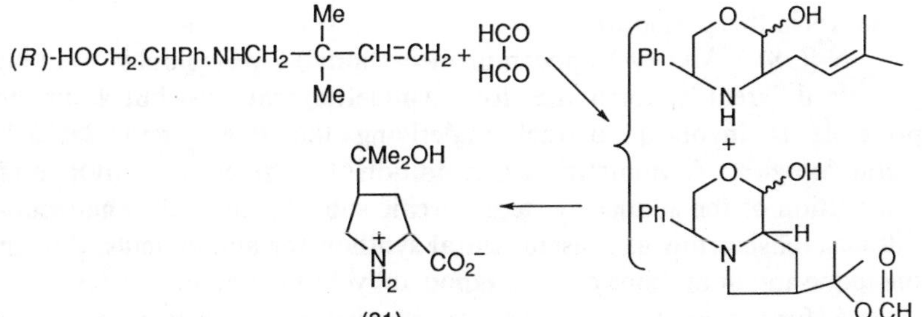

(30)

(*R*)-$HOCH_2.CHPh.NHCH_2$–C(Me)(Me)–$CH=CH_2$ + HCO–HCO

(31)

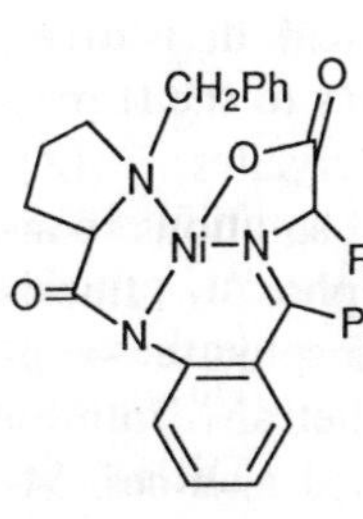

(32)

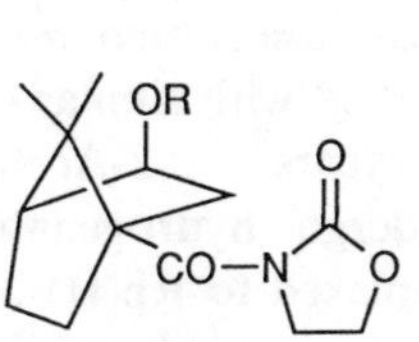

(33)

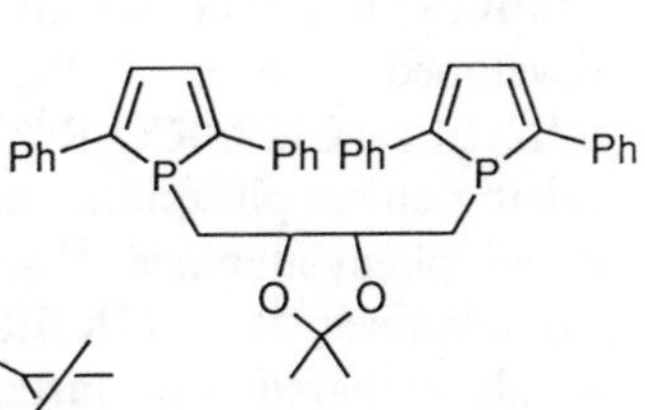

(34)

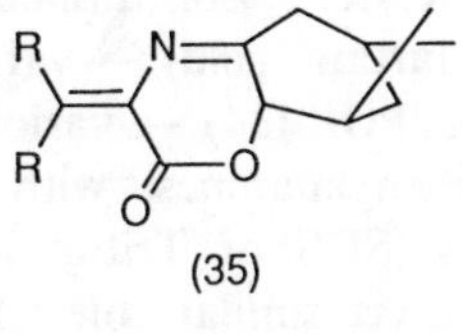

(35)

(32; R = H) has been indulged again, this year's literature describing its alkylation leading to syntheses of novel (S)-α-amino acids with phosphinate ester side-chains (e.g. -$CH_2CH_2P(Me)(O)OH$) with about 90% diastereoselectivity,[112] to (2R,3S)- and (2S,3R)-β-(4-methoxytetrafluorophenyl)serine and β-(pentafluorophenyl)serine,[113] and to propargylglycine (92% d.e.).[114] New variants of the process include the use of its β-alanine homologue for enantioselective β-amino acid synthesis [PhCHO → a mixture of (2S,3S)- and (2S,3R)-$PhCH(OH)CH(CH_2NH_2)CO_2H$].[115]

The "double asymmetric induction" principle in which both the Schiff base AND the ester moiety are chiral directors now has a considerable volume of literature behind it (Vol. 24, p.10). (R)-α-Amino acids form in 73 - 100% e.e. when (+)-camphorimines of (−)-menthylglycinate undergo alkylation,[116] and a wider range of e.e. values accompanies the exchange of the ester moiety for its (+)-enantiomer.[117] It is difficult to see any trends in the data, especially when faced with numerous other reports for alkylation of analogues containing achiral ester moieties; e.g. D-phenylalanine is formed in 99% e.e. when the imine formed between camphor-10-di-isopropylsulphonamide and glycine t-butyl ester is alkylated in the same protocol as for the preceding examples in this paragraph.[118]

N-[(1S)-2-Alkoxy-1-apocamphanecarbonyl)-2-oxazolones (33) provide different opportunities for asymmetric synthesis, but a similar principle is involved to that underlying the Evans and Seebach methodologies. Asymmetric cycloaddition to azines RN=NR and elaboration of the adduct by stereospecific substitution with organocuprates establishes this as a useful chiral synthon for amino acids, though the scope and limitations of the method are yet to be explored.[119]

A further approach to asymmetric synthesis, asymmetric hydrogenation and alkylation of αβ-unsaturated α-amino acid derivatives, continues to be thoroughly studied with similar objectives to the themes developed over the years. 2-Acetamidoacrylates (Z)-$R^1NH.C(=CHR^2)CO_2R^3$ undergo hydrogenation in solutions containing chiral phosphines complexed to Rh(II)-"Propraphos" → fluorinated phenylalanines,[120] -chiral aminophosphine - phosphinite → β-furylalanines,[121] -(3R,4R)-3,4-bis(diphenylphosphino)tetrahydrofuran (easily prepared from tartaric acid) → various substituted alanines, 54-97% e.e.,[122] -(R,R)-DIPPOP (34) → various substituted alanines,[123] → various substituted phenylalanines, with e.e. enhanced through the presence of detergents (SDS or Triton X-100).[124] Chiral phosphines complexed to Ru(II) exert similar roles; five-co-ordinate Ru(II)-binap complex → various substituted alanines, with a notable effect of temperature on e.e. (at 70°, <70% e.e.; at 50°, 96% e.e.);[125] (R,R)-[dipamp-Ru(II)(2-methylallyl)$_2$] for reduction of the ketone function in α-acetylglycine to allothreonine.[126]

The same objective can be approached through Pd/C-catalysed hydrogenation of chiral Schiff bases of acetamidoacrylates, e.g. (35), gives better than 95% diastereofacial discrimination in favour of the (S)-β-substituted alanine (the same result is obtained by reduction with L-Selectride).[127] While little (44%) or no d.e. is observed in mixed cuprate (PhMgBr/CuI) alkylation of homochiral esters (Z)-$R^1NHC(=CHR^2)CO_2R^3$ [R^3 = (−)-menthyl],[128] higher order cuprates give β-substituted tryptophans with the L-tryptophan-derived synthon (36).[129]

Enzyme-assisted aldol condensation between glycine and an aldehyde is one of the areas where an approach to asymmetric synthesis of amino acids can be contemplated, but hydroxymethyltransferase shows very little ability to determine homochirality at the side-chain chiral centre.[130]

4.3 *Synthesis of Protein Amino Acids and Other Naturally Occurring α-Amino Acids.*

Several instances of the use of naturally-occurring α-amino acids as synthesis targets have been described in the preceding sections, and this section deals with specifically-designed syntheses. The usual opening topic of this section, fermentative production of the protein amino acids, is given less thorough coverage this year, citations being restricted to a key review (particularly of the production of glutamic acid, lysine, phenylalanine, and aspartic acid),[131] and indicative primary papers (reductive amination of pyruvate by alanine dehydrogenase to give L-alanine;[132] production of phenylalanine analogues by amination of cinnamates catalysed by the yeast *Rhodotorula glutinis*;[133] and methods using free and immobilized *Corynebacterium glutamicum* cells for L-lysine production.[134]

Natural cyclopropane-based α-amino acids tackled recently, include 1-aminocyclopropanecarboxylic acid (ACC), obtained in near-quantitative yield by cyclopropanation of a dehydroalanine Schiff base $Ph_2C=NC(=CH_2)CO_2Me$ using CH_2N_2,[135] and (+)-(1R,2S)-allocoronamic acid obtained through similar cyclopropanation of a chiral alkylidenedioxopiperazine (37).[136] A different, but still classical, route leads to N-Boc-ACC benzyl ester, elaborated for a synthesis of the the *Coprinus atramentarius* toxins coprine and O-ethylcoprine from L-glutamic acid.[137] An asymmetric total synthesis of individual diastereo-isomers of hypoglycin A employs Sharpless oxidation of the appropriate

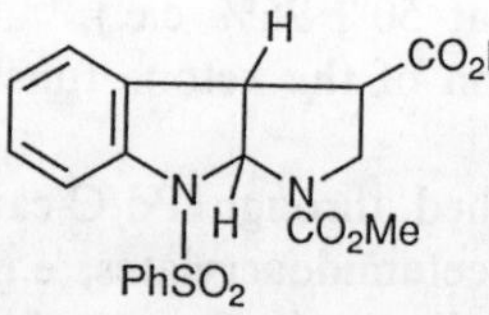

H CO_2Me N H N CO_2Me $PhSO_2$

(36)

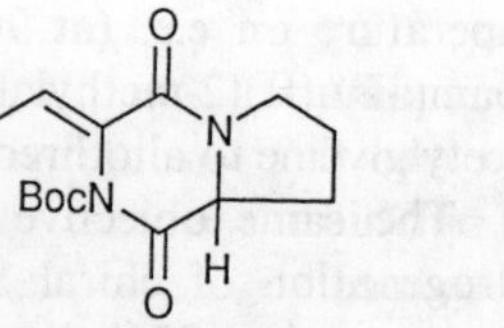

O N BocN H O

(37)

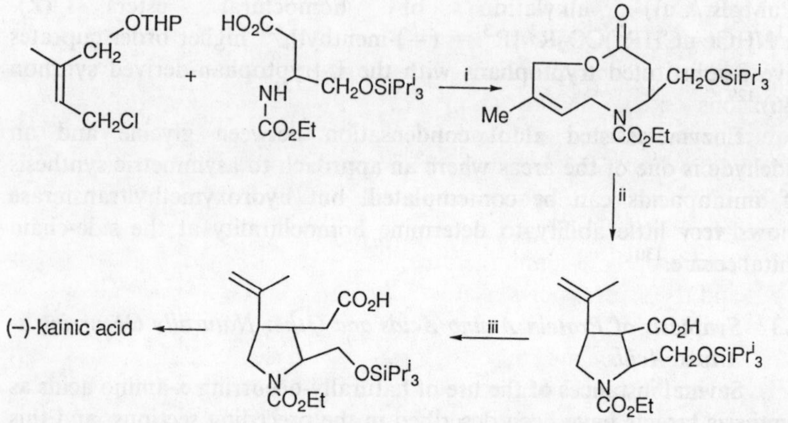

OTHP CH_2 CH_2Cl + HO_2C NH $CH_2OSiPr^i_3$ CO_2Et —i→ O O Me $CH_2OSiPr^i_3$ N CO_2Et

↓ ii

(−)-kainic acid ←— CO_2H N $OSiPr^i_3$ CO_2Et ←iii— CO_2H $CH_2OSiPr^i_3$ N CO_2Et

Reagents: i, BuLi/THF/N_2/−70 °C; then deprotection and intramolecular esterification; ii, LDA/BuLi/−100 °C→r.t.; iii, Arndt-Eistert homologation, then routine functional group manipulation

Scheme 11

HO_2C O CO_2H

(38)

CH_2 AcO CH_2 CH CH CH AcO–CH CH N CO_2Et Boc

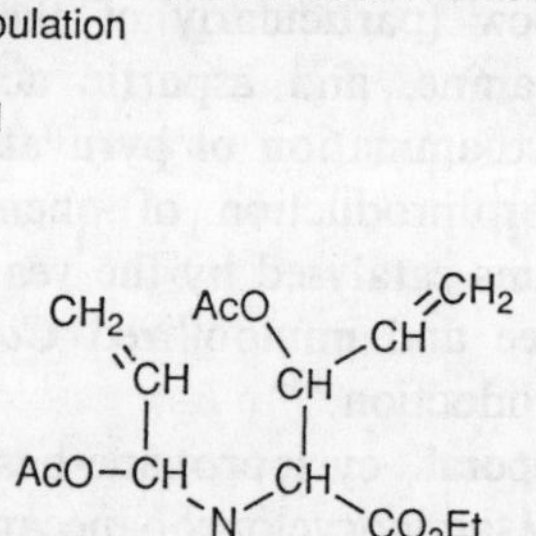

(39)

alkene to open up access to the chiral methylenecyclopropane moiety.[138] The readily-available (2R,3R)-epoxysuccinic acid (38) has been elaborated into the γ-azetidinyl-β-hydroxy-α-amino acid related to mugineic acid by Shioiri's group,[139] for whom (2R,3R)-2,3-epoxycinnamyl alcohol has seemed a better starting point in a later study culminating in syntheses of mugineic acid[140] 3-epi-hydroxymugineic acid, and distichonic acid (the azetidine-ring opened analogue).[141]

The kainoids and their relatives, like several of the preceding examples, have important, even spectacular, physiological properties that justify the unflagging interest in their synthesis.

Recent papers describing synthesis in this area are mostly extensions or consummations of earlier strategies. The stereochemical requirements as far as relative configurations are concerned, determine that the crop of recent papers, with one exception, exemplify the most obvious approach, in which the eventual substituted proline ring is formed from an acyclic precursor. The exception is the enolate Claisen route from a cyclic β-amino acid derived from L-aspartic acid (Scheme 11)[142] and the other reports relate to intramolecular ene-carbocyclization of (39), obtained by elaboration of N-Boc α-(acetoxyallyl)glycine ethyl ester[143] intramolecular Pauson-Khand ring formation (Scheme 12)[144] and tandem Michael addition (Scheme 13).[145] A previously-established route has been extended to provide a pair of acromelic acid A analogues (40), one of which ($n = 0$) is as potent in its physiological properties as kainic acid, while the other ($n = 1$) is inactive.[146]

A new synthesis of (−)-bulgecine uses the increasingly-popular chiral epoxide methodology (Scheme 14).[147] The first synthesis of the non-protein amino acid form the mushroom *Lycoperdon perlatum* involves the SmI_2-mediated formation of a spirolactone at C-4 of 4-hydroxyproline (Scheme 15).[148]

In spite of the vigorous activity surrounding the synthesis of the L-threonine derivative "MeBmt" [*alias* (2S,3R,4R,6E)-3-hydroxy-4-methyl-2-methylamino-6-octenoic acid], the unusual amino acid residue in cyclosporin, only one paper has appeared (Ref. 71 on p. 8, Vol. 20) reporting the synthesis of "MeBma", its C-9 analogue that is epimeric at the chiral centre carrying the hydroxy group. Another expedition along this path is shown in Scheme 16.[149]

Natural amino acids with unusual aromatic side-chains of the O-aryl ether variety, as are found in the ristomycins, offer formidable problems of synthesis. Continuing attempts to simplify the condensation of polyfunctional benzene derivatives through benzene moieties, include a study of the phenoxide-chloroarene-$Mn(CO)_3$ system, with aryl ether formation being followed by alkylation by a chiral glycine enolate to lead

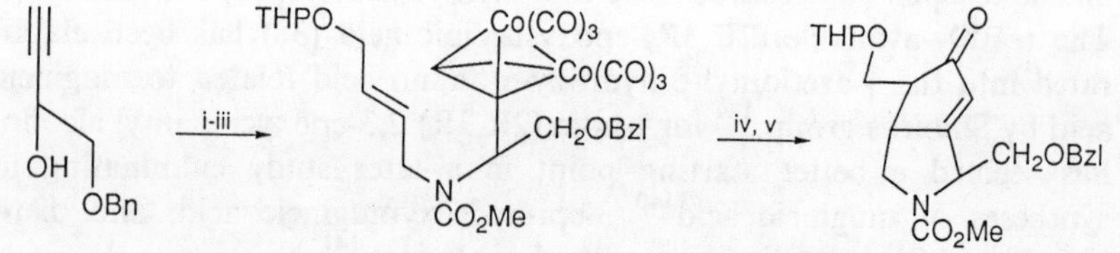

Reagents: i, PhtH, ($Pr^iO_2C.N$=)$_2$ /Ph_3P; ii, $NH_2.NH_2$ then $MeO_2C.Cl$;
iii, routine development; iv, Pauson-Khand reaction;
v, $LiAlH_4$, followed by iii→(–)-kainic acid

Scheme 12

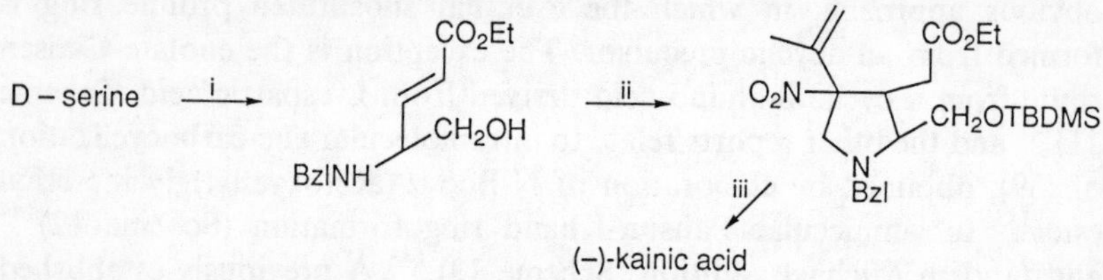

D – serine → (–)-kainic acid

Reagents: i, six steps; ii, $H_2C{=}CMe.C(NO_2){=}CH_2$;
iii, Pd(0)/$HCOO^-NH_4^+$, followed by routine steps→(–)-kainic acid

Scheme 13

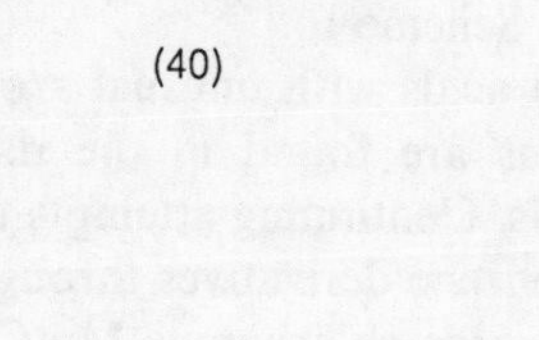

(40)

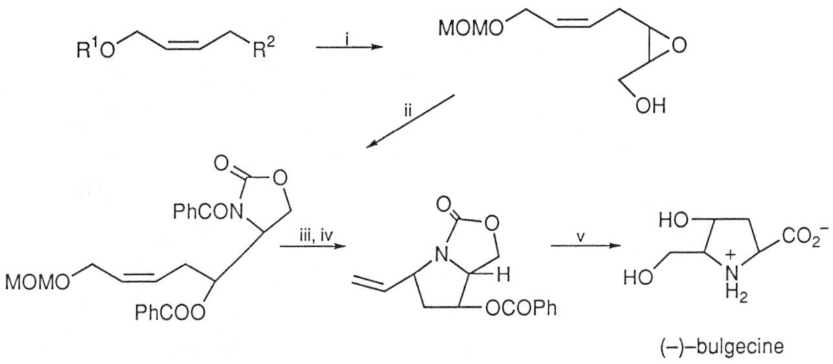

Reagents: i, several steps; ii, Ph.CO.NCO; iii, H_3O^+, deprotection; iv, β–elimination; v, O_3 and $KMnO_4$

Scheme 14

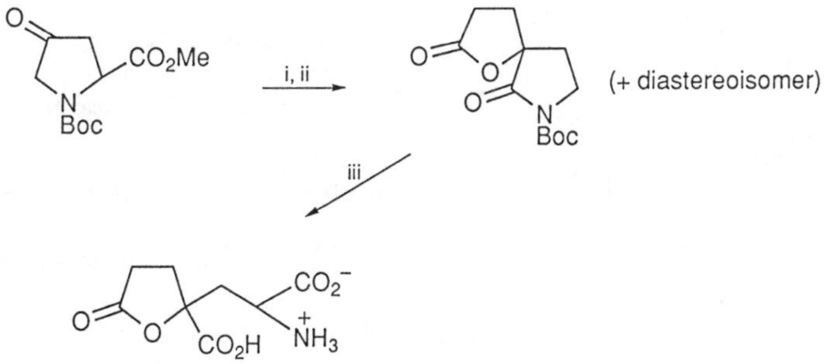

Reagents: i, $CH_2{=}CHCO_2Me/SmI_2$; ii, $RuO_2/NaIO_4$; iii, separate, reflux 6M HCl/12h

Scheme 15

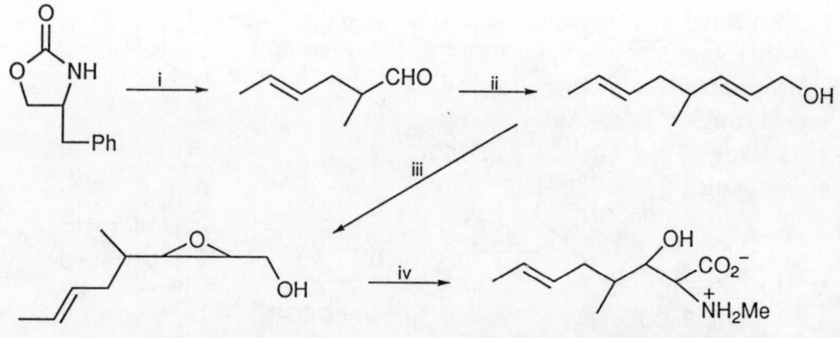

"MeBma"

Reagents: i, Evans methodology (*Tetrahedron Lett.*, 1987, **28**, 39);
ii, $(MeO)_2P(O)CH_2CO_2Me$/BuLi, then DIBAL;
iii, $Bu^tOOH/Ti(OPr^i)_4$/L-(+)-DET; iv, PDC then $MeNH_2$

Scheme 16

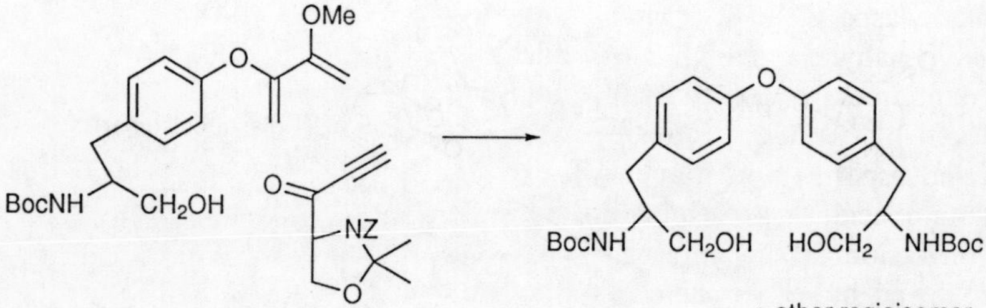

+ other regioisomer

Scheme 17

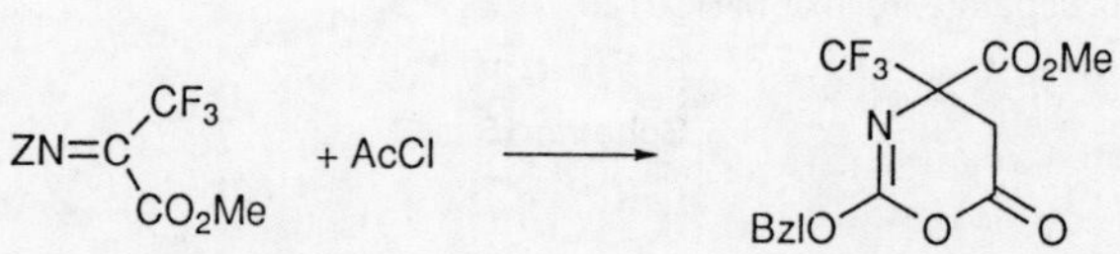

(41)

to protected ristomycinic acid derivatives in high optical purity.[150] Another approach avoids the condensation by starting with an appropriate alkenyl ether of Boc-L-tyrosine and forming the other arene ring through a Diels-Alder strategy (Scheme 17; see also Vol. 24, p.18). This leads to a 1:1-mixture of the required isomer, (S,S)-isodityrosinol accompanied by the the meta-isomer.[151]

4.4 *α-Alkyl Analogues of Protein Amino Acids.*

Important pharmaceutical roles are arising for these derivatives, associated with the irreversible enzyme inhibitory activity of some of them, and novel synthetic approaches are being established. Some material under this heading will be found in preceding sections, and elsewhere, in this Chapter. Thus, a general asymmetric α-amino acid synthesis based on alkylation of a homochiral imidazolinone has served to provide enantiomers of α-methylserine, elaboration through standard methods [$-CH_2OH \rightarrow -CH_2Cl \rightarrow -CH_2py^+$] giving the pyridiniomethyl analogue.[152] A quite different approach applied to the preparation of α-methylaspartic acid, is the alkylation with retention of configuration of the hexahydropyrrolo[2,3-*b*]indole easily obtainable from a tryptophan enantiomer, followed by degradation of the indole moiety ($NaIO_4$/$RuCl_3$).[153] A more fundamental approach to assembling the α-substituted aspartic acid framework exploits the dual categorization of this amino acid as a member of both α- and β-amino acid families, by conversion of the imine $ZN=C(CF_3)CO_2Me$ into the oxazinone (41) followed by nucleophilic ring-opening[154] (see also Vol. 22, p.22 for an exact precedent for this work).

An unexpected entry to α-[2-(phenylseleno)ethyl]-α-amino acids was discovered through attempted PhSeSePh/$NaBH(OMe)_3$/DMF reduction (thought of as "$NaPhSeBH_3$" reduction) of N-benzoyl homoserine lactone (Scheme 18).[155] There is no competing lactone reduction in this process.

Ornithine derivatives are crucially important in the context of the opening sentence of this section, and absolute configurations have been assigned by X-ray crystal analysis and by classical chemical correlation methods to enantiomers of irreversible inactivators of decarboxylases, namely, α-chlorofluoromethyl derivatives of this amino acid and of glutamic acid and m-tyrosine.[156] α-Methylornithine enantiomers have been converted into L-(+)- and D-(−)-α-methylarginines through standard side-chain functional group elaboration using ZN=C(SMe)NHZ.[157]

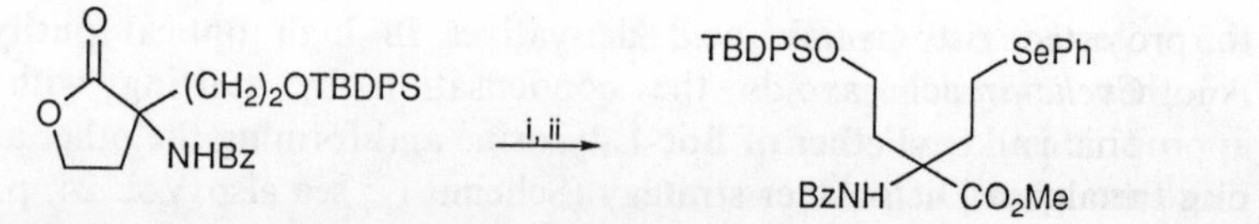

O
$(CH_2)_2OTBDPS$
O
NHBz
i, ii
TBDPSO
SePh
BzNH
CO_2Me

Reagents: i, $(PhSe)_2$, $NaBH(OMe)_3$/DMF/60 °C; ii, CH_2N_2, H_3O^+

Scheme 18

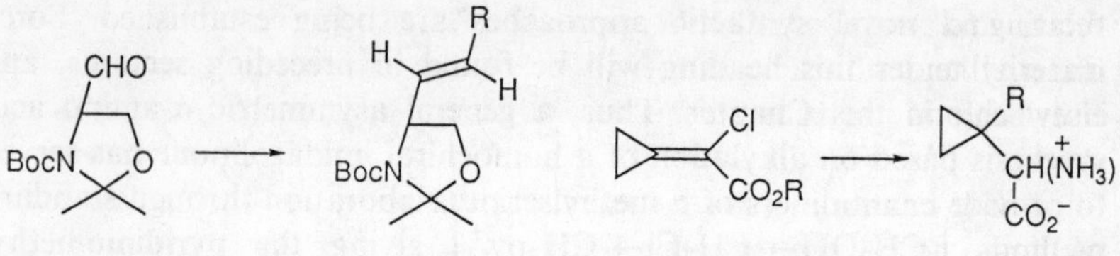

CHO
BocN
O
(42)
H
R
H
BocN
O
Cl
CO_2R
R
$\overset{+}{C}H(NH_3)$
CO_2^-
(43)

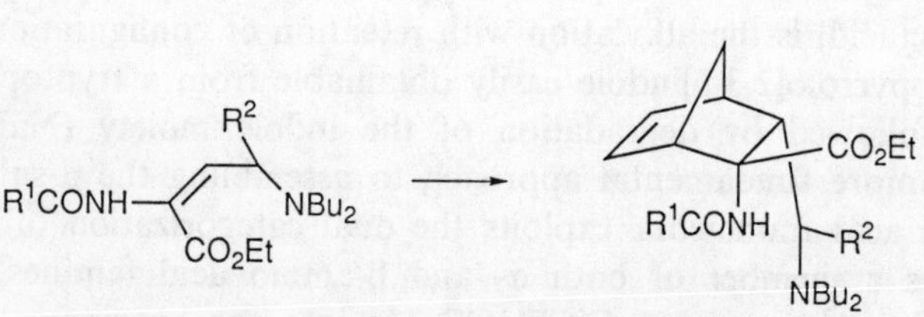

R^2
R^1CONH
NBu_2
CO_2Et
CO_2Et
R^1CONH
R^2
NBu_2
(44)

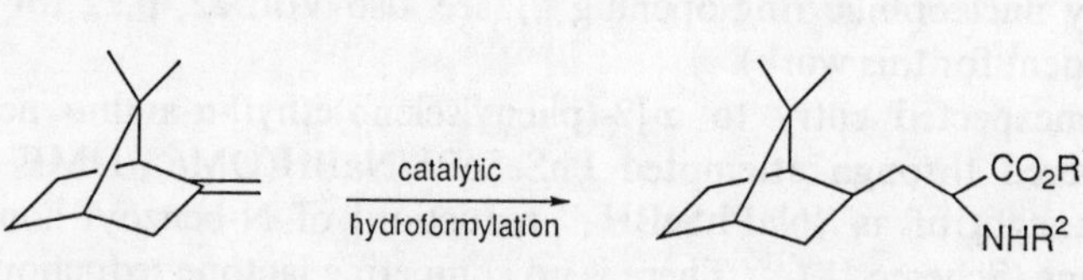

catalytic
hydroformylation
CO_2R^1
NHR^2
(45)

4.5 *Synthesis of C-Alkyl and Substituted C-Alkyl α-Amino Acids and Cyclic Analogues.*

This is a Section intended to give refuge to papers covering both close analogues of protein amino acids, and to more distant structural relatives that are of aliphatic, alicyclic, or saturated heterocyclic types. Again, some papers that might have been included here, because they are covered by this description will be found in a preceding section instead.

β-Iodo-alanine is becoming established as a useful synthon, when converted into its organozinc derivative, for development of side-chain aliphatic features, such as 4-oxo-alkyl (reaction with RCOCl) and 3-arylalkyl (reaction with an aryl iodide).[158] An uninformative abstract (*Chem.Abs.*, 1992, **117**, 112003) has to be blamed for lack of details relating to syntheses of DL-2-amino-5,5-dimethylhexanoic acid and -6,6-dimethylheptanoic acid as their N-acetyl methyl esters and employing t-butyl chloride (presumably a version of the acetamidomalonate synthesis).[159]

Side-chain alicyclic structures may be built on to acyclic side-chains of amino acid synthons in a variety of ways for a variety of purposes - ethyl diazoacetate to a protected D-allylglycine to give D-2-amino-4,5-methano-adipates that show NMDA receptor activity;[160] αβ-methano-valine, -phenylalanine, and -alanine by treatment of N-phthaloyl β-bromo amino acids with NaH,[161] DL-(E)- and (Z)-2,3-methano-m-tyrosines (potent L-aromatic amino acid decarboxylase inhibitors, by similarly classical methods from mono-ethyl malonate,[162] and (E)-α-[2-phenyl(or ethyl)cyclopropyl]glycines by diastereoisomeric dibromocyclopropanation of the alkene derived from the D-serine synthon (42; cf, 23) followed by reductive debromination.[163] A novel and efficient route to these cyclopropanated side-chains starts with the dehydro-alanine imine $(MeS)_2C{=}NC({=}CH_2)CO_2Me$ and involves its cyclopropanation with diazomethane.[164] The only recent study within this family of compound, involving amination of a substrate that already has the cyclopropane moiety built in to it, has been directed at syntheses of natural cyclopropylglycines (43; R = Me), including cleomin (43; R = OH).[165]

Cognate areas in which alicyclic side-chains are constructed analogously, are covered in cycloadditions to chiral αβ-dehydro-amino acid derivatives (e.g. 44), including diazomethane cyclopropanation,[166] and a synthesis of the (+)-fenchane derivative (45), which is 2200 times sweeter than sucrose.[167] Improved syntheses are reported for β-cyclohexyl-L-aspartic acid and γ-cyclohexyl-L-glutamic acid in protected forms.[168]

Alkylation of N-TBDMS-azetidinone-4-carboxylates (*alias* β-lactam-4-carboxylates; 46) through standard LDA/alkyl halide protocols

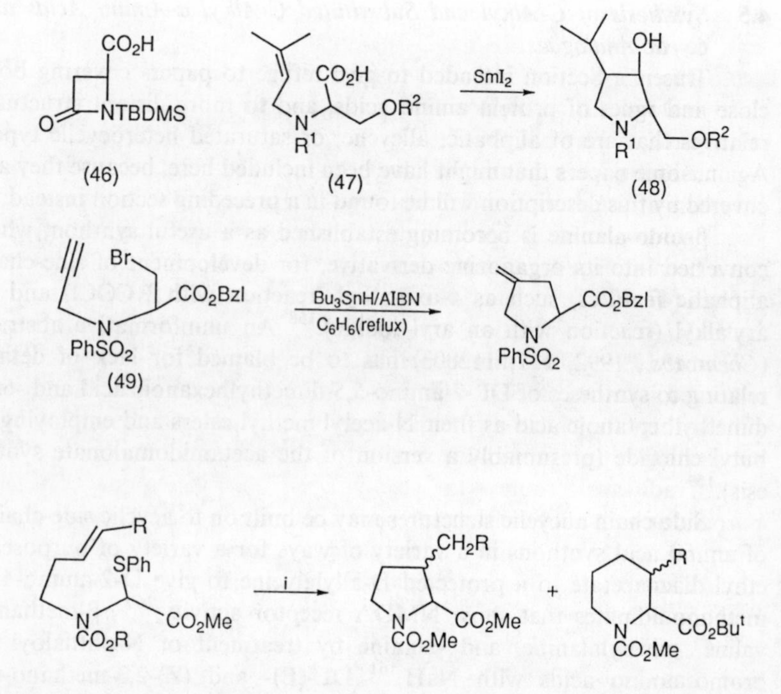

CO_2H NTBDMS O (46)

CO_2H N R^1 OR^2 (47) $\xrightarrow{SmI_2}$ OH N R^1 OR^2 (48)

Br CO_2Bzl N $PhSO_2$ (49) $\xrightarrow[C_6H_6(reflux)]{Bu_3SnH/AIBN}$ CO_2Bzl N $PhSO_2$

R SPh N CO_2R CO_2Me $\xrightarrow{i}$ CH_2R N CO_2Me CO_2Me + R N CO_2Me CO_2Bu^t

Reagents: i, Bu_3SnH, AIBN

Scheme 19

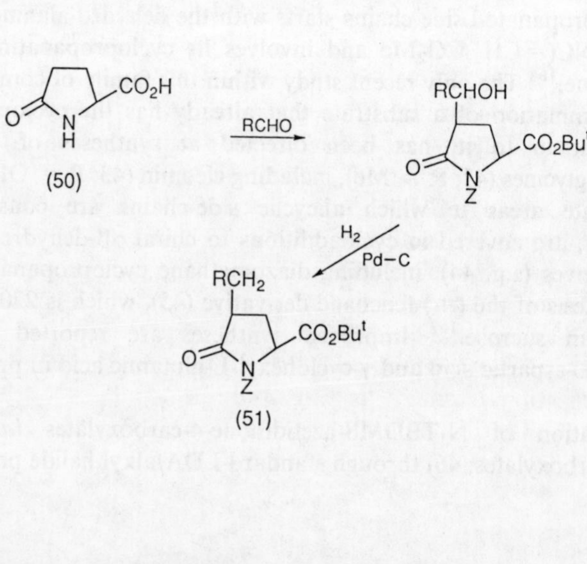

O N H CO_2H (50) $\xrightarrow{RCHO}$ RCHOH O N Z CO_2Bu^t $\xrightarrow[Pd-C]{H_2}$ RCH_2 O N Z CO_2Bu^t (51)

gives anti-oriented products from which (2S,3R)-3-alkylaspartic acids can be obtained.[169] (2RS,E)-3-Ethylidene azetidine-2-carboxylic acid (*alias* racemic polyoximic acid) has been prepared through $Rh(OAc)_4$-catalysed cyclization of $BocNHCH_2COCN_2CO_2Bu^t$ followed by Wittig alkenylation with $PhSCOCH=PPh_3$ and routine elaboration.[170]

α-Imino acids are of ever-increasing interest, and some syntheses of members of the proline and pipecolic acid families of natural origin have been mentioned in the preceding sections. N-Alkenylserine derivatives have been shown to undergo ring-closure with oxophilic samarium(II) species (47 → 48),[171] and analogous starting materials are employed in Bu_3SnH/AIBN cyclisation of N-alkenyl α-(phenylthio)glycine esters (Scheme 19).[172] The minor competing 6-endo-cyclization mode was observed in some cases in this study, leading to pipecolic acid isomers, but not in the 5-exo-dig radical cyclisation of the L-serine-derived substrate (49) put through the same reagent treatment.[173]

Traditional routes to substituted prolines include uses for pyroglutamic acid enantiomers (aldolization after Li enolate generation; 50 → 51),[174] Diels-Alder reactions of N-benzylimines (52 → 53),[175] and routes to bicyclic prolines, one exploiting the L-tyrosine functional groups (54 → 55),[176] and the other following the "magic meso" philosophy that is currently intriguing several research groups and has been used for the preparation of the ACE inhibitor Trandolapril (Scheme 20).[177]

Syntheses within the pipecolic acid family (see also Refs.110,111), that approach the target through conventional cycloaddition strategies employing imines, include the use of Danishefsky's diene with an N-alkylidene L-amino acid ester (56 → 57),[178] and a similarly-oriented study using $RCH=CR^1C(OSiMe_2R^2)=CH_2$ + $R^3N=C(CO_2R^4)_2$ in which the outcome of the cycloaddition contradicts earlier claims.[179] The cyclohexenones resulting from these reactions can be elaborated further through their C=C grouping to give trans-3-substituted-4-oxo-L-pipecolic acids.[180] The hitherto unsuspected propensity for Schiff bases of αβ-dehydroamino acids $ArCH=NC(CO_2Me)=CH_2$ to undergo dimerization has now become a reality (Scheme 21), with the benefit of leading to new types of mono- and bicyclic α-amino acids.[181]

A different pattern of substitution arises from the cyclisation of derivatized 1-aminobut-3-en-2-ols, obtained from N-Boc-L-alanine and providing the N-methyl-L-pipecolic acid analogue (58).[182] An alternative cyclization strategy is involved in rhodium(II) diacetate catalysed NH insertion of L-glutamic acid-based diazoketones (59 → 60).[183]

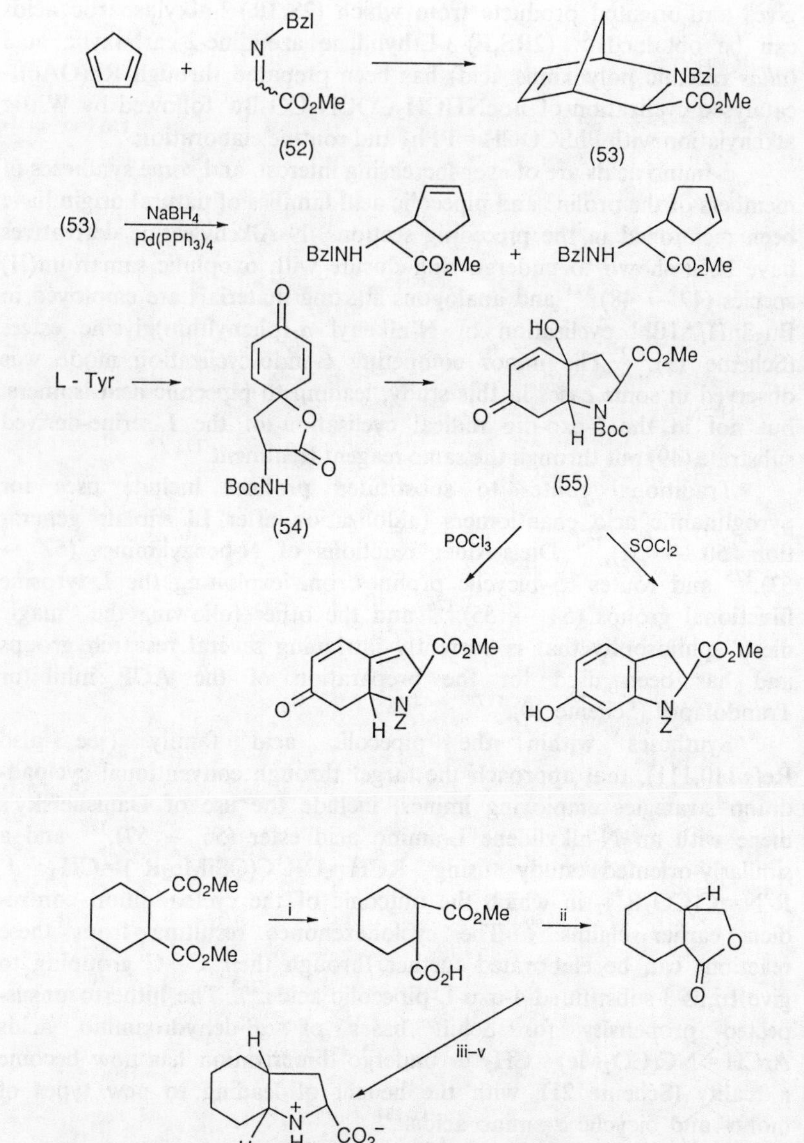

+ N–Bzl, CO_2Me (52) → NBzl, CO_2Me (53)

(53) $\xrightarrow[Pd(PPh_3)_4]{NaBH_4}$ BzlNH, CO_2Me + BzlNH, CO_2Me

L - Tyr → O, O, O, BocNH (54) → HO, CO_2Me, O, N, H, Boc (55)

$POCl_3$: CO_2Me, O, N, H, Z

$SOCl_2$: CO_2Me, HO, N, Z

CO_2Me, CO_2Me $\xrightarrow{i}$ CO_2Me, CO_2H $\xrightarrow{ii}$ H, O, O

iii–v: H, $\overset{+}{N}$, H_2, H, CO_2^-

Reagents: i, Pig liver esterase; ii, $NaAlH_2Et_2$/toluene
iii, isomerize ring junction stereochemistry (MeONa)
iv, NH_3, Beckmann rearrangement; v, HCHO/KCN then H_3O^+

Scheme 20

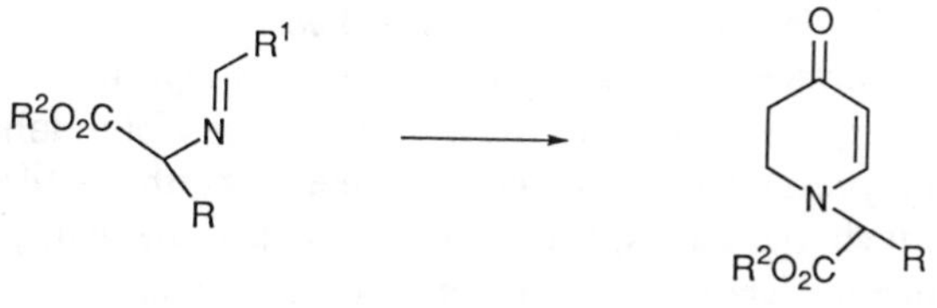

R^1 R^2O_2C N R

(56)

O N R^2O_2C R

(57)

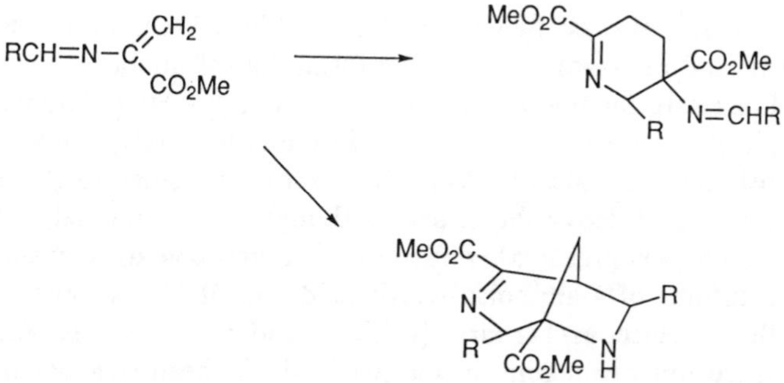

CH_2 RCH=N–C CO_2Me

MeO_2C N CO_2Me N=CHR R

MeO_2C N R R CO_2Me N H

Scheme 21

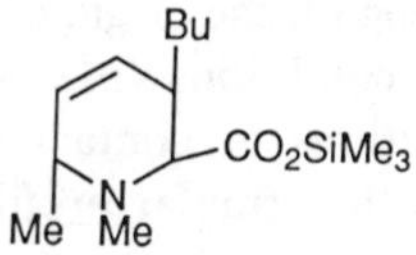

Bu CO_2SiMe_3 N Me Me

(58)

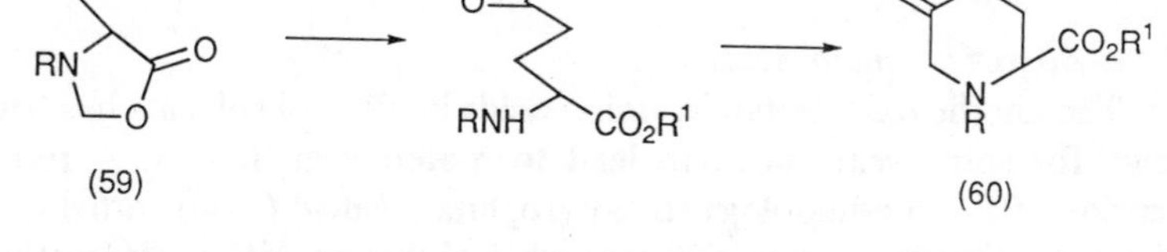

HO_2C RN O O

(59)

N_2 O RNH CO_2R^1

O CO_2R^1 N R

(60)

4.6 *Models for Prebiotic Synthesis of Amino Acids.*

It is many Volumes ago in this Series that the *deja-vu* nature was suggested, of papers still appearing then as now, describing laboratory modelling of the presumed chance environmental synthesis of protein amino acids from simple atmosphere components and an energy source. However, the picture has certainly taken a wider perspective, and has become more detailed, as the years have gone by, even though the general approach to the underlying chemistry has remained the same. In particular, the discovery of nucleic acid components[184] and carbohydrates[185] in the laboratory soups has been a useful outcome, as well as establishing the role of HCN and opening up its organic chemistry a little more.

The former[184] of these two studies has established the intermediacy of oxalic acid and/or oxamic acid in the formation of amino acids and nucleic acid bases during u.v. photolysis (<280 nm, $> -80°$) of aqueous $(NH_4)_2CO_3$ in the presence of Mg^{2+} salts. In the latter study, conditions mimicking ash-gaseous volcanic clouds were shown to generate glucose, mannose, ribose, and deoxyribose, and (although not mentioned in the abstract of this paper) presumably also traces of amino acids. Unusually high concentrations of 4-aminoisobutyric acid and of DL-isovaline are found at the Cretaceous/Tertiary (K/T) boundary at Sterns Klint, Denmark. Intrusion of magma into a coal bed has been speculated to give the simple gases needed to synthesize hydantoins from which these amino acids could have formed.[186]

Simulated hydrothermal conditions (submarine volcanic vents) are underlying experiments exploring the latest thinking on the origin of life, and the high temperature (150°) and reducing conditions (pH $\sim$ 7) applied to HCHO, NH_3, HCN, and H_2O mixtures in contact with mineral surfaces yield a range of amino acids with a similar profile to that obtained in traditional electric spark discharge experiments, but with higher relative yields of amino acids other than glycine.[187] Bombardment of $CO/N_2/H_2O$ mixtures with 2.8-40 MeV protons to yield HCN and amino acids is claimed to show that cosmic radiation is a more effective energy source in this area than electric spark discharge.[188]

4.7 *α-Alkoxy α-Amino Acids.*

The anodic oxidation of α-imino acids in alkanol solvents has been known for some years now, to lead to α-alkoxy analogues. A recent extension of this methodology to L-pyroglutamic acid (+)-menthyl ester and substitution of the α-methoxy group by reaction with allyltrimethylsilane and $TiCl_4$, gives a 2:1-mixture of diastereoisomers of spiro-compound (61).[189]

CO_2-(+)-menthyl
R^1
(R^1 = OMe → R^1 = allyl)
$TiCl_4$
(61)

NAr
Boc
(62)
ArN
OBzl
Boc
(63)
ArN
RO_2C
OBzl
β-hydroxy-Asp

Br
OH Br
(64)
HO OH
CO_2^-
H_2
(65)

Ph H
(66)
Ph
Ph
Ph
(67)
NH_2
$HO_2C(CH_2)_2$
$(CH_2)_2CO_2H$
NH_2
(68)

MeO_2C—L-Trp—OMe
85% H_3PO_4
Ac_2O/py.
CO_2Me
Ac
CO_2Me
(69)

α-Benzyloxy- and -methoxyglycines are more stable than corresponding bromides, but undergo analogous substitution reactions and are thus more convenient for such processing; e.g. reaction with Sn_2Bu_6 and a disulphide RSSR gives α-(alkanethio)glycines.[190]

4.8 *α-(Halogenoalkyl) α-Amino Acids.*

In noting that synthetic work starting in a routine way with β-bromoalanine and related compounds has been cited in preceding sections of this Chapter, this section covers papers more centrally concerned with the halogenoalkyl amino acids as final targets. 4-Fluoroglutamic acid is a representative illustration of the attraction of such compounds, not in their own right but as potential enzyme inhibitors or otherwise capable of disrupting certain metabolic processes. The stereospecific route to this compound requires the hydroxypyroglutamic acid as precursor, to be substituted with inversion by diethylaminosulphur trifluoride.[191] This work follows pioneering work by Hudlicky[192] (see Vol. 23, p.26) which is discussed in Ref. 191.

4.9 *α-(Hydroxyalkyl) α-Amino Acids.*

Several examples of aldol reactions with glycine carbanions or equivalents have been described in preceding pages, and these and others like them collected here, lead to β-hydroxy-α-amino acids. N^{τ}-Trityl β-hydroxy-L-histidine has been prepared by the oxazolid-2-one method using the trityl 4-formylimidazole[193] and β-hydroxyaspartates via β-lactam formation from the Garner oxazoline derived from L-serine (62 → 63).[194] An interesting trimethylsilylated homoserine lactone analogue has been prepared.[195]

The elaboration of simple monosaccharide derivatives into hydroxylated amino acids continues to offer particularly convenient access to prolines (64 from D-xylonolactone → 65).[196]

The versatile chiral glycine synthon (66) has recently shown even more versatility in the synthetic applications of the stable azomethine ylides (67) formed from it. The previously-established uses flow from cycloadditions, but if excess aldehyde is used in the preparation of the ylide, then adducts transformable into β-hydroxy-α-amino acids are formed in the absence of dipolarophile.[197]

4.10 *α-Amino Acids with Aminoalkyl Side-Chains.*

An increasing number of studies under this heading is reflected in entries in a number of other Sections in this Chapter (Refs. 86, 87, 104, 105, for example). Typical studies to be described here, must start with a heroic preparation of L-(+)-2,4-di-aminobutanoic acid, by refluxing L-glutamic acid with NaN_3 and H_2SO_4,[198] followed by a more controlled synthesis of (S,S)-2,3-di-aminobutanoic acid, the constituent of antimycins and cirratiomycins.[199] The latter synthesis starts with N-Boc-L-threonine, via the 1,3-diol and exploiting the propensity of TBSCl/DMAP to silylate the primary alcohol function selectively; mesylation of the other function, and displacement by azide, completes the process. An interesting preparation of di-amino di-acids (68) uses the known 1,4,6-tri-O-acetyl-2-azido-2,3-dideoxy-3-nitro-α-D-glucose.[200]

4.11 *α-Amino Acids with Unsaturated Side-Chains.*

A number of examples of alkenylation of α-functionalized glycine synthons has been reported in the current literature, viz. S_N1-substitution of N-alkoxycarbonyl α-methoxyglycinamide by allylsilanes mediated by BF_3,[201] and Wittig-Horner reaction of $Me_2O_3PCH(NHR)CO_2Me$ with R^1OCH_2CHO to give (E)-vinyl ethers[202] and protected (E)-αβ-didehydro-amino acids more generally.[203] Acetamidomalonate routes to propargylglycines, and the homologation of the products through $Pd(PPh_3)_4$-mediated reactions with aryl and vinyl halides or triflates ($-CH_2C{\equiv}CH \rightarrow -CH_2C{\equiv}CR$) without racemization, have been described.[204] A synthesis of 5-methoxy-N-methoxy-L-proline methyl ester and its reaction with bis(trimethylsilyl)acetylene/$TiCl_4$ to give cis- and trans-5-vinyl-L-proline, and analogous carbanion substitution reactions to give the 5-ethynyl analogues, as well as a synthesis of 4-methylene-L-proline from the 4-oxo-compound, have been described.[205]

A standard synthesis of αβ-dehydroalanine and its butyrine analogue through dehydration of protected serines and threonines respectively, has been improved through the use of diethylchlorophosphate [$(EtO)_2P(O)Cl$] and NaH in THF.[206]

The first of many potential uses for a new L-serine-derived Zn-Cu reagent, $IZn(CN)CuCH_2CH(NHBoc)CO_2Bzl$, prepared from the protected iodo-alanine by reaction first with the Zn-Cu couple, then with Knochel's soluble copper salt CuCN,2LiCl, has been chosen to be homologation by reaction with an allyl halide, or other allylic electrophile.[207]

Manipulation of the products of these syntheses is feasible in several ways, such as (E)- → (Z)-isomerization by acid, base or radical catalysis,[203] and other examples of re-location of the site of unsaturation described in a later Section (6.3: Specific Reactions of Amino Acids) extends the synthetic usefulness of these compounds.

4.12 *α-Amino Acids with Aromatic and Heteroaromatic Side-Chains.*

A major feature over the years, has been the collection of papers describing syntheses of near-relatives of the protein amino acid representatives conforming to the title of this Section. The significance of the work carried out lies only occasionally in the synthetic details, but more in the potential physiological properties, particularly towards the NMDA receptor, of the amino acids.

Standard general methods (e.g. bis-lactim ether alkylation,[208] Pd-catalysed cross-coupling of an aryl iodide with methyl 2-acetamidoacrylate[209]) have been illustrated for syntheses of N-Boc p-dimethylphosphonomethyl-L-phenylalanine[208] (see also[210] for this and for p-sulpho- and -carboxy and -N-hydroxycarboxamido-L-phenylalanines, and see[211] for o-, m-, and p-phosphonomethyl-, o-, m-, and p-sulphomethyl-, m-carboxymethyl- and m-N-hydroxycarboxamidomethyl-L-phenylalanines (and some cyclohexylalanine analogues), and see[212] for syntheses of 6-fluoroDOPA and its potential metabolites. Pd-Catalysed cross-coupling reactions have been explored, so as to develop methods for the connection of relevant aryl moieties, viz. (R)-4-hydroxyphenylglycine and (S)-3,5-dihydroxyphenylglycine, seen in vancomycin.[213]

Some general methods related to those of the preceding paragraph have proved satisfactory for syntheses of tryptophan analogues, viz. $EtAlCl_2$-catalysed Michael-type addition of indoles to Schiff bases of αβ-dehydroalanine, $Ph_2C{=}NC({=}CH_2)CO_2Me$,[214] and (−)-(R)-7-aza-tryptophan through alkylation [-100°/HMPA/THF/$KN(SiMe_3)_2$] of the (+)-camphor imine of glycine t-butyl ester.[215] A 5-cyano-L-tryptophan synthesis employs a protected L-tryptophan (69) as starting material and proceeds via the 5-bromo-analogue (NBS/AcOH).[216]

N^{τ}-Alkylation of L-histidine using 2-chloro-1-methoxymethylindole starts a short synthesis of the amino acid residue present in moroidin.[217] Non-protein natural amino acids with heteroaromatic side-chains often possess spectacular physiological properties, represented in this year's literature by analogues of homoibotenic acid and of its isoxazole variant,[218] that include potent agonists tested for NMDA receptor activity. Quisqualic acid analogues tested for, but ineffective in, sensitizing cell neurons to depolarization by L-2-amino-4-phosphonobutanoic acid, involve replacement of the isoxazole moiety by the maleimide ring and other simple five-membered heterocycles.[219] However, DL-[tetrazol-5-yl]glycine, prepared using a rarely-used but specifically appropriate glycine equivalent (Scheme 22),[220] is a highly potent NMDA agonist. (2S)-Nicotinylalanine (in Scheme 23) has been prepared through an unusual L-aspartic acid synthon in which the β-carboxy group is liberated for activation as the acid chloride. The product shows promise as a new neuroprotecting agent.[221]

Analogues of DOPA and methylDOPA in which the catechol moiety is replaced by analogous benzimidazole and benzotriazole group-

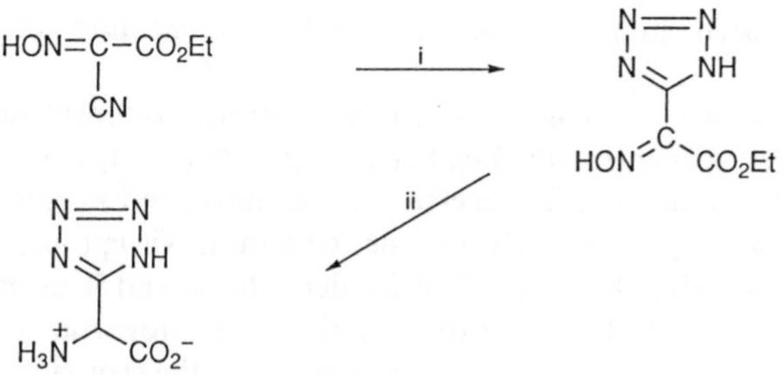

Reagents: i, NaN_3/DMF; ii, H_2/Pt

Scheme 22

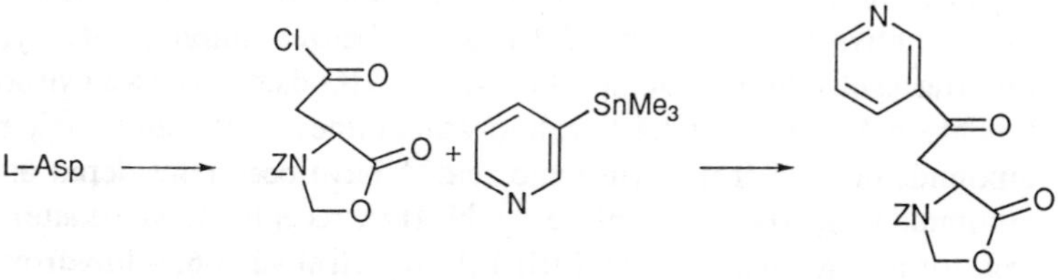

Reagents: i, $PdCl_2(PPh_3)_2$; ii, deprotection, H_3O^+

Scheme 23

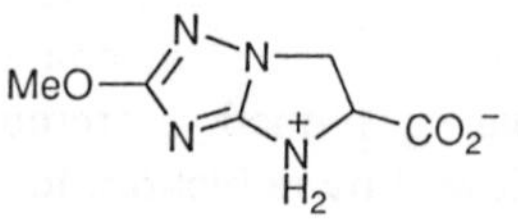

(70)

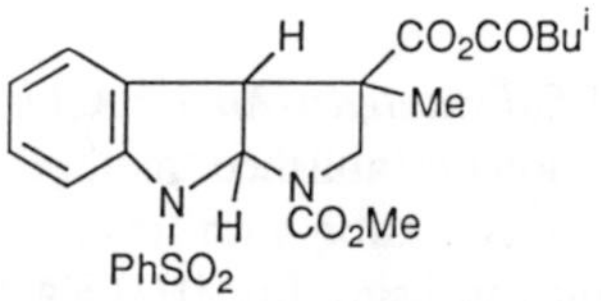

(71)

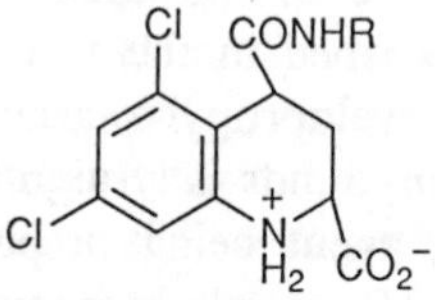

(72)

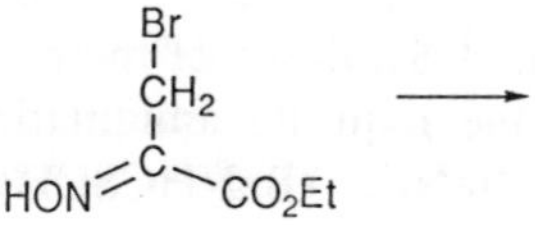

(73)

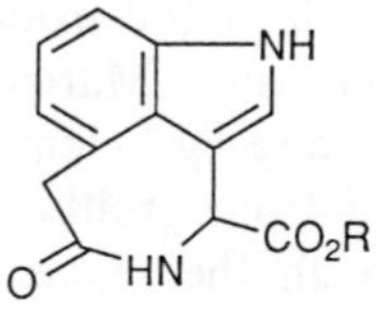

(74)

ings have been prepared through routine application of standard methods.[222]

Increasing interest is being shown in sterically-constrained analogues of these aromatic and heteroaromatic amino acids, in which the common alanyl moiety is connected through nitrogen to the aromatic or heteroaromatic grouping. From one structural viewpoint, these are proline, or occasionally pipecolic acid, derivatives, and it seems that the interest in them for their potential physiological properties may be the result of empirical thinking-compounds that fuse the structural details of two classes of neuroactive α-amino- and -imino acids might themselves show some valuable properties. Distant proline relatives include the triazole (70) formed from the 5-bromo-1,2,4-triazole and methyl N-benzylaziridinecarboxylate,[223] and the unusually stable mixed carboxylic carbonic anhydride (71) formed from N^{im}-benzenesulphonyl-L-tryptophan after cyclisation through established methods.[224] trans-2-Carboxy-5,7-dichloro-4-amido-1,2,3,4-tetrahydroquinolines (72)[225] and analogous compounds derived from kynurenic acid[226] have been considered useful compounds with which to explore the NMDA receptor (glycine site) in a search for antagonists. (R)- and (S)-1,2,3,4-Tetrahydro-6,7-dihydroxy-1-methylisoquinoline-1-carboxylic acids (metabolic precursors of (R)-salsinol in humans) have been synthesized through standard isoquinoline methodology, and their absolute configurations have been established.[227] The cyclic tryptophan derivative (74) has arisen from studies of tryptophan synthesis using indole and the β-bromoalanine synthon (73).[228]

4.13 *N-Substituted α-Amino Acids.*

A long-established method for preparing N-methyl amino acid esters, condensation with formaldehyde followed by reduction, has been accomplished using Et_3SiH/TFA for the reduction step and now understood to proceed via the iminium ion.[229] Side-chain N-alkylated arginines have become important again, this time in the context of potential nitric oxide synthase inhibitors, as described in the work of Olken and Marletta[230] on N^G-allyl- and -cyclopropyl-L-arginine (reviewed by Feldman).[231] Synthesis of these compounds is straightforward from ornithine, the requisite amidinating agent being prepared through the sequence BzNCS→BzNHCSCHR (R = allyl or cyclopropyl) → S-alkyl, followed by established methodology.

4.14 *Sulphur Containing α-Amino Acids.*

The most familiar of the α-amino acids under this heading, cysteine and its post-translationally modified derivatives, are featured in this year's crop of citations that are not without interest even though exploring familiar and otherwise thoroughly-studied amino acids.

A preparation of S-alkylcysteine homologues involving AIBN-catalysed addition of thiols to (R)- or (S)-allylglycine is accompanied by only a small degree of racemisation.[232] Enantioselective synthesis of (2R,3S)- and (2R,3R)-3-phenylcysteine derivatives has been accomplished in this general way.[233] Preparations of N-Boc-S-trifluoromethyl-DL-cysteine,[234] and of D-(β-ureidoethyl)-cysteine and -homocysteine, involve routine methodology.[235] The last-mentioned preparation involves hydantoin formation and *Agrobacterium radiobacter* resolution.

A synthesis of meso-lanthionine calls for a rather more subtle strategy than at first sight would appear to be needed. Temporary linking of two cysteine molecules of opposite configuration through their carboxy-groups, then disulphide formation, has been achieved through esterifying N-Z-D-cysteine with N-Boc-S-trityl-L-cysteine 2-hydroxyethyl ester (DCCI/DMAP) followed by $PhI(OAc)_2$ oxidation. The functional group transformation disulphide → sulphide was achieved in the usual way with $(Et_2N)_3P$ in DMF.[236] The bis-cysteine disulphide of meso-dimercaptosuccinic acid (75) may be the form in which dimercaptosuccinic acid (a metal-binding agent) is excreted by mammals.[237]

4.15 *Phosphorus Containing α-Amino Acids.*

There is a steady interest in amino acid analogues in which the carboxy group is replaced by a phosphorus oxy-acid group, but the policy for this Specialist Periodical Report continues to exclude these and other carboxy group substitutions. However, the presence of a phosphorus oxy-acid moiety in an amino acid side-chain does render that amino acid open to consideration for inclusion in this Section. Indeed, there is growing interest in such compounds, both for synthetic applications (see preceding Section 4.1) and for their potential physiological activity.

Treatment of the serine-derived β-lactone (76) with dimethyl trimethylsilylphosphite gives β-(dimethylphosphonyl)-L-alanine in a protected form suitable for use in peptide synthesis.[238] The homologous γ-(dimethylphosphonyl)butyrine has been approached by elaboration of the side-chain carboxy group of N-Boc-L-aspartic acid α-t-butyl ester → CH_2OH → CHO by TEMPO-catalysed NOCl oxidation, and dimethyl trimethylsilylphosphate treatment[239] and alternatively, by asymmetric amination of $Et_2O_3PCH_2CH_2COCO_2Et$ with (1S,2R)-$PhCH(OH)CHPhNH_2$.[240] L-Aspartic acid is the source of phosphinothricin, $MeP(O)(OH)CH_2CH_2CH(NH_2)CO_2H$, in protected form, based on conversion into an N-protected L-homoserine lactone, ring-opening

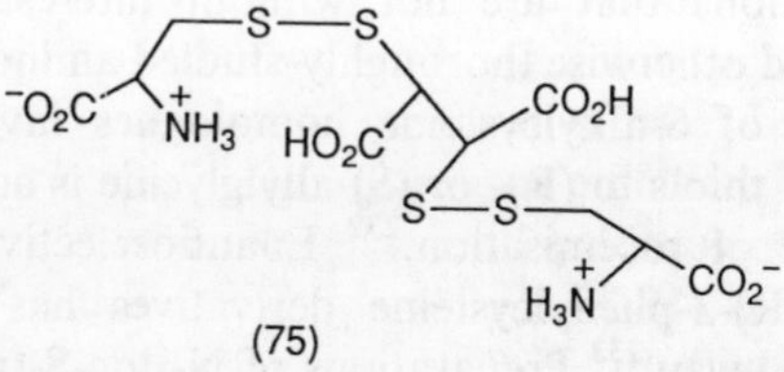

S–S
$^{-}O_2C$ $\overset{+}{N}H_3$ HO_2C CO_2H
S–S
$H_3\overset{+}{N}$ CO_2^-

(75)

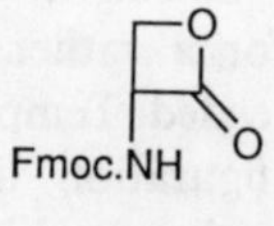

O
Fmoc.NH O

(76)

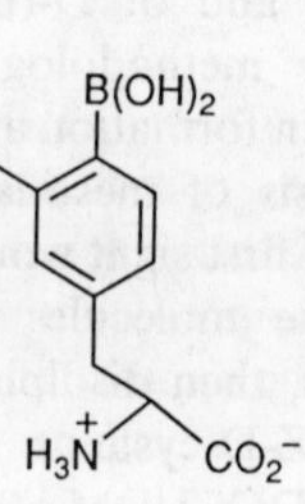

$B(OH)_2$
^{18}F
$H_3\overset{+}{N}$ CO_2^-

(77)

$PhCH_2$
Pr^iO_2C CO_2Pr^i
HO

i, ii

OAc
$PhCH_2$
O O O

iii

$PhCH_2$
H_2NC CO_2Pr^i
O AcO

iv, v

H_2N CO_2Pr^i
HO

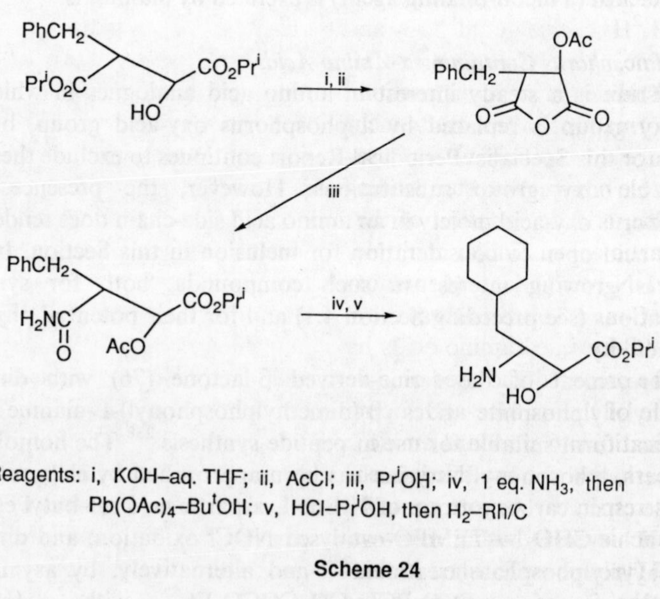

Reagents: i, KOH–aq. THF; ii, AcCl; iii, Pr^iOH; iv, 1 eq. NH_3, then $Pb(OAc)_4$–Bu^tOH; v, HCl–Pr^iOH, then H_2–Rh/C

Scheme 24

with HCl to give the β-chlorobutyrine, and reaction with $MeP(OEt)_2$,[241] An alternative starting point for this compound is a protected L-vinylglycine, prepared from L-methionine or L-glutamic acid in established ways.[242] The phosphinic acid analogue of kainic acid in which the γ-carboxy group is replaced, has been synthesized by Michael addition to $HC{\equiv}CPO_3Et_2$ of the (1S,2S,5S)-2-hydroxypinanone - glycine Schiff base, N-alkenylation with $BrCH_2CH{=}CMe_2$ an ene cyclisation through established methodology.[243]

4.16 *Synthesis of Labelled Amino Acids.*

The reports collected for this Section all concern α-amino acids, except for one citation for the preparation of ^{13}N-labelled GABA (from $^{13}NH_3$ + tri-isopropylbutenoate).[244] Taking the papers in order of increasing atomic number of the labelled atom(s), [4-$^2H_3$]-3-methylbutyraldehyde acetal has been prepared from crotonaldehyde via 3-bromobutanal and C^2H_3MgI, and used in preparations of [5-$^2H_3$]leucine and [4-$^2H_3$]valine.[245] Isotope exchange processes with amino acids themselves, selectively with ^{2}H-ethanoic acid/0.05 eq.PhCHO to give better than 99.5% α-exchange[246] and general exchange (solid L-alanine/$^3H_2$ with retention of configuration see Vol.24, p.31).[247] Selective deuteriation by ^{1}H-^{2}H exchange at C-3 and C-5 of a ^{15}N-labelled DL-γ-keto-ornithine, $H_2{}^{15}NCH_2COCH_2CH(NH_2)CO_2H$, gives multiply-labelled DL-histidine after construction of the imidazole ring.[248] Further transformations in this study include enzymic resolution, C-5′-^{2}H → ^{1}H-exchange by boiling water, use of $NaSC^{15}N$ as a variation in the imidazole contruction step leading to the multiply-labelled 2′-mercapto-histidine, and degradation of the labelled histidines into corresponding L-aspartic acids. Similar multiple labelling including an approach from methyl N-benzylidene (4-methoxyphenyl)alaninate to α-C^3H_3-tyrosine with ^{11}C- and ^{14}C-variants at the α-position, has been described.[249]

^{11}C-Labelled amino acids have important clinical applications but must be prepared and used on a very short timescale because of the short half-life of the isotope and the need for operator protection. Ingenious modifications to standard amino acid syntheses have been explored over the years, and the modified Bucherer-Strecker route has been improved in this respect by remote control equipment[250] and by reduced reaction times achieved by using microwave irradiation.[251] A synthesis of DL-[1-^{11}C]Tyrosine established in this study was achieved in about 1/15 to 1/20 of the previous best time, leading to 70-100% higher radioactive emission. α-[^{11}C]Methyl tryptophan and the corresponding methyl ester have been prepared from the N-benzylidene derivatives by LDA deprotonation and alkylation with [^{11}C]methyl iodide.[252] Within the 50-

minute preparation time reported for the [1-^{11}C]acetyl derivative of leukotriene E_4, *alias* (5S)-hydroxy-(6R)-(N-[1-^{11}C]acetylcysteinyl)-7,9-trans-11,14-cis-eicosatetraenoic acid, most of the initial radioactivity (originating in Me^{11}COCl) will have decayed.[253]

Preparations of the six possible N-Boc-[^{13}C]glycines, with and without [^{15}N]-labelling, have been achieved by amination of the corresponding alkyl bromoacetates.[254] Syntheses of L-[4-^{13}C]- and L-[3,4-$^{13}C_2$]aspartic acids from L-serine and L-[3-^{13}C]serine respectively, via K^{13}CN opening of the derived β-lactones, have been developed.[255] [1-^{14}C]-1-Aminocyclopropane-1-carboxylic acid is available from Ba$^{14}CO_3$ via the arylideneglycine 4-$BrC_6H_4CH=NCH_2{}^{14}CO_2H$ through cyclopropane ring construction with $BrCH_2CH_2Br$.[256] The burgeoning interest in carboranylation of amino acids for clinical investigations is illustrated in a synthesis of Me_3NBH_2CO-[^{14}C]phenylalanine methyl ester from the labelled amino acid ester with Me_3N-carboxyborane/$Ph_3P/CCl_4/Et_3N/MeCN$.[257]

A review has appeared of the preparation of 6-[^{18}F]fluoroDOPA, discussing the non-regioselective nature of electrophilic fluorination, and comparison with regioselective fluoro-demetallation and nucleophilic substitution routes.[258] Direct fluorination of 3-O-methylDOPA gives 2- and 6-[^{18}F]fluoroDOPAs.[259] The [^{18}F]DOPA analogue (77) shows good uptake in melanoma tissue, and is a promising positron emission tomography tracer for melanoma imaging.[260]

A description of upscaling an existing route (see Vol.20, p.26) to L-[^{75}Se]selenomethionine, and remote handling conditions to give a high level of radioactivity in the product, have been described.[261]

4.17 *Synthesis of β-Amino Acids, and Higher Homologous Amino Acids.*

The startling increase in the number of papers in this area is a reflection of the increasing level of interest in the amino acid field as a whole, but it certainly represents larger proportional growth in recent times. Taken with the growing interest in oligopeptide isosteres (covered later in this Specialist Periodical Report), compounds that can reasonably be categorized as higher homologous amino acids, the rate of development of interest seems remarkable.

Simple routes with original features, leading to β-amino acids, continue to be explored, such as α-amidoalkylation of esters [$PhCH_2CO_2Bu^t + PhCON=CHPh \rightarrow PhCONHCHPhCHPhCO_2Bu^t$],[262] and by an unusually facile oxidative C–N cleavage reaction of polyamines (the first example of such a cleavage reaction), e.g. diethylenetriamine $H_2N(CH_2)_2NH(CH_2)_2NH_2$, mediated by the rhenium(V) nitrido complex $ReNCl_2(PPh_3)_2$.[263] Buried within a new stereoselective nor-C-statine synthesis (Scheme 24)(the 15-year-old route is not suitable for large-scale operation) from malic acid esters, is the traditional Hofmann rearrangement approach to β-amino acids in a novel $Pb(OAc)_4$-mediated form.[264]

The other β-amino acid syntheses culled from the recent literature are developments of asymmetric syntheses based on methods or principles established relatively recently. Evans' methodology starting with an N-bromoacetyloxazolidin-2-one (cf. 24) leads to N-protected β-phenylisoserine, given as a representative application, in a highly stereoselective route.[265] An oxazolin-2-one intermediate arises in a new synthesis of the (2S,3R)-α-hydroxy-β-amino acid (−)-bestatin but not as a chiral synthon; it is formed by elaboration of diethyl L-maleate and its formation involves a Curtius rearrangement of the acyl azide of a β-hydroxy-acid.[266] Some other stereoselective β-amino acid syntheses use L-aspartic acid as chiral auxiliary, e.g. a route to the 3-aminohex-2-ene-1,6-dioic acid moiety present in the potent gastroprotective agent AI-77-B, via a protected L-aspartic α-semialdehyde,[267] but most of the current papers employ chiral heterocyclic synthons. The presently-favoured example is the (S)-tetrahydropyrimidin-4-one (78) from L-asparagine, that gives (R)-α-methyl-β-alanine as a result of highly stereoselective trans alkylation with MeI after lithiation.[268] The synthon is prepared via the dihydropyrimidin-4-one (79), which can also yield β-amino acids through Pd-catalysed conjugate addition of aryl iodides.[269] Chromatographic resolution of β-alanine cyclic acetals (78), and of glycine analogues (80) has been achieved over silica gel coated with a copolymer of ethyl N-acryloyl-L-phenylalaninate.[270] The 6-methyl homologue of (78), prepared from enantiomers of 3-aminobutanoic acid, has also been submitted to the same alkylation procedure for synthesis of (2R,3R)- and (2S,3S)-2-substituted 3-aminobutanoic acids.[271] The need to resolve DL-3-aminobutanoic acid to create the chiral 6-methyl synthon is avoided if the N-[(S)-phenylethyl] analogue of compound (78, lacking t-butyl), formed by Hg-mediated cyclization of βγ-unsaturated amidals,[272] is used (see Vol. 24, p.6). These synthons are also amenable to highly stereoselective trans-alkylation.[273]

A conversion of a protected β-alanine into α-(2-hydroxyalkyl) analogues through classical aldolization has been illustrated for the reaction of $BocNMeCH_2CH_2CO_2Et$ with benzaldehyde, to give an erythro/threo mixture of products.[274]

An oxygen near-analogue of (78) has been involved in an unusual asymmetric synthesis of β-amino acid esters (Scheme 25) starting with amidals of (R)- or (S)-phenylglycinol.[275] A Reformatzky reagent brings about ring-expansion and retains 60-92% of the initial homochirality in

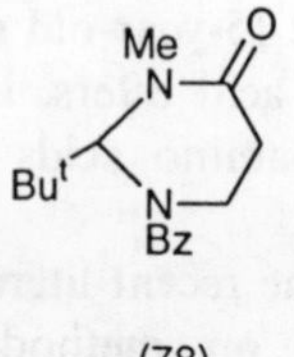

(78) 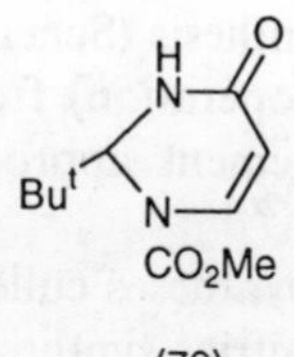 (79) 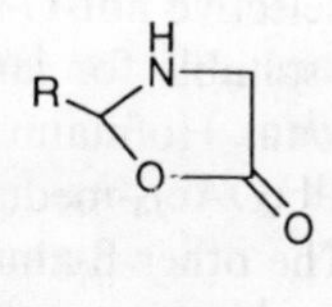 (80)

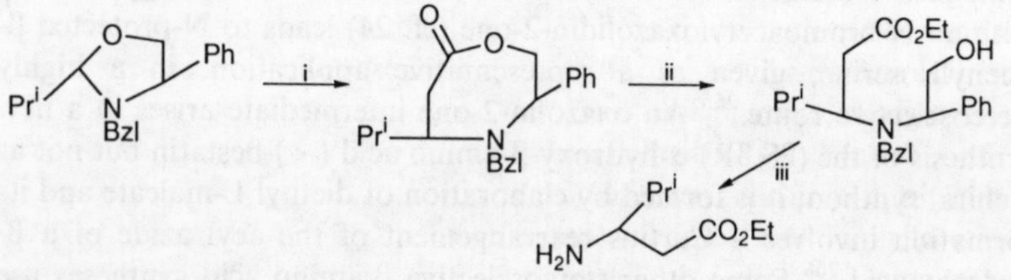

$$R^1CONH.CHR^3.CH(OH).CO_2R^2$$

Reagents: i, $BrZnCH_2CO_2Et/Et_2O/35\ ^\circ C$; ii, sat. HCl in EtOH;
iii, H_2–Pd/C

Scheme 25

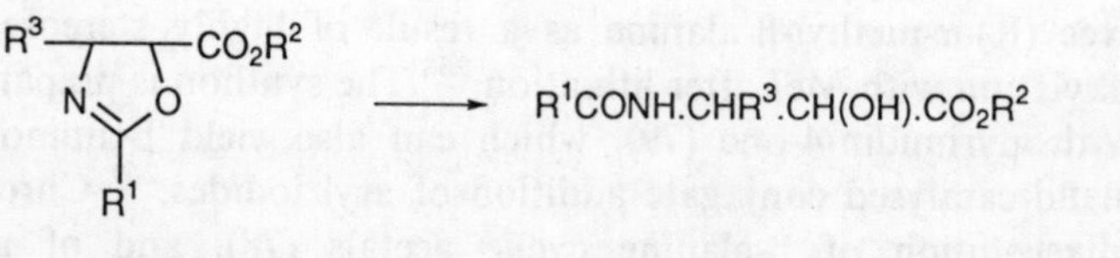

(81)

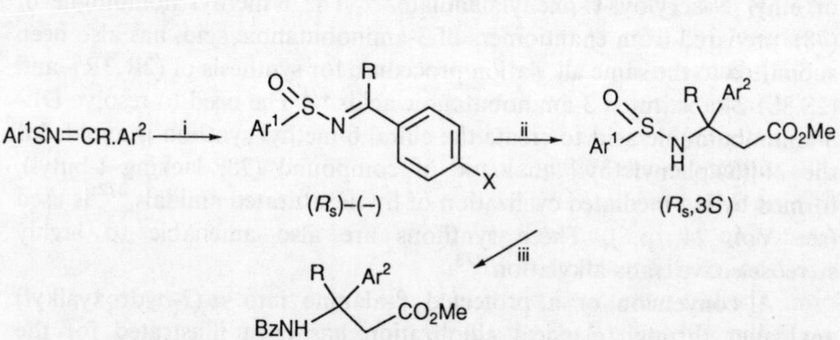

$Ar^1SN{=}CR.Ar^2$ → (R_s)–(−) → (R_s,3S)

Reagents: i, chiral oxaziridine; ii, $CH_2{=}C(OLi)OMe$;
iii, TFA, then PhCOCl

Scheme 26

leading to the product. Ring-opening of oxiranes by nitriles gives oxazolines (81) that are readily hydrolysed to provide a general synthesis of α-hydroxy-β-amino acids.[276] Corresponding oxazolines (81; R^1 = Ph; R^3 = CH_2OTs, H in place of CO_2Me) undergo nucleophilic substitution with CN^- in a variant of a standard route from α- to β-amino acids that is achieved by aqueous acid hydrolysis of the resulting β-aminonitriles.[277] Chiral oxaziridines are not themselves the synthons in an asymmetric synthesis developed to β-amino acids, but the means of converting prochiral sulphenimines into homochiral sulphinimines that provide the β-amino group through addition to a lithium enolate (Scheme 26).[278] The method, for which the diastereoselectivity was not high, was illustrated further in a synthesis of the (2R)-hydroxy-(3S)-amino-alkanoic acid moiety of taxol. Azetidinones that are obtainable from non-amino acid sources continue to be valuable sources of β-amino acids, and full details are available (see Vol.24, p.34) of the [2 + 2]cycloaddition of benzyloxyketene with a chiral imine from methyl (R)- or (S)-mandelates, exemplified in the preparation of (2R,3S)- and (2S,3R)-3-amino-2-hydroxybutanoic acids, components of a renin inhibitor, and of bestatin, respectively.[279] N-Tosylaldimines undergo $TiBr_4$-mediated condensation with ketene trimethylsilyl acetals $RCH{=}C(OSiMe_3)_2$ to give anti-β-amino acid derivatives with 92% diastereoselectivity.[280]

More conventional syntheses, against the background of established standard methods, include extensions of α-amino aldehydes in highly diastereoselective aldol reactions leading to the γ-hydroxy-β-amino acid “Aboa” ($p\text{-}BrC_6H_4CH{=}CHCMe{=}CHCH(OH)CH(NH_3^+)\text{-}CH_2CO_2^-$, a constituent of theonellamide F),[281] and (2R,3S)-3-amino-4-cyclohexyl-2-hydroxybutanoic acid and the 4-phenyl analogue, and their (2S,3R)-diastereoisomers.[282]

Diastereoselective synthesis of β-oxygenated γ-amino acids and γ-oxygenated δ-amino acids has been reviewed.[283] Hydroxylated γ-amino acids calling for a relatively simple stereoselective route, such as (S)-(−)-4-amino-2-hydroxybutanoic acid,[284] and (R)-erythro-4-amino-3-hydroxybutanoic acid,[285] have been prepared by Baker's yeast-catalysed reduction of the corresponding oxo-analogues in suitably protected form. The last-mentioned protected product was needed in syntheses of sperabillin C and (R)-GABOB (the β-hydroxylated version of the neurotransmitter, GABA, *alias* γ-aminobutyric acid). Another synthesis of (R)-(−)-GABOB is based on aminolysis in 89% enantiomeric excess, of the chiral oxirane (82) formed from $BrCH_2CHO$ and allyl(di-isocamphenyl)borane.[286] The GABA analogue synthesis field is very actively cultivated at the moment, with routes to (S)-γ-acetylenic-[287] and -trans-γ-butenyl-GABAs, using phthalimide for Mitsunobu amination,

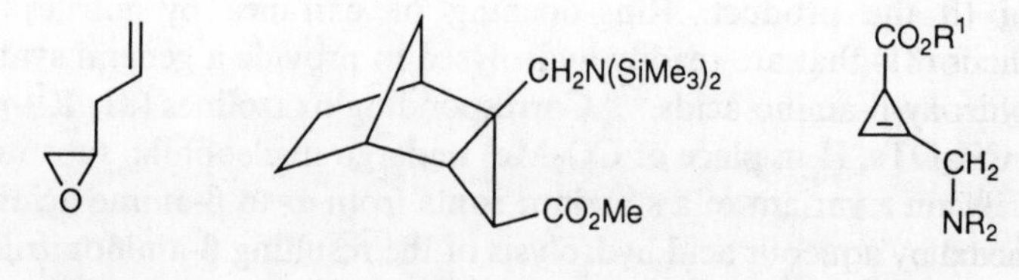

O (82)

$CH_2N(SiMe_3)_2$ CO_2Me (83)

CO_2R^1 CH_2 NR_2 (84)

N_3 CO_2Me $SiMe_2Ph$ —i→ BnO OMe N_3 Me CO_2Me

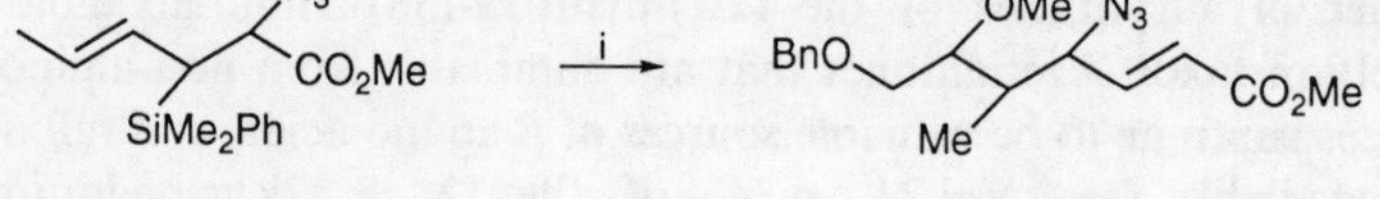

Reagents: i, $BnOCH_2CH(OMe)_2$, TMSOTf

Scheme 27

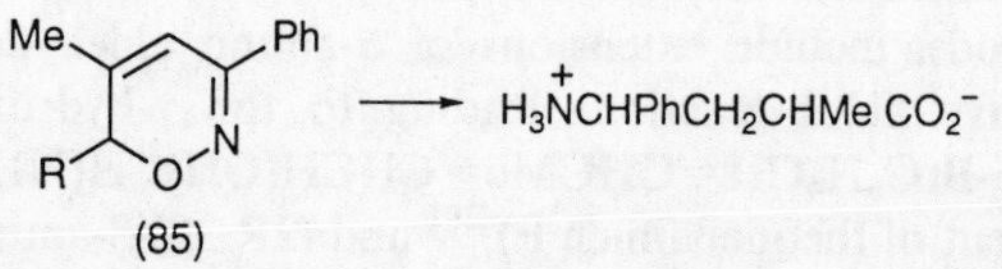

Me Ph R O N (85) → $H_3\overset{+}{N}CHPhCH_2CHMe\ CO_2^-$

L–Glu→ O O NHBoc → BocNH HO RO CO_2H (86)

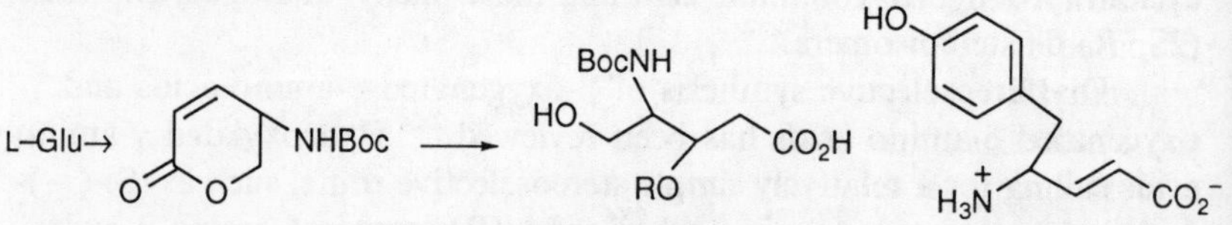

HO $H_3\overset{+}{N}$ CO_2^- (87)

e.g. of (R,E)-$Me_3SiC{\equiv}CCH{=}CHCH(OH)(CH_2)_3OSiPh_2Bu^t$ followed by functional group elaboration in the latter case.[288] VinylGABA (4-aminohex-5-enoic acid) has been obtained from L-glutamic acid (itself a γ-amino acid as well as an α-amino acid) in 33% overall yield in six relatively routine steps. One of the steps applied to the pyroglutamate intermediate in this synthesis involves a novel alkenyl protecting group to permit the conversion to (S)-2-oxopyrrolidine-5-carboxaldehyde.[289] α-Methylene-β-hydroxy-GABA derivatives are easily available by aldolization of an acrylate ester with an α-amino aldehyde.[290] Diels-Alder synthesis of the rather remote GABA analogue (83) uses the cyclopropene (84) formed by $Rh_2(OAc)_4$-catalysed cyclopropenation of $HC{\equiv}CCH_2N(SiMe_3)_2$ with an alkyl diazoacetate.[291]

Some new approaches to γ-amino acids have been studied. Stereospecific allylic azide isomerization offers a way of inter-relating the α- and γ-amino acid series, since this process accompanies the attempted alkylation of α-azido-βγ-unsaturated esters (Scheme 27) by acetals.[292] Only low diastereoselectivity is shown, however, in the outcome of hydrogenation (H_2/Pd), and functional group elaboration, of chiral 3-phenyl-6H-1,2-oxazines (85).[293]

Synthesis studies for other γ-amino acids with significant biological importance have been reported. "MeBmt" analogues (86), prepared starting with L-glutamic acid[294] are obtained in 53–100% yields with diastereoisomer ratios 73:27–99:1. Statine syntheses are being pursued currently, mainly as vehicles for exploring interesting synthetic methodologies, as in a new route to β-hydroxy-γ-amino acids starting with Sharpless oxidation of 3-hydroxy-5-methylhex-1-ene.[295] The 8 possible isostatine isomers [the natural compound is (3S,4R,5S)] have been prepared from the 4 isomeric isoleucinals/allo-isoleucinals by reaction with ethyl lithioacetate.[296] A similar use of an α-amino aldehyde arises in the synthesis of the D-tyrosine relative (87) that is a constituent of cyclotheonamide A.[297] Full details of the elaboration of (R)-2-hydroxy-3-phenylpropanoic acid into statine have been supplied.[298]

At least five syntheses have been reported for detoxinine (88), a constituent of (−)-detoxin D_1, to which is added a sixth new stereospecific synthesis (based on a modification of an earlier route).[299]

Syntheses from monosaccharide derivatives are appropriate in the (R)-carnitine area [two simple syntheses from D-galactono-1.4-lactone (see also Vol. 24, p.33)],[300] and analogues (89) of this compound have been synthesized as potential carnitine acyltransferase inhibitors.[301] An efficient (R)-carnitine synthesis employs microbial "resolution" of (RS)-2,3-dichloropropan-1-ol to the derived homochiral oxirane (as in Scheme 4, with Cl in place of OTf) followed by routine steps.[302]

Diacetone-D-glucose → N_3, HO, MsO (sugar intermediate)

→ → (88)

HO_2CCH_2, R, OH, Me_2, X^-

(89)

$Cl(OPr^i)_2Ti$ ← O

→ $PhCH_2$, BocNH, CO_2H, OH

(90)

HO, $H_3\overset{+}{N}$, ^-O_2C → $H_3\overset{+}{N}$, ^-O_2C

(91)

Aldolization of N-Boc-L-phenylalaninal with the chiral Ti-enolate (90) is the first step in a synthesis of δ-amino acids.[303] The cyclization of the γ-lactam of proclavaminic acid into (91) and the aminoethylidene analogue is an unusual though doubtless, non-general, conversion of one δ-amino acid into another.[304]

Monosaccharide elaboration is also involved in a synthesis of the γ-amino acid destomic acid (92) from Bu^tPh_2SiO-protected Z-L-serinal.[305] The use of the same starting material in the synthesis of anhydrogalantinic acid is also described in this paper.

4.18 *Resolution of DL α-Amino Acids.*

Again as in previous Volumes, this topic is separated into preparative resolution, details of which are carried here; and analytical resolutions (even though they use the same chromatographic and other separation principles), which are described in the later Analytical Sections 7.2.–7.5.

Preparative chromatographic resolution employing chiral stationary phases has been reviewed,[306] and a Symposium account touching on prebiotic and environmental resolution of DL-amino acids, has appeared.[307]

The rapidly growing area under this heading is the exploitation of physical chiral recognition principles in heterogeneous systems, for resolution of DL-amino acids and their derivatives. At its simplest level, a successful approach is the passage of a solution through a column of porous, insoluble material to which is adsorbed, or bonded, a homochiral amino acid derivative. (R)-N-(2-Naphthyl)alanine is such a derivative, thought to offer preferential hydrophobic interactions towards one enantiomer of a solute, rather than the other.[308] Other "Pirkle CSPs" include N-(3,5-dinitrobenzoyl)amino acid derivatives, and a series of 15 N-(3,5-dinitrobenzoyl)dipeptide esters,[309] and N-(3,5-dinitrobenzamides of α-aminophosphonates[310] have been resolved on four different CSPs. An L-amino acid-glutaraldehyde "condensate" introduced into a poly(sulphone) membrane matrix was permeable to D-phenylalanine in preference to its L-enantiomer.[311]

A recent fascinating development (see Vol.24, p.39) has been the use of "imprinted" polymers to promote chiral recognition, and a clever extension of this, combining two different strands of chiral recognition science, has been the generation of "chiral catalytic cavities" in silica gel, using Z-L-alanine N-benzylamide, then to use this material for selective 2,4-dinitrophenolysis of the L-enantiomer of benzoic N-Z-alanine anhydride.[312] The distantly-related principle of "replacing crystallization" has been illustrated for the seeding of a saturated solution or melt

(92)

(93)

(94)

(95)

(96)

of (RS)-thiazolidine-4-carboxylic acid with L-isoleucine or L-cysteine [to cause crystallization of the (R)-enantiomer, and with other L-amino acids to bring out the (S)-enantiomer].[313]

Classical examples of the conventional resolution approach based on differential solubility of diastereoisomeric salts or derivatives include 4-(tetrazolylmethyl)pipecolic acids (di-p-toluyltartaric acid)[314] DL-amino acids generally (N-protected aspartame; salts with the D-amino acids are less soluble),[315] N-Z-α-methoxyglycine (D- or L-phenylalaninol, or (1S,2S)-2-amino-1-phenylpropane-1,3-diol),[316] and salts with quinine or ephedrine.[162] 5,5-Di-alkylhydantoins derivatized with (S)-(−)-phenylethyl isocyanate have been separated by medium pressure liquid chromatography, followed by hydrolysis into homochiral α-alkyl-α-amino acids,[317] and DL-α-alkyl-α-amino acids have been resolved by coupling as benzoyl derivatives to L-phenylalanine NN-dimethylamides or L-phenylalanyl-L-phenylalanine NN-dimethylamides.[318]

Methods involving enantioselective transport across liquid interfaces, especially involving liquid membranes entrapping a chiral amino acid-complexing agent, are prominently represented in the recent literature. The crown ether (93) is central to one of these studies.[319] The passage of L-phenylalanine from a solution of racemate, into a liquid membrane containing a chiral cation complexing agent, is assisted by di-(2-ethylhexyl)phosphoric acid as carrier.[320] A variant of this approach is to entrap an enzyme in the liquid membrane. When such a membrane is adjacent to a solution of DL-phenylalanine ethyl ester, the separation of enantiomers through the hydrolysis of one of them has been demonstrated.[321]

Solid-phase versions of this principle are illustrated in the intercalation of layered α-$Zr(HPO_4)_2$ by the chiral selector cation 3,5-dinitrobenzoyl-L-Leu-$NHCH_2CH_2NMe_3^+$, which selectively binds methyl N-(2-naphthyl)-L-alaninate from a solution of the racemate in MeCN,[322] and in the use of microporous hollow fibres involving a poly(vinyl alcohol) barrier, that permits the preferential passage of L-leucine from a stationary aqueous solution of the racemate through to a flowing octan-1-ol solution of N-dodecyl-L-hydroxyproline.[323]

Solution studies involving familiar enantioselective chiral host-guest systems have continued. Natural chiral pool constituents, of course, offer convenient starting points, and complexes of amine salts with crown-type chiral ionophore receptors from the monensin family have been studied.[324] X-Ray analysis of complexation of tartaric acid-derived hosts (e.g. 94) with amino acid esters shows that hydrogen bonding interactions are important in the chiral selection.[325] Such studies of the origins of the chiral recognition are becoming more common, as in n.m.r. studies of the L-tryptophan - cyclodextrin complex that indicate that the protonated amino group does not contribute to the binding.[326] Unmodified and methylated cyclodextrins show the ability to complex selectively with enantiomers of N-dansyl amino acids, and this underlies the enantiomeric analysis of these derivatives by capillary zone electrophoresis.[327]

Conformationally homogeneous podand ionophores (95) have been thoroughly studied for enantioselective complex formation with enantiomers of amino acid ester cations, and other ammonium ions.[328] Those ionophores with hydroxy functions (95; X = OH) were much more effective than (95; X = H) with enantioselectivities expressed (in organic chemists' terms!) as 0.8 - 1.4 kCal mol^{-1}. The ion-encapsulating host (96) strongly (40%) favours complexation with L-enantiomers of phenylalanine and tryptophan from neutral solutions of racemates, but showed no discrimination towards valine enantiomers.[329]

Applications of enzymes for the delivery of particular amino acid enantiomers from racemates - the overall outcome is resolution, but the process can hardly be called resolution! - are expanding considerably, with the use of enzymes outside the protease and aminopeptidase areas for the purpose. Thus, microbial lipases catalyse the enantioselective hydrolysis of Z-DL-amino acid methyl esters, in the same manner as subtilisin,[330] and pancreatic lipase and a lipase from *Aspergillus niger* perform similarly with 2-phenyloxazol-5(4H)-ones.[331] DL-α-Methyl-α-amino acids can be "resolved" in this classical manner, using a lipase from *Candida lipolytica*[332] (see also Ref.62), and alcalase catalyses the hydrolysis of N-acetyl-DL-amino acid esters in aqueous organic solvents,[333] an enzyme used in the same way with the unsaturated methionine analogue (Z)-MeSCH = CHCH(NHAc)CO_2R.[334]

More conventional enzyme applications (see also Ref.248) are described in a larger number of papers than usual, in some of which the resolution step is an incidental routine detail while in others, there is some novelty to the study. Subtilisin Carlsberg esterase (see also Ref.210) has been used in a prototype liquid membrane-entrapped-in-hollow-fibre reactor (described above in this Section) for continuous flow enantioselective hydrolysis of isopropyl DL-phenylalaninate.[335] The enzymatic resolution of non-protein amino acids is attracting more interest, and serine proteases have been shown to be capable of mediating the kinetic resolution of DL-DOPA ethyl ester in organic solvents,[336] while in this area, the protease from *Bacillus subtilis* is more effective than that from *Aspergillus oryzae*.[337] A further example of the toleration by enzymes of unusual experimental conditions, now that their usefulness in organic media is well-established, is the conversion of DL-proline into L-proline in better than 98% yield by D-amino acid oxidase and $NaBH_4$.[338] The process involves the conversion of D-proline into Δ^1-pyrroline-2-carboxylic acid, which is immediately reduced to DL-proline, thus effecting a novel kinetic resolution process. For a use of L-amino acid oxidase, see Ref. 43, and for uses of acylases, see Refs. 39 and 167. Delivery of L-tyrosine in high optical purity by α-chymotrypsin-catalysed hydrolysis of DL-tyrosine ethyl ester in largely organic media is a significant observation.[339]

Whole-cell processes include *Candida maltosa*/DL-alanine → D-alanine,[340] *Rhodococcus rhodochrons* PA34/DL-α-aminonitriles → L-amino acids,[341] and immobilized enzyme studies [poly(acrylate)-amino-acylase/N-acetyl-DL-methionine].[342] Resolution of 5-alkylhydantoins using *Agrobacterium radiobacter* continues to be a convenient variation of these approaches.[234]

A review has appeared of chemo-enzymic synthesis of amino acids, and the stereoselective hydrolysis of DL-amino acid amides by an aminopeptidase from *Pseudomonas putida* and by an amidase from *Mycobacterium neoaurum*[343] (see also Ref.46).

Speculation on resolution mechanisms operating on DL-amino acids in prebiotic times (and, presumably, still operating) continues to be published when backed up by experimental observations. Familiar aspects represented this year concern the stereoselective bias in favour of L-isomers accompanying de-amination of aqueous aspartic and glutamic acids over sodium montmorillonite at pH 6,[344] and reversible redistribution of L- and D-valine in the surface layer of an aqueous solution under non-equilibrium conditions.[345] The exact context of this claim is not clear from the information in the abstract, and neither is its relevance to the present dominance of L-amino acids in life processes, in suggesting that the ocean-air interface is selective in favour of one enantiomer rather than the other.

5 Physico-chemical Studies of Amino Acids

5.1 *X-Ray Crystal Structure Analysis of α-Amino Acids.*

Most of the recent papers cover the protein amino acids in derivatized forms: bis(glycine)HBr (i.e., $^+H_3NCH_2CO_2^-$ $^+H_3NCH_2CO_2H$ Br^-; a re-investigation),[346] bis(N-Boc)glycine N′-methoxy N′-methylamide,[347] N-Boc-D-alanine benzylamide,[348] N^{β}-hydroxy-DL-asparagine monohydrate,[349] ammonium L-glutamate monohydrate,[350] sodium DL-glutamate monohydrate,[351] potassium L-glutamate monohydrate,[352] propanediamine complexed with L-glutamic and with DL-glutamic acids,[353] Z-L-prolinamide,[354] di-L-phenylalaninium sulphate monohydrate,[355] L-phenylalaninium formate,[356] and N-Boc-L-phenylalanine benzyl ester.[357]

Unusual amino acids subjected to X-ray crystal analysis include two αα-disubstituted glycines (as their N-carboxyanhydrides),[358] L-azetidine-2-carboxylic acid (as its N-Boc derivative),[359] and indospicine hydrochloride monohydrate.[360]

5.2 *Nuclear Magnetic Resonance Spectrometry.*

The role of n.m.r. spectrometry in revealing finer structural details is exemplified well in carefully-devised experiments reported in the current literature. The complex formation that occurs through adding L-Chirasil-Val to solutions of enantiomers of N-trifluoroacetylamino acid methyl esters in CCl_4 is revealed in chemical shift non-equivalence in both ^{1}H-n.m.r. and in ^{19}F-n.m.r.[361] Weak hydrogen bonding interactions involving the amide proton are identified for N-acetyl-DL-valine through ^{2}H-n.m.r. spectrometry of a series of isotopically-labelled compounds.[362] Less ambitious studies are represented in the collection of ^{1}H-n.m.r. data for mono- and di-fluorotryptophans,[363] and in ^{13}C-n.m.r. assessment through lanthanide shift studies, of conformational information for L-lysine in solution.[364] Similar conformational objectives for L-leucine in its cationic and anionic forms have been achieved through interpretation of n.m.r. spin-spin coupling constants and nuclear Overhauser effect data.[365]

CP/MAS ^{13}C-n.m.r. spectra of crystalline L-leucine and DL-leucine show considerable differences, including a different order of peaks in some cases, with well-known solution spectra of these amino acids.[366]

^{14}N-N.m.r. spectra for 19 amino acids in various protonated/deprotonated forms have been published.[367] This technique has shown its potential for providing structural information, in a measurement of rate constants for H-exchange involving glycine, proline and aspartic acid in aqueous solutions.[368]

Peak separation of 0.099 ppm is seen in ^{31}P-n.m.r. of diastereoisomeric phosphinamides formed between (S,S)-$(EtCHMeO)_2P(O)H$ and DL-alanine, permitting the establishment of a novel method of estimating enantiomer excesses for amino acids that compares well with results from analogous α-chloropropionyl chloride derivatization and polarimetric quantitation.[369]

5.3 *Optical Rotatory Dispersion and Circular Dichroism.*

These techniques, or at least, circular dichroism spectrometry (c.d.), are occasionally combined with other physical methods for routine structural and stereochemical assignments to amino acids, as in an exploration of interactions between β-cyclodextrin as host and 2,4-dinitrophenyl-D- or -L-valine as guest.[370]

Pioneering studies of vibrational Raman optical activity continue to use homochiral amino acids as models, and more data have been accumulated of the back-scattered absorption features of L-alanine as a function of pH.[371]

5.4 *Mass Spectrometry of α-Amino Acids and Related Gas Phase Studies.*

This and the neighbouring Sections of this Chapter have been steadily declining in scope in relative terms, as some of the techniques become established and therefore routine in nature. The newer ionization techniques are yielding new information, as in the positive plasma desorption mass spectra for the 20 common amino acids,[372] and corresponding positive and negative ion mass spectra for N-acetylcysteine and a number of its biologically-important S-alkylated derivatives.[373]

Collection of gas-phase proton affinity data interests several research groups, one study relating to arginine by determining dissociation rates of protonated dimers,[374] another[375] relating accumulated rate data with adiabatic ionization energies obtained from photoelectron spectra.[376] A theoretical study relating to glycine[377] has appeared. Fourier transform ion cyclotron resonance spectrometry provides proton affinities for the 20 common amino acids that point to intramolecular hydrogen bonding in the gas phase in the cases of lysine and glutamic acid.[378]

5.5 *Other Spectrometric Studies of α-Amino Acids.*

Excluding routine spectrometric measurements that are incidental to the main thrust of amino acid studies, there are relatively few papers calling for discussion here.

Fourier transform-i.r. studies of N-Boc sarcosine methylamide show that equilibria between seven-membered ring intramolecular hydrogen bonded structures, and extended conformations are very sensitive to solvent characteristics, e.g. hexane versus CCl_4.[379] A similar study with N-Boc proline methylamide [380] shows the greater propensity in this case towards the intramolecularly hydrogen-bonded conformation.

Prominent bands in polarized Raman spectra, claimed to be useful markers for the amino acids phenylalanine, tyrosine and tryptophan, have been studied further.[381] In a similar vein, new bands and revisions of earlier assignments are reported for Raman spectra of glycine in the solid state, in water, and in ^2H_2O.[382] Vibrational spectra (i.r. absorption and Raman scattering) of L-amino acids have been measured over the range 10 - 400 cm^{-1}.[383]

Continuation of sophisticated laser-induced fluorescence spectrometry of phenylalanine and tyrosine in a supersonic jet has been justified by the important results already obtained, since spectral features can be assigned by this technique to each of the several conformations present. No support was obtained of intramolecular exciplex formation in these samples, in contrast to earlier claims for tryptophan.[384] A single crystal electron spin-echo envelope modulation study of copper(II)-doped L-histidine hydrochloride has provided values for ^{14}N hyperfine and quadrupole coupling tensors for this compound.[385]

5.6 *Other Physico-Chemical Studies.*

The ability of low levels of amino acids to modify considerably, the structure of aqueous solutions, is well-known. They play a natural cryoprotectant role, by their ability to depress the freezing point of such solutions which is crucially important biologically (and also useful in food processing), and studied recently for glutamine in aqueous solutions modelling cell fluids.[386] Glycine, sarcosine, NN-dimethylglycine, and glycine betaines have been studied for their role in preventing leakage from liposomes after freezing and thawing.[387]

Further unusual properties of solutions of amino acid derivatives are coming to light, in the gel-like nature seen in certain cases, for example. Fibrous aggregates are a feature of the gels formed at low temperatures by N-acyl-L-aspartic acids with long-chain (C_{12}-C_{18}) alkanoyl groups in neutral aqueous solutions, and by N-dodecanoyl-β-alanine (but not by N-dodecanoyl-L-glutamic acid).[388] A most extraordinary example in this area is the thermoreversible hard gel that forms on cooling a 1% solution of Z-L-alanine 4-hexadecanoyl-2-nitrophenyl ester, in methanol or in cyclohexane.[389] The nature of the intramolecular interactions involved in such cases is assumed to be predominantly hydrogen-bonding, but N-dodecanoylhistidine methyl ester forms a complex with dodecyl dihydrogen phosphate in $CHCl_3$ that must involve both hydrogen bonding and ion-pair interactions.[390] Efficient hydrogen bonding occurs between N-acetyl amino acids and a model (97) for the vancomycin-D-alanyl-D-alanine interaction.[391]

Calorimetric studies with amino acids include glycine dissociation in aqueous glucose over the temperature range 5–45°,[392] enthalpy of protonation of glycine in water,[393] thermogravimetry and differential scanning calorimetry of amino acids,[394] and further measurements of enthalpies of dilution of solutions of glycine in binary solvent

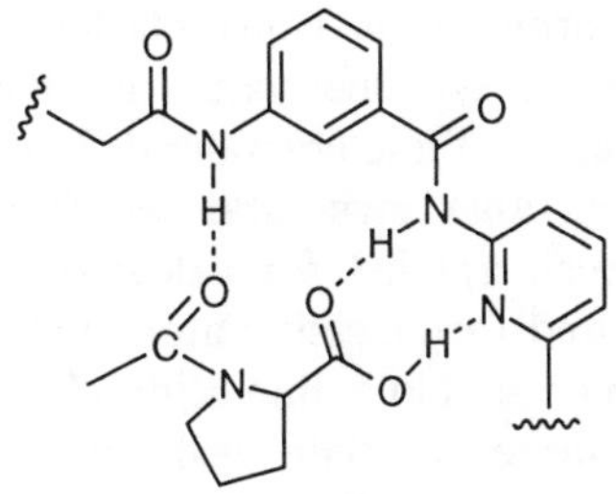
O
N
H
O
C
N
O
H
N
O
H
N
O

(97)

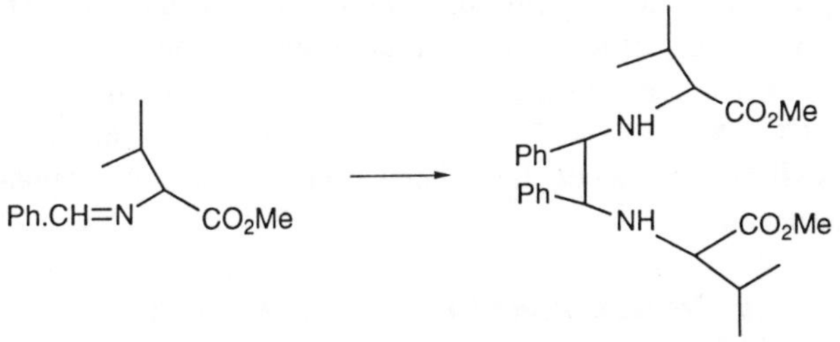
Ph.CH=N
CO2Me
NH
CO2Me
Ph
Ph
NH
CO2Me

(98)

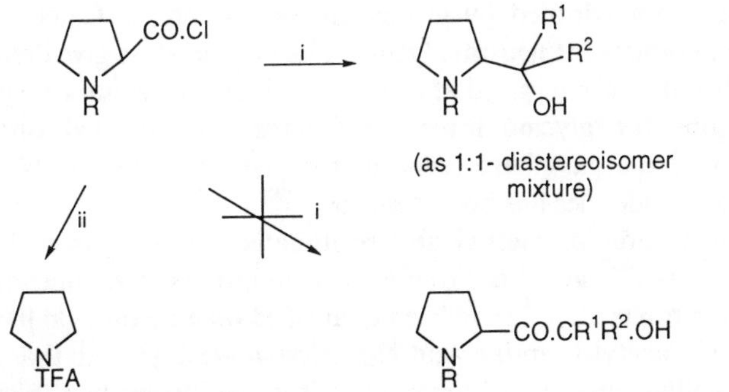
CO.Cl
N
R
i
R1
R2
N
R
OH
(as 1:1- diastereoisomer
mixture)
ii
i
N
TFA
CO.CR1R2.OH
N
R

Reagents: i, R^1.CO.R^2 /SmI_2/THF; ii, conditions as in i, but R = TFA

Scheme 28

mixtures.[395] Microscopic protonation constants have been determined for 10 amino acids related to tyrosine; values for the basicities of the amino and phenolate groups in these compounds have been interpreted in terms of the roles of other structural features.[396] A molecular connectivity model has been applied for calculating the isoelectronic points of amino acids,[397] and for assessing atomic charges on atoms of second row elements when present in amino acids.[398]

Apparent molar volumes correlate well with computed van der Waals and molecular volumes for 17 amino acids.[399] Dipole moment data for L-threonine in water confirm the existence of a solvent - solute interaction,[400] and corresponding data for L-tyrosine have shown the existence of the dihydrated zwitterion in aqueous solutions.[401]

2,4-Dinitrophenyl-L-amino acid derivatives R-X-$OCH_2C{\equiv}CC{\equiv}CCH_2O$-X-R (R-X = dnp-L-amino acyl), otherwise viewed as chiral acetylenes, have been prepared as potential non-linear optical materials.[402]

5.7 *Molecular Orbital Calculations for α-Amino Acids.*

A relatively larger proportion of papers than usual under this heading deals with calculations for the amino acids themselves, rather than for their simple derivatives.

Calculations for glycine that attend to both charge distribution and conformational features[403] establish the point that a zwitterionic structure is not adopted by glycine in the gas phase. Calculations for vibrational modes for neutral glycine in its ground state give deformation frequencies that are in good agreement with experimental i.r. spectra.[404] Calculations for glycine force fields have been carried out,[405] and extension of the gas-phase scenario for glycine to its C-, N-, and O-methylated analogues has been described.[406]

The motion of methyl and protonated amino groups in the L-alanine crystal,[407] and the broader conformational dynamics of the L-tryptophan molecule,[408] have been quantified on a theoretical basis.

The N-acetyl-L-amino acid N-methylamide approach that has been such a familiar part of this topic, since it is used to probe the behaviour of an amino acid as a residue in a protein, continues with comparisons of calculations for the L-alanine representative for the gas phase compared with aqueous solutions, to assess solvent-solute interactions.[409] Calculations aimed at the assessment of helicogenic properties, or otherwise, of the side-chains of the protein amino acids, have been reviewed.[410] Conformational deductions based on calculations previously (1983) presented for N-acetyl-L-phenylalanine N′-(4-acetylphenyl)amide have been revised on the basis of current MM2 methodology.[411]

Many of the studies described in the preceding paragraphs are aimed at useful extensions of understanding of the ways in which amino acids perform in chemical and physical contexts, but two further papers indicate the two extremes encompassed by the topic of this Section. In one, a very specific interest is addressed by conformational calculations for N-acryloyl-L-phenylalanine methyl ester and its alanine analogue, to explain the observed asymmetric induction in their Diels-Alder additions to cyclopentadiene;[412] in the other study, an outcome of much broader relevance arises from calculations suggesting the existence of a weak hydrogen bond between a methyl group and a negatively-charged oxygen atom in creatinine and in its competitive inhibitor, N-carbamoylsarcosine.[413]

6 Chemical Studies of Amino Acids

6.1 *Racemization.*

The scope of recent studies encompasses familiar themes, though with rather fewer racemization studies taking place overall.

Reviews have appeared covering the dating of fossils through the measurement of the extent of racemization of indigenous amino acids.[414,415] A similar use of racemization data in the broader area of geochronology is covered in one of these reviews.[415] Indigenous aspartic acid has racemized relatively faster than the other protein amino acids in ancient mollusc shells, within the range 2–5% per century (depending on ambient temperature and other environmental factors) and enantiomeric analysis of the aspartic acid content has been advocated for dating purposes in this fossil area.[416] However, isoleucine epimerization continues in favour for the purpose, especially for workers with interests in fossils at the older end of the age range; ostrich egg shells retain proteinaceous material over more than 10 million years,[417] the date given to samples from the Border Cave in South Africa. A salutary warning has been given that estimates of isoleucine epimerization may be open to a systematic error. A standard amino acid derivatization reaction (o-phthaldialdehyde-β-mercaptoethanol; see Section 7.4) occurs at different rates with isoleucine and allo-isoleucine (Ile > alloIle), and incorrect ratios can be obtained by incomplete derivatization.[418]

Laboratory studies of amino acid racemization are usually performed for specific preparative purposes, such as the asymmetric transformation of L-histidine into its enantiomer by salicylaldehyde Schiff base formation using (2R,3R)-tartaric acid in acetic acid, followed by hydrolysis,[419] and in assessing the causes of epimer formation in peptide synthesis. In the latter category, racemization of an N-substituted L-proline phenacyl ester by 1-hydroxybenzotriazole[420] contravenes the standard dogma that this acidic compound is generally protective against racemization when used as an additive in the aminolysis of active esters (the other compound in this study is a member of that class).

More fundamental aspects are explored in the determination of absolute rate constants for acid- and base-catalysed racemization of representative amino acids in the pH range 0.53 - 10.35 (i.e. the rate constants for each of the three, or more, ionic species of amino acids).[421] Results - a massive arithmetical problem solved with the help of simplex optimization computer analysis - are consistent with the classic Neuberger mechanism requiring α-proton abstraction. Racemization rates for N-acetyl-L-tyrosine, phenylalanine, 4-hydroxyphenylglycine, phenylglycine and alanine using (RS)-α-methylbenzylamine as base catalyst have been determined.[422] An interesting but subsidiary aspect of this study is the demonstration of asymmetric transformation of 4-hydroxyphenylglycine by (R)-α-methylbenzylamine, a subtle weak acid-weak base interaction presumably being the underlying explanation of this phenomenon.

6.2 *General Reactions of Amino Acids.*

Reactions involving the amino and carboxy functions, and the α-carbon atom, of the amino acids are covered in this Section, while reactions involving mainly the side chain are covered in the following Section 6.3: Specific Reactions of Amino Acids.

Though amino acids are generally regarded as stable towards routine handling, chemical changes can be inflicted on amino acids thermally, and such changes are subject to catalysis. The extraordinary ease of polymerization brought about in aqueous solutions by copper(II) salts (see Vol. 24, p. 47) is given continuing attention.[423] $CuCl^+(H_2O)_n$ ions acting in concert with $Na^+(H_2O)_n$ ions (ions with incomplete hydration shells), are considered to be responsible for accepting water liberated in the self-condensation of amino acids in 0.5M copper(II) chloride in aqueous 5M NaCl. Corresponding self-condensation of L-aspartic acid in the presence of metal ions at 200° during 2 hours is claimed to lead to polymer mixtures.[424] Thermal decarboxylation of L-ornithine to give putrescine is brought about by heating in aqueous solution in the absence of oxygen.[425]

Reliable protocols for the preparation of tetra-n-butylammonium salts of amino acids that are soluble in organic solvents (e.g. dichloromethane), have been published.[426]

Reactions at the amino group are often pursued with an eye on the biological roles of the amino acids, as well as on the uses in synthesis of N-protected amino acids. The biological context frequently implies condensation with aldehydes and the ensuing reactions of the resulting Schiff bases; in the food science area this means primarily the Maillard reaction, with its multitudes of eventual condensation products. Amino- and hydroxyethylpyridines, imidazoles and cyanopyrroles are formed through condensation of aspartic acid or asparagine and glucose,[427] and more than 50 different indoles are formed between tryptophan and glucose, xylose, or furfurals at 220°.[428] Fructose-methionine Amadori compounds give methioninal on heating, with pyridines, pyrazoles, pyrroles, and furans, some retaining the methylthioethyl side chain.[429] Similar cysteine - carbohydrate systems have been studied under milder conditions, to establish the primary formation of the corresponding thiazolidine carboxylic acid, which decomposes through two pathways, one via the classical Amadori compound; and that the level of browning and the development of meat flavour are in proportion to the concentration of the thiazolidinecarboxylic acid.[430] Also on a more descriptive note, but probably of considerable significance in nutritional contexts, is the observation that glucose-amino acid reaction products have anti-oxidant properties.[431] L-Lysine reacts with the lipid oxidation product, (E)-4,5-epoxyheptanal, under microwave irradiation to give Schiff bases, thence to browning reaction products; a role is thus demonstrated for oxidized lipids in non-enzymic browning processes.[432] With importance in another biological context, the Schiff bases formed between retinal and amino acids continue to stimulate chemical studies, and their demonstrated formation in reversed micelles also reflects the growing interest in performing organic reactions in unusual media.[433] A curious reductive dimerization process has been illustrated with N-benzylidene-L-valine methyl ester (Zn/MsOH/THF/$-50°$) to induce mainly the (R,R)-stereochemistry in the newly-created chiral centres (98).[434] The substantial body of papers already published on the kinetics of N-chlorination in aqueous solutions, is added to this year with demonstrations of a rate-determining Cl transfer to nitrogen from hypochlorite ion in alkaline media.[435] Decomposition of N-halogeno-amino acids follows first order kinetics.[436,437]

Replacements of the α-amino group by other nitrogen-containing groups are described for the preparation of isocyanates (amino acid esters with phosgene in toluene under mild conditions),[438] and of azides via α-hydroxy esters and their O-(p-nitrobenzenesulphonyl) derivatives without racemization.[439]

The replacement of an N-Z protecting group by Boc is accomplished by catalytic transfer hydrogenation of the Z-amino acid in the presence of $(Boc)_2O$,[440] while the one-pot conversion of an amino acid into its N-Boc methyl ester has been accomplished by Fischer esterification followed without purification by reaction with $(Boc)_2O$.[441] A new base-labile N-protecting group, N-2-(2,4-dinitrophenyl)ethoxycarbonyl, has been advocated.[442]

Functional group modifications to the carboxy group of amino acids can be divided into the topics that are familiar to readers of this section over the years. Reduction of amino acids to amino alcohols, once seeming so difficult, is accomplished by $NaBH_4$-H_2SO_4 (1:1),[443] presumably relying on the generation of B_2H_6 *in situ*. $LiAlH_4$ in THF, followed by py-SO_3, is an effective sequence for overall reduction of Z-amino acids to Z-amino alcohols.[444] An alternative way of converting α-amino alcohols to α-amino aldehydes without racemization, employs TEMPO oxidation (NaOCl/2,2,6,6-tetramethylpiperidin-N-oxyl).[445] Little or no racemization accompanies electrochemical reduction leading to α-amino aldehydes.[446] In this method, a Z-L-amino acid is electrolysed at -30° in the presence of PPh_3 under nitrogen with $HPPh_3^+$ ClO_4^-. Another illustration of a standard preparation of α-amino aldehydes [R.CON(OMe)Me → RCHO with $LiAlH_4$] has been published in a preparation of the N-Boc β-cyclohexylalanine aldehyde for use in statine syntheses.[447] Aldehydes prepared in these ways have become important intermediates for carbon-carbon bond-forming processes mentioned in a number of contexts earlier in this Chapter, and used outside the amino acid field e.g. to prepare chiral allylsilanes [BocNHCHRCHO → (Z)-BocNHCH=CHMe → (Z)-BocNHCHRCH=$CHCH_2SiMe_3$].[448] N-Acyl-L-prolinols prepared from corresponding esters using $NaBH_4$-MeOH also represents standard practice.[449]

Thiazol-2-yl ketones BocNZCH(CH_2Ph)COR show further uses in synthesis.[450] The thiazolylketone moiety can be converted into an α-hydroxyaldehyde [-COR → -CH(OH)CHO], from which by standard C-C bond-forming operations, an overall synthesis of δ-amino acid esters can be defined.

N-Protected α-amino acid chlorides and fluorides are proving useful in synthesis after evidence has accumulated, showing that they have sufficient stability to survive the procedures involved in their preparation. Oxalyl chloride has been used for this purpose, with N-ethoxycarbonyl α-amino acids,[451] and N-TFA- or -Fmoc-L-prolyl chlorides have been useful in chain extension to alkanols (Scheme 28) by reductive condensation with ketones mediated by SmI_2.[452]

Amide formation from N-protected amino acids with amines has been accomplished, using standard peptide coupling reagents BOPCl and

TBTU[453] and with crystalline ammonium salts of 3-hydroxy-1,2,3-benzotriazin-4(3H)-one with N-hydroxysuccinimide.[454]

N-Protected α-amino acid amides can be O-alkylated with trialkyloxonium tetrafluoroborates to give iminium salts that undergo thiohydrolysis to give thionesters, $R^1NHCHR^2CSOR^3$, without racemization.[455] Corresponding dithioesters are prepared from N-protected α-amino nitriles in the time-honoured way [$RCN \rightarrow RCSNH_2 \rightarrow RC(SMe){=}NH_{2^+}\ X^- \rightarrow RCSSR'$].[456]

There are relatively few novel esterification studies, but an increasing number of enantioselective hydrolysis projects. L-Aspartic acid gives the di(trimethylsilyl) di-ester by refluxing with excess trimethylsilazane, and ways of preparing N-protected derivatives of the diester have been established.[457] An interesting study of relative rates of acylation of 5′-AMP by N-acetyl-D-amino acids reveals faster rates than for L-enantiomers, particularly so for β-branched amino acids valine and isoleucine, with generally decreasing rate differences as the hydrophobicity of the side-chain decreases.[458]

The simplest catalyst system so far studied, in which rate differences are seen for the hydrolysis of enantiomeric amino acid esters, is a combination of an L- or D-amino acid with a metal oxide (ZnO, γ-Al_2O_3, or SiO_2).[459] Further examples (see Vol.24, p.52) involving chiral catalysts within micelles, N-Z-L-Phe-L-His-L-Leu-OH for example,[460,461] for such studies with L-phenylalanine p-nitrophenyl ester, have been published, and the chiral ligand (99) contained in mixed micelles has been applied to the same purpose.[462] In the last-mentioned study, the presence of metal ions Zn, Co, Cu, or Ni, only slightly enhances the catalytic effect of the organic ligand. Outside the enantioselective hydrolysis area, transition metal ions are good catalysts for the hydrolysis of amino acid esters, and their effect is enhanced by certain ligands.[463] Reactivity studies of conventional types have been applied to determine rates of aminolysis of Z-amino acid N-hydroxysuccinimide esters[464] and a series of N-phthaloyl-L-phenylalanine 2-substituted indan-1,3-dione enol esters [highest acylation reactivity was shown by the 2-(4-fluorophenyl) compound].[465]

Oxidation studies continue for common amino acids with conventional oxidants [serine/Tl-$HClO_4$ oxidation catalysed by Ru(III) > Os(VIII) > Nd(III)],[466] [aspartic acid/Bi(V)-$HClO_4$ oxidation → HO_2CCH_2CHO].[467] Other oxidative decarboxylative processes leading to aldehydes have been established for oxidation at the Pt anode of aqueous solutions at pH = 1 and pH = 13, of α-, β-, and γ-aminobutanoic acids.[468] Peroxydisulphate ($S_2O_8^-$) oxidation of 2,2(phenylcyclopropyl)glycine catalysed by silver picolinate gives 2-hydroxy-5-phenylte-

N N CH_2 NMe R N CH_2OH

(99)

HO $NHCH_2CO_2H$ → HO_2C CHO NH CH_2CO_2H + HO N HO_2C NH CH_2CO_2H

(100) (101)

R^2 R^1N CO O CO

(102)

R^1 O N R^2 N H O

(103)

R^1 N R^1 O O

(104)

NMe_2 N Me NHCOPh Ph N Ph O

(105)

PhCONH Me O O NR O

(106)

trahydrofuran, via an α-amino radical intermediate.[469] Further studies of oxidative de-aldolization and oxazolopyrroloquinoline formation through reactions of co-enzyme PQQ with amino acids have been reported.[470] More extensive changes accompany the tyrosinase-catalysed oxidation of N-(4-hydroxyphenyl)glycine (→ 100 + 101).[471]

Heterocyclic compounds feature substantially in every year's literature dealing with the general reactions of amino acids, and one of the classes given much attention in earlier times, the N-carboxylic amino acid anhydrides (NCAs, 102) has been revisited by several groups. One valuable new synthesis starts with N-Boc-L-amino acids, using PCl_3.[472] Others use triphosgene/NEt_3,[473] $SOCl_2$ (with Z-α-trifluoromethyl-α-amino acids),[474] or PyBrOP or PyClOP with N-Boc-N-methyl-L-amino acids[475] for the cyclization. N,N-Bis(Boc)- or N-Boc-N-Z-L-amino acids react with $SOCl_2$/DMF to give N-Boc- or N-Z-L-NCAs, respectively, but react with cyanuric fluoride to give the corresponding acid fluorides.[476] NCAs react with N-Boc-anthranilic acid to give 1,4-benzodiazepine-2,5-diones (103),[472] and in addition to the well-known exploitation of their acylation reactivity in peptide synthesis, they participate in Friedel-Crafts reactions without racemization under standard conditions.[477]

Oxazol-5(4H)-ones (104) represent another class of five-membered heterocyclic compound that is readily accessible from N-acyl-α-amino acids, and their reactions usually preserve intact, the α-aminoacyl residue -NHCHRCO-. They are prepared using simple cyclization reagents, though the synthesis of an optically-active oxazol-5(4H)-one by treatment of L-tryptophan with trifluoroacetic anhydride is an interesting exception to a rule, precedents suggesting that the oxazol-5(2H)-one tautomer is usually favoured by the 2-trifluoromethyl substituent.[478] This work was published previously in preliminary form.[479] Diazadienes react with 4-methyl-2-phenyloxazol-5(4H)-one to give dihydropyrimidin-6-ones (105),[480] and 3-(aryliminomethyl)chromones undergo cycloaddition with oxazolones to give azetidinones (106).[481] 4-Alkylidene-oxazol-5-ones are implicated in reactions of N-acetyl αβ-dehydro-α-amino acids, which give poor yields during attempted formation of amides through standard peptide coupling methodologies.[482]

6.3 *Specific Reactions of Amino Acids.*

This Section deals with reactions involving primarily the side-chains of the more common amino acids. The outcome is often a synthetic route to another amino acid (but biosynthetic studies are not covered here), and many of the examples discussed in earlier "Synthesis" sections (in Sections 4.1–4.16) could have been located here instead.

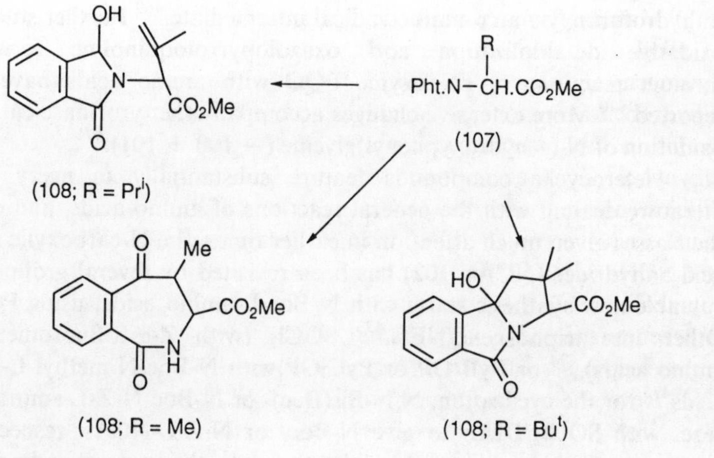

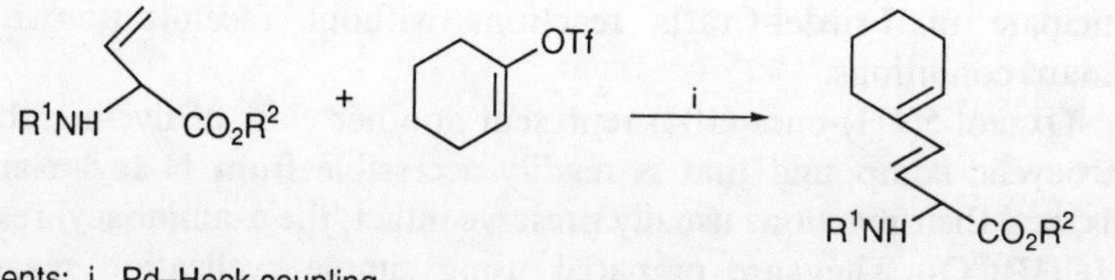

(107)

(108; R = Pr^i)

(108; R = Me)

(108; R = Bu^t)

Reagents: i, Pd–Heck coupling

Scheme 29

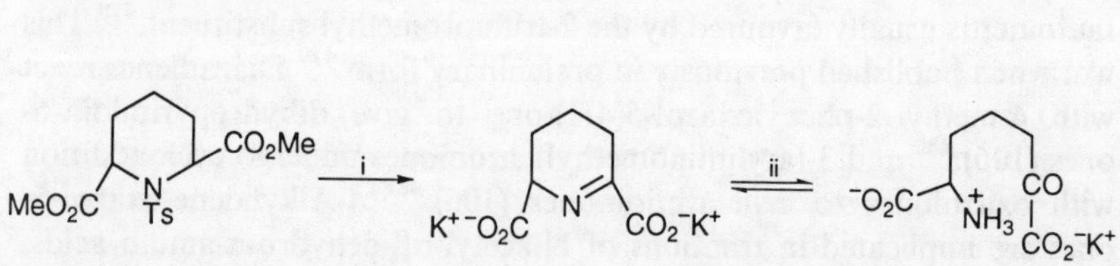

Reagents: i, H_3O^+, bring to near-neutral pH; ii, spontaneous ring-opening in aq. soln.

Scheme 30

Manipulations of aliphatic side-chains include photo-isomerization of alkyl α-phthalimido-alkanoates involving an unprecedented double hydrogen transfer (107→108) accompanied by photocyclization to dihydrobenzazepinones,[483] and the use of unsaturated aliphatic side-chains to build heterocyclic systems. N-Protected α-trifluoromethyl-α-alkynyl-α-amino acids (and their α-hydroxy-analogues) add to 1,3-dipoles (nitrile oxides, diazoalkanes, PhN_3) to give corresponding five-membered heterocycles,[484] while αβ-di-amino-αβ-unsaturated acids $Me_2NCH{=}C(NHBz)CO_2Me$ condense with pyrazolones to give (Z)-β-(pyrazolon-5-yl)-αβ-unsaturated-α-amino acids.[485] Heck couplings between protected L-vinylglycine and aryl and vinyl triflates lead to regioselective, racemization-free, γ-alkenylation (Scheme 29).[486]

Reversible ring-opening follows toluene-p-sulphinic acid elimination from dimethyl meso-N-toluene-p-sulphonylpiperidine-2,6-dicarboxylic acid to give 2,3,4,5-tetrahydrodipicolinic acid, isolated as the potassium salt (Scheme 30), giving an *in vitro* model for the L-lysine biosynthetic pathway followed in bacteria and in fungi.[487] The biosynthesis of avicin (109; R = H) and its 4-hydroxy-analogue (109; R = OH) involves ornithine as primary precursor, rather than glutamic acid or glutamine.[488] Enzyme-catalysed conversion of 5-aminolevulinic acid into porphobilinogen has been contrasted with the formation of pyrazine (110) and pyrrole (111) in the absence of enzyme in some laboratory procedures.[489] A well-known procedure[490] for selective carbamylation of the side-chain amine groups of L-ornithine and L-lysine as their copper(II) complexes, has been fully described.[491]

Ring-opening processes are also a feature of reports of reactions of aromatic and heteroaromatic side-chains, 2,3-bond-cleavage to give (112) occurring when potassium superoxide acts on N-acyl-L-tryptophans [but not when it acts on corresponding tryptophanamides, when dioxindoles (113) are produced].[492] Products of Udenfriend or Fenton hydroxylation of tryptophan (viz., four hydroxytryptophans and oxindole-3-alanine, with N-formylkynurenine) are similar to those obtained by the effects of ionizing radiation on the amino acids in aqueous solutions,[493] while differences arise, comparing Fenton hydroxylation with radiation, with phenylalanine (Fenton → 2-, 3-, and 4-hydroxylation, etc) and tyrosine (Fenton → 2,3- and 3,4-di-hydroxylated products).[494]

Pictet-Spengler reactions of aldehydes with N-diphenylmethyltryptophan isopropyl ester[495] and with di-iodo- or dibromo-L-tyrosines[496] give indolo- (114) and benzo-pipecolic acids (115), respectively. From the point of view of the research interest in preparing these compounds, they can be seen as conformationally-constrained analogues of their parent amino acids. The now notorious L-tryptophan contaminant responsible

(109) (110) (111)

(112) (113)

(114) (115)

(116) (117)

(118)

for an outbreak of eosinophilia myalgia syndrome, 1,1′-ethylidene bis(L-tryptophan), is accompanied by another impurity (UV-5; 0.002g from 150g), "(+)-3-anilino-L-alanine", i.e. (S)-2-amino-3-phenylaminopropanoic acid, presumably introduced during the production process (see Vol.24, p.58).[497]

Protein modification at histidine residues by 4-hydroxynon-2-enal, a lipid peroxidation breakdown product, has been modelled by reactions with N-acetylhistidine,[498] leading after $NaBH_4$ reduction, to N^{π}- and N^{τ}-1,4-dihydroxynonanylhistidines. Lipid hydroperoxides react with tyrosine to give fluorescent 3,3′-dityrosine via radical intermediates.[499]

N-Alkoxycarbonyl derivatives of aspartic acid and glutamic acid give internal anhydrides through carboxy-group activation (using dicyclohexylcarbodi-imide or an alkyl chloroformate), and in no case was a pyroglutamate formed.[500] Pyroglutamic acid derivatives are easily prepared, however, and in protected form this compound is an increasingly valuable synthon, undergoing regioselective nucleophilic ring-opening without racemization.[501] The α-diazoketone of β-t-butyl Z-L-aspartate has been subjected to Wolff rearrangement using silver benzoate, to give the corresponding β-amino acid methyl t-butyl di-ester from which, by selective elaboration at either ester (e.g. by Curtius rearrangement at the methyl ester), (3S)- or (3R)-3,4-di-aminobutanoic acid mono-esters have been obtained.[502] ω-(9-Fluorenylmethyl) esters of aspartic and glutamic acids are formed between the acid and alkanol reactants in THF using HBF_4-OEt_2.[503] Preparations of N-Boc β-aspartyl and γ-glutamyl fluorides have been detailed.[504]

Hydroxyalkyl side-chains of N,C-protected serines and threonines provide sites for O-glycosylation (by a disaccharide derivative, α-D-Xylp-(1→3)-β-D-Glcp,[505] and transglycosylation (from raffinose and lactose, catalysed by α- and β-galactosidases, respectively).[506] Analogous reactions leading to ether phospholipids have been established using a novel phosphite coupling procedure.[507] Methyl N-Z-β-iodoalaninate can be coupled to electron-deficient alkenyl carbohydrates using the Zn/Cu alloy.[508] Cyclization of benzyl N-trityl-L-serine or threonine esters with sulphuryl chloride gives benzyl (2S)-1-trityl-2-aziridinecarboxylate esters or (2S,3S)-1-trityl-3-methyl-2-aziridinecarboxylate esters in excellent yields, something of a breakthrough in achieving this transformation, which seems to be facilitated by the bulky N-substituent.[509] N,O-Acetal formation of L-serine through the hydroxy group, and reduction of the carboxy group of the acetal, provides[510] the Garner aldehyde (23) that has been mentioned in several contexts in this Chapter, all opening up valuable synthetic routes. Another example is reductive amination leading to a product (116) that can be categorized as both an α- or a β-amino acid derivative.[510] Methyl Z-L-threoninate has been converted into a β-keto-ester analogue through straightforward elaboration.[511] Preparations of t-butyldimethylsilyl ethers of serine and threonine have been described.[512]

Cleavage of t-butyl ethers and esters of hydroxyalkyl amino acids by TFA/CH_2Cl_2 is accompanied by fewer artefacts if Et_3SiH is added to the reaction mixture as carbocation scavenger; this deprotection strategy does not affect Bu^tS, Z, Fmoc, or O- and S-benzyl protection.[513] S-Trifluoromethylation of L-homocysteine can be accomplished using CF_3I in liquid NH_3 under u.v. irradiation,[514] and new S-protection strategies include S-(2,4,6-trimethoxybenzyl)ation (removable with 30% $TFA/CH_2Cl_2/PhOH$),[515] and S-[2-(2,4-dinitrophenyl)ethyl]ation (removable by base).[516]

L-Cysteine easily undergoes autoxidation in solution, and electrochemical reduction of L-cystine has been advocated for its clean preparation in 96% yield.[517] Alkaline aqueous solutions of cystine or homocystine contain various sulphur functional groups (RS^-, RSS^-, and $RSSO_3^-$; R = β-alanyl, γ-butyrinyl, respectively),[518] and slow oxidation of methionine to methionine sulphoxide occurs in solutions containing glucose and copper(II) salts.[519]

6.4 *Effects of Electromagnetic Radiation on Amino Acids.*

This Section deals with non-routine reactions and properties of the more common amino acids when subjected to irradiation. Changes brought about in amino acids as a result of pulse radiolysis in aqueous solutions have been reviewed,[520] and it has been noted that aspartic acid complexed with $Al(OH)_3$ is an effective γ-radiation trap.[521] These reports, and the observation that sunlight photolysis of 0.1M aqueous solutions of aspartic and glutamic acids at pH 7 yields mainly malonic and succinic acids respectively,[522] are relevant to prebiotic and environmental aspects of amino acid science.

Laboratory studies concern the aromatic and heteroaromatic protein amino acids, almost exclusively as in previous years. These studies often call for sophisticated spectrometric techniques, and are aimed more at extending knowledge in general of the interaction of energy with matter, than with interests in the reaction products obtained. Fluorescence studies with N-acetyltyrosinamide in the presence of proton acceptors,[523] and time-resolved fluorescence studies with L,L-dityrosine in aqueous solutions,[524] have been reported. Anthrapyrazoles are novel anti-cancer compounds that are effective agents for photosensitized oxidation of DOPA into o-semiquinone radicals.[525] Tyrosine usefully attenuates the 4-iodophenol-enhanced chemiluminescence assay consistent with its competition with luminol for the aryloxy radicals for which this becomes a sensitive assay system.[526]

Fluorescence studies of phenylalanine, tyrosine, and tryptophan and N-acetyltryptophan amide (193nm and 248nm laser fluorescence spectrometry),[527] of tryptophan,[528] and of the constrained tryptophan (117),[529] follow familiar lines. The last-mentioned study is particularly informative in demonstrating a ^{2}H-isotope effect on the fluorescence yield and its variation as a function of pH and temperature. Flash photolysis of tryptophan in 2,2,2-trifluoroethanol results in transient formation of a cyclohexadienyl-type cation (118) due to photoprotonation of the indole moiety at C-4.[530]

7 Analytical Studies with Amino Acids

7.1 *General.*

Results of a 1989 collaborative study of 43 core amino acid analysis facilities have been reported, in an attempt to measure accuracy and precision in amino acid analysis by various methods.[531] A similar objective has been shared with five laboratories who have co-operated in an independent two-part study to try to improve understanding of factors influencing accuracy and precision.[532]

7.2 *Gas-Liquid Chromatography.*

Many of the derivatization protocols that have been described over the years in this Chapter have been in use again this year. Several papers make little of the routine procedures involved, for preparing and using N- and O-trifluoroacetyl n-butyl esters,[533] N- and O-i-butoxycarbonyl t-butyldimethylsilyl esters,[534] and N-methoxycarbonyl 2′,3′,4′,5′,6′-pentafluorobenzyl esters.[535] The particular analytical context in some of these papers is unusually interesting, e.g. the estimation of hydroxyproline in a few fibres of ancient leather after vapour phase hydrolysis (1:1-HCl/$EtCO_2H$ at 150° during 1 hour)[533] by g.l.c. with mass-spectrometric detection, an instrumental combination also used for 3-methylhistidine estimations in physiological samples.[536]

G.l.c. analysis of cleavage products from peptide sequencing remains a research interest for several groups, especially for ever-lower levels of analyte. Derivatization remains obligatory, 2-anilinothiazol-5(4H)-ones from the Edman degradation procedure being converted by aminolysis into N-phenylthiocarbamoylamino acid 2,2,2-trifluoroethylamides for 180 femtomole level analysis by g.l.c./electron capture detection.[537] N- and O-t-Butyldimethylsilylated methylthiohydantoins prepared using $^tBuSiMe_2CNMeCOCF_3$ have been shown to be of suitable volatility for routine g.l.c. analysis.[538]

Enantiomer ratios for amino acid samples can be estimated by g.l.c., either by passage of samples derivatized as above, over a chiral stationary phase, or by diastereoisomer formation using a chiral derivatization reagent; a new example in the latter category, 2-methoxy-2-trifluoromethylpropanoic acid, has been advocated.[539]

7.3 *Thin-Layer Chromatography.*

Leaving aside the routine use of t.l.c. in support of synthetic and reaction studies, an effective use of two-dimensional t.l.c. to estimate enantiomer ratios (yellow spots after derivatization with Marfey's reagent) of naturally-occurring amino acids, can be cited as an example of out-of-the-ordinary work.[19]

7.4 *Ion-Exchange Chromatography.*

Standard techniques are applicable for the analysis of unusual amino acids, viz. mature and immature crosslinking amino acids in collagen and elastin,[540] and 2,2′-di-aminopimelic acid.[541]

7.5 *High-Performance Liquid Chromatography.*

The predominant approach to amino acid analysis by h.p.l.c. currently involves derivatization of mixtures by one of a number of protocols, followed by separation of the resulting derivatives (the "pre-column derivatization approach"). Reviews of derivatization procedures and appropriate h.p.l.c. methods for the different approaches have been published,[542,543] together with specific coverage of the analysis of amino acids in foods.[544]

The conversion of amino acids into iso-indoles using o-phthaldicarboxaldehyde (OPA) in combination with an alkanethiol continues to be represented in most of the current papers. Although the reaction fails with prolines and other N-alkylamino acids, a "double derivatization approach" in which the OPA-thiol derivatization reaction is followed by Fmoc-chloride derivatization,[545-547] seems very reliable. [Of course, Fmoc-chloride derivatization of amino acid mixtures would convert every amino AND imino acid into derivatives that can be readily separated and quantitated on the basis of their fluorescence, but this approach has not gained popularity]. Among studies using the standard OPA-2-mercaptoethanol protocol, the amino acid content of glycoconjugates[548] describes conventional acid hydrolysate analysis, while identification of di-aminopimelic acid in physiological fluids is aided by the presence of an ion-pairing agent in the mobile phase, sharpening peaks of nearby co-eluting compounds.[549] An h.p.l.c. analysis explored with 17 of the protein amino acids using ion-pairing additives has been described.[550] Electrochemical detection methods offer lower detection limits (at about the 0.66 picomole level).[551-553] It has been claimed that recent improvements to OPA - thiol derivatization put it on a par with classical ion-exchange separation and post-column ninhydrin analysis of amino acids. An advantage for this method accompanies the use of a chiral alkanethiol, which yields diastereoisomeric mixtures with D/L-amino acid samples and permits enantiomer ratio analysis. This has been seized upon in studies employing D-3-mercaptopropanoic acid[554,555] and other chiral thiols,[556] but particularly N-acetyl-L-cysteine[557] and N-Boc-L-cysteine.[558] The last two studies have particular aspects of interest; one explores automated derivatization by flow injection analysis as well as demonstrating the greater stability of the derivatives prepared using N-acetylcysteine, while the other establishes substantial levels of D-serine in rat brain, but only trace levels of D-alanine and D-aspartic acid.

Cyanobenz[*f*]isoindoles formed analogously using naphthalene-2,3-dicarboxaldehyde with 18 amino acids have been assessed for their electrochemical oxidation characteristics, that permit voltammetric detection.[559] These derivatives are capable of detection at an extremely low level, and are mentioned in this respect in the later Section 7.5. Another recent derivatization method that has become widely used, involves reaction with phenyl isothiocyanate to give N-phenylthiocarbamoyl derivatives (PTC-amino acids). As with other pre-column derivatization methods, it compares well with classical ion exchange analysis, and the derivatives are more stable (unchanged during at least 32 hours) than OPA-thiol condensation products.[560] An application has been described that is routine in terms of the amino acids identified as their PTC-derivatives, but in a spectacular context. Microgram samples from a Renaissance painting (Cosimo Tura's "Annunciation with St Francis and St Maurelius of Toulouse", 1475?) were hydrolysed in acid vapour and analysed to determine whether the painting had an egg tempera base.[561] Amino acids presenting difficulties as far as phenyl isothiocyanate derivatization is concerned, constitute the other papers cited here: N^G-methyl-, N^GN^G-dimethyl- and $N^GN'^G$-dimethylarginines in physiological fluids,[562] hydroxyproline,[563] but particularly cysteine and cystine.[564,565] In acid hydrolysates containing dithioglycollic acid, some problems of identification of cysteine are inevitable; the PTC-derivatives prepared from cysteine and cystine have the same retention time, probably indicating cleavage of the disulphide bond by phenyl isothiocyanate.[565] It is more likely to be due to the formation of the same phenyldithiocarbamate from both amino acids, after disulphide bond cleavage of cystine; or, possibly, to be due to the formation of the same β-elimination product from both amino acids – see next paragraph.

Phenylthiohydantoins formed through Edman sequencing (they can also be formed from PTC-amino acids) are prone to elimination when the amino acid from which they are derived carries a β-heteroatom. They are stabilised by allyloxycarbonyl side-chain protection, and such PTHs show clean h.p.l.c. traces in the cases of cysteine, lysine, threonine and tyrosine, but serine, aspartic acid, glutamic acid and arginine show some deprotection while histidine is completely deprotected during the manipulations involved.[566] The PTH of S-(β-amidoethyl)cysteine has been prepared as a standard to assign artefactual h.p.l.c. peaks that have been found to originate in un-polymerized acrylamide present in polyacrylamide gels.[567]

Alternative established derivatization protocols that have been illustrated further this year include dansylation[568] and the related dabsylation (used for estimation of the protein crosslinking amino acid, hypusine, down to 500 femtomole levels),[569] and the more distantly-related thiohydantoins, DABTHs.[570] The N-γ-lysine Schiff base with malondialdehyde has been detected in urine and in enzymic digests of certain foods, and can be assayed at 280nm after Diels-Alder adduct formation with diethyl ethoxymethylene malonate (50°/50 min).[571] A procedure for selenocysteine estimation is based on the fluorescence of its N-(iodoacetylaminoethyl)-5-naphthylamine-1-sulphonyl derivative.[572] Further unfamiliar fluorescence-generating procedures include conversion of N-acetylamino acids into esters with 9-anthryldiazomethane,[573] and formation of the NN-diethyl-2,4-dinitro-5-fluoroaniline-hydroxyproline adduct,[574] Edman cleavage products, the 2-anilinothiazolin-5(4H)-ones, can be detected through fluorescence measurement at the 100 attomole level after aminolysis by 4-aminofluorescein,[575] and almost as effective is aminolysis by *N*-(4-aminohexyl)-*N*-ethylisoluminol and chemiluminescence detection.[576]

Post-column derivatization is a feature of the identification and quantitation of a huge range of amino acids in a test of the Pickering lithium ion gradient mobile phase (ninhydrin colorimetry).[577] ^{3}H-labelling of proteins is assessed by hydrolysis and separation of the resulting amino acids by classical ion-exchange chromatography, then the various fractions are quantitated by OPA fluorimetry and scintillation counting.[578] N^G-Methylarginine content of myelin basic protein has been estimated by the same post-column fluorimetry protocol,[579] also used for the quantitation of enantiomer mixtures by separation through the ligand exchange principle using copper(II)-complexed amino acids passing over N-alkyl-L-proline-coated stationary phases.[580,581] The same principle but using a commercial chiral stationary phase (Chiral-ProCu = S,100) has successfully separated all four isomers of threonine and of β-phenylserine and its o- and p-fluoro-substituted analogues.[582]

Enantiomer quantitation is achieved for N-Boc-amino acids over hydroxypropylated β-cyclodextrin[583] and for dansylamino acids through a switchable two-column system, with one of the columns operating with β-cyclodextrin as a chiral mobile phase additive.[584] Another example of the alternative diastereoisomer-forming derivatization approach has been described, in which chiral monochloro-s-triazines are the reagents,[585] prepared by replacing one chlorine atom of s-trichlorotriazine with 4-aminoazobenzene, and another chlorine atom with L-alanine amide.

Ion-exchange h.p.l.c. procedures are appropriate for certain highly-polar analytes, such as taurine, cysteine-sulphinic acid and cysteic acid,[586] the cyanogen bromide-selenomethionine reaction product,[587] protein cross-linking amino acids pyridinoline and deoxypyridinoline in urine or tissue hydrolysates,[588] and S-adenosyl-L-methionine (after the use of a strong cation exchange sulphonic acid resin for its isolation with other cations from mixtures).[589]

Quantitative h.p.l.c. analysis without derivatization is a feature of a number of studies in the preceding paragraph, and an assay of phenylalanine in dried blood spots by u.v. absorption detection,[590] of photodegradation products of aromatic amino acids by electrochemical detection,[591] and of amino acid pentafluorobenzyl esters at sub-picogram levels by negative ion mass spectrometry.[592]

The expected relationship exists between reversed phase h.p.l.c. retention and structure for a homologous series of aliphatic amino acids, with retention times increasing with molecular mass, while each member of this series shows a lower retention index than its cyclic analogue.[593]

7.6 *Other Analytical Methods.*

Appreciation of the superior separation characteristics of capillary zone electrophoresis (CZE) and related techniques is clear from the growing interest in their use, based on similar derivatization protocols to those established for h.p.l.c.

CZE Analysis of fluoresceamine-derivatized proline and hydroxyproline, (these imino acids give non-fluorescent amino-enones),[594] and of dansylated lysine and valine,[595] have been reported. However, a stunning demonstration has been published, of the sensitivity of naphthalene-2,3-dicarboxaldehyde derivatization (0.8 attomole detection limit for the leucine derivative), using CZE, micellar electrokinetic chromatography (MEKC), and cyclodextrin-modified MEKC (CD-MEKC; 30 minutes analysis time), with laser-induced fluorescence detection.[596] Chiral separations were established in some cases in this study, by the CD-MEKC technique.

7.7 *Determination of Specific Amino Acids.*

Whereas general analytical methods are, of course, applicable to specific cases, there are uniquely beneficial short cuts available when knowledge is needed of the level of a particular amino acid in a complex mixture. There are limits to this complexity when using specific functional group chemistry, as in an assay of cysteine, homocysteine and cystine in urine based on cyanide - nitroprusside colorimetry (ca. 524nm),[597] and in a specific condensation of lysine with furfuraldehyde that allows this amino acid to be determined in the presence of other di-amino acids,[598] but an assay of 1-aminocyclopropane-1-carboxylic acid in apple juice can be safely based on ethene evolution caused by NaOCl-$HgCl_2$.[599]

As usual, the predominant theme for this Section lies in the exploitation of enzyme selectivity. A good example of the more recently developed technology is found in estimation of L-leucine in blood, using tRNA/acyl-tRNA synthetase competing with L-[U-^{14}C]leucine and measuring the radioactivity of the acid-insoluble tRNA fraction.[600]

More traditional approaches are being studied, for lysine oxidation to 2-oxo-6-aminocaproate, NH_3, and H_2O_2 catalysed by *Trichoderma viride* L-lysine α-oxidase (O_2 uptake and H_2O_2 generation can be easily measured electrochemically),[601] and a similar L-lysine oxidase-based procedure (chicken kidney tissue) linked to an oxygen electrode.[602] A recently-isolated L-lysine dehydrogenase immobilized on a platinum electrode by gelatin entrapment forms the basis for an amperometric biosensor for L-lysine.[603]

L-Glutamine determination, through the use of glutamine oxidase immobilized by cross-linking to bovine serum albumin deposited on pre-activated nylon,[604] and through the recycling of pink rose petals ("Sonia"; *Rosa hybrida* Hort.: see Vol. 23, p. 70) by attachment to an NH_3 sensor,[605] has been explored further. Immobilized glutamate dehydrogenase acts on L-glutamic acid to liberate the reduced form of NADH that can be assayed through the luminescence generated in a separate sensor, a nylon coil carrying immobilized bacterial bioluminescence-generating enzymes.[606]

References

1. R.O.Duthaler, *GIT Fachz.Lab.*, 1992, **36**, 479, 485, 488.
2. M.T.Reetz, *Angew.Chem.Int.Edit., 1991*, **30**, 1531.

3. D.L.Hatfield, I.S.Choi, B.J.Lee, and J.E.Jung, *Transfer RNAs in Protein Synthesis*, Eds. D.L.Hatfield, B.J.Lee, and R.M.Pirtle, CRC Press, Boca Raton, Florida, 1992, p.269.
4. G.C. Barrett, *Chem.Brit.*, 1992, **28**, 783.
5. V.F. Selemenev, N.Ya.Korenman, and V.Yu.Khokhlov, *Zh.Prikl.Khim.*, 1991, **64**, 2707.
6. L.Handojo, W.Degener, and K.Schuegerl, *Process Metall.*, 1992, **7B** (Solvent Extraction), 1785; T.Hano, T.Ohtake, M.Matsumoto, D.Kitayama, F.Hori, and F.Nakashio, *Ibid.*, p.1881.
7 P.Kaur, G.L.Mundhara, H.S.Kar, and J.-S.Tiwari, Proceedings of the 3rd International Conference on Fundamentals of Adsorption, 1989, Eds. A.B.Mersmann and S.E.Scholl, A.I.Ch.E., New York, 1991, p.379 (*Chem.Abs.*, 1992, **117**, 119108).
8. S.Kaneko, S.Ohmori, M.Mikawa, T.Yamazaki, M.Nakamura, and S.Yamagiwa, *Chem.Lett.*, 1992, 2249.
9. V.A.Prokhorenko, V.V.Kolov, T.I.Struchalina, and V.V.San, *Izv.Akad.Nauk Kirg.S.S.R., Khim.-Tekhnol.Biol.Nauki.*, 1990, 35 (*Chem.Abs.*, 1992, **117**, 91688).
10. I.Iliev and B.Chorbanov, *Biotechnol.Appl.Biochem.*, 1992, **16**, 29.
11. T.C.Pochapsky and Q.Gopen, *Protein Sci.*, 1992, **1**, 786.
12. S.Hatanaka, *Prog.Chem.Org.Nat.Prod.*, 1992, **59**, 1.
13. J.Thompson and J.A.Donkersloot, *Ann.Rev.Biochem.*, 1992, **61**, 517.
14. A.H.Van Gennip, S.Busch, E.G.Schotten, L.E.Stroomer and N.G.Abeling, *Adv.Exp.Med.Biol.*, 1991, **309B**, 15.
15. E.Heidemann, J.Seba, I.Tanay, C.Leutze, M.Rohn, and T.Kisselbach, *Leder*, 1992, **43**, 75 (*Chem.Abs.*, 1992, **117**, 92582).
16. S.Kim, M.Ubukata, K.Kobayashi, and K.Isono, *Tetrahedron Lett.*, 1992, **33**, 2561.
17. F.Itagaki, H.Shigemori, M.Ishibashi, T.Nakamura, T.Sasaki, and J.Kobayashi, *J.Org.Chem.*, 1992, **57**, 5540.
18. G.M.Lee and T.F.Molinski, *Tetrahedron Lett.*, 1992, **33**, 7671.
19. Y.Nagata, K.Yamamoto and T.Shimojo, *J.Chromatogr.*, 1992, **575**, 147.
20. M.Minagawa, S.Egawa, Y.Kabaya, and K.Karasawa-Tsuru, *Shitsuryo Bunseki*, 1992, **40**, 47 (*Chem.Abs.*, 1992, **117**, 86129).
21. J.A.Silfer, M.H.Engel, and S.A.Macko, *Chem.Geol.*, 1992, **101**, 211.
22. M.Ijima, T.Masuda, H.Nakamura, H.Naganawa, S.Kurasawa, Y.Okami, M.Ishizuka, T.Takeuchi, and Y.Iitaka, *J.Antibiot.*, 1992, **45**, 1553.
23. B.C.Van Wegenen, R.Larsen, J.H.Cardellina, D.Randazzo, Z.C.Lidert, and C.Swithenbank, *J.Org.Chem.*, 1993, **58**, 335.
24. S.Fushiya, S.Sato, Y.Kera, and S.Nozoe, *Heterocycles*, 1992, **34**, 1277.
25. K.Yamano and H.Shirahama, *Tetrahedron*, 1992, **48**, 1457.
26. K.Yamano, K.Hashimoto, and H.Shirahima, *Heterocycles*, 1992, 34, 445.
27. S.Hakoda, S.Tsubotani, T.Iwasa, M.Suzuki, M.Kondo, and S.Harada, *J.Antibiot.*, 1992, **45**, 854.
28. S.W.Ayer, B.G.Isaac, K.Luchsinger, N.Makkar, M.Tran, and R.J.Stonard, *J.Antibiot.*, 1991, **44**, 1460.
29. J.M.Pezzuto, W.Mar, L.Z.Lin, G.A.Cordell, A.Neszmelyi, and H.Wagner, *Phytochemistry*, 1992, **31**, 1795.
30. F.Nakamura, K.Yamazaki, and K.Suyama, *Biochem.Biophys.Res.Commun.*, 1992, **186**, 1533.
31. K.Nakamura, T.Hasegawa, Y.Fukunaga, and K.Ienaga, *J.Chem.Soc., Chem.Commun.*, 1992, 992.

32. F.E.Koehn, R.E.Longley, and J.K.Reed, *J.Nat.Prod.*, 1992, **55**, 613.
33. K.Ueda, J.-Z.Xiao, N.Doke, and S.Nakatsuka, *Tetrahedron Lett.*, 1992, **33**, 5377.
34. Y.Koiso, M.Natori, S.Iwasaki, S.Sato, R.Sonoda, Y.Fujita, H.Yaegashi, and Z.Sato, *Tetrahedron Lett.*, 1992, **33**, 4157.
35. G.Ivanova, *Tetrahedron*, 1992, **48**, 177.
36. A.P.Combs and R.W.Armstrong, *Tetrahedron Lett.*, 1992, **33**, 6519.
37. Z.Du, X.Zhou, Y.Shi, and H.Hu, *Chin.J.Chem.*, 1992, **10**, 82 (*Chem.Abs.*, 1992, **117**, 90238).
38. H.Watanabe, Y.Hashizume, and K.Uneyama, *Tetrahedron Lett.*, 1992, **33**, 4333.
39. U.Groeger, K.Drauz, and H.Klenk, *Angew.Chem.Int.Edit.*, 1992, **31**, 195.
40. M.Kobow, W.D.Sprung, and E.Schultz, *Pharmazie*, 1992, **47**, 55.
41. P.Pachaly, H.S.Kang, and D.Wahl, *Arch.Pharm.*, 1991, **324**, 989.
42. L.Petrus and J.N.BeMiller, *Carbohydr.Res.*, 1992, **230**, 197.
43. M.C.Pirrung and N.Krishnamurthy, *J.Org.Chem.*, 1993, **58**, 957.
44. S.Kotha, D.Anglos, and A.Kuki, *Tetrahedron Lett.*, 1992, **33**, 1569; S.Kotha and A.Kuki, *Ibid.*, p.1565.
45. J.P.Genet, J.Uziel, M.Port, A.M.Touzin, S.Roland, S.Thorimbert, and S.Tanier, *Tetrahedron Lett.*, 1992, **33**, 77.
46. B.Kaptein, W.H.J.Boesten, Q.B.Broxterman, H.E.Schoemaker, and J.Kamphuis, *Tetrahedron Lett.*, 1992, **33**, 6007.
47. A.De Nicola, J.Einhorn, and J.-L.Luche, *Tetrahedron Lett.*, 1992, **33**, 646. 48. T.R.Hoye, S.R.Duff, and R.S.King, *Tetrahedron Lett.*, 1985, **26**, 3433.
49. J.M.McIntosh, R.Thangarasa, D.J.Ager, and B.Zhi, *Tetrahedron*, 1992, **48**, 6219.
50. S.Knapp, S.Naudan, and L.Resnick, *Tetrahedron Lett.*, 1992, **33**, 5485.
51. W.O.Moss, A.L.Jones, A.Wisedale, M.F.Mahon, K.C.Molloy, R.H.Bradbury, N.J.Hales, and T.Gallagher, *J.Chem.Soc. Perkin Trans. 1*, 1992, 2615.
52. W.O.Moss, E.Wakefield, M.F.Mahon, K.C.Molloy, R.H.Bradbury, N.J.Hales, and T.Gallagher, *Tetrahedron*, 1992, **48**, 7551.
53. E.C. Roos, H. Hiemstra, W.N. Speckamp, B. Kaptein, J. Kamphuis, and H.E. Schoemaker, *Recl.Trav.Chim.Pays-Bas*, 1992, **111**, 360.
54. E.C.Roos, H.Hiemstra, W.N.Speckamp, B.Kaptein, J.Kamphuis, and H.E. Schoemaker, *SynLett.*, 1992, 451.
55. K.Burger, E.Heistradner, R.Simmerl, and M.Eggersdorfer, *Z.Naturforsch, B: Chem.Sci.*, 1992, **47**, 424.
56. C.Cativiela, M.P.Lopez, and J.A.Mayoral, *SynLett.*, 1992, 121.
57. J.Barker, S.L.Cook, M.E.Lasterra-Sanchez, and S.E.Thomas, *J.Chem.Soc. Chem.Commun.*, 1992, 830.
58. D.Albanese, D.Landini, and M.Penso, *J.Org.Chem.*, 1992, **57**, 1603.
59. M.Nasreddine, S.Szonyi, F.Szonyi, and A.Cambon, *Synth.Commun.*, 1992, **22**, 1547.
60. F.Degerbeck, B.Fransson, L.Grehn, and U.Ragnarsson, *J.Chem.Soc., Perkin Trans.1*, 1992, 245.
61. K.Burgess and K.-K.Ho, *Tetrahedron Lett.*, 1992, **33**, 5677.
62. K.Burgess and K.-K.Ho, *J.Org.Chem.*, 1992, **57**, 5931.
63. M.Mehmandoust, Y.Petit, and M.Larcheveque, *Tetrahedron Lett.*, 1992, **33**, 4313.
64. E.J.Corey, J.O.Link, and Y.Shao, *Tetrahedron Lett.*, 1992, **33**, 3435.
65. E.J.Corey and J.O.Link, *J.Am.Chem.Soc.*, 1992, **114**, 1906.
66. J.M.Chong and S.B.Park, *J.Org.Chem.*, 1992, **57**, 2220.
67. E.Vedejs and H.Sano, *Tetrahedron Lett.*, 1992, **33**, 3261.
68. J.Legters, L.Thijs, and B.Zwanenburg, *Recl.Trav.Chim.Pays.-Bas*, 1992, **111**, 16.

69. J.Legters, J.G.H.Willems, L.Thijs, and B.Zwanenburg, *Recl.Trav.Chim.Pays.-Bas*, 1992, **111**, 59.
70. R.S.Atkinson, B.J.Kelly, and J.Williams, *Tetrahedron*, 1992, **48**, 7713.
71. Y.Ohfune, *Acc.Chem.Res.*, 1992, **25**, 360.
72. R.M.Williams and J.A.Hendrix, *Chem.Rev.*, 1992, **92**, 889.
73. C.G.Kruse, J.Brusse, and A.Van der Gen, *Spec.Chem.*, 1992, **12**, 184, 186, 189, 192 (*Chem.Abs.*, 1992, **117**, 150273).
74. T.Inaba, I.Kozono, M.Fujita, and K.Ogura, *Bull.Chem.Soc.Japan*, 1992, **65**, 2359.
75. P.Zandbergen, J.Brusse, A.Van der Gen, and C.G.Kruse, *Tetrahedron: Asymmetry*, 1992, **3**, 769.
76. T.Durst and K.Koh, *Tetrahedron Lett.*, 1992, **33**, 6799.
77. E.J.Corey, D.-H.Lee, and S.Choi, *Tetrahedron Lett.*, 1992, **33**, 6735.
78. C.-N.Hsaio and T.Kolasa, *Tetrahedron Lett.*, 1992, **33**, 2629.
79. D.B.Berkowitz and W.B.Schweizer, *Tetrahedron*, 1992, **48**, 1715.
80. M.J.O'Donnell and S.Wu, *Tetrahedron: Asymmetry*, 1992, **3**, 591.
81. D.Guillerm and G.Guillerm, *Tetrahedron Lett.*, 1992, **33**, 5047.
82. U.Groth, W.Halfbrodt, and U.Schöllkopf, *Liebigs Ann.Chem.*, 1992, 351.
83. M.S.Allen, L.K.Hamaker, A.J.La Loggia, and J.M.Cook, *Synth.Commun.*, 1992, **22**, 2077.
84. S.Kotha and A.Kuki, *J.Chem.Soc., Chem.Commun.*, 1992, 404.
85. J.E.Rose, P.D.Leeson, and D.Gani, *J.Chem.Soc. Perkin Trans. 1*, 1992, 1563.
86. G.Bold, T.Allmendinger, P.Herold, L.Moelsch, H.P.Schaer, and R.O.Duthaler, *Helv.Chim.Acta*, 1992, **75**, 865.
87. A.R.Jurgens, *Tetrahedron Lett.*, 1992, **33**, 4727.
88. M.Orena, G.Porzi, and S.Sandri, *J.Org.Chem.*, 1992, **57**, 6532.
89. K.Busch, U.M.Groth, W.Kuehnle and U.Schöllkopf, *Tetrahedron*, 1992, **48**, 5607.
90. D.P.G.Hamon, R.A.Massy-Westropp, and P.Razzino, *Tetrahedron*, 1992, **48**, 5163.
91. H.G.Chen, V.G.Beylin, M.Marlatt, and O.P.Goel, *Tetrahedron Lett.*, 1992, **33**, 3293.
92. R.Dharanipragada, K.Van Hulle, A.Bannister, B.Soaring, L.Kennedy, and V.J.Hruby, *Tetrahedron*, 1992, **48**, 4733.
93. M.F.Beatty, C.Jennings-White, and M.A.Avery, *J.Chem.Soc. Perkin Trans. 1*, 1992, 1637.
94. D.A.Evans and K.M.Lundy, *J.Am.Chem.Soc.*, 1992, **114**, 1495.
95. L.W.Boteju, K.Wegner, and V.J.Hruby, *Tetrahedron*, 1992, **48**, 1457.
96. K.J.Hale, V.M.Delisser, and S.Manaviazu, *Tetrahedron Lett.*, 1992, **33**, 7613.
97. W.Oppolzer, O.Tamura, and J.Deerberg, *Helv.Chim.Acta*, 1992, **75**, 1965.
98. L.S.Hegedus, E.Lastra, Y.Narukawa, and D.C.Snustad, *J.Am.Chem.Soc.*, 1992, **114**, 2991.
99. J.R.Miller, S.R.Pulley, L.S.Hegedus, and S.DeLombaert, *J.Am.Chem.Soc.*, 1992, **114**, 5602.
100. J.M.Vernier, L.S.Hegedus, and D.B.Miller, *J.Org.Chem.*, 1992, **57**, 6914.
101. R.M.Williams, W.Zhai, D.J.Aldous, and S.C.Aldous, *J.Org.Chem.*, 1992, **57**, 6527.
102. R.M.Williams and G.J.Fegley, *Tetrahedron Lett.*, 1992, **33**, 6755.
103. R.M.Williams and C.Yuan, *J.Org.Chem.*, 1992, **57**, 6519.
104. Z.Dong, *Tetrahedron Lett.*, 1992, **33**, 7725.
105. J.E.Baldwin, V.Lee, and C.J.Schofield, *SynLett.*, 1992, 249.
106. M.J.Crossley and C.W.Tansey, *Aust.J.Chem.*, 1992, **45**, 479.
107. G.Keum, Y.J.Chung, and B.H.Kim, *Bull.Korean Chem.Soc.*, 1992, **13**, 343 (*Chem.Abs.*, 1992, **117**, 171978).

108. W.Mueller, D.A.Lowe, H.Neist, S.Urwyler, P.L.Herrling, D.Blaser, and D.Seebach, *Helv.Chim.Acta*, 1992, **75**, 855.
109. C.Andres, A.Maestro, R.Pedrosa, A.Perez-Encabo, and M.Vicente, *SynLett.*, 1992, 45.
110. C.Agami, F.Conty, and M.Poursoulis, *SynLett.*, 1992, 847; C.Agami, F.Conty, M.Poursoulis, and J.Vaissermann, *Tetrahedron*, 1992, **48**, 431.
111. P.D.Bailey, G.R.Brown, F.Korber, A.Reid, and R.D.Wilson, *Tetrahedron: Asymmetry*, 1992, **3**, 161.
112. V.A.Soloshonok, Yu.N.Belokon, N.A.Kuzmina, V.I.Maleev, N.Yu.Svistunova, V.A.Solodenko, and V.P.Kukhar, *J.Chem.Soc. Perkin Trans. 2*, 1992, 1525.
113. V.A.Soloshonok, N.Yu.Svistunova, V.P.Kukhar, N.A.Kuzmina, and Yu.N.Belokon, *Izv.Akad.Nauk S.S.S.R., Ser.Khim.*, 1992, 687.
114. V.I.Losikina and V.I.Sokolov, *Doklady Akad.Nauk, S.S.S.R.*, 1992, **320**, 339.
115. V.A.Soloshonok, S.Mociskite, V.I.Maleev, S.A.Orlova, N.S.Ikonnikov, E.R.Shamuratov, A.S.Batsanov, and Yu.T.Struchkov, *Mendeleev Commun.*, 1992, 89.
116. G.Liu, C.Zhou, H.Piao, L.Wu, A.Mi, and Y.Jiang, *Huaxue Xuebao*, 1992, **50**, 89 (*Chem.Abs.*, 1992, **116**, 214861).
117. Y.Jiang, C.Zhou, H.Piao, G.Liu, Z.Ma, amd A.Mi, *Youji Huaxue*, 1992, **12**, 77 (*Chem.Abs.*, 1992, **117**, 234475).
118. S.S.Lu and B.T.Uang, *J.Chin.Chem.Soc.*, 1992, **39**, 245 (*Chem.Abs.*, 1992, **117**, 131521).
119. H.Matsunega, T.Ishizuka, N.Marubayashi, and T.Kunieda, *Chem.Pharm.Bull.*, 1992, **40**, 1077.
120. H.W.Krause, H.J.Kreuzfeld and C.Doebler, *Tetrahedron: Asymmetry*, 1992, **3**, 555.
121. H.W.Krause, F.W.Wilcke, H.J.Kreuzfeld and C.Doebler, *Chirality*, 1992, **4**, 110.
122. A.Terfort, *Synthesis*, 1992, 951.
123. J.J.Brunet, H.Hajouji, J.C.Ndjanga, and D.Weibecker, *J.Mol.Catal.*, 1992, **72**, L21.
124. G.Oehme, E.Paetzold, and R.Selke, *J.Mol.Catal.*, 1992, **71**, L1.
125. M.Saburi, M.Ohnuki, M.Ogasawara, T.Takahashi, and Y.Uchida, *Tetrahedron Lett.*, 1992, **33**, 5783.
126. J.P.Genet, C.Pinel, S.Mallart, S.Juge, N.Cailhol, and J.A.Lafitte, *Tetrahedron Lett.*, 1992, **33**, 5343.
127. C.Cativiela, M.D.Diaz-de-Villegas, and J.A.Galvez, *Tetrahedron: Asymmetry*, 1992, **3**, 567.
128. C.Cativiela, M.D.Diaz-de-Villegas, and J.A.Galvez, *Can.J.Chem.*, 1992, **70**, 2325.
129. M.Bruncko and D.Crich, *Tetrahedron Lett.*, 1992, **33**, 6251.
130. A.Saeed and D.W.Young, *Tetrahedron*, 1992, **48**, 2507.
131. S.Kumon and T.Kawakita, in "Biotechnology of Food Ingredients", Eds. I.Goldberg and R.A.Williams, Van Nostrand-Reinhold, New York, 1991, p.125.
132. S.Suye, M.Kawagoe, and S.Inuta, *Can.J.Chem.Eng.*, 1992, **20**, 306.
133. G.Renard, J.C.Guilleux, C.Bore, V.Malta-Valette, and D.Lerner, *Biotechnol.Lett.*, 1992, **14**, 673.
134. S.G.Velizarov, E.I.Rainina, A.P.Sinitsyn, S.D.Varfolomeev, V.I.Lozinskii, and A.L.Zubov, *Biotechnol.Lett.*, 1992, **14**, 291.
135. C.Cativiela, M.Diaz-de-Villegas, and A.I.Jimenez, *Synth,Commun.*, 1992, **22**, 2955.
136. C.Alvarez, A.Herrero, J.L.Marco, E.Fernandez-Alvarez, and M.Bernabe, *Tetrahedron Lett.*, 1992, **33**, 5605.
137. T.Kienzler, P.Strazewski, and C.Tamm, *Helv.Chim.Acta*, 1992, **75**, 1078.

138. J.E.Baldwin, R.M.Adlington, D.Bebbington, and A.T.Russell, *J.Chem.Soc. Chem.Commun.*, 1992, 1249.
139. F.Matsuura, Y.Hamada, and T.Shioiri, *Tetrahedron: Asymmetry*, 1992, **3**, 1069.
140. F.Matsuura, Y.Hamada, and T.Shioiri, *Tetrahedron Lett.*, 1992, **33**, 7917.
141. F.Matsuura, Y.Hamada, and T.Shioiri, *Tetrahedron Lett.*, 1992, **33**, 7921.
142. J.Cooper, D.W.Knight, and P.T.Gallagher, *J.Chem.Soc. Perkin Trans. 1*, 1992, 553.
143. M.Kininata, T.Kaziwara, Y.Kawashima, and I.Ichimoto, *Agric.Biol.Chem.*, 1991, **55**, 3033.
144. S.Takano, K.Inomata, and K.Ogasawara, *J.Chem.Soc. Chem.Commun.*, 1992, 169.
145. A.Barco, S.Benetti, G.Spalluto, A.Casolari, G.P.Pollini, and V.Zanirato, *J.Org.Chem.*, 1992, **57**, 6729.
146. K.Konno, K.Hashimoto, and H.Shirahama, *Heterocycles*, 1992, **33**, 303.
147. Y.Hirai, T.Terada, Y.Amemiya, and T.Momose, *Tetrahedron Lett.*, 1992, **33**, 7893.
148. M.Kaname and S.Yoshifuji, *Tetrahedron Lett.*, 1992, **33**, 8103.
149. J.P.Genet, J.O.Durand, M.Savignac, and D.Pons, *Tetrahedron Lett.*, 1992, **33**, 2497.
150. A.J.Pearson and H.Shin, *Tetrahedron*, 1992, **48**, 7527.
151. X.Feng and R.K.Olsen, *J.Org.Chem.*, 1992, **57**, 5811.
152. H.Pervez and C.J.Suckling, *Sci.Int. (Lahore)*, 1992, **4**, 49.
153. C.-O.Chan, D.Crich, and S.Natarajan, *Tetrahedron Lett.*, 1992, **33**, 3405.
154. N.Sewald, J.Reide, P.Bissinger, and K.Burger, *J.Chem.Soc. Perkin Trans. 1*, 1992, 267.
155. M.L.Pedersen and D.B.Berkowitz, *Tetrahedron Lett.*, 1992, **33**, 7315.
156. D.Schirlin, J.B.Ducep, S.Baltzer, P.Bey, F.Piriou, J.Wagner, J.M.Hornsperger, J.G.Heydt, M.J.Jung, C.Danzin, R.Weiss, J.Fischer, A.Mitschler, and A.De Cian, *J.Chem.Soc. Perkin Trans. 1*, 1992, 1053.
157. Z.Tian, P.Edwards, and R.W.Roeske, *Int.J.Pept.Protein Res.*, 1992, **40**, 119.
158. R.F.W.Jackson, N.Wishart, A.Wood, K.Janes, and M.J.Wythes, *J.Org.Chem.*, 1992, **57**, 3397.
159. Z.Shen, J.Qiau, W.Qiang, and Y.Zhang, *Chin.Chem.Lett.*, 1992, **3**, 237.
160. R.Pellicciari, B.Natalini, M.Marinozzi, B.M.Sadeghpour, A.A.Cordi, T.H. Lanthorn, W.F.Hood, and J.B.Monahan, *Farmaco*, 1991, **48**, 1243.
161. C.J.Easton, E.W.Tan, and C.M.Ward, *Aust.J.Chem.*, 1992, **48**, 395.
162. S.Ahmad, R.S.Phillips, and C.H.Stammer, *J.Med.Chem.*, 1992, **35**, 1410.
163. M.Pilar de Frutos, M.D.Fernandez, and M.Bernabe, *Tetrahedron*, 1992, **48**, 1123.
164. E.Bunnet, C.Cataviela, M.D.Diaz-de-Villegas, and A.I.Jiminez, *SynLett.*, 1992, 579.
165. L.Wessjohann, N.Krass, D.Yu, and A.De Meijere, *Chem.Ber.*, 1992, **125**, 867.
166. M.T.Reetz, F.Kayser, and K.Harms, *Tetrahedron Lett.*, 1992, **33**, 3453.
167. Y.Yuasa, T.Watanabe, A.Nagakura, H.,Tsuruta, G.A.King, J.G.Sweeney, and G.A.Iacobucci, *Tetrahedron*, 1992, **48**, 3473.
168. G.K.Toth and B.Penke, *Synthesis*, 1992, 361.
169. S.Hanessian, K.Sumi, and B.Vanasse, *SynLett.*, 1992, 33.
170. G.Emmer, *Tetrahedron*, 1992, **48**, 7165.
171. J.E.Baldwin, S.C.M.Turner, and M.G.Moloney, *Tetrahedron Lett.*, 1992, **33**, 1517.
172. P.M.Esch, H.Hiemstra, R.F.De Boer, and W.N.Speckamp, *Tetrahedron*, 1992, **48**, 4659.
173. R.M.Adlington and S.J.Mantell, *Tetrahedron*, 1992, **48**, 6529.
174. D.K.Dikshit and S.K.Panday, *J.Org.Chem.*, 1992, **57**, 1920.
175. A.Bourgeois-Cury, D.Doan, and J.Gore, *Tetrahedron Lett.*, 1992, **33**, 1277.
176. P.Wipf and Y.Kim, *Tetrahedron Lett.*, 1992, **33**, 5477.

177. F.Brion, C.Marie, P.Mackiewicz, J.M.Roul, and J.Buendia, *Tetrahedron Lett.*, 1992, **33**, 4889.
178. H.Waldmann and M.Braun, *J.Org.Chem.*, 1992, **57**, 4444.
179. C.Herdeis and W.Engel, *Arch.Pharm.*, 1992, **325**, 411.
180. C.Herdeis and W.Engel, *Arch.Pharm.*, 1992, **325**, 419.
181. G.Wulff and H.T.Klinken, *Tetrahedron*, 1992, **48**, 5985.
182. S.R.Angle, J.G.Breitenbucher, and D.O.Arnaiz, *J.Org.Chem.*, 1992, **57**, 5947.
183. K.-K.Ko, K.-I.Lee, and W.-J.Kim, *Tetrahedron Lett.*, 1992, **33**, 6651.
184. S.Kihara, M.Samada, S.Kuwada, Y.Sohrin, O.Shirai, H.Kokusen, M.Suzuki, and M.Matsui, *Anal.Sci.*, 1991, **7** (Suppl.Proc.Int.Congr.Anal.Sci., 1991), 663.
185. G.A.Lavrent'ev, *Zh.Evol.Biokhim.Fiziol.*, 1991, **27**, 395 (*Chem.Abs.*, 1992, **116**, 209828).
186. B.Knight, *Nature*, 1992, **357**, 202.
187. R.J.C.Hennet, N.G.Holm, and M.H.Engel, *Naturwissenschaften*, 1992, **79**, 361.
188. K.Kobayashi, T.Kaneko, T.Kobayashi, L.Hui, M.Tsuchiya, T.Saito, and T.Oshima, *Anal.Sci.*, 1991, **7** (Suppl.Proc.Int.Congr.Anal.Sci., 1991), 921.
189. K.D.Moeller and L.D.Rutledge, *J.Org.Chem.*, 1992, **57**, 6360.
190. C.J.Easton and S.C.Peters, *Tetrahedron Lett.*, 1992, **33**, 5581.
191. A.G.Avent, A.N.Bowler, P.M.Doyle, C.M.Marchand, and D.W.Young *Tetrahedron Lett.*, 1992, **33**, 1509.
192. M.Hudlicky and J.S.Merola, *Tetrahedron Lett.*, 1990, **31**, 7403; M.Hudlicky and J.S.Merola, *Tetrahedron Lett.*, 1991, **34**, 3134.
193. D.L.Boger and R.F.Menezes, *J.Org.Chem.*, 1992, **57**, 4331.
194. C.Palomo, F.Cabré, and J.Ontoria, *Tetrahedron Lett.*, 1992, **33** 4819.
195. S.Ebeling, D.Matthies, and D.McCarthy, *J.Prakt.Chem./Chem.Ztg.* 1992, **334**, 361.
196. M.Bols and I.Lundt, *Acta Chem.Scand.*, 1992, **46**, 298.
197. L.M.Harwood, J.Macro, D.Watkin, C.E.Williams, and L.F.Wong *Tetrahedron: Asymmetry*, 1992, **3**, 1127.
198. W.Yang, X.Pi, Q.Huang, and Y.Liu, *Huaxue Shije*, 1991, **32**, 516 (*Chem.Abs.*, 1992, **117**, 29163).
199. Y.Nakamura and C.G.Shin, *Chem.Lett.*, 1992, 49.
200. M.A.R.C.Bulusu and P.Waldstatten, *Tetrahedron Lett.*, 1992, **33** 1859.
201. E.C.Roos, H.H.Mooiweer, H.Hiemstra, W.N.Speckamp, B.Kaptein H.J.W.Boesten, and J.Kamphuis, *J.Org.Chem.*, 1992, **57**, 6769.
202. M.Daumas, L.Vo-Quang, and F.Le Goffic, *Tetrahedron*, 1992, **48** 2373.
203. U.Schmidt, H.Griesser, V.Leitenberger, A.Lieberknecht, R.Mangold R.Meyer, and B.Reidl, *Synthesis*, 1992, 487.
204. G.T.Crisp and T.A.Robertson, *Tetrahedron*, 1992, **48**, 3239.
205. F.Manfre, J.-M.Kern, and J.-F.Biellmann, *J.Org.Chem.*, 1992, **57** 2060.
206. F.Berti, C.Ebert, and L.Gardossi, *Tetrahedron Lett.*, 1992, **33** 8145.
207. M.J.Dunn and R.F.W.Jackson, *J.Chem.Soc. Chem.Commun.*, 1992, 319.
208. M.Cushman and E.-S.Lee, *Tetrahedron Lett.*, 1992, **33**, 1193.
209. J.H.Dygos, E.E.Yonan, M.G.Scaros, O.J.Goodmonson, D.P.Getman R.A.Periana, and G.R.Beck, *Synthesis , 1992, 741.*
210. C.Garbay-Jaureguiberry, I.McCort-Tranchepain, B.Barbe, D.Ficheux and B.P.Roques, *Tetrahedron: Asymmetry*, 1992, **3**, 637.
211. A.Dorville, I.McCort-Tranchepain, D.Vichard, W.Sather, R.Maroun P.Ascher, and B.P.Roques, *J.Med.Chem.*, 1992, **35**, 2551.
212. J.Y.Nie and K.L.Kirk, *J.Fluorine Chem.*, 1992, **55**, 259.
213. A.V.Rama Rao, T.K.Chakraborty, and S.P.Joshi, *Tetrahedron Lett.* 1992, **33**, 4045.

214. G.Spadoni, C.Balsamini, A.Bedini, E.Duranti and A.Tontini *J.Heterocycl.Chem.*, 1992, **29**, 305.
215. R.Sanchez-Obregon, A.G.Fallis, and A.G.Szabo, *Can.J.Chem.*, 1992 **70**, 1531.
216. P.K.Dua and R.S.Phillips, *Tetrahedron Lett.*, 1992, **33**, 29.
217. M.F.Comber and C.J.Moody, *Synthesis*, 1992, 731.
218. I.T.Christensen, B.Ebert, U.Madsen, B.Nielsen, L.Brehm, and P.Krogsgaard-Larsen, *J.Med.Chem.*, 1992, **35**, 3512.
219. N.Subasinghe, M.Schulte, R.T.Roon, J.F Koerner, and R.L.Johnson *J.Med.Chem.*, 1992, **35**, 4602.
220. W.H.W.Lunn, D.D.Schoepp, D.O.Calligaro, R.T.Vasileff, L.J.Heinz C.R.Salhoff, and P.J.O'Malley, *J.Med.Chem.*, 1992, **35**, 4608.
221. R.Pellicciari, M.A.Gallo-Mezo, B.Natalini, and A.M.Amer, *Tetrahedron Lett.*, 1992, **33**, 3003.
222. M.Loriga, G.Paglietti, F.Sparatore, G.Pinna, and A.Sisini, *Farmaco*, 1992, **47**, 439.
223. S.Farooq, W.E.Swain, R.Daeppu, and G.Rihs, *Tetrahedron: Asymmetry* 1992, **3**, 51.
224. C.O.Chan, C.J.Cooksey, and D.Crich, *J.Chem.Soc. Perkin Trans. 1* 1992, 777.
225. P.D.Leeson, R.W.Carling, K.W.Moore, A.M.Moseley, J.D.Smith G.Stevenson, T.Chan, R.Baker, A.C.Foster, et al., *J.Med.Chem.* 1992, **35**, 1954.
226. R.W.Carling, P.D.Leeson, A.M.Moseley, R.Baker, A.C.Foster S.Grimwood, J.A.Kemp, and G.R.Marshall, *J.Med.Chem.*, 1992, **35** 1942.
227. P.Dostert, M.Varasi, A.Della Torre, C.Monti, and V.Rizzo, *Eur.J.Med.Chem.*, 1992, **27**, 57.
228. J.Y.L.Chung, J.T.Wasicak, and A.M.Nadzan, *Synth.Commun.*, 1992, **22** 1039.
229. T.Beulshausen, U.Groth, and U.Schollköpf, *Liebigs Ann.Chem.*, 1992 523.
230. N.M.Olken and M.A.Marletta, *J.Med.Chem.*, 1992, **35**, 1137.
231. P.L.Feldman, *Chemtracts: Org.Chem.*, 1992, **5**, 217.
232. Q.B.Broxterman, B.Kaptein, J.Kamphuis, and H.E.Schoemaker *J.Org.Chem.*, 1992, **57**, 6286.
233. M.A.Lago, J.Samancu, and J.D.Elliott, *J.Org.Chem.*, 1992, **57** 3493.
234. V.V.Ryakhovskii, S.V.Agafonov, O.B.Vinogradova, Yu.M.Kosyrev, M.V. Kiselevskii, and V.S.Dobryanski, *Khim.Farm.Zh.*, 1992, **26**, 46.
235. K.Drauz, M.Kottnhahn, and H.Klenk, *J.Prakt.Chem./Chem.Ztg.*, 1992 **334**, 214.
236. F.Cavelier-Frontin, J.Dannis, and R.Jacquier, *Tetrahedron: Asymmetry*, 1992, **3**, 85.
237. R.Polt, Y.Li, Q.Fernando, and M.Rivera, *Tetrahedron Lett.*, 1992 **33**, 2961.
238. J.P.E.Hutchinson and K.E.B.Parkes, *Tetrahedron Lett.*, 1992, **33** 7065.
239. J.W.Perich, *SynLett.*, 1992, 595.
240. X.Jiao, W.Chen, and B.Hu, *Synth,Commun.*, 1992, **22**, 1179.
241. M.G.Hoffmann and H.-J.Zeiss, *Tetrahedron Lett.*, 1992, **33**, 2669.
242. H.-J.Zeiss, *Tetrahedron*, 1992, **48**, 8263.
243. M.Tabcheh, L.Pappalardo, and M.L.Roumestant, *Amino Acids*, 1992, **2** 191.
244. G.W.Kabalka, Z.Wang, J.F.Green, and M.M.Goodman *Appl.Radiat,Isot.*, 1992, **43**, 389.
245. S.S.Yuan and A.M.Ajami, *Huaxue*, 1991, **49**, 257 (*Chem.Abs.*, 1992 **117**, 251722).
246. S.T.Chen, C.C.Tu, and K.T.Wang, *Biotechnol.Lett.*, 1992, **14**, 269.
247. Yu.A.Zolotarev, V.S.Kozik, E.M.Dorokhova, V.Yu.Tatur S.G.Rosenberg, and N.F.Myasoedov, *J.Labelled Compd.Radiopharm.* 1992, **31**, 71.
248. T.Furuta, M.Katayama, H.Shibasaki, and Y.Kasuya, *J.Chem.Soc. Perkin Trans. 1*, 1992, 1643.
249. R.Rajagopal, T.K.Venkatachalan, T.Conway, and M.Diksic *Appl.Radiat.Isot.*, 1992, **43**, 979.

250. J.D.Fissekis, C.Nielsen, S.Tirelli, A.F.Knott, and J.R.Dahl *Appl.Radiat.Isot.*, 1992, **42**, 1169.
251. J.O.Theorell, S.Stone-Elander, and N.Elander, *J.Labelled Compd.Radiopharm.*, 1992, **31**, 207.
252. M.Suehiro, H.T.Ravert, A.A.Wilson, U.Scheffel, R.F.Dannals, and H.N.Wagner, *J.Labelled Compd.Radiopharm.*, 1992, **31**, 151.
253. F.Oberdorfer, T.Siegel, A.Guhlmann, D.Keppler, and W.Maier-Borst *J.Labelled Compd.Radiopharm.*, 1992, **31**, 903.
254. L.Grehn, U.Bondesson, T.Pekh, and U.Ragnarsson, *J.Chem.Soc. Chem.Commun.*, 1992, 1332.
255. S.N.Lodwig and C.J.Unkefer, *J.Labelled Compd.Radiopharm.*, 1992 **31**, 95.
256. L.Li and F.Liu, *Hejishu*, 1992, **15**, 43 (*Chem.Abs.*, 1992, **117** 70280).
257. M.C.Miller, S.D.Wyrick, I.H.Hall, A.Sood, and B.F.Spielvogel *J.Labelled Compd.Radiopharm.*, 1992, **31**, 595.
258. A.Luxen, M.Guillaume, W.P.Melega, V.W.Pike, O.Solin, and R.Wagner *Nucl.Med.Biol.*, 1992, **19**, 149.
259. M.J.Adam and S.Jivan, *J.Labelled Compd.Radiopharm.*, 1992, **31**, 39.
260. K.Ishiwata, T.Ido, C.Honda, M.Kawamura, M.Ichihashi, and Y.Mishima, *Nucl.Med.Biol.*, 1992, **19**, 311.
261. J.Roemer, P.Maeding, and F.Roesch, *Appl.Radiat.Isot.*, 1992, **43** 495.
262. A.Dobrev, V.Benin, and L.Nechev, *Liebigs Ann.Chem.*, 1992, 863.
263. R.Bernardi, M.Zanotti, G.Bernardi, and A.Duatti, *J.Chem.Soc. Chem.Commun.*, 1992, 1015.
264. R.W.Dugger, J.L.Ralbovsky, D.Bryant, J.Commander, S.S.Massett N.A.Sage, and J.R.Selvidio, *Tetrahedron Lett.*, 1992, **33**, 6763.
265. A.Commercon, D.Bezard, F.Bernard, and J.D.Bourzat, *Tetrahedron Lett.*, 1992, **33**, 5185.
266. B.H.Norman and M.L.Morris, *Tetrahedron Lett.*, 1992, **33**, 6883.
267. Q.A.Ward and G.Procter, *Tetrahedron Lett.*, 1992, **33**, 3359.
268. E.Juaristi and D.Quintana, *Tetrahedron: Asymmetry*, 1992, **3**, 723.
269. K.S.Chu, G.R.Negrete, J.P.Konopelski, F.J.Lakner, N.T.Woo, and M.M.Olmstead, *J.Am.Chem.Soc.*, 1992, **114**, 1800.
270. J.N.Kinkel, U.Gysel, D.Blaser, and D.Seebach, *Helv.Chim.Acta* 1991, **74**, 1622.
271. E.Juaristi, J.Escalante, B.Lamatsch, and D.Seebach, *J.Org.Chem.* 1992, **57**, 2396.
272. R.Amoroso, G.Cardillo, and C.Tomasini, *Heterocycles*, 1992, **34** 349.
273. R.Amoroso, G.Cardillo, and C.Tomasini, *Tetrahedron Lett.*, 1992 **33**, 2725.
274. T.Kurihara, Y.Matsubara, S.Harusawa, and R.Yoneda, *J.Chem.Soc. Perkin Trans. 1*, 1992, 3177.
275. C.Andres, A.Gonzalez, R.Pedrosa, and A.Perez-Encabo, *Tetrahedron Lett.*, 1992, **33**, 2895.
276. J.Legters, E.van Dienst, L.Thijs, and B.Zwanenburg *Rec.Trav.Chim.Pays-Bas*, 1992, **111**, 69.
277. A.Atmani, A.El Hallaoui, S.El Hajji, M.L.Roumestant, and P.Viallefont, *Synth.Commun.*, 1992, **21**, 2383.
278. F.A.Davis, R.T.Reddy, and R.E.Reddy, *J.Org.Chem.*, 1992, **57**, 6387.
279. Y.Kobayashi, Y.Takemoto, T.Kamijo, H.Harada, Y.Ito, and S.Terashima, *Tetrahedron*, 1992, **48**, 1853.
280. S.Shimada, K.Saigo, M.Abe, A.Sudo, and M.Hasegawa, *Chem.Lett.* 1992, 1445.
281. K.Tohdo, Y.Hamada, and T.Shioiri, *Tetrahedron Lett.*, 1992, **33** 2031.

282. F.Matsuda, T.Matsumoto, M.Ohsaki, Y.Ito, and S.Terashima *Bull.Chem.Soc.Jpn.*, 1992, **65**, 360.
283. T.Yokomatsu, Y.Yuasa, and S.Shibuya, *Heterocycles*, 1992, **33** 1051.
284. K.J.Harris and C.J.Sih, *Biocatalysis*, 1992, **5**, 195.
285. S.Hashiguchi, A.Kawada, and H.Natsugari, *Synthesis*, 1992, 403.
286. Yu.N.Bubnov, L.I.Lavrinovich, A.Yu.Zykov, and A.V.Ignatenko *Mendeleev Commun.*, 1992, 86.
287. A.B.Holmes, A.B.Tabor, and R.Baker, *J.Chem.Soc. Perkin Trans. 1* 1991, 3301.
288. A.B.Holmes, A.B.Tabor, and R.Baker, *J.Chem.Soc. Perkin Trans. 1* 1991, 3307.
289. T.W.Kwon, P.F.Keusenkothen, and M.B.Smith, *J.Org.Chem.*, 1992, **57** 6169.
290. T.Manickum and G.Rous, *Synth.Commun.*, 1992, **21**, 2269.
291. K.Paulini and H.U.Reissig, *SynLett.*, 1992, 505.
292. J.S.Panek, M.Yang, and I.Muler, *J.Org.Chem.*, 1992, **57**, 4063.
293. R.Zimmer, M.Hoffmann, and H.U.Reissig, *Chem.Ber.*, 1992, **125** 2243.
294. M.Yoda, T.Shirai, T.Katagiri, K.Takabe, and K.Hosoya *Chem.Express*, 1992, **7**, 477.
295. M.Bessodes, M.Saiah, and K.Antonakis, *J.Org.Chem.*, 1992, **57** 4441.
296. K.L.Rinehart, R.Sakai, K.Vishore, D.W.Sullins, and K.M.Li, *J.Org.Chem.*, 1992, **57**, 3007.
297. P.Roth and R.Metternich, *Tetrahedron Lett.*, 1992, **33**, 3993.
298. F.J.Urban and B.S.Moore, *J.Heterocycl.Chem.*, 1992, **29**, 431.
299. W.-R.Li, S-Y.Han, and M.M.Joullie, *Tetrahedron Lett.*, 1992, **33** 3595.
300. M.Bols, I.Lundt, and C.Pedersen, *Tetrahedron*, 1992, **48**, 319.
301. R.D.Gandour, N.L.Blackwell, W.T.Colucci, C.Chung, L.L.Bieber R.R.Ramsay, E.P.Brass, and F.R.Fronczek, *J.Org.Chem.*, 1992, **57** 3426.
302. N.Kasai and K.Sakaguchi, *Tetrahedron Lett.*, 1992, **33**, 1211.
303. J.D.Armstrong, F.W.Hartner, A.E.DeCamp, R.P.Volante, and I.Shinkai, *Tetrahedron Lett.*, 1992, **33**, 6599.
304. J.E.Baldwin, R.M.Adlington, J.S.Bryans, M.D.Lloyd, T.J.Sewell C.J.Schofield, K.H.Baggaley, and R.Cassels, *J.Chem.Soc. Chem.Commun.*, 1992, 877.
305. A.Golebiowski, J.Kozak, and J.Jurczak, *J.Org.Chem.*, 1991, **56** 7344.
306. E.Francotte and A.Junker-Buchheit, *J.Chromatogr.*, 1992, **576**, 1.
307. S.F.Mason, *Ciba Foundation Symposium*, 1991, **162** (Biol. Asymmetry Handedness), 3.
308. W.H.Pirkle, J.P.Chang, and J.A.Burke, *J.Chromatogr.*, 1992, **598** 1.
309. M.H.Hyun, J.J.Ryoo, C.S.Min, and W.H.Pirkle, *Bull.Korean Chem.Soc.*, 1992, **13**, 407 (*Chem.Abs.*, 1993, **118**, 39362).
310. W.H.Pirkle and J.A.Burke, *J.Chromatogr.*, 1992, **598**, 159.
311. T.Masawaki, M.Sasai, and S.Tone, *J.Chem.Eng.Jpn.*, 1992, **25**, 33 (*Chem.Abs.*, 1992, **116**, 152334).
312. K.Morihara, M.Kurokawa, Y.Kamata, and T.Shimada, *J.Chem.Soc. Chem.Commun.*, 1992, 358.
313. T.Shiraiwa, M.Yamauchi, T.Tasumi, and H.Kurokawa *Bull.Chem.Soc.Jpn.*, 1992, **65**, 267.
314. P.L.Ornstein, D.D.Schoepp, M.B.Arnold, N.D.Jones, J.B.Deeter D.Lodge, and J.D.Leander, *J.Med.Chem.*, 1992, **35**, 3111.
315. K.Oyama, M.Itoh, and Y.Nonaka, *Bull.Chem.Soc.Jpn.*, 1992, **65** 1751.
316. M.Kawai, Y.Omori, H.Yamamura, and Y.Butsugan, *Tetrahedron: Asymmetry*, 1992, **3**, 1019.
317. W.H.Pirkle, R.Heire, and M.H.Hyun, *Chirality*, 1992, **4**, 302.

318. D.Obrecht, C.Spiegler, P.Schoenholzer, K.Mueller, H.Heimgartner and F.Stierli, *Helv.Chim.Acta*, 1992, **75**, 1666.
319. T.Shinbo, T.Yamaguchi, K.Sakaki, H.Yamagishita, D.Kitamoto, and M.Sugiura, *Chem.Express*, 1992, **7**, 781.
320. S.A.Hong, H.J.Choi, and S.W.Nam, *J.Membr.Sci.*, 1992, **70**, 225.
321. H.Y.Ha and S.A.Hong, *Process Metall.*, 1992, **7B**, 1899 (*Chem.Abs.* 1992, **117**, 88661).
322. G.Cao, M.E.Garcia, M.Alcala, L.F.Burgess, and T.E.Mallonk *J.Am.Chem.Soc.*, 1992, **114**, 7574.
323. H.B.Ding, P.W.Carr, and E.L.Cussler, *A.I.Ch.E.*, 1992, **38**, 1493.
324. K.Maruyama, H.Sohmiya, and H.Tsukube, *Tetrahedron*, 1992, **48**, 805.
325. F.Toda, A.Sato, L.R.Nassimbeni, and M.L.Nive, *J.Chem.Soc. Perkin Trans. 1*, 1991, 1971.
326. K.B.Lipkowitz, S.Raghothama, and J.A.Yang, *J.Am.Chem.Soc.*, 1992 **114**, 1554.
327. M.Tanaka, M.Yoshinaga, S.Asano, Y.Yamashoji, and Y.Kawaguchi *Fresenius' J.Anal.Chem.*, 1992, **343**, 896.
328. A.Armstrong and W.C.Still, *J.Org.Chem.*, 1992, **57**, 4581.
329. A.Galan, D.Andreu, A.M.Echavarren, P.Prados, and J.De Mendoza *J.Am.Chem.Soc.*, 1992, **114**, 1511.
330. A.J.Chiou, S.H.Wu, and K.T.Wang, *Biotechnol.Lett.*, 1992, **14**, 461.
331. R.-L.Gu, I.-S.Lee, and C.J.Sih, *Tetrahedron Lett.*, 1992, **33** 1953.
332. C.Yee, T.A.Blythe, T.J.McNabb, and A.E.Watts, *J.Org.Chem.*, 1992 **57**, 3525.
333. S.T.Chen, S.Y.Chen, S.C.Hsaio, and K.T.Wang, *Biotechnol.Lett.* 1991, **13**, 773.
334. V.Alks, D.D.Keith, and J.R.Sufrin, *Synthesis*, 1992, 623.
335. E.E.Ricks, M.C.Estrada-Valdes, T.L.McLean, and G.A.Iacobucci *Biotechnol.Prog.*, 1992, **8**, 197.
336. K.Ohshima, Y.Tomiuchi, and H.Kise, *Yukagaku*, 1992, **41**, 1141 (*Chem.Abs.*, 1993, **118**, 81371).
337. T.Miyazawa, H.Iwanaga, T.Yamada, and S.Kuwata, *Chirality*, 1992, **4** 427.
338. J.W.Huh, K.Yokoigawa, N.Esaki, and K.Soda, *J.Ferment.Bioeng.* 1992, **74**, 189.
339. Y.Tomiuchi, K.Ohshima, and H.Kise, *Bull.Chem.Soc.Jpn.*, 1992, **65** 2599.
340. I.Umemura, K.Yanagiya, S.Komatsubara, T.Sato, and T.Tosa, *Appl.Microbiol. Biotechnol.*, 1992, **36**, 722.
341. T.K.Bhall, A.Miura, A.Wakamoto, Y.Ohba, and K.Furuhashi, *Appl.Microbiol. Biotechnol.*, 1992, **37**, 184.
342. S.Maxim, A.Fondor, I.Bunia, and A.Kayacsa, *J.Mol.Catal.*, 1992, **72**, 351.
343. H.E.Schoemaker, W.H.J.Boesten, B.Kaptein, H.F.M.Hermes, T.Sonke, Q.B.Broxterman, W.J.J.van den Tweel, and J.Kamphuis, *Pure Appl.Chem.*, 1992, **64**, 1171.
344. B.Siffert and A.Naidja, *Clay Mineral*, 1992, **27**, 109.
345. V.A.Tverdislov, M.R.Kuznetsova, and H.V.Yakovenko, *Biofizika* 1992, **37**, 391 (*Chem.Abs.*, 1992, **117**, 192296).
346. S.Natarajan and E.Zangrando, *Proc.-Indian Acad.Sci., Chem.Sci.* 1992, **104**, 483 (*Chem.Abs.*, 1992, **117**, 161377).
347. J.Lubkowski, L.Lankiewicz, and Z.Kosturkiewicz, *Acta Crystallogr. Sect.C, Cryst.Struct.Commun.*, 1992, **C48**, 1616.
348. G.Valle, M.Crisma, and C.Toniolo, *Z.Kristallogr.*, 1992, **199**, 293.
349. G.Tadeusz and B.Kurzak, *J.Crystallogr.Spectrosc.Res.*, 1992, **22** 95.
350. N.Nagashima, C.Sano, T.Kawakita, and Y.Iitaka, *Anal.Sci.*, 1992, **8** 119.
351. N.Nagashima, C.Sano, T.Kawakita, and Y.Iitaka, *Anal.Sci.*, 1992, **8** 267.

352. N.Nagashima, C.Sano, T.Kawita, and Y.Iitaka, *Anal.Sci.*, 1992, **8** 115.
353. S.R.Ramaswamy and M.R.N.Murthy, *Acta Crystallogr., Sect.B, Struct.Sci.*, 1992, **B48**, 488.
354. A.M.Piazzesi, R.Bardi, F.Formaggio, M.Crisma, and C.Toniolo *Z.Kristallogr.*, 1992, **200**, 93.
355. N.Nagashima, C.Sano, T.Kawakita, and Y.Iitaka, *Anal.Sci.*, 1992, **8** 723.
356. C.H.Goerbitz and M.C.Etter, *Acta Crystallogr., Sect.C Cryst.Struct.Commun.*, 1992, **C48**, 1317.
357. M.Gikas, G.Agelis, J.Matsoukas, G.J.Moore, L.Dupont, and S.Englebert, *Acta Crystallogr., Sect.C, Cryst.Struct.Commun.* 1992, **C48**, 216.
358. G.Valle, M.Crisma, F.Formaggio, C.Toniolo, A.S.Redlinski K.Kaczmarek, and M.T.Leplawy, *Z.Kristallogr.*, 1992, **199**, 229.
359. V.Nastopoulos, M.Gikas, J.Matsoukas, and O.Dideberg, *Z.Kristallogr.*, 1992, **199**, 149.
360. M.P.Hegarty, C.H.L.Kennard, K.A.Byriel, and G.Smith, *Aust.J.Chem.* 1992, **45**, 1021.
361. B.Koppenhoefer and M.Hummel, *Z.Naturforsch., B: Chem.Sci.*, 1992 **47**, 1034.
362. R.Gerald, T.Bernhard, U.Haeberln, J.Rendell, and S.Opella *J.Am.Chem.Soc.*, 1993, **115**, 777.
363. M.Lee and R.S.Phillips, *Magn.Reson.Chem.*, 1992, **30**, 1035.
364. J.Wang, X.Wang, Y.Ouyang, Y.You, and A.Dai, *Wuli Huaxue Xuebao* 1992, **8**, 647 (*Chem.Abs.*, 1992, **118**, 22586).
365. Z.Dzakula, A.S.Edison, W.M.Westler, and J.L.Markley *J.Am.Chem.Soc.*, 1992, **114**, 6200.
366. P.S.Belton, A.M.Gil, and S.F.Tanner, *Solid State Nucl.Magn.Reson.* 1992, **1**, 67.
367. P.A.Manovik, S.I.Tyukhtenko, K.B.Yatsimirski, M.A.Fedorenko, and E.I.Bliznyukova, *Doklady Akad.Nauk S.S.S.R.*, 1991, **320**, 108 (*Chem.Abs.*, 1992, **116**, 106714).
368. S.I.Tyukhtenko and V.G.Shtyrlin, *Zh.Fiz.Khim.*, 1992, **66**, 964.
369. R.Hulst, N.K.de Vries, and B.L.Feringa, *Angew.Chem.Int.Edit.* 1992, **31**, 1092.
370. S.Li and W.C.Purdy, *Anal.Sci.*, 1992, **64**, 1405.
371. L.D.Barron, A.R.Gargaro, L.Hecht, and P.L.Polavarapu, *Spectrochim.Acta*, 1992, **48A**, 261.
372. S.Bouchonnet, J.P.Denhez, Y.Hoppilliard, and C.Mauriac, *Anal.Chem.*, 1992, **64**, 743.
373. E.Pittenauer, A.Pachinger, G.Allmeier, and E.R.Schmid, *Org.Mass.Spectrom.*, 1991, **26**, 1065.
374. Z.Wu and C.Fenselau, *Rapid Commun.Mass Spectrom.*, 1992, **6**, 403.
375. R.A.J.O'Hair, J.H.Bowie, and S.Gronert, *Int.J.Mass Spectrom.* 1992, **117**, 23.
376. S.Campbell, J.L.Beauchamp, M.Rempe, and D.L.Lichtenberger, *Int.J.Mass Spectrom.*, 1992, **117**, 83.
377. S.Bouchonnet and Y.Hoppilliard, *Org.Mass Spectrom.*, 1992, **27**, 71.
378. G.S.Gorman, J.P.Speir, C.A.Turner, and I.J.Amster *J.Am.Chem.Soc.*, 1992, **114**, 3986.
379. J.Parmentier, C.Samyn, and Th.Zeegers-Huyskens, *Spectrochim.Acta* 1992, **48A**, 1091.
380. J.Parmentier, K.De Wael, and Th.Zeegers-Huyskens, *J.Mol.Struct.* 1992, **270**, 217.
381. W.B.Fischer and H.U.Eysel, *Spectrochim.Acta*, 1992, **48A**, 725.
382. K.Furic, V.Mohacek, M.Bonifacic, and I.Stefanic, *J.Mol.Struct.* 1992, **267**, 39.
383. T.L.Botte, G.I.Dovbeshko, and G.S.Litvinov, *Biopolim.Kletka*, 1991 **7**, 48 (*Chem.Abs.*, 1992, **116**, 124227).

384. S.J.Martinez, J.C.Alfano, and D.H.Levy, *J.Mol.Spectrosc.*, 1992 **156**, 421.
385. M.J.Colaneri and J.Peisach, *J.Am.Chem.Soc.*, 1992, **114**, 5335.
386. J.Kruuv and D.J.Glofcheski, *Cryobiology*, 1992, **29**, 291.
387. A.W.Lloyd, J.A.Baker, G.Smith, C.J.Ollitt, and K.J.Rutt, *J.Pharm.Pharmacol.*, 1992, **44**, 507.
388. T.Imae, Y.Takahashi, and H.Muramatsu, *J.Am.Chem.Soc.*, 1992, **114** 3414.
389. K.Hanabusa, K.Okui, K.Karaki, T.Koyama, and H.Shirai, *J.Chem.Soc. Chem.Commun.*, 1992, 1371.
390. J.Verma and D.Sahal, *Bull.Chem.Soc.Jpn.*, 1992, **65**, 1719.
391. C.Vicent, E.Fan, and A.D.Hamilton, *Tetrahedron Lett.*, 1992, **33** 4269.
392. J.Z.Yang, J.Wang, and H.Li, *J.Solution Chem.*, 1992, **21**, 1131.
393. R.M.Izatt, J.L.Oscarson, S.E.Gillespie, H.Grimsrud J.A.R.Renuncio, and C.Pando, *Biophys.J.*, 1992, **61**, 1394.
394. F.Rodante, G.Marrosu, and G.Catalini, *Thermochim.Acta*, 1992, **194** 197.
395. J.Fernandez and T.H.Lilley, *J.Chem.Soc. Faraday Trans. 1*, 1992 **88**, 2503.
396. P.Sipos, G.Peintler, and G.Toth, *Int.J.Pept.Protein Res.*, 1992 **39**, 207.
397. L.Pogliani, *J.Pharm.Sci.*, 1992, **81**, 334.
398. L.Pogliani, *J.Pharm.Sci.*, 1992, **81**, 967.
399. W.J.Spillane, G.G.Birch, M.G.B.Drew, and I.Bartolo, *J.Chem.Soc. Perkin Trans. 2*, 1992, 497.
400. L.M.Cordoba, S.A.Brandan, and M.A.Acuna Molina, *An.Asoc.Quim.Argentina*, 1992, **79**, 151.
401. S.A.Brandan, L.M.Cordoba, and M.A.Acuna Molina, *An.Asoc.Quim.Argentina*, 1992, **79**, 189.
402. E.C.Bolton, G.A.Thomson, and G.H.W.Milburn, *J.Chem.Res.Synop*. 1992, 210.
403. D.Yu, D.A.Armstrong, and A.Rauk, *Can.J.Chem.*, 1992, **70**, 1762; A.G.Csaszar, *J.Am.Chem.Soc.*, 1992, **114**, 9568.
404. A.Vijay and D.N.Sathyanarayana, *J.Phys.Chem.*, 1992, **96**, 10735.
405. J.T.Lopez-Navarrete, J.J.Quirante, and F.J.Ramirez, *J.Mol.Struct*. 1992, **268**, 249.
406. F.Jensen, *J.Am.Chem.Soc.*, 1992, **114**, 9533.
407. J.Garen, M.J.Field, G.Kneller, M.Karplus, and J.Smith, *J.Chem.Phys,, Phys-Chim.Biol.*, 1991, **88**, 2587.
408. H.L.Gordon, H.C.Larrell, A.G.Szabo, K.J.Willis, and R.L.Somorjai, *J.Phys.Chem.*, 1992, **96**, 1915.
409. D.J.Tobias and C.L.Brooks, *J.Phys.Chem.*, 1992, **96**, 3864; see also I.R.Gould and P.A.Kollmann, *Ibid.*, p.9255.
410. E.Alvira, *Amino Acids*, 1992, **2**, 97.
411. P.Ivanov and M.Ivanova, *J.Mol.Struct.*, 1992, **269**, 223.
412. F.Torrens and M.Ruiz-Lopez, *Tetrahedron*, 1992, **48**, 5209.
413. P.L.A.Popelier and R.F.W.Bader, *Chem.Phys.Lett.*, 1992, **189**, 542.
414. V.R.Meyer, *Chem.Tech.*, 1992, **22**, 412.
415. D.S.Kaufman and G.H.Miller, *Comp.Biochem., Physiol., B*, 1992 **102B**, 199.
416. G.A.Goodfriend, *Nature*, 1992, **357**, 399.
417. G.H.Miller, P.B.Beaumont, A.J.T.Jull, and B.Johnson, *Philos.Trans.Roy.Soc., Ser.B*, 1992, **337**, 149.
418. M.W.Meyer, V.R.Meyer, and S.Ramseyer, *Chirality*, 1991, **2**, 471.
419. T.Shiraiwa, K.Shinjo, Y.Masui, A.Ohta, H.Natsuyama, H.Miyazaki and H.Kurokawa, *Bull.Chem.Soc.Jpn.*, 1991, **64**, 3741.
420. H.Kuroda, S.Kubo, N.Clino, T.Kimura, and S.Sakakibara, *Int.J.Pept.Protein Res.*, 1992, **40**, 114.

421. R.Valcarce and G.G.Smith, *Chemom. Intell.Lab.Syst.*, 1992, **16**, 61.
422. T.Shiraiwa, S.Sakata, H.Natsuyama, K.Fujishima, H.Miyazaki S.Kubo, T.Nitta, and H.Kurokawa, *Bull.Chem.Soc.Jpn.*, 1992, **65** 965.
423. B.M.Rode, *J.Phys.Chem.*, 1992, **96**, 4170.
424. K.Okamoto, H.Oohata, T.Konno, and J.Hidaka, *Viva Origino*, 1991 **19**, 133 (*Chem.Abs.*, 1993, **118**, 54508).
425. C.Wong, J.C.Santiago, L.Rodriguez-Paez, M.Ibenez, and I.Baeza, *Origins Life Evol.Biosphere*, 1991, **21**, 145.
426. L.A.Andreeva, L.Yu.Alfeeva, V.N.Potaman, and V.N.Nezavibatko, *Int.J.Pept.Protein Res.*, 1992, **39**, 493.
427. C.Bohnenstengl and W.Baltes, *Z.Lebensm.-Unters.Forsch.*, 1992, **194** 366.
428. E.Knoch and W.Baltes, *Food Chem.*, 1992, **44**, 243.
429. G.Vernin, J.Metzger, C.Bonifac, M.H.Murello, A.Siouffi J.L.Larice, and C.Parkanyi, *Carbohydr.Res.*, 1992, **230**, 15.
430. K.B.De Roos, *A.C.S.Symp.*(Flavour Precursors), 1992, **490**, 203.
431. N.Yamaguchi, *Shokuhin Kogyo*, 1992, **35**, 26 (*Chem.Abs.*, 1992, **116** 233967).
432. R.Zamora and F.J.Hildago, *J.Agric.Food Chem.*, 1992, **40**, 2269.
433. A.V.Rani and A.K.Singh, *Tetrahedron Lett.*, 1992, **33**, 1379.
434. T.Shono, N.Kise, H.Oike, M.Yoshimoto, and E.Okazaki, *Tetrahedron Lett.*, 1992, **33**, 5559.
435. X.L.Armesto, M.Canle, and J.A.Santabella, *Tetrahedron*, 1993, **49** 275; J.M.Antelo, F.Arce, and J.C.Perez-Moure, *Int.J.Chem.Kinet.* 1992, **24**, 1093.
436. M.Losada, J.A.Santabella, X.L.Armesto, and J.M.Antelo, *Acta Chim.Hung.*, 1992, **129**, 535.
437. J.M.Antelo, F.Arce, J.L.Armesto, J.Crugeiras, and J.Franco, *Afinidad*, 1992, **49**, 45.
438. J.S.Nowick, N.A.Powell, T.M.Nguyen, and G.Noronha, *J.Org.Chem.* 1992, **57**, 7364.
439. R.V.Hoffmann and H.O.Kim, *Tetrahedron*, 1992, **48**, 3007.
440. J.S.Bajwa, *Tetrahedron Lett.*, 1992, **33**, 2955.
441. J.McNulty and I.W.J.Still, *Synth.Commun.*, 1992, **22**, 979.
442. M.Acedo, F.Albericio, and R.Eritja, *Tetrahedron Lett.*, 1992, **33** 4989.
443. A.Abiko and S.Masamune, *Tetrahedron Lett.*, 1992, **33**, 5517.
444. Y.Guindon, A.Slassi, E.Ghiro, G.Bantle, and G.Jung, *Tetrahedron Lett.*, 1992, **33**, 4257.
445. M.R.Leanna, T.J.Sowin, and H.E.Morton, *Tetrahedron Lett.*, 1992 **33**, 5029.
446. H.Maeda, T.Maki, and H.Ohmori, *Tetrahedron Lett.*, 1992, **33**, 1347.
447. M.A.Poss and J.Reid, *Tetrahedron Lett.*, 1992, **33**, 1411.
448. M.Franciotti, A.Mordini, and M.Taddei, *SynLett.*, 1992, 137.
449. S.B.Mandal, B.Achari, and S.Chattopadhyay, *Tetrahedron Lett.* 1992, **33**, 1647.
450. A.Dondini and D.Perrone, *Tetrahedron Lett.*, 1992, **33**, 7259.
451. T.Ye and M.A.McKervey, *Tetrahedron*, 1992, **48**, 8007.
452. J.Collin, J.L.Namy, G.Jones, H.B.Kagan, *Tetrahedron Lett.*, 1992 **33**, 2973.
453. M.Boomgaarden, G.Heinrich, and P.Henklein, *Pharmazie*, 1992, **47** 710.
454. C.Somlai, G.Szokan, and L.Balaspiri, *Synthesis*, 1992, 285.
455. A.Brutsche and K.Hartke, *Liebigs Ann.Chem.*, 1992, 921.
456. A.Kohrt and K.Hartke, *Liebigs Ann.Chem.*, 1992, 595.
457. A.M.Castano and A.M.Echevarren, *Tetrahedron*, 1992, **48**, 3377.
458. J.C.Lacey, N.S.M.D.Wickramasinghe, and R.S.Sabatini, *Experientia* 1992, **48**, 379.
459. T.Moriguchi, Y.G.Guo, S.Yamamoto, Y.Matsubara, M.Yoshihara, and T.Maeshima, *Chem.Express*, 1992, **7**, 625.

460. R.Ueoka, Y.Matsumoto, K.Goto, and Y.Kato, *Process Metall.*, 1992 **7B**, 1649 (*Chem.Abs.*, 1992, **117**, 172011).
461. R.Ueoka, Y.Yano, H.Uchiyama, H.Oyama, A.Sakoguchi, and Y.Kato *Nippon Kagaku Zaisshi*, 1992, 255 (*Chem.Abs.*, 1992, **116**, 236114).
462. J.G.J.Weijnen, A.Kondijs, and J.F.J.Engbersen, *J.Mol.Catal.*, 1992, **73**, L5.
463. P.Scrimin, P.Tecilla, and V.Tonellato, *J.Phys.Org.Chem.*, 1992, **5** 619.
464. P.Stefanowicz and I.Z.Siemion, *Pol.J.Chem.*, 1992, **66**, 111.
465. Kh.Neder, S.Minchev, and N.Stoyanov, *Dokl.Bulg.Akad.Nauk*, 1991 **44**, 61 (*Chem.Abs.*, 1992, **117**, 112001).
466. S.Srivastava, S.Agrawal, and K.C.Nand, *Oxid.Commun.*, 1992, **15** 43.
467. I.Rao, S.Jain, and P.D.Sharma, *Int.J.Chem.Kinet.*, 1992, **24**, 963.
468. I.G.N.Wylie and S.G.Roscoe, *Bioelectrochem.Bioenerg.*, 1992, **28** 367.
469. Y.Zelechonok and R.B.Silverman, *J.Org.Chem.*, 1992, **57**, 5787.
470. S.Itoh, M.Mure, A.Suzuki, H.Murao, and Y.Ohshiro, *J.Chem.Soc. Perkin Trans. 1*, 1992, 1245.
471. D.Mascagna, C.Costantini, M.D'Ischia, and G.Prota, *Tetrahedron* 1992, **48**, 8309.
472. M.Akssira, H.Kasmi, A.Dahdouh, and M.Boumzebra, *Tetrahedron Lett.* 1992, **33**, 1887.
473. R.Wilder and S.Mobashery, *J.Org.Chem.*, 1992, **57**, 2755.
474. C.Schierlinger and K.Burger, *Tetrahedron Lett.*, 1992, **33**, 193.
475. E.Frerot, J.Coste, J.Poncet, P.Jouin, and B.Castro, *Tetrahedron Lett.*, 1992, **33**, 2815.
476. J.Savrda and M.Wakselman, *J.Chem.Soc. Chem.Commun.*, 1992, 812.
477. O.Itoh, T.Honnami, A.Amano, K.Murata, Y.Koichi, and T.Sugita, *J.Org.Chem.*, 1992, **57**, 7334.
478. J.Bergman and G.Lidgren, *Bull.Soc.Chim.Belg.*, 1992, **101**, 643.
479. J.Bergman and G.Lidgren, *Tetrahedron Lett.*, 1989, **30**, 4597.
480. B.Sain, S.P.Singh, and J.S.Sandhu, *Tetrahedron*, 1992, **48**, 4567.
481. D.Prajapati, A.R.Mahajan, and J.S.Sandhu, *J.Chem.Soc. Perkin Trans. 1*, 1992, 1821.
482. N.L.Subasinghe and R.L.Johnson, *Tetrahedron Lett.*, 1992, **33**, 2649.
483. A.G.Griesbeck, H.Mauder, and I.Mueller, *Chem.Ber.*, 1992, **125** 2467; A.G.Griesbeck and H.Mauder, *Angew.Chem.Int.Ed.*, 1992, **31** 73.
484. N.Sewald and K.Burger, *Liebigs Ann.Chem.*, 1992, 947.
485. B.Stanovnik, L.Golic, P.Kmecl, B.Ornik, J.Svete, and M.Tisler, *J.Heterocycl.Chem.*, 1992, **28**, 1961.
486. G.T.Crisp and P.T.Glink, *Tetrahedron*, 1992, **48**, 3541.
487. L.Couper, D.J.Robins, and E.J.T.Chrystal, *Tetrahedron Lett.*, 1992 **33**, 2717.
488. S.J.Gould and S.Ju, *J.Am.Chem.Soc.*, 1992, **114**, 10166.
489. A.R.Butler and S.George, *Tetrahedron*, 1992, **48**, 7879.
490. J.Meienhofer, in "Chemistry and Biochemistry of the Amino Acids" ed. G.C.Barrett, Chapman and Hall, London, 1985, p.302.
491. E.Masiukiewicz, B.Rzeszotarska, and J.Szczerbaniewicz, *Org.Prep.Proced.Int.*, 1992, **24**, 191.
492. K.Itakura, K.Uchida, and S.Kawakishi, *Tetrahedron Lett.*, 1992, **33** 2567.
493. Z.Maskos, J.D.Rush, and W.H.Koppenol, *Arch.Biochem.Biophys.*, 1992 **296**, 514.
494. Z.Maskos, J.D.Rush, and W.H.Koppenol, *Arch.Biochem.Biophys.*, 1992 **296**, 521.
495. K.M.Czerwinski, L.Deng, and J.M.Cook, *Tetrahedron Lett.*, 1992, **33** 4721.
496. K.Verschueren, G.Toth, D.Tourwe, M.Lebl, G.van Binst, and V.Hruby, *Synthesis*, 1992, 458.

497. Y.Goda, J.Suzuki, T.Maitani, K.Yoshihira, M.Takeda, and M.Uchiyama, *Chem.Pharm.Bull.*, 1992, **40**, 2236.
498. K.Uchida and E.R.Stadtman, *Proc.Natl.Acad.Sci.U.S.A.*, 1992, **89**, 4544.
499. K.Kikugawa, T.Kato, and A.Hayasaka, *Lipids*, 1992, **26**, 922.
500. F.M.F.Chen and N.L.Benoiton, *Int.J.Pept.Protein Res.*, 1992, **40**, 13.
501. J.Ezquerra, J.de Mendoza, C.Pedregal, and C.Ramirez, *Tetrahedron Lett.*, 1992, **33**, 5589.
502. D.Misiti, M.Santaniello, and G.Zappia, *Synth.Commun.*, 1992, **22**, 883.
503. P.J.Belshaw, J.G.Adamson, and G.A.Lajoie, *Synth.Commun.*, 1992, **22**, 1001.
504. L.A.Carpino and E.S.M.E.Mansour, *J.Org.Chem.*, 1992, **57**, 6371.
505. B.Luening, T.Norberg, and J.Tejbrant, *J.Carbohydr.Chem.*, 1992, **11** 933.
506. D.Cantacuzune, S.Attal, and S.Bay, *Biomed.Biochim.Acta*, 1991, **50** S231.
507. A.B.Kazi and J.Hajdu, *Tetrahedron Lett.*, 1992, **33**, 2291.
508. P.Blanchard, M.S.El Kortbi, J.-L.Fourrey, and M.Robert-Gero, *Tetrahedron Lett.*, 1992, **33**, 3319.
509. E.Kuyl-Yeheskiely, M.Lodder, G.A.van der Marel, and J.H.van Boom, *Tetrahedron Lett.*, 1992, **33**, 3013.
510. M.S.Stanley, *J.Org.Chem.*, 1992, **57**, 6421.
511. J.M.Delacotte and H.Galone, *Synth.Commun.*, 1992, **48**, 3075.
512. P.M.Fischer, *Tetrahedron Lett.*, 1992, **33**, 7605.
513. A.Mehta, R.Jaouhari, T.J.Benson, and K.T.Douglas, *Tetrahedron Lett.*, 1992, **33**, 5441.
514 V.Soloshonok, V.Kukhar, Yu.Pustovil and V.Nazaretyan, *SynLett.* 1992, 657.
515. M.C.Munson, C.Garcia-Echeverria, F.Albericio, and G.Barany, *J.Org.Chem.*, 1992, **57**, 3013.
516. M.Royo, C.Garcia-Echeverria, E.Giralt, R.Eritja, and F.Albericio, *Tetrahedron Lett.*, 1992, **33**, 2391.
517. Z.Yang, B.Li, W.Cao, Y.Hu, Y.Tong, and D.Gu, *Hebei Shifan Daxue Xuebao, Ziran Kexueban*, 1992, 47 (*Chem.Abs.*, 1992, **117**, 221962).
518. R.Stendel and A.Albertson, *J.Chromatogr.*, 1992, **606**, 260.
519. P.K.Hall and R.C.Roberts, *Biochim.Biophys,Acta*, 1992, **112**, 325.
520. N.Getoff, *Amino Acids*, 1992, **2**, 195.
521. T.A.Himdan, J.Nothig-Laslo, and H.Bihinski, *Radiat.Res.*, 1992, **131**, 266.
522. T.G.Waddell and T.J.Miller, *Origins Life Evol.Biosphere*, 1992, **21**, 219.
523. J.K.Lee, R.T.Ross, S.Thampi, and S.Leurgans, *J.Phys.Chem.*, 1992, **96**, 9158.
524. A.G.Kungl, G.Laudl, A.J.W.G.Visser, M.Breirenbach, and H.F.Kauffmann, *J.Fluoresc.*, 1992, **2**, 63.
525. K.Reszka, J.W.Lown, and C.F.Chignell, *Photochem.Photobiol.*, 1992, **55**, 359.
526. T.E.G.Candy, D.Mantle, and P.Jones, *J.Biolumin.Chemilumin.*, 1991, **6**, 245.
527. D.N.Nikogosyan and H.Goerner, *J.Photochem.Photobiol., B*, 1992, **13**, 219.
528. N.Vekshin, M.Vincent, and J.Gallay, *Chem.Phys.Lett.*, 1992, **199**, 459.
529. L.P.McMahon, W.J.Colucci, M.L.McLaughlin, and M.D.Barkley *J.Am.Chem.Soc.*, 1992, **114**, 8442.
530. F.Cozens, R.A.McClelland, and S.Steenken, *Tetrahedron Lett.*, 1992, **33**, 173.
531. L.H.Ericsson, D.Atherton, R.Kutny, A.J.Smith, and J.W.Crabb, 8th International Conference on Methods of Protein Sequence Analysis, Eds. H.Joernvall, J.-O.Hoeoeg, and A.-M.Gustavsson, Birkhaeuser Basel, 1991, p.143.
532. R.L.Davies, D.R.Baigent, M.S.Levitt, Y.Mollah, C.J.Rayner, and A.B.Frensham, *J.Sci.Food Agric.*, 1992, **59**, 423.

533. P.Richardin, S.Bonnassies, and C.Chahine, *Leder*, 1991, **42**, 201 (*Chem.Abs.*, 1992, **116**, 193240).
534. K.R.Kim, J.H.Kim, C.H.Oh, and T.J.Mabry, *J.Chromatogr.*, 1992, **605**, 241.
535. T.Zee, F.Stellaard, and C.Jakobs, *J.Chromatogr.*, 1992, **574**, 335.
536. J.A.Rathmacher, G.A.Link, P.J.Flakoll, and S.L.Nissen, *Biol.Mass Spectrom.*, 1992, **21**, 560.
537. C.S.Jone, M.Kamo, M.Sano, and A.Tsugita, *Chem.Lett.*, 1992, 929.
538. K.L.Woo, *Han'guk Nonghwha Hakhoechi*, 1992, **35**, 132 (*Chem.Abs.* 1992, **117**, 151304).
539. F.Yasuhara, H.Takeda, Y.Ochiai, S.Miyano, and S.Yamaguchi, *Chem.Lett.*, 1992, 251.
540. T.J.Sims and A.J.Bailey, *J.Chromatogr.*, 1992, **582**, 49.
541. N.T.Faithful, *Lab.Pract.*, 1992, **41**, 37.
542. G.A.Qureshi, in "H.p.l.c. of Proteins, Peptides and Polynucleotides", Ed. M.T.W.Hearn, VCH, New York, 1991, p.625.
543. J.A.White and R.J.Hart, *Food Sci.Technol.*, 1992, **52** (Food Analysis by H.p.l.c.), 53.
544. J.A.White and R.J.Hart, *Food Sci.Technol.*, 1992, **52** (Food Analysis by H.p.l.c.), 75.
545. H.G.Worthen and H.Liu, *J.Liq.Chromatogr.*, 1992, **15**, 3323.
546. G.R.Nathans and D.R.Gere, *Anal.Biochem.*, 1992, **202**, 262.
547. W.D.Beinert, A.Meisner, M.Fuchs, E.Riedel, M.Luepke, and H.Brueckner, *G.I.T.Fachz.Lab.*, 1992, **36**, 1018 (*Chem.Abs.*, 1993 **118**, 15646).
548. F.Altmann, *Anal.Biochem.*, 1992, **204**, 215.
549. M.E.R.Dugan, W.C.Sauer, K.A.Lien, and T.W.Fenton, *J.Chromatogr.*, 1992, **582**, 242.
550. K.H.Horz and J.Reichling, *Bioforum*, 1992, **15**, 155.
551. L.Canevari, R.Vieira, M.Aldegunde, and F.Dagani, *Pharmacol.Res.* 1992, **25**, 115.
552. L.Canevari, R.Vieira, M.Aldegunde, and F.Dagani, *Anal.Biochem.* 1992, **205**, 137.
553. V.R.Roettger and M.D.Goldfinger, *J.Neurosci.Methods*, 1991, **39**, 263.
554. S.Palmero, M.De Marchis, M.Prati, and E.Fugassa, *Anal.Biochem.* 1992, **202**, 152.
555. A.L.L.Duchateau, H.Knuts, J.M.M.Boesten, and J.J.Guns, *J.Chromatogr.*, 1992, **623**, 237. 556. L.Kang and R.H.Buck, *Amino Acids*, 1992, **2**, 1403.
557. M.J.Medina Hernandez, S.Sagrado Vives, and M.C.Garcia Alvarez-Coque, *Mikrochim.Acta*, 1992, **108**, 293.
558. A.Hashimoto, T.Nishikawa, T.Oka, K.Takahashi, and T.Harashi, *J.Chromatogr.*, 1992, **582**, 41.
559. M.A.Nussbaum, J.E.Przedwiecki, D.U.Staerk, S.M.Lunte, and C.M.Riley, *Anal.Chem.*, 1992, **64**, 1259.
560. A.S.Feste, *J.Chromatogr.*, 1992, **574**, 23.
561. S.M.Halpine, *Stud.Conserv.*, 1992, **37**, 22.
562. S.Ueno, A.Sano, K.Kotani, K.Kondoh, and Y.Kakimoto, *J.Neurochem.*, 1992, **59**, 2012.
563. G.D.Green and K.Reagan, *Anal.Biochem.*, 1992, **201**, 265.
564. J.G.Hoogerheide and C.M.Campbell, *Anal.Biochem.*, 1992, **201**, 146.
565. I.Molnar-Perl and M.Morvai, *Chromatographia*, 1992, **34**, 132.
566. C.G.Fields, A.Loffet, S.A.Kates, and G.B.Fields, *Anal.Biochem.*, 1992, **203**, 245.
567. D.C.Brune, *Anal.Biochem.*, 1992, **207**, 285.
568. F.Zezza, J.Kerner, M.R.Pascale, R.Giannini, and A.Arrigoni-Martelli, *J.Chromatogr.*, 1992, **593**, 99.
569. D.Bartig and F.Klink, *J.Chromatogr.*, 1992, **606**, 43.

570. S.Terada, S.Fujimura, K.Noda, and E.Kimoto, *Fukuoka Daigaku Rigaku Shuho*, 1992, **22**, 35 (*Chem.Abs.*, 1992, **117**, 103273).
571. J.Giron, M.Alaiz, and E.Vioque, *Anal.Biochem.*, 1992, **206**, 155.
572. W.C.Hawkes and M.A.Kutnink, *J.Chromatogr.*, 1992, **576**, 263.
573. Y.Kawakami, T.Ohga, C.Shimamoto, N.Satoh, and S.Ohmori, *J.Chromatogr.*, 1992, **576**, 63.
574. R.Paroni, E.De Vecchi, I.Fermo, C.Arcelloni, L.Diomede, F.Magni, and P.A.Bonini, *Clin.Chem.*, 1992, **38**, 407.
575. A.Tsugita and M.Kamo, *8th International Conference on Methods of Protein Sequence Analysis*, Eds. H.Joernvall, J.-O.Hoeoeg, and A.-M.Gustavsson, Birkhaeuser, Basel, 1991, p.123.
576. Y.Hasegawa, D.C.Jette, A.Miyamoto, H.Kawasaki, and H.Yuki *Anal.Sci.*, 1991 (Suppl.Proc.Int.Congr.Anal.Sci.), **7**, 945.
577. J.A.Grunan and J.M.Swiader, *J.Chromatogr.*, 1992, **594**, 165.
578. M.Zhou and G.Wang, *Hejishu*, 1992, **15**, 117.
579. N.Rawal, Y.J.Lee, W.K.Paik, and S.Kim, *Biochem.J.*, 1992, **287**, 929.
580. S.H.Lee, T.S.Oh, and W.H.Lee, *Bull.Korean Chem.Soc.*, 1992, **13**, 280 (*Chem.Abs.*, 1992, **117**, 131529).
581. S.H.Lee, T.S.Oh, and W.H.Lee, *Bull.Korean Chem.Soc.*, 1992, **13**, 285 (*Chem.Abs.*, 1992, **117**, 192276).
582. S.V.Galushko, I.P.Shishkina, and V.A.Soloshonok, *J.Chromatogr.*, 1992, **592**, 345.
583. S.C.Chang, L.R.Wang, and D.W.Armstrong, *J.Liq.Chromatogr.*, 1992, **15**, 1411.
584. M.A.Rizzi, P.Briza, and M.Breitenbach, *J.Chromatogr.*, 1992, **582**, 35; see also K.Shimadu and K.Hirakata, *J.Liq.Chromatogr.*, 1992, **15**, 1763.
585. H.Brueckner and B.Strecker, *Chromatographia*, 1992, **33**, 586.
586. M.H.Stipanuk, L.L.Hirschberger, and P.J.Bagley, *Adv.Exp.Med.Biol.*, 1992, **315**, 429.
587. W.R.Wolf, D.E.La Croix, and M.E.Slagt, *Anal.Lett.*, 1992, **25**, 2165.
588. D.A.Pratt, Y.Daniloff, A.Duncan, and S.P.Robins, *Anal.Biochem.*, 1992, **207**, 168.
589. J.Lagendijk, J.B.Ubbink, and W.J.H.Vermaak, *J.Chromatogr.*, 1992 **576**, 95.
590. S.J.Standing and R.P.Taylor, *Ann.Clin.Biochem.*, 1992, **29**, 668.
591. L.Dou and I.S.Krull, *Electroanalysis (N.Y.)*, 1992, **4**, 381.
592. H.J.Leis, P.T.Ozand, A.Al Odaib, and U.Gleispach, *J.Chromatogr.*, 1992, **578**, 116.
593. S.Kagabu, S.Kawamura, T.Asano, and M.Kanoh, *Biosci.Biotechnol.Biochem.*, 1992, **56**, 729.
594. N.A.Guzman, J.Moschera, K.Iqbal, and A.W.Malick, *J.Liq.Chromatogr.*, 1992, **15**, 1163.
595. C.Fujimoto, T.Fujikawa, and K.Jinuo, *J.High Resolut.Chromatogr.*, 1992, **15**, 201.
596. T.Ueda, R.Mitchell, F.Kitamura, T.Metcalf, T.Kuwana, and A.Nakamoto, *J.Chromatogr.*, 1992, **593**, 265.
597. J.T.Wu, L.W.Wilson, and S.Christensen, *Ann.Clin.Lab.Sci.*, 1992, **22**, 18.
598. I.Molnar-Perl and M.Pinter-Szakacs, *Anal.Chim.Acta*, 1992, **257**, 209.
599. E.G.Slkova, V.G.Kartvelishvili, and A.A.Bezzubov, *Prikl.Biokhim.Mikrobiol.*, 1992, **28**, 275.
600. E.Latres, J.A.Fernandez-Lopez, X.Remesar, and M.Alemany, *J.Biochem.Biophys.Methods*, 1992, **24**, 39.
601. E.Weber, K.Siegler, B.Weber, K.Tonder, K.Unverhau, H.Weide, and H.Aurich, *G.B.F. Monogr.*, 1992, **17**(Biosensors: Fundamentals and Technological Applications), 303.

602. J.Kong, H.He, and J.Deng, *Fenxi Huaxue*, 1992, **20**, 1265 (*Chem.Abs*. 1993, **118**, 76331).
603. E.Dempsey, J.Wang, U.Wollenberger, M.Ozsoz, and M.R.Smyth *Biosens.Bioelectron*., 1992, **7**, 323.
604. M.V.Cattaneo, J.H.J.Luong, and S.Mercille, *Biosens.Bioelectron., 1992,* **7**, 329.
605. G.S.Ihn, B.W.Kim, S.T.Woo, and Y.G.Jeon, *Anal.Sci.Technol*., 1992 **4**, 139 (*Chem.Abs*., 1993, **118**, 3167).
606. S.Girotti, S.Ghini, R.Budino, A.Pistillo, G.Carrea, R.Bovara S.Piazzi, R.Merighi, and A.Roda, *Anal.Lett*., 1992, **25**, 637.

2
Peptide Synthesis

BY D.T. ELMORE

1 Introduction

The format of this report is the same as that used last year[1]. Readers should note that symposium communications and patents are not cited. Three relevant books[2-4] and many reviews[5-26] have been published. Some of the latter are rather specialized or not easily available or both. There should be something of interest, however, to all peptide chemists.

2 Methods

2.1 *α-Amino group protection.*

A method has been described for the one-pot conversion of Z-derivatives into Boc-derivatives[27]; catalytic transfer hydrogenation using 1,4-cyclohexadiene and 10% Pd on charcoal is carried out in the presence of Boc_2O. Asymmetric syntheses of amino acids in the form of their Fmoc derivatives can be achieved[28]. The attachment of a specially designed α-*N*-protecting group at the end of the cycles for chain growth during solid-phase peptide synthesis (SPPS) but before cleavage from the matrix can permit simple purification of the product by affinity or immunoaffinity chromatography[29]. Biotin-based protecting groups or α-DNP have been most widely used. A new candidate (1) is the tetrabenzo[a,c,g,i]fluorenyl-17-methoxycarbonyl (Tbfmoc) group (**1**). Tbfmoc derivatives of peptides were retained on porous graphitised carbon columns whereas incomplete and capped chains were rapidly eluted. The Tbfmoc group could be removed on the column and the peptide eluted by treating with 20% piperidine in dioxan/water (2:1). In a related study[30], the carboxyl group in 9-hydroxymethyl-fluorene-4-carboxylic acid (**2**) was coupled to one of several amines [$NH_2CH(C_8H_{17})CO_2Me$, $NH_2CH(C_8H_{17})CONHCH(C_8H_{17})CO_2Me$, H-Lys-Lys-Lys-OMe, H-Glu-Glu-Glu(OEt)-OEt] to provide lipophilic or charged probes. The modified Fmoc group is attached to the last amino acid to be coupled and the peptide is cleaved from the polystyrene resin using HF. The peptide carrying its probe could be purified by reverse-phase hplc or ion-exchange chromatography and finally the *N*-terminal protecting group could be removed with base.

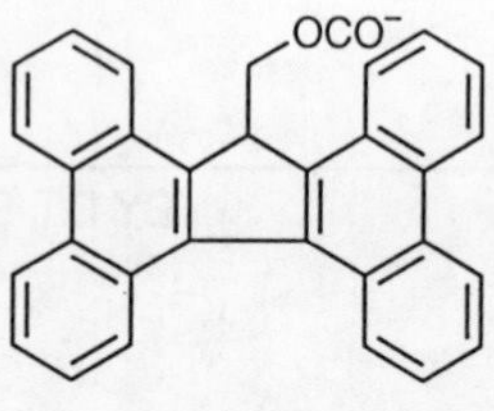

OCO^-

(1)

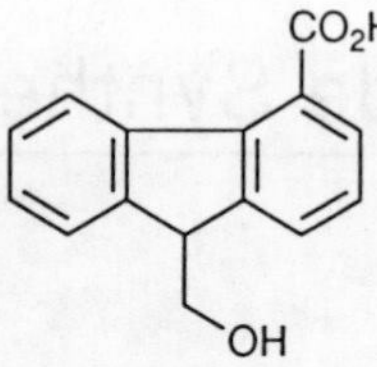

CO_2H

OH

(2)

NO_2 OCOCl NO_2

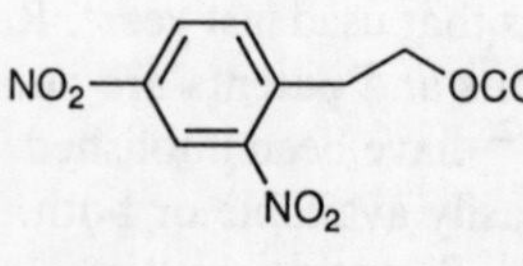

(3)

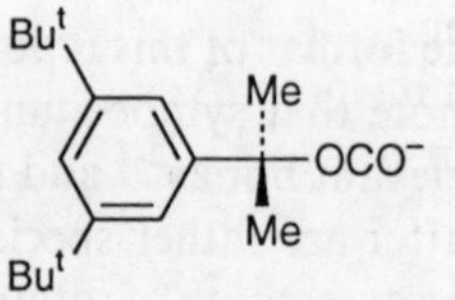

Bu^t Me OCO^- Me Bu^t

(4)

R O O^- NH_3^+ —i–iii→ R O OMe ONs —iv→ R O OMe N_3

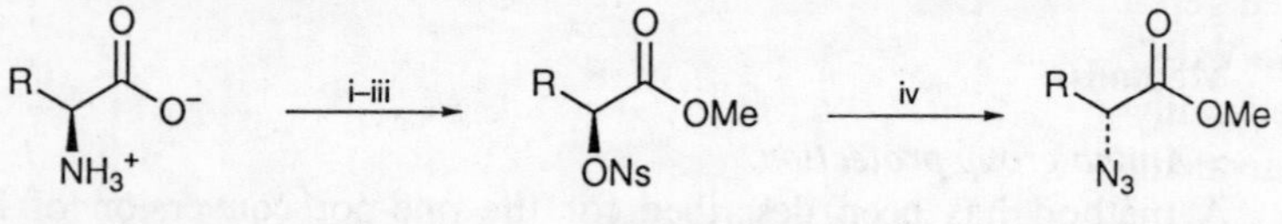

Reagents: i, $NaNO_2$, N-H_2SO_4; ii, K_2CO_3, MeI, MeCOMe;
iii, 4-$NO_2C_6H_4SO_2Cl$(NsCl), Et_3N, DMAP, CH_2Cl_2; iv, NaN_3, MeSOMe, 55 °C

Scheme 1

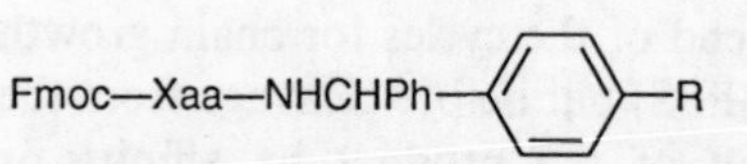

Fmoc—Xaa—NHCHPh—C_6H_4—R

(5; R = SMe or SOMe)

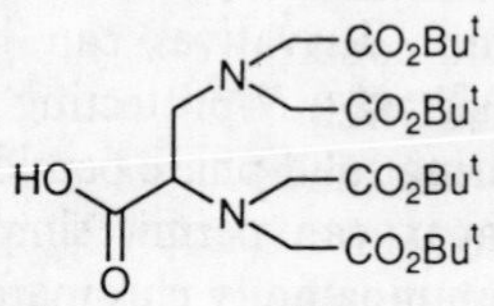

HO O N CO_2Bu^t CO_2Bu^t N CO_2Bu^t CO_2Bu^t

(6)

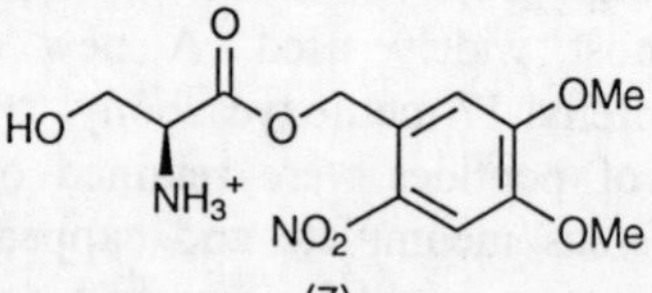

HO NH_3^+ O O OMe NO_2 OMe

(7)

BocNH BocN SMe + H_2N O OH ZNH

BocNH BocN N H O OH ZNH

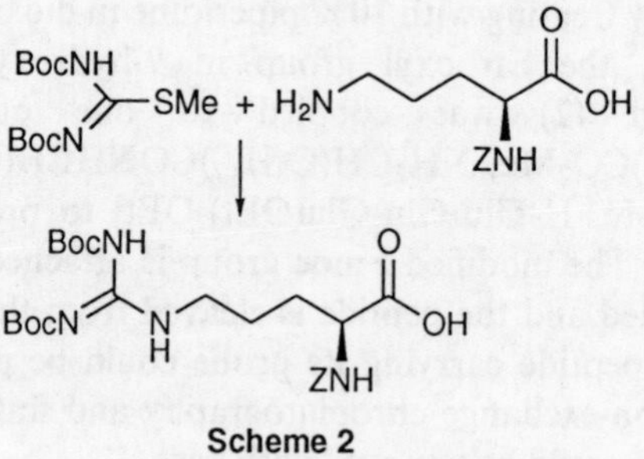

Scheme 2

A new base-labile protecting group has been reported[31]. The 2-(2′,4′-dinitrophenyl)ethoxycarbonyl (Dnpeoc) group can be introduced by silylating an amino acid and then treating with the chloroformate (**3**). Yields range from 84-95%. Removal of the Dnpeoc group is complete in <1 min in either piperidine/$CHONMe_2$ (1:4) or 0.5 M DBU in an aprotic solvent. Voelter[32] has revived his brainchild, the 1-(3′,5′-di-t-butylphenyl)-1-methylethoxycarbonyl (t-Bumeoc) group (**4**) for SPPS. It is cleaved in 1% CF_3CO_2H/CH_2Cl_2 in 1-3 min. It is doubtful, however, if the t-Bumeoc group will achieve any greater popularity now than when it was first introduced since Boc chemistry has become so firmly established where an acid-labile group is required.

P4-Phosphazene can be used to mediate the *N*-perbenzylation of Boc- or Z-peptides by BzlBr[33]. Amide nitrogens are also substituted. Deprotection is effected by sodium in liquid ammonia. The perbenzylated peptides are very soluble in hydrophobic solvents; the product from Boc-Val-Gly-Leu-OH is very soluble in pentane! Finally, 2-azidoesters are readily prepared (Scheme 1) and can be used as protected amino acid equivalents for the synthesis of peptides[34]. No racemization was detected.

2.2 *Carboxyl group protection.*

N-Protected peptides can be converted into their benzyl esters by interaction with *p*-hydroxyphenylbenzylmethylsulfonium chloride under mild conditions with no risk of racemization[35]. A peptide containing -Glu(OBzl)- groups was found to undergo readily trans-esterification with MeOH under conditions as mild as those obtaining during elution with MeOH from a chromatographic column[36]. It would be valuable to know if this is a general phenomenon with peptides containing -Glu(OBzl)- and also if -Asp(OBzl)-peptides are similarly affected. Carboxylate groups can be esterified with alkyl iodides in the presence of CsF in $CHONMe_2$ at room temperature[37]. Only one example in amino acid chemistry was reported, but it is probably quite general. Unfortunately, the reaction is rather slow.

Acetolyl esters can be prepared easily from Boc- or Z-protected amino acids and acetol in the presence of DCCI and $4\text{-}Me_2NC_5H_4N$ (DMAP). Acetolyl esters are stable to acids and to hydrogenolysis but are deprotected by exposure to tetrabutylammonium fluoride trihydrate in tetrahydrofuran at room temperature in 5-45 min[38]. Benzyl and t-butyl esters are unaffected by this treatment. Succinimidoalkyl esters have also been used for carboxyl group protection[39], but the method does not appear to offer any particularly attractive features. *p*-Azobenzenecarboxamidomethyl esters, $RCO_2CH_2CONHC_6H_4N{=}NC_6H_5$, have also been proposed[40], mainly because they are hydrolysed by K_2CO_3 solution in 15-20 min at room temperature, but there is no information about possible racemization. Finally, the bis(trimethylsilyl) ester of aspartic acid is recommended as an intermediate for the protection of the amino group with e.g. Z, Troc or Aloc, but only two acylation experiments indicated that it could be used in peptide synthesis[41].

A method of protecting amide groups[42] is placed here because it does not just apply to side chains of Gln and Asn. A new type of benzhydryl protective protective group (**5**) employs the safety-catch principle. When R = -SMe, the group is labile to acids, but when R = -SOMe, the group is resistant to conditions that remove Boc groups. The amide protecting group can be removed after reduction of the sulfoxide moiety. The method was tested by synthesizing H-Pro-Leu-Gly-NH_2 by the solution method.

2.3 *Side chain protection.*

A one-pot synthesis of α-*N*-Z-ω-*N*-Boc-ornithine and -lysine has been described[43]. Since the presence of a charged side chain facilitates the removal of impurities from a peptide, the temporary removal of ε-*N*-protecting groups such as Fmoc and its replacement by a trifluoroacetyl group after purification permits further manipulation such as fragment condensation[44]. It is worth noting that the authors report that some commercial samples of CF_3CH_2OH used as solvent contain an impurity that effects trifluoro-acetylation. Attachment of the acid (**6**) to the ε-amino group of Fmoc-Lys-OH permits the synthesis of peptides with a novel metal-ligating group[45]. The synthesis of (**6**) started from αβ-diaminopropionic acid (presumably the L-form although this was not specified in the paper).

The synthesis of the ω-cyclohexyl esters of Asp and Glu has been rather laborious. It has now been reported[46] that direct esterification is suitable provided that strict control of temperature and time of reaction are maintained. Reaction of cyclohexanol with the amino acid is carried out in $Et_2O/C_6H_{11}OH$ in the presence of H_2SO_4 at 70 °C for 2h in a rotary evaporator under vacuum. After work up, the product requires removal of Na_2SO_4. The ω-9-fluorenyl-methyl esters of Asp and Glu have also been prepared by direct esterification[47] in the presence of tetrafluoroboric acid diethyletherate.

Protection of the hydroxyl group of Ser, Thr and Tyr by silylation continues to be investigated[48,49]. Saponification of Z-Ser($SiBu^tPh_2$)-OMe gave a rather poor yield of the corresponding acid[48]. Avoidance of alkaline conditions gave satisfactory results and the ease of deprotection can be adjusted by choosing an appropriately substituted silyl group. The lability of *O*-silylated Tyr derivatives appears to be rather different from those of Ser and Thr[49]. An interesting method of deprotecting silyl ethers using 2,3-dichloro-5,6-dicyano-*p*-benzoquinone in MeCN/H_2O (9:1) has been reported[50], but has not so far been applied to peptide synthesis. The deprotection probably occurs by single electron transfer and this could cause problems with some peptides. The synthesis of photodeprotectable serine derivatives[51] such as (7) allows the light-controlled generation of free serine for enzymes that use this as a substrate, but this chemistry has yet to be applied for protection purposes to peptide synthesis. The safety-catch principle has been applied to the selective sulfation of a Tyr hydroxyl group in a peptide that also contains Ser[52]. Using Fmoc chemistry, a peptide containing *O*-*p*-(methylsulfinyl)-benzylserine [Ser(Msib)] and Tyr(Bu^t) was assembled. The latter can be deprotected using CF_3CO_2H without affecting the former. After detachment from the resin, the Tyr residue was sulfated with $CHONMe_2$-SO_3 complex and then the Msib group was reduced using $HSCH_2CH_2SH$ rendering it cleavable by CF_3CO_2H.

Earlier reports that the *S*-(3-nitro-2-pyridinesulfenyl) protecting group is partially cleaved by HOBt have been rechecked using both soluble and polymeric forms of Cys(Npys)[53]. It transpires that the fear was unfounded; the *S*-Npys group is quite stable to HOBt. A novel protecting group for the thiol group of cysteine has been described in detail[54]. The *S*-2,4,6-trimethoxybenzyl (Tmob) group is introduced by treating cysteine with 2,4,6-trimethoxybenzyl alcohol in the presence of CF_3CO_2H. The amino group is protected with Fmoc. Recommended conditions for the removal of the Tmob group involve treatment with either 30% CF_3CO_2H in CH_2Cl_2 in the presence of PhOH, PhSMe and H_2O (5% of each) or 6% CF_3CO_2H in CH_2Cl_2 in the presence of 0.5% of either Et_3SiH or $Pr^i{}_3SiH$. The *S*-ethylcarbamoyl group has been recommended for protection of the cysteine side chain[55] and it is stable to acidolysis and long-term storage. It is removed by NaOH, conditions which may deter potential users. It has been used for the SPPS of more than 100 peptides in order to prepare peptide-protein conjugates in which the presence of a small amount of racemized peptide is unlikely to be crucially important. The *S*-(2,4-dinitrophenyl)ethyl (Dnpe) group, which has been used in nucleotide chemistry, has also been recommended for thiol group protection in SPPS[56]. *N*-Boc-Cys-OH reacts with 2,4-dinitrophenyl-ethyl bromide and the product can be used in SPPS. The Dnpe group is stable to acid but is removed by β-elimination on treatment with base. After removal of the Dnpe group it is possible to form cyclic disulfides while the precursor peptide is still attached to the support. The method was tested by synthesizing oxytocin and deaminooxytocin.

A method for preparing the ωω′-bis-Boc derivatives of Fmoc-and Z-Arg-OH has been described[57]. It involves the preparation of *NN*′-bis-Boc-*S*-methyl isothiourea followed by the guanidination of the appropriate α-protected ornithine (Scheme 2). Parallel SPPS routes to α-MSH using Fmoc-Arg$(Boc)_2$-OH, Fmoc-Arg(Pmc)-OH and Fmoc-Arg(Mtr)-OH revealed that the first two gave good yields of crude product whereas the last gave a lot of sulfonated byproducts. It has been reported that the Pmc and Mtr groups are not always cleanly removed by reagents involving CF_3CO_2H particularly if there are several Arg residues present[58]. The authors found that only hard-acid deprotection with Me_3SiBr reliably removed these groups. Like many other 1,2-bis-carbonyl compounds, diketopinic acid (**8**) forms adducts with the guanidino group of arginine derivatives[59]. The conditions for decomposing the adduct (0.2 M phenylenediamine at pH 8–9) are too mild for the reagent to find much use in peptide synthesis, but it might be useful for temporarily blocking selected guanidino groups in a trypsin-catalysed fragment condensation.

The 4-methyltrityl group has been recommended for protecting the carboxamide side chains of Asn and Gln because it is more readily cleaved than the Trt group by acid[60]. It is not clear why the Me group should make such a large difference. The use of a Trt group for protecting the amide group of Asn can cause problems if this group is *N*-terminal[61]. Deprotection with CF_3CO_2H was found to be incomplete under normal conditions. This difficulty does not arise if the Asn group is not *N*-terminal and deprotection of *N*-terminal Gln does not cause problems. The authors' suggest that acidolysis of a Trt group involves an intermediate in which the amide nitrogen is protonated. This would be opposed by the adjacent α-NH_3^+- group especially with the short distance involved with Asn. Kinetic measurements of the rate of deprotection of Fmoc-Asn(Trt)-OH and H-Asn(Trt)-OH support this view. Use of the methyltrityl protecting group avoids this difficulty with *N*-terminal Asn residues. Alkylation of Trp occurs during removal of Tmob groups from the side chains of Asn and Gln[62]. It would be useful to know if this is a serious problem when Tmob is used to protect the cysteine thiol group as described above.

2.4 *General deprotection.*

6M-HCl is a satisfactory alternative to CF_3CO_2H in CH_2Cl_2 for removing Boc groups during SPPS[63]. The use of Et_3SiH as a carbocation scavenger in conjunction with CF_3CO_2H in CH_2Cl_2 leads to improved yields, decreased reaction times and better selectivity in the presence of other acid-labile groups[64]. The use of CF_3CO_2H for deprotection can

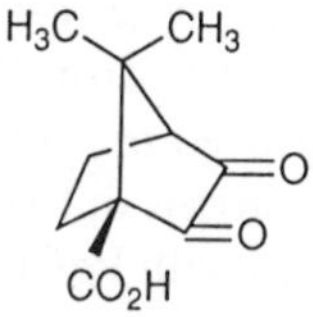

(8)

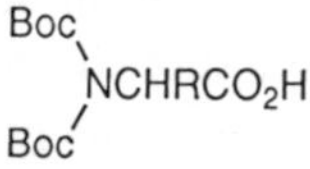

(9)

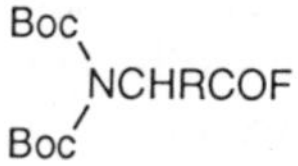

(10)

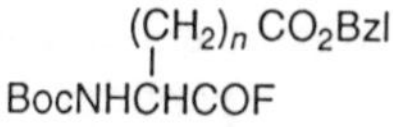

(11)

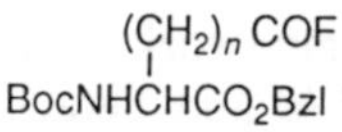

(12)

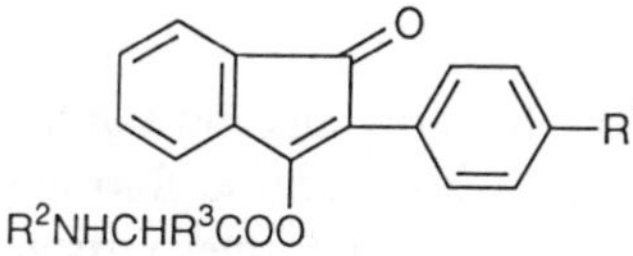

(13; R^1 = H, F)

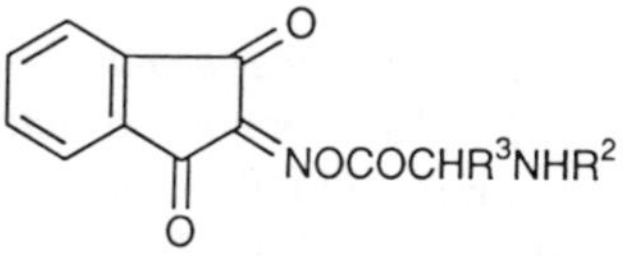

(14)

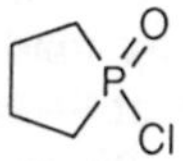

(15)

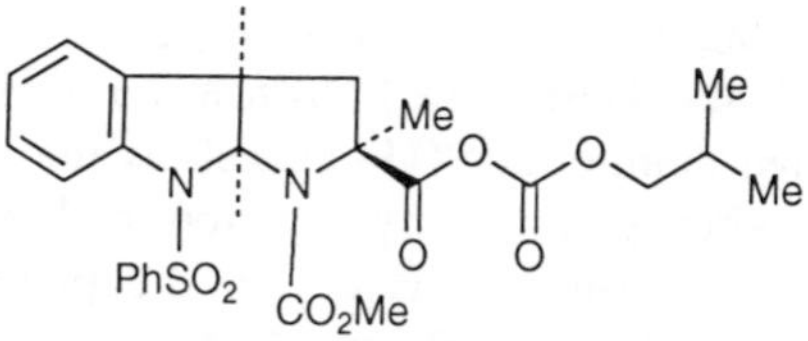

(16)

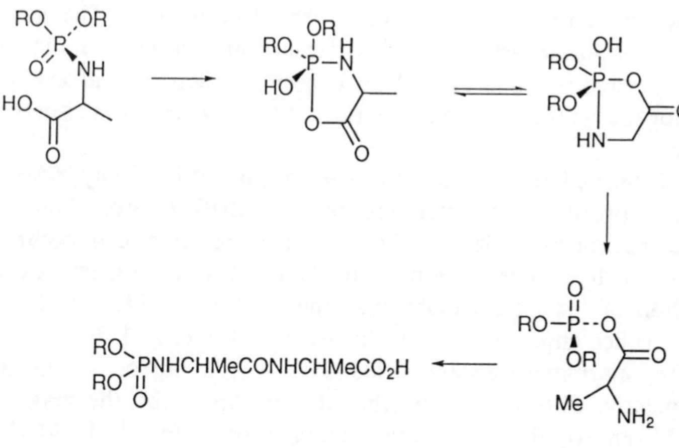

Scheme 3

cause trifluoroacetylation of the α-amino group if Thr is *N*-terminal[65]. It was suggested that trifluoroacetylation of the side-chain hydroxyl group is followed by $O \rightarrow N$ acyl migration. There is an improved procedure for deprotection by hydrogenolysis[66]. The procedure is carried out in dilute aqueous ethanol in a milky suspension which is shaken with $PdCl_2$ under hydrogen (5 atm) until the turbidity clears. The main advantages are the speed of reaction and the visually recognizable endpoint of reaction. Finally, a robotic work-station has been constructed which mechanizes the deprotection and detachment of synthetic peptides by CF_3CO_2H from the resin after solid-phase synthesis[67]. The apparatus can be used with 50–500 mg of peptide-resin every 2 h affording yields of 50–80%.

2.5 *Peptide bond formation.*

NN-Bis-Boc-amino acids (**9**) gave the hitherto unknown *NN*-bis-Boc-amino acid fluorides (**10**) when treated with cyanuric fluoride in CH_2Cl_2 at −30°C. In contrast, $SOCl_2/CHONMe_2$ (Vilsmeier reagent) converts (**9**) into the *N*-carboxyanhydride[68]. The acyl fluorides (**10**) are good acylating agents as expected and, for example, Boc_2-Phe-Leu-OBzl was obtained in 80% yield. The previously described derivatives of Asp and Glu (**11**) are now known to be the isomeric compounds (**12**)[69]. The desired compounds (**11**) as well as the Fmoc analogues are made from the ω-benzyl esters by reaction with cyanuric fluoride at −20°C to −30°C in $MeCN/CH_2Cl_2$ in the presence of pyridine. Some of the corresponding *N*-carboxyanhydride was also formed.

The kinetics of interaction of *N*-hydroxysuccinimide esters of Z-amino acids with *p*-anisidine in Me_2SO were studied in detail and the enthalpy and entropy of activation were determined[70]. The authors propose that the transition state is tetrahedral and zwitterionic. New coloured, reactive esters (**13,14**) have been synthesized and evaluated briefly for the synthesis of small peptides[71,72]. They are about as active as Pfp esters.

N-Protected amino acids give rise to symmetrical anhydrides in a convenient procedure involving reaction with dialkyl pyrocarbonates in the presence of a tertiary amine[73]. A side reaction can occur, for example, with a pyrocarbonate derived from a primary alcohol. Formation of the intermediate unsymmetrical anhydride, e.g. RCOO-COOEt, is accompanied by the liberation of EtOH and this can react with the unsymmetrical anhydride to give the unreactive RCO_2Et. Consequently, di-t-butyl pyrocarbonate appears to be the reagent of choice. Even so, these can give t-butyl esters using $4\text{-}Me_2NC_5H_4N$, pyridine and Bu^tOH in EtOAc or dioxan. The previously described 1-oxo-1-chlorophospholane (**15**) forms peptide bonds through an intermediate phospholanic-carboxylic unsymmetrical anhydride[74]. The formation, stability and reactivity of such anhydrides has been studied using ^{31}P nmr spectroscopy. Addition of $CuCl_2$, which is known to suppress racemization in peptide bond formation involving carbodiimides (see below), is now reported to be equally effective in couplings mediated by $ClCO_2Bu^i$, BOP-Cl or EEDQ[75]. $CuCl_2$ is reported to be much more effective than $ZnCl_2$, ZnF_2 or even HOBt. Preparation of the *N*-carboxyanhydrides from H-Glu(OMe)-OH or H-Glu(OBzl)-OH produces a tenacious impurity, the γ-alkyl-L-glutamate ester hydrochloride, which could not easily be removed by recrystallization[76]. Retreatment with $COCl_2$ in tetrahydrofuran at 47-48C for 3 h is recommended. *N*-Phosphorylated amino acids on warming to 40–60°C in solution undergo cyclization and rearrangement (Scheme 3)[77]. Condensation of the cyclized and rearranged products gives an *N*-phosphoryl-dipeptide. These reactions do not occur if R = Pr^i. In a special case, an unsymmetrical carbonic anhydride (**16**) was found to be sufficiently stable to isolate and determine its structure by X-ray crystallography[78].

Suppression of racemization during peptide-bond generation involving carbodiimides continues to evoke interest. For example, when Z-Gly-L-Val-OH and H-L-Val-OMe were coupled in $CHONMe_2$ using $C_6H_{11}N{=}C{=}NC_6H_{11}$ (DCCI), a yield of 69% which was 44% racemized was obtained[79]. With one equivalent of $CuCl_2$, the yield was marginally improved, but racemization was diminished to $<0.1\%$. Similar results were obtained when Z-Ala-Val-OH was coupled to a wide range of amino acids and peptides. Addition of HOBt and $CuCl_2$, the latter in smaller quantities, effectively eliminated racemization and the yield was improved[80]. The use of either 1-hydroxy-4,6-dimethyl-2(1H)-pyrimidone (**17**) or 1-hydroxy-5,6-dimethyl-2(1H)-pyrazinone (**18**) can replace HOBt with consequent improvement of yield of peptide[81]. The use of a polymer-bound 4-carboxy-1-(4-pyridyl)-piperidine has been recommended for markedly decreasing racemization during peptide-bond formation using DCCI[82]. Development of a method for separating diastereoisomers of Z-Gly-Xaa-Xbb-OMe by reversed phase hplc facilitated another extensive investigation of racemization during peptide synthesis using DCCI. The well known penalty associated with the use of a very polar solvent such as $CHONMe_2$ has been underlined[83]. It was also shown that the nature of the side chain of the attacking nucleophile influences the extent of racemization with Ile and Val being the worst offenders. Again, when the attacking nucleophile was the phenacyl ester of either L-Pro or *O*-Bzl-*trans*-4-hydroxy-L-proline in a coupling effected by a water-soluble carbodiimide, extensive racemization occurred at the α-carbon atom of Pro or Hyp[84]. Moreover, racemization was increased

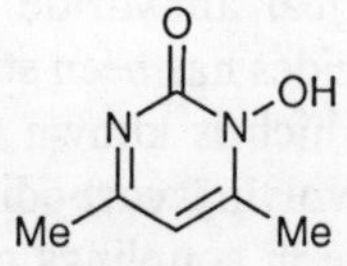

(17) 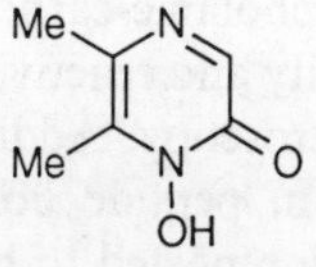 (18)

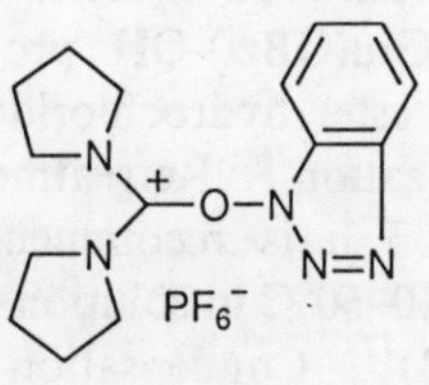

(19) 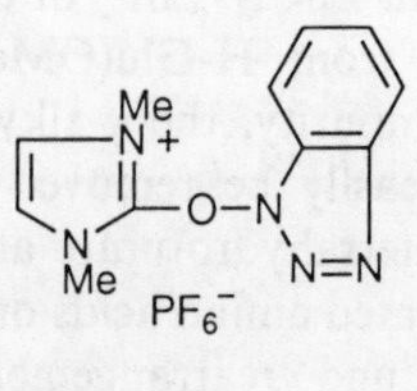 (20)

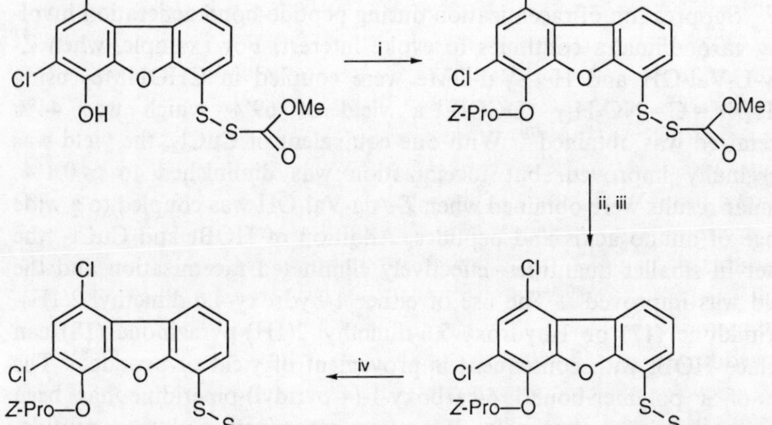

Reagents: i, $ClCO_2Bu^t/Et_3N/Z$-Pro—OH; ii, $Et_3P/CF_3CHOHCF_3/CH_2Cl_2$
ii, Boc-Cys(Scm)—OMe; iv, CF_3CO_2H

Peptide assembled: H–RPDFCLEPPYTGP–CRAKRNNFKSADECMRTCGGA–OH

Scheme 4

by the addition of HOBt. Almost no racemization occurred when coupling was mediated by an unsymmetrical anhydride or DCCI minus HOBt. The authors postulate intermediate formation of a carbinolamine between the -NH- of Pro or Hyp and the $>C=O$ group of the phenacyl ester. Clearly, this is a special case, but it serves to remind us that synthetic protocols are only secure within the conditions that have been investigated. The synthesis of the phenacyl esters of some protected peptides can be complicated by low solubility when removal of the phenacyl group is required. A mixture of CF_3CH_2OH or $(CF_3)_2CHOH$ and either $CHCl_3$ or CH_2Cl_2 has been recommended[85]. The use of fluorinated alcohols as solvents during coupling has been investigated. $(CF_3)_2CHOH$ alone was unsatisfactory in reactions effected by DCCI, a result attributed to the low pK_a of $(CF_3)_2CHOH$[86]. Although BOP gave better yields than DCCI, substantial racemization occurred. Eventually, a solvent system comprising $(CF_3)_2CHOH/CHONMe_2/C_5H_5N$ (2:1:1 v/v) was recommended with DCCI/HOBt as the coupling agent.

A new uronium coupling agent, benzotriazol-1-yloxy-bis-(pyrrolidino)-carbonium hexafluorophosphate (**19**) is proposed[87]. It is stable, not hygroscopic and can be stored at room temperature. Good yields of peptides are obtained with minimal racemization. Another new coupling reagent, 2-(benzotriazol-1-yl)oxy-1,3-dimethylimidazolinium hexafluorophosphate (**20**) has been recommended for SPPS[88]. The authors used dilute $MeSO_3H$ for deprotecting the α-amino group. The synthesis of porcine brain ANP indicates the efficacy of this methodology. The value of uronium coupling reagents has been further underlined in the synthesis of peptides containing highly sterically hindered amino acids such as α-*N*-methylaminoisobutyric acid[89]. Moderate yields were obtained when HBTU or PyBroP were used, but other coupling procedures examined were unsatisfactory.

Several methods were compared for coupling Z-Gly-Xaa-OH and H-Val-OBzl in $CHONMe_2$ or CH_2Cl_2 at 5 °C and the extent of racemization was determined by chromatography of the products[90]. DCCI alone gave the worst results but was best when HOBt was present. A novel method has been developed[91] to check for racemization of the *N*-terminal amino acid after synthesis of a dipeptide. A chiral isocyanate, α-methylbenzyl isocyanate was coupled to the dipeptide ester. The resulting carbamoylated dipeptide ester undergoes *N*-terminal degradation with $SOCl_2/MeOH$ similar to Edman sequencing and a mixture of diastereoisomeric α-methylbenzyl-carbamoyl derivatives of the methyl ester of the *N*-terminal amino acid is produced if racemization has occurred during peptide synthesis. The product is analysed by hplc.

The remainder of this section comprises a miscellaneous collection of topics. Esters of *N*-protected amino acids and carbohydrates such as 1,2-*O*-isopropylidene-α-D-xyluronamide or 5-deoxy-5-dimethylamino-1,2-isopropylidene-α-D-xylofuranose, which are not particularly reactive towards nucleophiles, become much more reactive in the presence of Na^+ ions[92]. Kinetic studies revealed that the enhanced reactivity stems from a diminished entropy of activation. The difficulty of forming a Pro-Cys bond by Kemp's thiol capture method has been greatly reduced by the use of a dichlorodibenzofuran template[93] (Scheme 4). A one-pot synthesis of a peptide involving two redox reactions[94] starts from an *N*-protected amino acid or peptide, Ph_2Se_2 and Bu_3P which give a selenophenyl ester (**21**) and PhSeH at room temperature (Scheme 5). The *C*-terminal component of the final peptide is added as an azidoester which is reduced by the PhSeH formed in the foregoing step to the amino acid ester. The PhSeH is concomitantly reoxidized to Ph_2Se_2 to recycle. The amino acid ester reacts with the selenophenyl ester formed in the first step to give the protected peptide and PhSeH. With hindered azides such as 1-azido-1-ethoxy-carbonylcyclopropane, the yield can be improved by the addition of HOBt. Tests for racemization were encouraging. Young's peptide (Bz-Leu-Gly-OEt) was 4% racemized and Anderson's peptide (Z-Gly-Phe-Gly-OEt) was optically pure. Based on earlier work which showed that metal carboxylates derived from Ti and Zr were prone to nucleophilic attack by amines albeit under rather vigorous conditions, it has been found[95] that tantalum(V) complexes are much more reactive. For example, cyclopentadienyltantalum(V) amino acid carboxylate can be formed from $CpTaCl_4$ and an *N*-protected amino acid and this reacts at low temperature with an amino acid ester to give the expected peptide derivative. Little racemization occurred even when starting from Ac-Ala-OH. We now need data on the synthesis of peptides from sterically unfavourable precursors. Tetrabutylammonium salts of amino acids have been used in peptide synthesis on account of their high solubility in organic solvents[96]. There is a considerable acceleration resulting from the presence of the Bu_4N^+ ion. Coupling of an *N*-protected amino acid and an amine can be effected by a polymer-based phosphine derivative in presence of CCl_4 as an oxidizing agent and *N*-methylmorpholine[97]. In contrast to the results of earlier experiments, racemization was not serious. Although poor yields of dehydroaspartyl peptides originated from Ac-Δ^zAsp(OBu^t)-OH using various coupling procedures, Z-Δ^zAsp(OBu^t)-OH gave good yields[98]. The difference was attributed to the formation of an oxazolone derivative in the former case. Finally, dipeptides have been synthesized by the photolytic coupling of chromium-aminocarbene complexes with α-amino acid esters[99] (Scheme

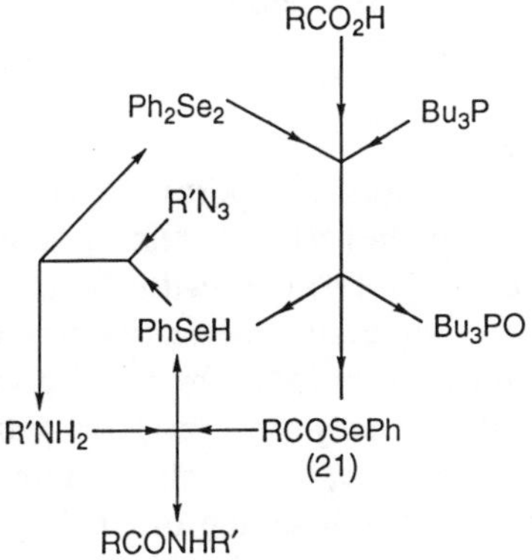

RCO$_2$H, Ph$_2$Se$_2$, Bu$_3$P, R′N$_3$, PhSeH, Bu$_3$PO, R′NH$_2$, RCOSePh (21), RCONHR′

Scheme 5

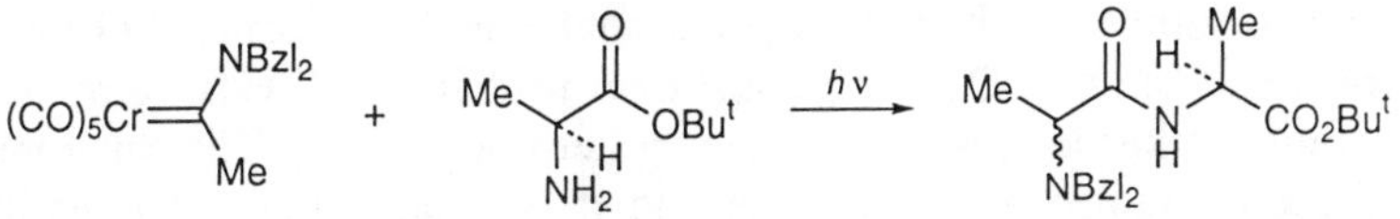

$(CO)_5Cr{=}C(NBzl_2)Me$ + Me–CH(NH$_2$)–CO–OBu^t $\xrightarrow{h\nu}$ Me–CH(NBzl$_2$)–CO–NH–CH(Me)–CO$_2$Bu^t

Scheme 6

H—Cys(Acm)-Gly-Thr-Tyr-Glu-Cys—OH
|
Ac—Cys(Acm)-Leu-Ser-Ala-Lys-Cys-Ala-Phe-Gly-Ser-Lys—OH

(22)

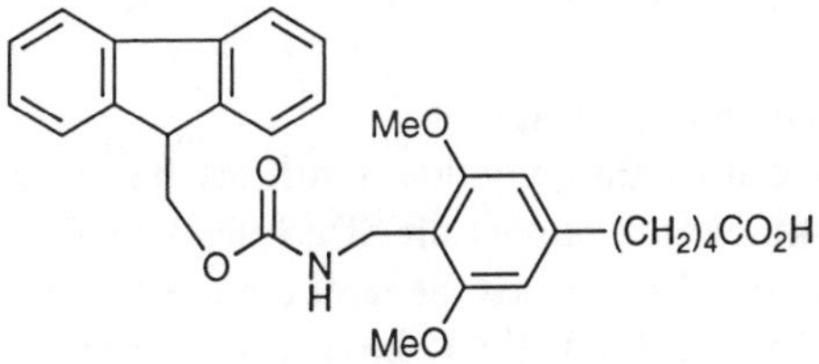

MeO, MeO, (CH$_2$)$_4$CO$_2$H, O, O, N, H

(23)

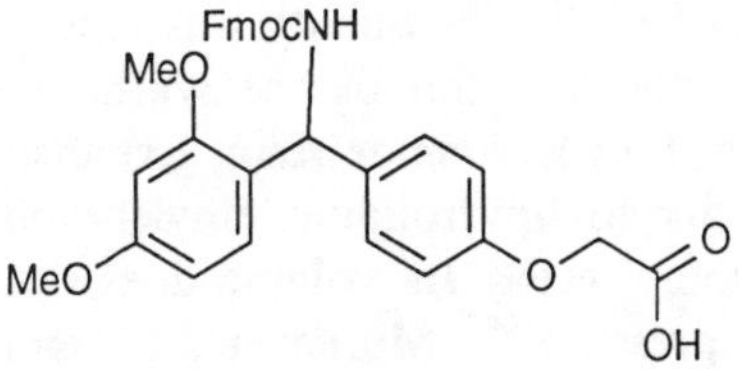

FmocNH, MeO, MeO, O, O, OH

(24)

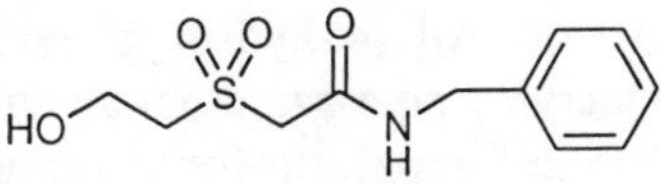

HO, O, O, S, O, N, H

(25)

6). The reaction has high diastereoselectivity, especially if the temperature is kept low.

2.6 *Disulfide bond formation.*

There is very little new work to report in this area. Fmoc-Cys(Boc-Cys-OH)-OBzl has been synthesized by treating a mixture of Fmoc-Cys(Acm)-OBzl and Boc-Cys(Trt)-OH with I_2 in MeOH at room temperature[100]. The foregoing cystine derivative was incorporated into the central position of an undecapeptide by SPPS using Boc chemistry. The *N*-terminal residue [Cys(Acm)] was acetylated and the Fmoc group was removed and a pentapeptide, which also terminated in Cys(Acm) was constructed on the other half of the original cystine moiety. The resultant peptide (**22**) was then treated with I_2 to generate a second disulfide bond. A new reagent, cyanogen iodide, has been used to generate disulfide bonds in a fragment analogue of calcitonin, EGF and a fragment of the DNA-binding *Cro* protein[101]. Yields, although sometimes rather low, were routinely better than with I_2. The calcitonin analogue contained a Met residue and it is notable that CNI, unlike CNBr, did not cause chain cleavage. Finally, $MeSiCl_3$ in CF_3CO_2H with admixed PhSOPh removes various *S*-protecting groups (Acm, Tacm, Bam, MeOBzl) and forms disulfides directly in not more than 30 min[102]. Trp residues, if present, must be *N*-formylated to prevent chlorination. Using this technique, oxytocin, human brain natriuretic peptide and somatostatin were synthesized.

2.7 *Solid phase peptide synthesis.*

Despite the claims that peptide synthesis has become a button-pushing process, numerous papers on SPPS describe new advances and the solution of some old problems. Several commercial resins, some sold with the first amino acid attached, have been compared for swelling properties, uniformity of beads, and level of substitution for those resins with an amino acid attached[103]. The quality was reported to vary from batch to batch and the effects of this on the synthesis of peptides was evaluated. A commercial polyacrylamide resin, Expansin™, is made by copolymerization of *N*-acryloyl-pyrrolidine, ethylene bisacrylamide and *N*-acryloyl-β-alanine methyl ester. Its volume does not alter much on change of solvent or pressure[104]. Moreover, ^{1}H nmr measurements revealed that attached amino acids have considerable mobility. A polymer (Copolymer Q) of *N*-[2(4′-hydroxyphenyl)ethyl]acrylamide cross-linked with *NN*′-diacryloylpiperazine is an ultra-high loading support[105]. A manually operated apparatus for continuous flow synthesis has been constructed and it has performed well in test syntheses. Various resins have been tested as media for the synthesis of peptide amides using Fmoc/Bu^t chemistry[106]. Only the methylbenzylhydrylamine (MBHA) resin containing the 4-hydroxymethylbenzoic acid linker, which was cleaved by NH_3, gave an acceptable yield of Boc-Tyr-Gly-Gly-Phe-Arg(Pmc)-NH_2. A previously described resin, α-(4-bromomethyl-3-nitrobenzamido)benzyl polystyrene has been found to be suitable for the synthesis of peptide *N*-alkylamides[107]. The electron-withdrawing effects of the -NO_2 and -CONH- groups facilitate nucleophilic cleavage by e.g. $MeNH_2$. A copolymer of polyethylene glycol and dimethylacrylamide has been prepared[108] and it passed the almost compulsory test for synthesizing ACP(65-74). The well-known Kaiser oxime resin has been used to synthesize fragments of β-amyloid protein for subsequent fragment coupling on the resin[109]. Moreover, peptide amides can be made on the Kaiser resin by cleaving the final peptide sequence with $CH_3CO_2NH_4$ (10-20 equivalents) in the presence of Pr^i_2EtN in $CHONMe_2$. Negligible epimerization of the *C*-terminal residue occurred during this cleavage. A cellulose-based support has been made[110] by treating beaded cellulose with CH_2:CHCN in the presence of NaOH. After reduction of the nitrile groups with B_2H_6, the resultant amino-propyl groups could be converted into glycolamido linkers by treatment with $(BrCH_2CO)_2O$ followed by reaction with the Cs salt of a Boc amino acid. A peptide can be constructed without further modification since the glycolamido moiety is labile to base. Alternatively, the aminopropyl group could be extended with a 4-hydroxymethylphenoxyacetyl moiety. Several small peptides including ACP(65-74) were synthesized satisfactorily. Several papers describe new linkers and improvements to old ones. The tertiary alcohol linker, $4\text{-}HOCMe_2(CH_2)_2C_6H_4OCH_2CO_2H$, has been used[111] to synthesize the mosquito oocytic hormone which has 6 Pro residues at the *C*-terminus. This is a tiresome molecule to make because diketopiperazine formation and consequent detachment tend to occur when the second Pro residue is deprotected. A way round was found by attaching the first residue followed by the next two together as Fmoc-Pro-Pro-OH using BOP as the coupling agent. For the synthesis of [Ala^8]-dynorphin A with Gln at the *C*-terminus, the PAL linker (**23**) was selected since it is recommended for synthesizing *C*-terminal amides. For this problem, the *C*-terminal amino acid was attached through its side chain[112]. An improved synthesis of 4-[(R,S)-1-[1-(9H-fluoren-9-yl)methoxy-carbonylamino]-(2′,4′-dimethoxybenzyl)]phenoxyacetic acid (**24**) for the SPPS of peptide amides has been described[113]. A new base-labile linker (**25**) has been synthesized for attachment to aminomethylpolystyrene[114]. Leu-enkephalin and its [D-Ala^2] analogue were synthesized using Boc protection in good yields and with no detectable racemization.

A new fluorene-based linker (**26**) has been synthesized[115] and attached to MBHA resin. Boc/Bzl chemistry was used and the peptide was preferably cleaved with *N*-methylmorpholine. The first enzyme-scissile linker for SPPS (**27**) uses phosphodiesterase to detach the peptide still containing part of the linker[116]. Since this part of the linker forms a benzyl ester of the peptide, it can be removed by hydrogenolysis, converted into hydrazide for an azide fragment coupling or retained if the peptide is intended for use as a substrate for a proteolytic enzyme possessing esterase activity. For the specific objective of synthesizing β-mercaptoethylamide derivatives of peptides a simple disulphide type of linker, $BocNH(CH_2)_2SSCHMeCO_2H$, has been proposed[117]. Boc/Bzl chemistry can be used and the product is detached by thiolysis with dithiothreitol.

There have been a few advances in coupling methodology. Two new related reagents are proposed for coupling the first Fmoc amino acid to a support[118]. Either 2-chloro-1,3-dimethylimidazolidinium hexafluorophosphate or the product (**20**) obtained from its reaction with HOBt are good esterification reagents. Yields compared satisfactorily with other methods on the whole. Two groups report that *in situ* neutralization is advantageous when using Boc chemistry[119,120]. After acidolytic deprotection, the base for neutralization and the reagents for coupling the next amino acid residue are added simultaneously. Operations are accelerated and the risk of racemization is decreased. The tests employed could hardly have been more severe. On the one hand[119], interleukin 8 (77 residues) was assembled while the other group[120] synthesized HIV-proteinase (99 residues) and two large fragments (1-50, 53-99). Coupling efficiency in SPPS can be improved by carrying out the reaction in a microwave oven[121]. The results are most notable when a sterically hindered amino acid is being coupled. For example, formation of a Val-Gly bond was improved from a yield of 58.5% under standard conditions to 99.8% with microwave irradiation. The temperature used was typically 55 °C and a coupling time of *ca.* 6 min was usually adequate. No significant racemization was detected. The lowest yield recorded during the synthesis of ACP(65-74) was 99.5%, a result which surely demands an exhaustive assessment of the methodology. Two coupling reagents, HOBt esters and the symmetrical anhydride from Fmoc amino acids, were used. Fmoc amino acid chlorides[122,123] and 2-(fluorenylmethoxy)-5-(4H)-oxazolones[122] have been recommended for the coupling step when the target peptide contains sequences of secondary amino acids.

There is another report of difficulties in synthesizing ACP(65-74)[124]. As well as the desired peptide, the des-Ile[72] and des-Asn[73] analogues were isolated. Fortunately, the Sheppard group in particular

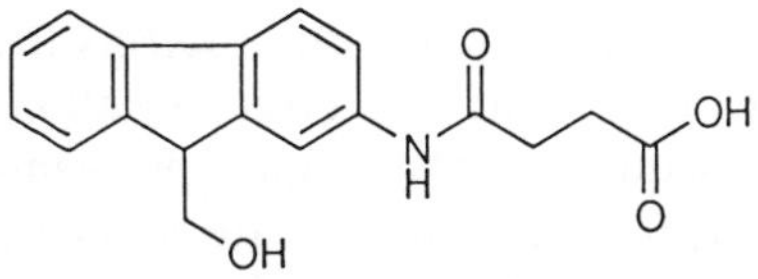

(26)

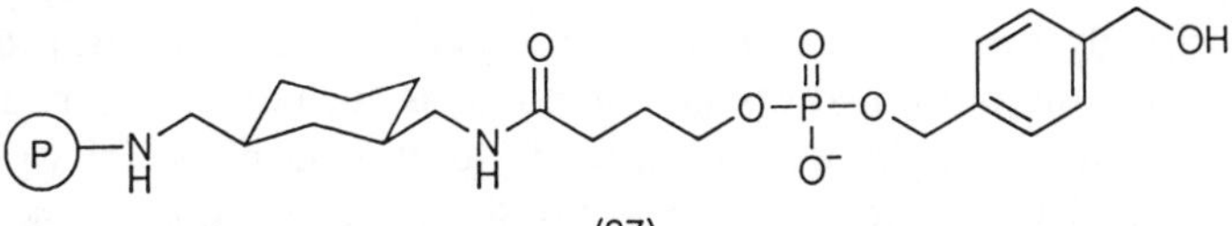

(27)

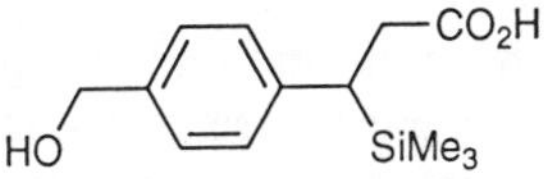

(28)

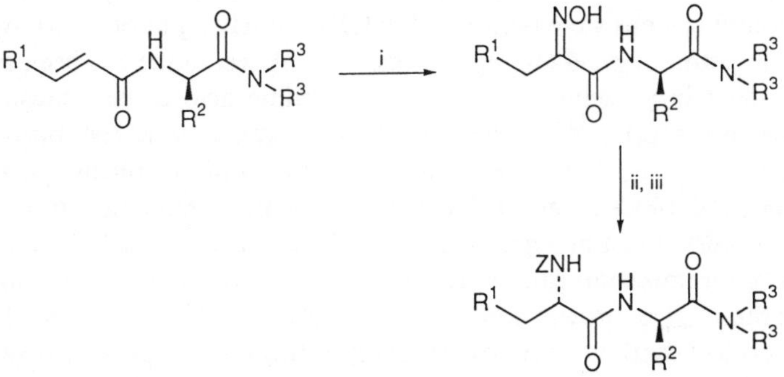

Reagents: i, $BuNO/R^3H$ + trace of (29) ii, $SmI_2/MeOH$; iii, ZCl/pyridine

Scheme 7

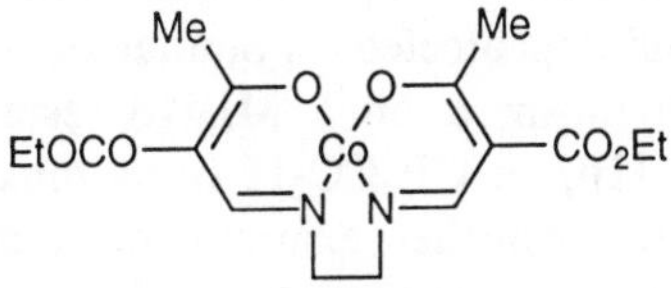

(29)

continues to grapple successfully with problems that arise in SPPS. The use of MeSOMe as solvent for both acylation and deacylation is recommended for sequences which cause problems because of internal aggregation[125]. Polyalanine sequences are notoriously troublesome and so the effect of other guest residues in the sequence, H-(Ala)$_n$-X-X-(Ala)$_3$-Val-polydimethylacrylamide, was determined using Fmoc chemistry[126]. Ala, Val, Ile and Sar cause serious problems, but perhaps surprisingly Leu and Phe are less troublesome. The choice of protecting group for the more hydrophilic amino acids such as Ser, Thr, Tyr and Cys is important. Bulky hydrophobic groups such as Trt can cause serious difficulties. On the other side of the coin, a comparison of the use of Boc and Fmoc protecting groups when assembling (Val)$_n$ or (Ala)$_n$ on MBHA-polystyrene led to the conclusion that Boc was preferable[127]. Doubtless the controversy will continue. Repeated coupling of Val was necessary in the synthesis of an undecapeptide using the Sheppard approach and the product gave three peaks on hplc[128]. One peak, which lacked the four *C*-terminal residues, was attributed to esterification of residual hydroxyl groups on the resin catalysed by His. In a synthesis of prepro-GnRH and three fragments thereof using Boc/Bzl protection, recoupling steps were performed in $CF_3CHOHCF_3$ in order to overcome aggregation of pendant peptide chains[129]. A technique has been reported for detecting racemization of the first amino acid to be attached to an insoluble support[130]. After acidolytic cleavage of a test portion, the amino acid is derivatized with 1-fluoro-2,4-dinitrophenyl-5-L-alanine amide (Marfey's reagent). The diastereoisomers resulting if racemization has occurred can be separated readily by reversed phase hplc. As little as 0.1% racemization can be detected with a lower limit of 100 pmol. The synthesis of a protein kinase C substrate (KRAKAKTTKKR) by the Merrifield method gave low recovery using HF for peptide detachment and deprotection[131]. The *N*-terminal Lys carried two Fmoc groups and it transpired that these had been attacked by carbocations.

A hexapeptide was constructed by attaching the *C*-terminal Asp residue through its β-CO_2H to the support. After deprotection of the *N*-terminal amino group and the *C*-terminal carboxyl group, head-to-tail cyclization was effected on the support using TBTU. The cyclic peptide was then detached and deprotected[132]. Similar side-chain attachment was used to assemble thymosin α on a MBHA-resin[133]. Detachment was effected with 0.1 M HBr in CF_3CO_2H containing pentamethylbenzene and 5.5% thioanisole. Another example of a modification achieved before detachment of a peptide from the support involved the biotinylation of the ε-amino group of a Lys residue in an analogue of NPY[134]. After attachment of an amino acid to a support, the deprotected amino group can be converted into an isocyanato group and that in turn is allowed to react with $RNHNH(CH_2)_2CONH_2$. The resultant peptide contains an azaglutamine residue[135] and a peptide chain can be constructed. Acidolytic detachment then affords a peptide with a *C*-terminal aldehyde group, a potentially useful route to proteinase inhibitors[136].

It is now possible to carry out SPPS of short peptides as spots on filter paper down to the 0.02 μmole scale[137]. This technique is intended for epitope work so that the presence of peptides with omissions in the sequence and the retention of a part of a linker is not a severe disadvantage. Clearly, this method appears to be ideal for the simultaneous production of vast numbers of peptides and the author uses computer-controlled pipetting operations for which commercial software is available. An alternative method uses a membrane sandwiched between two 96-well templates as the support[138]. This has been used to make peptides containing more than 20 residues. On the larger scale, fully automatic synthesizers have been described which can produce 36-48 peptides simultaneously[139-141]. The production of peptide libraries is now a major activity and a collection of 1900 peptides has been synthesized[142] based on the sequence: H-Tyr-Gly-Arg-Gly-X^1-X^2-X^3-OH.

This section is concluded by reference to some papers on the detachment of peptides from insoluble supports. Readers are reminded of one method[116] described above. The other three methods use tetrabutylammonium salts but the first of these has a quite different mechanism from the others. A new silicon-based linker (**28**) has been developed[143] which permits detachment of the peptide with $Bu_4N^+F^-$. A 35-residue fragment of ubiquitin was synthesized using this technique. The other two methods use tetrabutylammonium salts as phase-transfer catalysts. Tetrabutylammonium carbonate, prepared *in situ* from $Bu_4N^+HSO_4^-$ and K_2CO_3, releases peptides from Merrifield resin with the *N*-Boc group still attached[144]. Phase-transfer catalysed detachment of peptides from the support has been used in the synthesis of cyclic peptides[145]. Although peptide detachment by this method is rather slow, racemization is $<2\%$.

2.8 *Enzyme-mediated synthesis and semi-synthesis.*

There are nearly 70 references to this section and space does not permit detailed discussion of each paper. The material reviewed is considered under five main areas. First, the effect of physical conditions including the state of the enzyme, effect of organic solvents, temperature, pH, pressure, and the use of chemically modified or genetically engineered enzymes are considered. Readers should be aware that many papers cover several aspects of the subject.

There is no consensus concerning the most favourable state of the enzyme. It can be used in solution, in the solid state when the water concentration in the reaction medium is low, adsorbed on an inert support, covalently attached to an insoluble support or as a reversed micelle[146-152]. The use of various solvents is discussed in numerous papers[153-166]. Water-miscible solvents used include MeCN, $CHONMe_2$, $(CH_2CH_2OH)_2$, MeSOMe and $CF_3CH(OH)CF_3$ and mixtures of these. Two-phase systems seem to be rather less popular. Perhaps the most revolutionary idea is to use no solvent at all, but to triturate the acyl component, amine nucleophile and $Na_2CO_3.10H_2O$ in presence of a proteinase in a round-bottomed glass rotary homogenizer[166]. Very high yields of protected dipeptide were obtained.

The influence of other conditions has also been studied but less extensively. Kinetically-controlled syntheses catalysed by α-chymotrypsin in 50% $CHONMe_2$ at an apparent pH of 8.5 revealed that yields were better at 25 °C than at 35 °C because hydrolysis of the ester substrate was relatively more important at the higher temperature[158]. Similar studies led to the formulation of an equation relating log k, the first-order rate constant, to the absolute temperature, the percentage of $CHONMe_2$ in the solvent and the partition coefficient of the *N*-protecting group moiety in octanol/H_2O[167]. Pepsin-catalysed coupling of Z-X-Phe-OH (X = Ala, Phe, Trp, Tyr) and H-Phe-OMe was studied in biphasic systems and the effects of substrate and buffer concentrations, pH and the ratio of the volumes of the two phases on yields were examined[163]. A dramatic 6-fold increase in yield of product from *N*-furylacryloyl-Phe-OEt and either H-Phe-NH_2 or H-Gly-NH_2 under catalysis by carboxypeptidase Y resulted when the pressure was increased to 200 MPa[168].

Chemically modified or genetically engineered enzymes have been seldom used in peptide synthesis. α-Chymotrypsin in which the ε-amino groups had been blocked by Z-groups gave higher yields of peptides than did the native enzyme in both 1- and 2-phase systems[169]. That this may be the result of increasing the hydrophobicity of the enzyme stems from the observation that the esterification of Ac-Tyr-OH by α-chymotrypsin which had been modified by coupling with neutral liposaccharides, lipocarboxylic amphiphiles or octanal was accelerated[170]. Moreover, the last form of modified enzyme was more effective when adsorbed on a hydrophobic support such as chitin or octyl-Sepharose than when attached to the more hydrophilic Celite. Although anhydro-subtilisin is devoid of hydrolytic activity, it is reported to be a good catalyst for peptide synthesis[171]. This is surprising since aryl esters were used as acylating substrates and it would be interesting to know more about the mechanism of this process. Some genetically engineered variants of subtilisin have been developed for segment coupling[172]. Some of the variants were found to be stable and active in anhydrous $CHONMe_2$.

Comparison and explanation of the two mechanisms of enzyme-catalysed peptide synthesis continues[173-178]. It is recognized that once the substrate is bound to the enzyme the catalytic machinery works as well in low concentrations of water as it does in wholly aqueous solution. A decrease in rate is due to a combination of a high K_s value and a low concentration of added nucleophile compared to that of the competing nucleophile, water[173-174]. In trypsin-catalysed reactions, equilibrium-controlled synthesis from Bz-Arg-OH and kinetically controlled synthesis from Bz-Arg-OEt with various added nucleophiles have been compared[175]. In the former method, a high concentration of apolar solvent greatly inhibits adsorption of nucleophile, so an excess of the latter is recommended. The presence of salts such as $(NH_4)_2SO_4$ as a water-ordering agent greatly reduces the rate of hydrolysis of acyl-enzyme. In reactions catalysed by α-chymotrypsin, it was shown that in 2-phase systems, the rate of mass transfer between phases is rate-determining whereas in a homogeneous solution, peptide synthesis was kinetically controlled[176].

A wider range of proteinases is now being used and there are papers describing the use of angiotensin I converting enzyme[156], bacterial proteinases[179-181], a fungal carboxypeptidase[182] and chymosin[183]. A comparative kinetic study using five *N*-protecting groups led to Boc and Z being recommended for syntheses catalysed by α-chymotrypsin[167]. In high concentrations of organic solvents, α-chymotrypsin loses its stereospecificity at the S′ site since Ac-L-Tyr-OEt and D-Phe-NH_2 gave a very high yield of peptide[184]. A study of the specificity at the S′ site of papain has also been reported[185]. There has been a fairly detailed study[159] of the specificity of carboxypeptidase Y in organic solvents using *N*-[3-(2-furyl)acryloyl]-X-OMe as one of the substrates. The best yields were obtained when X = Phe and the worst when X = Ile or Val.

Examples of the synthesis of particular peptides are legion and the foregoing literature should be consulted. Other small peptides have been synthesized[186-189]. Perhaps the most interesting of these[189] describes the continuous synthesis of an 'Aspartame' precursor, Z-Asp-Phe-OMe, in a pulsed extraction column reactor. Other examples of syntheses achieved include *N*-acylenkephalin amides involving three enzyme-catalysed steps[190], a (5+4) semisynthesis of a 9-residue insect neuropeptide[146,191], a

(8 + 5) semisynthesis of α-MSH (the *N*-terminal octapeptide was fully deprotected apart from the *N*-Ac group before trypsin-catalysed coupling was effected)[162], a conversion of [Ala^{15,29}]-GRF(4-29)-OH into [Ala^{15}]-GRF-NH_2 by incubation with H-Arg-NH_2 in presence of carboxypeptidase Y[192], a (11 + 18) semisynthesis of GRF amide (protecting groups were retained on Lys^{12}, Arg^{20}, Lys^{21} and Arg^{29} until after the trypsin-catalysed step)[193], a trypsin-catalysed coupling of H-Leu-NH_2 to hGRF(1-43)-OH, which had been made by SPPS, to give hGRF(1-44)-NH_2[194], a (3 + 7) semisynthesis of GnRH[195], a further synthesis of GnRH involving several enzyme-catalysed steps[196], a synthesis of alamethicin[197], which has an unfavourable amino-acid composition for enzyme-catalysed synthesis, and a (15 + 9) semisynthesis of a fragment (17-40) of HbA[198], a process in which the driving force is the tendency by the solvent, Pr^nOH, to induce helical formation in the product. A thionodipeptide has been extended by one residue at the *C*-terminus using α-chymotrypsin[199]. Peptides or proteins can be labelled with a [^{3}H]-amino acid at the *C*-terminus by transpeptidation using carboxypeptidase Y[200]. This is a useful technique in sequence determination. There are two novel methods for introducing an amide group at the *C*-terminus. Firstly, a protected peptide ester can be coupled to 1,3,5-trimethoxybenzylamine using either papain or subtilisin[201]. The substituted benzyl substituent can be removed with CF_3CO_2H. Alternatively, the peptide ester can be coupled to 2-nitrobenzylamine or an α-substituted derivative of this using carboxypeptidase Y[202]. The nitrobenzyl group is then removed by photolysis under a medium pressure Hg lamp.

There are some examples of the use of enzymes to deprotect peptide derivatives after enzymic or chemical synthesis. *N*-Phenylacetyl groups or *p*-substituted analogues were removed by a penicillium amidase[203,204]. Ester groups have been removed from the *C*-terminus of peptide derivatives. Large groups can be removed by a lipase[204,205]. Because of the lability of Fmoc groups to bases, the enzymic removal of ester groups is a useful technique[206,207]. Thermitase has been used to cleave t-butyl esters of glycopeptides without disturbing the carbohydrate moiety[208]. A *C*-terminal amide group can be hydrolysed by a peptidase amidase from the flavedo of orange peel[209,210]. A somewhat more ambitious experiment involved the removal of the *N*-terminal tripeptide from the analogue of substance P, RLRRPKPQQFFRLR-NH_2 by trypsin[211]. This is a model of the processing of a peptide hormone precursor at a bond between two basic amino acid residues.

Although they do not involve the use of proteolytic enzymes, mention must be made of two syntheses where peptides have been glycosylated, one by a β-galactosidase[212], the other by an oligosaccharyl transferase[213].

2.9 *Miscellaneous reactions related to peptide synthesis.*

The researches reported in this section are unrelated to one another but have been selected on purely personal grounds for having evoked special interest in the Reporter. A new stereoselective route[214] to dipeptides involves the oximation of αβ-unsaturated amides (Scheme 7) with BuONO/R_3SiH in the presence of a catalytic amount of [*NN'*-bis(2-ethoxycarbonyl-3-oxybutylidene)-ethylenediaminato]cobalt(II) complex [Co(eobe)] (**29**). The oxime is reduced by SmI_2/MeOH and *N*-protected *in situ.* Another route to dipeptides[215] involves the addition of *N*-(nitroacetyl)amino acid derivatives to Michael acceptors followed, for example, by reductive acetylation (Scheme 8). A more specialized synthesis[216] of lactam-based dipeptide derivatives (**30**) has been reported. *C*-Terminal amino acids in peptides can be converted into taurine derivatives via *S*-acetylthio- or halogeno intermediates[217]. The terminal ester group is reduced to -CH_2OH with $LiBH_4$ and then converted into the alkyl bromide with CBr_4/Ph_3P. Finally, the bromide is treated with KSAc. Peptide alcohols can be converted into amines[218] by mesylation followed by reaction with NaN_3. Hydrogenation of the resultant azides gives the corresponding amines. Coupling reactions involving the Z- or Fmoc-derivatives of *N*-methylamino acids proceed straight-forwardly when PyBroP or PyCloP are used[219]. Boc-*N*-Methyl amino acids, on the other hand, form the corresponding *N*-methyl-*N*-carboxyanhydrides as a substantial byproduct. *N*-Alkoxycarbonyl derivatives of aminodicarboxylic acids give crystalline internal anhydrides when treated with $EtN{=}C{=}N(CH_2)_3NEt_2$ or with $ClCO_2Me$ in the presence of *N*-methylmorpholine[220]. The side chains of Asp residues in peptides also cyclize to the succinimide derivatives in aqueous solution especially at pH4-5[221]. This might be a nuisance in coupling reactions catalysed by aspartic proteinases. Reaction of 2-Z-imino-3,3,3-trifluoro-propionate with CH_3COCl/Et_3N gives an anionic intermediate by nucleophilic attack of the acetyl chloride anion[222]. The intermediate cyclizes by a 6-*exo-trig* process according to Baldwin's rules to give methyl 2-benzyloxy-6-oxo-4,5-dihydro-4-trifluoromethyl-1,3-oxazine-4-carboxylate (**31**), a β-activated equivalent for 2-trifluoromethylaspartic acid. This affords a convenient route to its peptides. Selective hydrogenolysis of the *N*-Z-group in (**32**; R = Z) occurred when using Pd/H_2 at 2-3 atm in $CHONMe_2$[223]. The phenoxy group was unaffected but, in addition, some of the ε-methylated compound was formed. The latter byproduct was not formed when hydrogenolysis was carried out in aqueous acetic acid

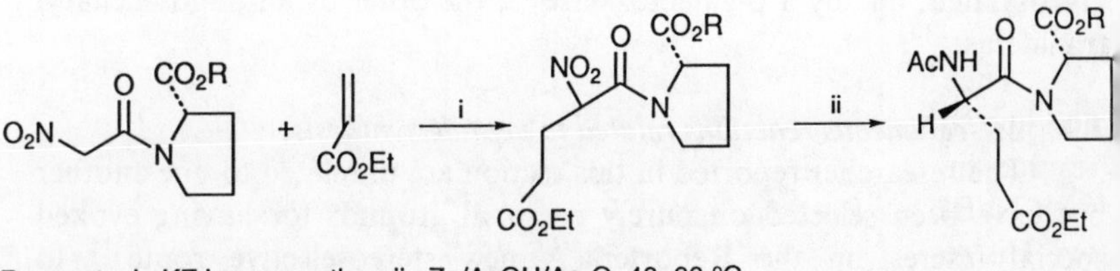

Reagents: i, KF in crown ether; ii, Zn/AcOH/Ac_2O, 40–60 °C

Scheme 8

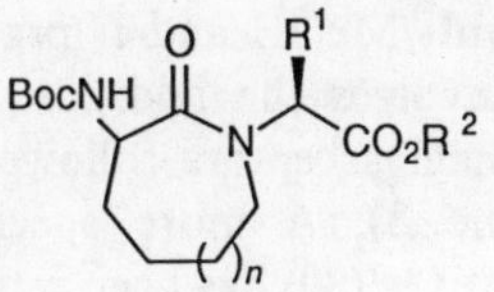

(30)

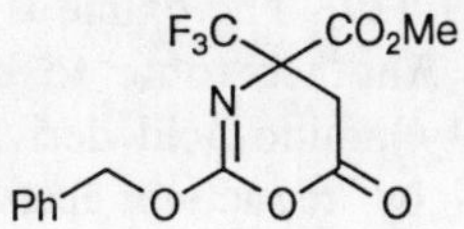

(31)

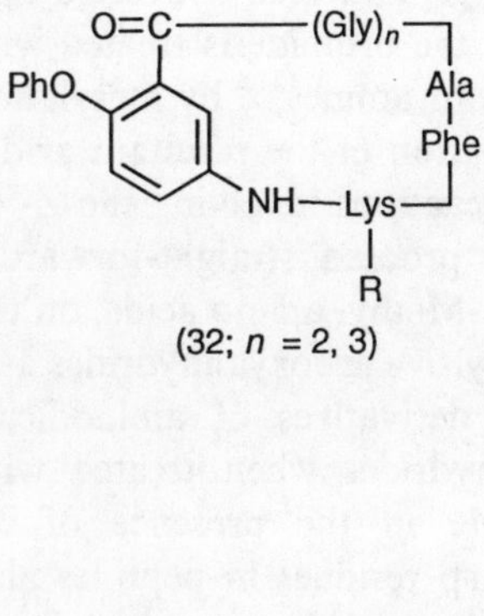

(32; n = 2, 3)

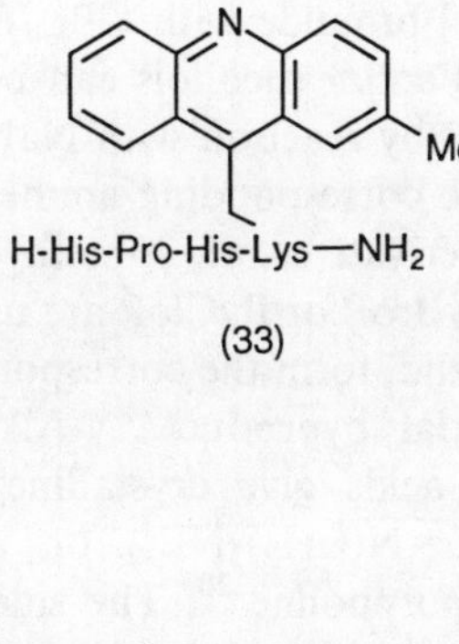

H-His-Pro-His-Lys—NH_2

(33)

during 30 min, but prolonged reaction removed the *O*-phenyl group as well.

The Schiff base produced by the condensation of 2-methyl-9-acridinecarboxaldehyde with the ε-amino group of the *C*-terminal residue in the peptide Fmoc-His-Pro-His-Lys-NH_2 has been reduced with $NaBH_3CN$ and then deprotected with piperidine to give (**33**). The two His residues are designed to simulate the active site of RNase A and the acridine ring is able to intercalate in and bind rRNA so that the latter is hydrolysed albeit in a random manner[224]. Analogues of proteins can be constructed by a site-specific condensation of two fragments[225]. One fragment is required to have Ser or Thr at its *N*-terminus and this is oxidized with periodate to the keto acyl derivative. The other peptide fragment is converted into an acyl hydrazide by a proteinase-catalysed reaction, although this could just as easily arise from a nonenzymic reaction. The two fragments are then allowed to interact and the resultant hydrazone is reduced with $NaBH_3CN$ to stabilize it. This approach has considerable potential in peptide-drug design since one fragment can function as a homing device while the other contains some moiety to destroy the receptor. Another possible approach to peptide-based drugs is based on the synthesis of peptoids[226,227], which are defined as oligomers of *N*-substituted Gly residues where the *N*-substituents simulate the side chains of amino acids in a natural peptide. The peptoids often have an affinity for the receptor similar to that of the native peptide. This is an important discovery since the peptoids are likely to be rather resistant to proteolysis.

3 Selected examples of peptide syntheses

As noted in last year's Report[1], only a few references are cited here and length of sequence synthesized is not necessarily a qualification for inclusion. Most of the citations concerning the synthesis of particular peptides will be found in the appendix.

As a demonstration that solution-phase synthesis still justifiably finds favour, the synthesis of 40 g batches of an anticoagulant decapeptide (MDL 28050): H-Suc-Tyr-Glu-Pro-Ile-Pro-Glu-Glu-Ala-Cha-D-Glu-OH is impressive[228]. Two pentapeptide segments were synthesized and then coupled to give an overall yield of 20% of 98% pure product. The construction of hEGF (53 residues)[229,230] and elafin[231], an elastase inhibitor, are more ambitious examples of solution-phase syntheses with fragment coupling.

Apart from the current clinical and social interest in Alzheimer's disease, the synthesis of the 42-residue βA4 amyloid protein presents considerable synthetic difficulties because of its insolubility. SPPS has given products that contained amino-acid deletions and that were difficult to purify. A new synthesis[232] commenced with the construction of the *C*-terminal pentapeptide by the Merrifield method and the coupling to it of other fragments which were prepared on the Kaiser oxime resin. As with Alzheimer's disease, the HIV proteinase continues to be the focus of attention for peptide chemists. A fully active engineered form, with Cys^{67} and Cys^{95} replaced by Aba and with the peptide bond between residues 52 and 53 replaced by a thiolester moiety, has been synthesized[233] from the fragments:

$$H\text{-}Pro^{1}...Ile^{50}NHCH_2COSH \text{ and } BrCH_2CO\text{-}Phe^{53}...Phe^{99}\text{-}OH.$$

$[Aba^{67,95}]$HIV proteinase has also been completely constructed from D-amino acids and the product was fully active but only on all-D-substrates[234]. A similar enzyme, the Simian immunodeficiency virus proteinase, has been synthesized[235] with an octapeptide extension found in the SIV *gag/pol* precursor polyprotein. Biotin was attached to the *N*-terminus to facilitate affinity chromatography. The extension peptide could be removed autocatalytically by cleavage of the Ala^{-1}-Pro^{1} bond. Horse heart apocytochrome c has been constructed[236] from fragment (1–65), with Met^{65} replaced by homoserine lactone, and fragment (66–104) in the presence of fragment (1–25). The interesting point here concerns the presence of the last fragment which forms a complex with the two larger fragments and thereby facilitates the formation of the peptide bond from the homoserine lactone.

4 Appendix: A list of syntheses reported in 1992

The syntheses are listed under the name of the peptide/protein to which they relate, but no arrangement is attempted under the subheading. In some cases, closely related peptides are listed together.

Peptide/protein *Ref.*

4.1 *Natural peptides, proteins and partial sequences.*

ACP

ACP(65–74) 108,110,121,124

ACTH

SPPS of fragment (4-11) on cellulose support 110

Acyl-CoA oxidase

C-Terminal peptides as minimal peroxisome-targeting signal 237

Agelenin

Synthesis to confirm structure 238

Alamethecin	
Fragment (7–17) involving enzyme-catalysed coupling	197
Anaphylatoxin C5a	
C-terminal octapeptide analogues	239
Angiotensin	
SPPS on cellulose support	110
New antagonist	240
Analogue containing 2,4-methanoproline	241
Annexin-II	
Nonapeptide fragment of putative site for binding F actin	242
Antibiotic peptides	
Bactenecin and several analogues	243
SPPS of defensin NP-1	244
Analogues and hybrids of cecropins and melittin	245
Antifreeze peptides	
Ala-rich antifreeze peptides	246
Atrial natriuretic peptide (factor), ANP, ANF, atriopeptin	
Brain natriuretic peptides	88,102
Analogues containing 3 identified essential regions	247
13-Residue analogue	248
Importance of *C*-terminal region	249
Bombesin	
Analogue of *C*-terminal sequence	250
Analogues with -CH_2NH- pseudopeptide bonds	251
Pseudononapeptide antagonists	252
Bombesin receptor antagonists with irreversible alkylating properties	253
C-Terminal nonapeptide with a thionopeptide bond	254
Bombyxin	
SPPS of A and B chains of bombyxin-IV	255
Bombyxin-II and its disulfide isomers	256
Bradykinin	
Antagonists containing bissuccinimidoalkane moiety	257
Antagonists with bulky substituents	258
(D-Phe^7,Leu^8)bradykinin	259
Analogue containing 2,4-methanoproline	241
Brain peptides	
N-terminal fragment of β-amyloid protein	260
γ-L-Glutamyltaurine	261
Caerins	
Structure and synthesis	262
Caliseptine	

4.2 *Sequential oligo- and poly-peptides.*

Polypeptides containing azetidine-2-CO_2H and Pro	398
Polypeptide formation from Asn dipeptides	399
α-*N*-Polyacyl-ε-poly-L-lysine derivatives	400
Synthesis of Ser polymers	401
Synthesis and size-exclusion chromatography of poly-L-amino acids	402

4.3 *Enzyme substrates and inhibitors.*

2-Aminobenzoyl-peptide-2,4-dinitroanilinoethylamides as fluorogenic substrates for endopeptidases	403
Intramolecularly quenched substrates of proteinases	404,405
N- and *C*-Terminal extensions of peptidyl-αα-difluoroalkyl ketones	406
Tripeptide trifluoromethylketones as HLE inhibitors	407
Peptidyl pentafluoroethyl ketones as HLE inhibitors	408
HLE inhibitors containing α,α-difluorostatone	409
Peptidyl α-ketobenzoxazoles as elastase inhibitors	410
2-(Peptidamido)cyclobutananones as elastase inhibitors	411
Nazumamide A, a tetrapeptide thrombin inhibitor	412
Fragment of bovine panreatic trypsin inhibitor	93
SPPS of bovine pancreatic trypsin inhibitor	413
Disulphide fragment of Bowman-Birk inhibitor	414
N-Terminal tetrapeptide analogues of fibrin α-chain as thrombin inhibitors	415
Mimetics of thrombin-bound segment of fibrinopeptide A	416
N-Terminal peptides from fibrinogen Lille as thrombin substrates	417
Tos-Phe(F)-Arg-OMe as substrates of thrombin	418
Cyclic analogues of hirudin (55–65)	419
Hirutonins, analogues of hirudin as thrombin inhibitors	420
Phosphonopeptides as thrombin inhibitors	421
A peptide model for the heparin-binding site of anti-thrombin III	422
RGD peptides as antiplatelet thrombotics	423,424
Fragment of platelet receptor site for thrombin	425
Synthesis of anticoagulant decapeptide MDL 28050	228
Thrombin inhibitors containing *C*-terminal boronic analogue of arginine	426
Ketomethylene pseudopeptide thrombin inhibitors	427
α-Hydroxy and of α-ketoester functionalized thrombin inhibitors	428

21 peptides derived from factor X sequence as potential inhibitors of prothrombin activation	429
Agonists for activation of thrombin receptor	430,431
Analogues of kininogen fragment as kallikrein inhibitors	432
Peptides of ε-aminocaproic acid as potential antifibrinolytic agents	433
Substrates for HIV-1 proteinase	434
Inhibitors of HIV-1 proteinase	435
HIV-1 proteinase inhibitors containing allophenylnorstatine	436
Slow-binding transition-state inibitor of HIV proteinase	437
A tripeptidyl epoxide as an irreversible inhibitor of HIV-1 proteinase	438
HIV-1 proteinase inhibitors containing a hydroxyethyl secondary amine transitions state isostere	439-443
Kinetic and modelling studies of S_3-S'_3 subsites of HIV proteinase	444
C_2-Symmetric phosphinic acid inhibitors of HIV proteinase	445
Conformationally restricted analogues of pepstatin	446
Potent, orally active renin inhibitors	447
Renin inhibitors containing conformationally restricted isosteres of Phe	448
Renin inhibitors containing a phenylcyclopropanecarboxamide moiety	449
Renin inhibitors containing heterocyclic derivatives of Phe	450
Renin inhibitors containing 2-amino-4-thiazolyl moiety	451
Substrate-analogue inhibitors of canine renin	452
Human renin inhibitors with *C*-termini derived from amides and esters of α-mercaptoalkanoic acids	453
Renin inhibitors containing homostatine	454,455
Retro-inverse tripeptide renin inhibitors	456
Inhibitors of penicillopepsin containing difluoro-statine and -statone	457
Inhibitors of Rhizopus pepsin	458
Prodrug inhibitors of enkephalin-degrading enzymes	459
Inhibitors of enkephalin-degrading aminopeptidase	460,461
Thiorphan and analogues	462
A mechanism-based inactivator of ACE	463
Zabicipril, a potent proinhibitor of ACE	464
Collagenase substrate	116
Hydroxamic acid derivatives of peptides as inhibitors of collagenase	465

References

1. D.T. Elmore, *Specialist Periodical Report: Amino Acids and Peptides,* 1993, **24**, 81.
2. G.A. Grant (ed.), 'Synthetic Peptides: a User's Guide', W.H. Freeman & Co. New York, 1992.
3. T. Wieland and M. Bodanszky, 'The World of Peptides. a Brief History of Peptide Chemistry', Springer-Verlag, Berlin, 1991.
4. A.R. Rees, M.J.E. Sternberg and R. Wetzel, "Protein Engineering: a Practical Approach", IRL Press at Oxford University Press, Oxford, 1992.
5. M. Bodanszky, *Pept.Res.*, 1992, **5**, 134.
6. D. Wang, *Youji Huaxue*, 1992, **12**, 345.
7. S. Aimoto, *Tanpakushitsu Kakusan Koso*, 1992, **37**, 410.
8. S. Tian and M. Cai, *Huaxue Tongbao*, 1992, 17.
9. G. Jung and A.G. Beck-Sickinger, *Angew.Chem.,Int.Ed.*, 1992, **31**, 367.
10. G. Jung, *Angew.Chem., Int.Ed.*, 1992, **31**, 1457.
11. S. Birnbaum and K. Mosbach, *Curr.Opinion Biotechnol.*, 1992, **3**, 49.
12. P. Hermann, *Biomed.Biochim.Acta*, 1991, **50**, S19.
13. H.D. Jakubke, *Kontakte (Darmstadt), 1992,* **1**, 46.
14. R. Matsuno, Y. Kimura, S. Adachi and K. Nakanishi, *Ann.N.Y.Acad.Sci.*, 1992, **672**, 363.
15. S. Yagisawa, S. Watanabe and Y. Sato, *Biomed.Biochim.Acta*, 1991, **50**, S187.
16. Y. Nakahara and T. Ogawa, *Yuki Gosei Kagaku Kyokaishi*, 1992, **50**, 410.
17. M. Sisido, *Prog.Polym.Sci.*, 1992, **17**, 699.
18. M. Shishido, *Kobunshi Kako*, 1991, **40**, 587.
19. T.J. Blacklock, R.F. Shuman, J.W. Butcher, P. Sohar, T.R. Lamanec, W.E. Shearin and E.J.J. Grabowski, *Chem.Ind. (Dekker)*, 1992, **47**, 45.
20. F. Jacquemotte, *Chim.Nouv.*, 1992, **10**, 1081.
21. F. Jacquemotte, *Chim.Nouv.*, 1992, **10**, 1109.
22. C. Di Bello, C. Vita and L. Gozzini, *Biochem.Biophys.Res.Commun.*, 1992, **183**, 258.
23. E.E. Buellesbach, *Kontakte (Darmstadt)*, 1992, **1**, 21.
24. S.M. Hecht, *Acc.Chem.Res.*, 1992, **25**, 545.25. D.P. Weiner, *Genet.Eng.Biotechnol.*, 1992, **12**, 9.
26. S. Takayuki, *Kagaku Zokan (Kyoto)*, 1992, **121**, 129.
27. J.S. Bajwa, *Tetrahedron Lett.*, 1992, **33**, 2955.
28. W. Oppolzer and P. Lienard, *Helv.Chim.Acta*, 1992, **75**, 2572.
29. R. Ramage and R. Raphy, *Tetrahedron Lett.*, 1992, **33**, 385.
30. H.L. Ball and P. Mascagni, *Int.J.Peptide Protein Res.*, 1992, **40**, 370.
31. M. Acedo, F. Albericio and R. Eritja, *Tetrahedron Lett.*, 1992, **33**, 4989.
32. W. Voelter, G. Breipohl, C. Tzougraki and E. Jungfleisch-Turgut, *Coll. Czech.Chem.Commun.*, 1992, **57**, 1707.

33. T. Pietzonka and D. Seebach, *Angew.Chem., Int.Ed.*, 1992, **31**, 1481.
34. R. Hoffman and H.-O. Kim, *Tetrahedron*, 1992, **48**, 3007.
35. M. Nakatani, T. Nakata, N. Mukaiyama, K. Kouge and H. Okai, *Chem. Express*, 1992, **7**, 761.
36. W. Zeng, B. Hemmasi and E. Bayer, *Tetrahedron Lett.*, 1992, **33**, 5945.
37. T. Sato, J. Otera and H. Nozaki, *J.Org.Chem.*, 1992, **57**, 2166.
38. B. Kundu, *Tetrahedron Lett.*, 1992, **33**, 3193.
39. Z. Wang, E. Wang, W. Li and R. Liu, *Jilin Daxue Ziran Kexue Xuebao*, 1991, 118.
40. V.G. Zhuravlev, A.A. Mazurov and S.A. Andronati, *Coll.Czech.Chem. Commun.*, 1992, **57**, 1495.
41. A. M. Castaño and A.M. Echavarren, *Tetrahedron.*, 1992, **48**, 3377.
42. M. Pátek and M. Lebl, *Coll.Czech.Chem.Commun.*, 1992, **57**, 508.
43. E. Masiukiewicz, B. Rzeszotarska and J. Szczerbaniewicz, *Org.Prep. Proced.Int.*, 1992, **24**, 191.
44. J. Rizo, F. Albericio, E. Giralt and E. Pedroso, *Tetrahedron Lett.*, 1992, **33**, 397.
45. T.M. Rana, M. Ban and J.E. Hearst, *Tetrahedron Lett.*, 1992, **33**, 4521.
46. G.K. Tóth and B. Penke, *Synthesis*, 1992, 361.
47. P.J. Belshaw, J.G. Adamson and G.A. Lajoie, *Synth.Commun.*, 1992, **22**, 1001.
48. J.S. Davies, C.L. Higginbotham, E.J. Tremeer, C. Brown and R.C. Treadgold, *J.Chem.Soc., Perkin Trans. 1*, 1992, 3043.
49. P.M. Fischer, *Tetrahedron Lett.*, 1992, **33**, 7605.
50. K. Tanemura, T. Suzuki and T. Horaguchi, *J.Chem.Soc., Perkin Trans. 1*, 1992, 2997.
51. M.C. Pirrung and D.S. Nunn, *Bioorg.Med.Chem.Lett.*, 1992, **2**, 1489.
52. S. Futaki, T. Taike, T. Akita and K. Kitagawa, *Tetrahedron*, 1992, **48**, 8899.
53. R. Matsueda, S. Higashida, F. Albericio and D. Andreu, *Pept.Res.*, 1992, **5**, 262.
54. M.C. Munson, C. García-Echeverría, F. Albericio and G. Barany, *J.Org. Chem.*, 1992, **57**, 3013.
55. J. Blake, B.A. Woodworth, L. Litzi-Davis and W.L. Cosand, *Int.J.Peptide Protein Res.*, 1992, **40**, 62.
56. M. Royo, C. García-Echeverría, E. Giralt, R. Eritja and F. Albericio, *Tetrahedron Lett.*, 1992, **33**, 2391.
57. A.S. Verdini, P. Lucietto, G. Fossati and C. Giordani, *Tetrahedron Lett.*, 1992, **33**, 6541.
58. P.M. Fischer, K.V. Retson, M.I. Tyler and M.E.H. Howden, *Int.J.Peptide Protein Res.*, 1992, **40**, 19.
59. C.S. Pande, K.D. Bassi, N. Jain, A. Dhar and J.D. Glass, *J.Biosci.*, 1991, **16**, 127.
60. B. Sax, F. Dick, R. Tanner and J. Gosteli, *Pept.Res.*, 1992, **5**, 245.
61. M. Friede, S. Denery, J. Neimark, S. Kieffer, H. Gausepohl and J.P. Briand, *Pept.Res.*, 1992, **5**, 145.
62. D. Shah, A. Schneider, S. Babler, R. Gandhi, E. Van Noord and E. Chess, *Pept.Res.*, 1992, **5**, 241.
63. H. Naharissoa, V. Sarrade, M. Follet and B. Calas, *Pept.Res.*, 1992, **5**, 293.
64. A. Mehta, R. Jaouhari, T.J. Benson and K.T. Douglas, *Tetrahedron Lett.*, 1992, **33**, 5441.
65. G. Huebener, W. Goehring, H.J. Musiol and L. Moroder, *Pept.Res.*, 1992, **5**, 287.
66. A.J. Pallenberg, *Tetrahedron Lett.*, 1992, **33**, 7693.
67. R.N. Zuckermann and S.C. Banville, *Pept.Res.*, 1992, **5**, 169.
68. J. Savrda and M. Wakselman, *J.Chem.Soc., Chem.Commun.*, 1992, 812.
69. L.A. Carpino and E.-S.M.E. Mansour, *J.Org.Chem.*, 1992, **57**, 6371.

70. P. Stefanowicz and I.Z. Siemion, *Pol.J.Chem.*, 1992, **66**, 111.
71. Kh. Nedev, S. Minchev and N. Stoyanov, *Dokl.Bulg.Akad.Nauk*, 1991, **44**, 61.
72. Kh. Nedev and S. Minchev, *Izv.Khim.*, 1991, **24**, 363.
73. V.F. Pozdnev, *Int.J.Peptide Protein Res.*, 1992, **40**, 407.
74. C. Poulos, C.P. Ashton, J. Green, O.M. Ogunjobi, R. Ramage and T. Tsegenidis, *Int.J.Peptide Protein Res.*, 1992, **40**, 315.
75. T. Miyazawa, T. Donkai, T. Yamada and S. Kuwata *Int.J Peptide Protein Res.*, 1992, **40**, 49.
76. L.C. Dorman and W.R. Shiang and P. Meyers, *Synth.Commun.*, 1992, **22**, 3257.
77. Y.-M. Li, Y.-W. Yin and Y.-F. Zhao, *Int.J.Peptide Protein Res.*, 1992, **39**, 375.
78. C.-O. Chan, C.J. Cooksey and D. Crich, *J.Chem.Soc., Perkin Trans. 1*, 1992, 777.
79. T. Miyazawa, T. Otomatsu, Y. Fukui, T. Yamada and S. Kuwata, *Int.J. Peptide Protein Res.*, 1992, **39**, 237.
80. T. Miyazawa, T. Otomatsu, Y. Fukui, T. Yamada and S. Kuwata, *Int.J. Peptide Protein Res.*, 1992, **39**, 308.
81. A. Katoh, J. Ohkanda, Y. Itoh and K. Mitsuhashi, *Chem.Lett.*, 1992, 2009.
82. F.C. Frontin, F. Guendouz, R. Jacquier and J. Verducci, *Bull.Soc.Chim. Fr.*, 1992, **129**, 463.
83. T. Miyazawa, T. Otomatsu, T. Yamada and S. Kuwata, *Int.J.Peptide Protein Res.*, 1992, **39**, 229.
84. H. Kuroda, S. Kubo, N. Chino, T. Kimura and S. Sakakibara, *Int.J. Peptide Protein Res.*, 1992, **40**, 114.
85. H. Kuroda, Y.-N. Chen, T. Kimura and S. Sakakibara, *Int.J.Peptide Protein Res.*, 1992, **40**, 294.
86. N. Nishino, H. Mihara, Y. Makinose and T. Fujimoto, *Tetrahedron Lett.*, 1992, **33**, 7007.
87. S. Chen and J. Xu, *Tetrahedron Lett.*, 1992, **33**, 647.
88. Y. Kiso, Y. Fujiwara, T. Kimura, A. Nishitani and K. Akaji, *Int.J. Peptide Protein Res.*, 1992, **40**, 308.
89. J.R. Spencer, V.V. Antonenko, N.G.J. Delaet and M. Goodman, *Int.J. Peptide Protein Res.*, 1992, **40**, 282.
90. N.L. Benoiton, Y.C. Lee, R. Steinaur and F.M.F. Chen, *Int.J.Peptide Protein Res.*, 1992, **40**, 559.
91. J.S. Davies, C. Enjalbal and G. Llewellyn, *J.Chem.Soc., Perkin Trans. 2*, 1992, 1225.
92. H. Kunz and R. Kullmann, *Tetrahedron Lett.*, 1992, **33**, 6115.
93. N. Fotouhi, B.R. Bowen and D.S. Kemp, *Int.J.Peptide Protein Res.*, 1992, **40**, 141.
94. S.K. Ghosh, U. Singh and V.R. Mamdapur, *Tetrahedron Lett.*, 1992, **33**, 805.
95. K. Joshi, J. Bao, A.S. Goldman and J. Kohn, *J.Amer.Chem.Soc.*, 1992, **114**, 6649.
96. L.A. Andreeva, L.Y. Alfeeva, V.N. Potaman and V.N. Nezavibatko, *Int.J. Peptide Protein Res.*, 1992, **39**, 493.
97. J.J. Landi and H.R. Brinkman, *Synthesis*, 1992, 1093.
98. N.L. Subasinghe and R.L. Johnson, *Tetrahedron Lett.*, 1992, **33**, 2649.
99. J.R. Miller, S.R. Pulley, L.S. Hegeduo and S. DeLombaert, *J.Amer.Chem. Soc.*, 1992, **114**, 5602.
100. E.E. Büllesbach and C. Schwabe, *Tetrahedron Lett.*, 1992, **33**, 5881.
101. P. Bishop and J. Chmielewski, *Tetrahedron Lett.*, 1992, **33**, 6263.
102. K. Akaji, T. Tatsumi, M. Yoshida, T. Kimura, Y. Fujiwara and Y. Kiso, *J.Amer.Chem.Soc.*, 1992, **114**, 4137.
103. K.C. Pugh, E.J. York and J.M. Stewart, *Int.J.Peptide Protein Res.*, 1992, **40**, 208.
104. C. Mendre, V. Sarrade and B. Calas, *Int.J.Peptide Protein Res.*, 1992, **39**, 278.

105. A.F. Coffey and T. Johnson, *Int.J.Peptide Protein Res.*, 1992, **39**, 419.
106. S.C. Story and J.V. Aldrich, *Int.J.Peptide Protein Res.*, 1992, **39**, 87.
107. E. Nicolás, J. Clemente, M. Perelló, F. Albericio, E. Pedroso and E. Giralt, *Tetrahedron Lett.*, 1992, **33**, 2183.
108. M. Meldal, *Tetrahedron Lett.*, 1992, **33**, 3077.
109. J.C. Hendrix, J.T. Jarrett, S.T. Anisfeld and P.T. Lansbury, *J.Org. Chem.*, 1992, **57**, 3414.
110. D.R. Englebretsen and D.R.K. Harding, *Int.J.Peptide Protein Res.*, 1992, **40**, 487.
111. J. Kochansky and R.M. Wagner, *Tetrahedron Lett.*, 1992, **33**, 8007.
112. N.A. Solé and G. Barany, *J.Org.Chem.*, 1992, **57**, 5399.
113. S. Sawada, K. Yasui and S. Takahashi, *Biosci.Biotechnol.Biochem.*, 1992, **56**, 1506.
114. S.B. Katti, P.K. Misra, W. Haq and K.B. Mathur, *J.Chem.Soc.,Chem. Commun.*, 1992, 843.
115. F. Rabanal, E. Giralt and F. Albericio, *Tetrahedron Lett.*, 1992, **33**, 1775.
116. D.T. Elmore, D.J.S. Guthrie, A.D. Wallace and S.R.E. Bates, *J.Chem. Soc., Chem.Commun.*, 1992, 1033.
117. J. Mery, J. Brugidou and J. Derancourt, *Pept.Res.*, 1992, **5**, 233.
118. K. Akaji, N. Kuriyama, T. Kimura, Y. Fujiwara and Y. Kiso, *Tetrahedron Lett.*, 1992, **33**, 3177.
119. J. Sueiras-Diaz and J. Horton, *Tetrahedron Lett.*, 1992, **33**, 2721.
120. M. Schnölzer, P. Alewood, A. Jones, D. Alewood and S.B.H. Kent, *Int.J. Peptide Protein Res.*, 1992, **40**, 180.
121. H.-M. Yu, S.-T. Chen and K.-T. Wang, *J.Org.Chem.*, 1992, **57**, 4781.
122. D.S. Perlow, J.M. Erb, N.P. Gould, R.D. Tung, R.M. Freidinger, P.D. Williams and D.F. Veber, *J.Org.Chem.*, 1992, **57**, 4394.
123. K.M. Sivanandaiah, V.V.S. Babu and C. Renukeshwar, *Int.J.Peptide Protein Res.*, 1992, **39**, 201.
124. C.-Y. Wu, W.-C. Chan, H.-M. Yu, S.-T. Chen and K.-T. Wang, *J.Chin. Chem.Soc.(Taipei)*, 1992, **39**, 195.
125. C. Hyde, T. Johnson and R.C. Sheppard, *J.Chem.Soc., Chem.Commun.*, 1992, 1573.
126. J. Bedford, C. Hyde, T. Johnson, W. Jun, D. Owen, M. Quibell and R.C. Sheppard, *Int.J.Peptide Protein Res.*, 1992, **40**, 300.
127. M. Beyermann and M. Bienert, *Tetrahedron Lett.*, 1992, **33**, 3745.
128. A. Pessi, V. Mancini, P. Filtri and L. Chiappinelli, *Int.J.Peptide Protein Res.*, 1992, **39**, 58.
129. S.C.F. Milton, W.F. Brandt, M. Schnölzer and R.C. deL. Milton, *Biochemistry.*, 1992, **31**, 8799.
130. J.G. Adamson, T. Hoang, A. Crivici and G.A. Lajoie, *Anal.Biochem.*, 1992, **202**, 210.
131. S.H. Grode, D.S. Strother, T.A. Runge and P.J. Dobrowolski, *Int.J. Peptide Protein Res.*, 1992, **40**, 538.
132. A. Trzeciak and W. Bannwarth, *Tetrahedron Lett.*, 1992, **33**, 4557.
133. S.S. Wang, B.S.H. Wang, J.L. Hughes, E.J. Leopold, C.R. Wu and J.P. Tam, *Int.J.Peptide Protein Res.*, 1992, **40**, 344.134. C.R. Baeza and A. Undén, *Int.J.Peptide Protein Res.*, 1992, **39**, 195.
135. C.J. Gray, M. Quibell, N. Baggett and T. Hammerle, *Int.J.Peptide Protein Res.*, 1992, **40**, 351.
136. A.M. Murphy, R. Dagnino, P.L. Vallar, A.J. Trippe, S.L. Sherman, R.H. Lumpkin, S.Y. Tamura and T.R. Webb, *J.Amer.Chem.Soc.*, 1992, **114**, 3156.
137. R. Frank, *Tetrahedron*, 1992, **48**, 9217.

138. Z. Wang and R.A. Laursen, *Pept.Res.*, 1992, **5**, 275.
139. R.N. Zuckermann, J.M. Kerr, M.A. Siani and S.C. Banville, *Int.J.Peptide Protein Res.*, 1992, **40**, 497.
140. J.E. Fox, *Biochem.Soc.Trans.*, 1992, **20**, 851.
141. H. Gausepohl, C. Boulin, M. Kraft and R.W. Frank, *Pept.Res.*, 1992, **5**, 315.
142. G.L. Hortin, W.D. Staatz and S.A. Santoro, *Biochem.Int.*, 1992, **26**, 731.
143. R. Ramage, C.A. Barron, S. Bielecki, R. Holden and D.W. Thomas, *Tetrahedron*, 1992, **48**, 499.
144. M.K. Anwer and A.F. Spatola, *Tetrahedron Lett.*, 1992, **33**, 3121.
145. A.F. Spatola, M.K. Anwer and M.N. Rao, *Int.J.Peptide Protein Res.*, 1992, **40**, 322.
146. N. Xaus, X. Jorba, S. Calvet, F. Albericio, P. Clapes, J.L. Torres and G. Valencia, *Biotechnol.Tech.*, 1992, **6**, 69.
147. M.E. Paulaitis, M.J. Sowa and J.H. McMinn, *Ann.N.Y.Acad.Sci.*, 1992, **672**, 278.
148. R.M. Blanco, J.L.L. Rakels, J.M. Guisán and P.J. Halling, *Biochim. Biophys.Acta*, 1992, **1156**, 67.
150. Yu.L. Khmetnitskii, A.K. Gladilin, V.L. Rubailo, K. Martinek and A.V. Levashov, *Eur.J.Biochem.*, 1992, **206**, 737.
151. C.A.A. Malak, I.Y. Filippova, E.N. Lysogorskaya, V.V. Anisimova, G.I. Lavrenova and V.M. Stepanov, *Int.J.Peptide Protein Res.*, 1992, **39**, 443.
152. Yu.L. Khmelnitskii, I.N. Neverova, A.V. Gedrovich, V.A. Polyakov, A.V. Levashov and K. Martinek, *Eur.J.Biochem.*, 1992, **210**, 751.
153. M.I. Vidal, M.L.M. Serralheiro and J.M.S. Cabral, *Biotechnol.Lett.*, 1992, **14**, 1041.
154. Y. Kimura, Y. Tari, S. Adachi and R. Matsuno, *Ann.N.Y.Acad.Sci.*, 1992, **672**, 458.
155. M.Y. Gololobov, T.L. Voyushina, V.M. Stepanov and P. Adlercreutz, *FEBS Lett.*, 1992, **307**, 309.
156. K. Tadasa, I. Shimoda and H. Kayahara, *Biosci.Biotechnol.Biochem.*, 1992, **56**, 804.
157. N. Nishino, M. Xu, H. Mihara and T. Fujimoto, *Chem.Lett.*, 1992, 327.
158. R.G. Whittaker, K.J. Bryant, E.A. Hamilton, L.T. McVittie and P.A. Schober, *Ann.N.Y.Acad.Sci.*, 1992, **672**, 387.
159. S. Kunugi, N. Suzuki, M. Yokoyama and A. Nomara, *Ann.N.Y.Acad.Sci.*, 1992, **672**, 323.
160. S.-T. Chen and K.-T. Wang, *J.Chin.Chem.Soc.*, 1992, **39**, 683.
161. P. Kuhl, S. Säuberlich and H.D. Jakubke, *Monatsh.Chem.*, 1992, **123**, 1015.
162. N. Nishino, M. Xu, H. Mihara and T. Fujimoto, *Tetrahedron Lett.*, 1992, **33**, 3137.
163. M.P. Bemquerer, F.C. Theobaldo and M. Tominaga, *Biomed.Biochim.Acta*, 1991, **50**, S94.
164. M. Sarra, G. Caminal, G. Gonzalez and J. Lopez-Santin, *Biocatalysis*, 1992, **7**, 49.
165. S. Tawaki and A.M. Klibanov, *J.Amer.Chem.Soc.*, 1992, **114**, 1882.
166. V. Cerovsky, *Biotechnol.Tech.*, 1992, **6**, 155.
167. S. Calvet, P. Clapés, J.P. Vigo, N. Xaus, X. Jorba, R.M. Mas, J.L. Torres, G. Valencia, M.L. Serralheiro, J.M.S. Cabral and J.M.A. Empis, *Biotechnol.Bioeng.*, 1992, **39**, 539.
168. S. Kunugi, *Ann.N.Y.Acad.Sci.*, 1992, **672**, 293.
169. Y. Kawasaki, M. Murakami, S. Dosako, I. Azuse, T. Nakamura and H. Okai, *Biosci.Biotechnol.Biochem.*, 1992, **56**, 441.
170. D. Cabaret, S. Bourcier, S. Maillot and M. Wakselman, *Biocatalysis*, 1992, **6**, 191.
171. K. Tanizawa, A. Sugimura and Y. Kanaoka, *FEBS Lett.*, 1992, **296**, 163.
172. C.H. Wong, K.K.C. Liu, T. Kajimoto, L. Chen, Z. Zhong, D.P. Dumas, J.L.C. Liu, Y. Ichikawa and G.J. Shen, *Pure Appl.Chem.*, 1992, **64**, 1197.
173. S. Chatterjee and A.J. Russell, *Biotechnol.Bioeng.*, 1992, **40**, 1069.

174. S.-T. Chen, S.-Y. Chen and K.-T. Wang, *J.Org.Chem.*, 1992, **57**, 6960.
175. R.M. Blanco, G. Alvaro, J.C. Tercero and J.M. Guisan, *J.Mol.Catal.*, 1992, **73**, 97.
176. A. Nadim, I.B. Stoineva, B. Galunsky, V. Kasche and D.D. Petkov, *Biotechnol.Tech.*, 1992, **6**, 539.
177. M.Y. Gololobov, E.V. Kozlova, I.L. Borisov, U. Schellenberger, V. Schellenberger and H.D. Jakubke, *Biotechnol.Bioeng.*, 1992, **40**, 432.
178. T. Nagashima, A. Watanabe and H. Kise, *Enzyme Microb.Technol.*, 1992, **14**, 842.
179. K.H. Lee, P.M. Lee, Y.S. Siaw and K. Morihara, *Biotechnol.Lett.*, 1992, **14**, 779.
180. K. Peek, S.-A. Wilson, M. Prescott and R.M. Daniel, *Ann.N.Y.Acad.Sci.*, 1992, **672**, 471.
181. T. Ohshiro, K. Mochida and T. Uwajima, *Biotechnol.Lett.*, 1992, **14**, 175.
182. F. Dal Degan, B. Ribadeau-Dumas and K. Breddam, *Appl.Environ. Microbiol.*, 1992, **58**, 2144.
183. C.A. Abdel Malak, *Biochem.J.*, 1992, **288**, 941.
184. D.H.G. Crout, D.A. MacManus, J.M. Ricca, S. Singh, P. Critchley and W.T. Gibson, *Pure Appl.Chem.*, 1992, **64**, 1079.
185. M. Schuster, V. Kasche and H.-D. Jakubke, *Biochim.Biophys.Acta*, 1992, **1121**, 207.
186. K. Aso and H. Kodaka, *Biosci.Biotechnol.Biochem.*, 1992, **52**, 755.
187. K. Aso, H. Kodaka, H. Fukushi and H.H. Li, *Biotechnol.Lett.*, 1992, **14**, 451.
188. B. Deschrevel, J.D. Dugast and J.C. Vincent, *C.R.Acad.Sci., Ser. III*, 1992, **314**, 519.
189. A. Hirata, M. Hirata and N. Hond, *Biochem.Eng. 2001*, 1992, 463.
190. I. Gill and E.N. Vulfson, *J.Chem.Soc., Perkin Trans. 1*, 1992, 667.
191. N. Xaus, F. Albericio, X. Jorba, S. Calvet, P. Clapés, J.L. Torres and G. Valencia, *Int.J.Peptide Protein Res.*, 1992, **39**, 528.
192. J. Bongers, T. Lambros, W. Liu, M. Ahmad, R.M. Campbell, A.M. Felix and E.P. Heimer, *J.Med.Chem.*, 1992, **35**, 3934.
193. N. Nishino, M. Xu, H. Mihara and T. Fujimoto, *J.Chem.Soc., Chem.Commun.*, 1992, 648.
194. J. Bongers, R.E. Offord, A.M. Felix, R.M. Campbell and E.P. Heimer, *Int.J.Peptide Protein Res.*, 1992, **40**, 268.
195. M. Schuster, A. Aaviksaar and H.-D. Jakubke, *Tetrahedron Lett.*, 1992, **33**, 2799.
196. U. Slomczynska, T. Leplawy and M.T. Leplawy, *Z.Naturforsch. B, Chem. Sci.*, 1992, **47**, 1424.
197. U. Slomczynska, J. Zabrochi, K. Kaczmarek, M.T. Leplawy, D.D. Beusen and G.R. Marshall, *Biopolymers*, 1992, **32**, 1461.
198. R.P. Roy, K.M. Khandke, B.N. Manjula and A.S. Acharya, *Biochemistry*, 1992, **31**, 7249.
199. C. Unverzagt, A. Geyer and H. Kessler, *Angew.Chem., Int.Ed.*, 1992, **31**, 1229.
200. P.-F. Berne, S. Blanquet and J.-M. Schmitter, *J.Amer.Chem.Soc.*, 1992, **114**, 2603.
201. J. Green and A.L. Margolin, *Tetrahedron Lett.*, 1992, **33**, 7759.
202. D.B. Henriksen, K. Breddam, J. Muller and O. Buchardt, *J.Amer.Chem. Soc.*, 1992, **114**, 1876.
203. I.B. Stoineva, B.P. Galunsky, V.S. Lazanov, I.P. Ivanov and D.D. Petkov, *Tetrahedron*, 1992, **48**, 1115.
204. H. Waldmann, A. Heuser, P. Braun and H. Kunz, *Indian J.Chem., Sect. B*, 1992, **31B**, 799.
205. P. Braun, H. Waldmann and H. Kunz, *Synlett*, 1992, 39.
206. S. Reissmann and G. Greiner, *Int.J.Peptide Protein Res.*, 1992, **40**, 110.
207. S.-T. Chen, S.-C. Hsiao, C.-H. Chang and K.-T. Wang, *Synth.Commun.*, 1992, **22**, 391.

208. M. Schultz, P. Hermann and H. Kunz, *Synlett*, 1992, 37.
209. C. Wandrey, A. Fischer, B. Joksch and A. Schwarz, *Ann.N.Y.Acad.Sci.*, 1992, **672**, 528.
210. A. Schwarz, C. Weandrey, D. Steinke and M.R. Kula, *Biotechnol. Bioeng.*, 1992, **39**, 132.
211. V. Schellenberger, W. Tegge, K.-D. Klöppel and R. Franck, *Int.J.Peptide Protein Res.*, 1992, **39**, 472.
212. E.W. Holla, M. Schudok, A. Weber and M. Zulauf, *J.Carbohydr.Chem.*, 1992, **11**, 659.
213. J. Lee and J.K. Coward, *J.Org.Chem.*, 1992, **57**, 4126.
214. T. Mukaiyama, K. Yorozu, K. Kato and T. Yamada, *Chem.Lett.*, 1992, 181.
215. A. Thomas, S.G. Marijunatha and S. Rajappa, *Helv.Chim.Acta.*, 1992, **75**, 715.
216. J. Aubé and M.S. Wolfe, *Biorg.Med.Chem.Lett.*, 1992, **2**, 925.
217. K. Higashiura and K. Ienaga, *J.Chem.Res.(S)*, 1992, 250.
218. G. Kokotos and V. Constantinou-Kokotou, *J.Chem.Res.(S)*, 1992, 391.
219. E. Frérot, J. Coste, J. Poncet and P. Jouin, *Tetrahedron Lett.*, 1992, **33**, 2815.
220. F.M.F. Chen and N.L. Benoiton, *Int.J.Peptide Protein Res.*, 1992, **40**, 13.
221. S. Capasso, L. Mazzarella, F. Sica, A. Zagari and S. Salvadori, *J.Chem. Soc., Chem.Commun.*, 1992, 919.
222. N. Sewald, J. Riede, P. Bissinger and K. Burger, *J.Chem.Soc., Perkin Trans. 1*, 1992, 267.
223. J.-P. Mazaleyrat, J. Xie and M. Wakselman, *Tetrahedron Lett.*, 1992, **33**, 4301.
224. C.-H. Tung, Y. Ebright, X. Shen and S. Stein, *Bioorg.Med.Chem.Lett.*, 1992, **2**, 303; C.-H. Tung, Z. Wei, M.J. Leibowitz and S.Stein, *Proc. Natl.Acad.Sci., U.S.A.*, 1992, **89**, 7114.
225. H.F. Gaertner, K. Rose, R. Cotton, D. Timms, R. Camble and R.E. Offord, *Bioconjugate Chem.*, 1992, **3**, 262.
226. R.J. Simon, R.S. Kania, R.N. Zuckermann, V.D. Huebner, D.A. Jewell, S. Banville, S. Ng, L. Wang, S. Rosenberg, C.K. Marlowe, D.C. Spellmeyer, R. Tan, A.D. Frankel, D.V. Santi, F.E. Cohen and P.A. Bartlett, *Proc. Natl.Acad.Sci.,U.S.A.*, 1992, **89**, 9367.
227. R.N. Zuckermann, J.M. Kerr, S.B.H. Kent and W.H. Moos, *J.Amer.Chem.Soc.*, 1992, **114**, 10646.
228. W.J. Hoekstra, S.S. Sunder, R.J. Cregge, L.A. Ashton, K.T. Stewart and C.-H.R. King, *Tetrahedron*, 1992, **48**, 307.
229. S.Y. Shin, Y. Kaburaki, M. Watanabe and E. Munekata, *Biosci.Biotechnol. Biochem.*, 1992, **56**, 399.
230. S.Y. Shin, Y. Kaburaki, M. Watanabe and E. Munekata, *Biosci.Biotechnol. Biochem.*, 1992, **56**, 404.
231. M. Tsunemi, H. Kato, Y. Nishiuchi, S.-i. Kumagaye and S. Sakakibara, *Biochem.Biophys.Res.Commun.*, 1992, **185**, 967.
232. J.C. Hendrix, K.J. Halverson and P.T. Lansbury, *J.Amer.Chem.Soc.*, 1992, **114**, 7930.
233. M. Schnözler and S.B.H. Kent, *Science*, 1992, **256**, 221.
234. R.C.deL. Milton, S.C.F. Milton and S.B.H. Kent, *Science*, 1992, **256**, 1445.
235. A.G. Tomasselli, C.A. Bannow, M.R. Deibel, J.O. Hui, H.A. Zurcher-Neely, I.M. Reardon, C.W. Smith and R.L. Heinrikson, *J.Biol.Chem.*, 1992, **267**, 10232.
236. C. Vita, L. Gozzini and C. Di Bello, *Eur.J.Biochem.*, 1992, **204**, 631.
237. S. Miura, I. Kasuya-Arai, H. Mori, S. Miyazawa, T. Osumi, T. Hashimoto and Y. Fujiki, *J.Biol.Chem.*, 1992, **267**, 14405.

238. T. Inui, K. Hagiwara, K. Nakajima, T. Kimura, T. Nakajima and S. Sakakibara, *Pept.Res.*, 1992, **5**, 140.
239. M. Kawai, D.A. Quincy, B. Lane, K.W. Mollison, Y.-S. Or, J.R. Luly and G.W. Carter, *J.Med.Chem.*, 1992, **35**, 220.
240. K. Plucinska, W. Gumulka and E.I. Wisniewska, *Pol.J.Chem.*, 1991, **65**, 1251.
241. P. Juvvadi, D.J. Dooley, C.C. Humblet, G.H. Lu, E.A. Lunney, R.L. Panek, R. Skeean and G.R. Marshall, *Int.J.Peptide Protein Res.*, 1992, **40**, 163.
242. P.G. Jones, G.J. Moore and D.M. Waisman, *J.Biol.Chem.*, 1992, **267**, 13993.
243. Y. Uchida and M. Shindo, *Bull.Chem.Soc.Jpn.*, 1992, **65**, 615.
244. A.G. Rao, T. Rood, J. Maddox and J. Duvick, *Int.J.Peptide Protein Res.*, 1992, **40**, 507.
245. D. Wade, D. Andreu, S.A. Mitchell, A.M.V. Silveira, A. Boman, H.G. Boman and R.B. Merrifield, *Int.J. Peptide Protein Res.*, 1992, **40**, 429.
246. D. Wen and R.A. Laursen, *J.Biol.Chem.*, 1992, **267**, 14102.
247. T.W.von Geldern, T.W. Rockway, S.K. Davidsen, G.P. Budzik, E.N. Bush, M.Y. Chu-Moyer, E.M. Devine, W.H. Holleman, M.C. Johnson, S.D. Lucas, D.M. Pollock, J.M. Smital, A.M. Thomas and T.J. Opgenorth, *J.Med.Chem.*, 1992, **35**, 808.
248. E.M. Devine, A.M. Buko, S.P. Cepa, S.K. Davidsen, W.H. Holleman, C.A. Marselle, T.J. Opgenorth, T.W. Rockway and T.W. von Geldern, *Int.J. Peptide Protein Res.*, 1992, **40**, 532.
249. Y. Shimekake, T. Kawabata, M. Nakamura and K. Nagata, *FEBS Lett.*, 1992, **309**, 185.
250. J.A. Malikayil, J.V. Edwards and L.R. McLean, *Biochemistry*, 1992, **31**, 7043.
251. F.C. Kull, J.J. Leban, A. Landavazo, K.D. Stewart, B. Stockstill and J.D. McDermed, *J.Biol.Chem.*, 1992, **267**, 21132.
252. R.Z. Cai, S. Radulovic, J. Pinski, A. Nagy, T.W. Redding, D.B. Olsen and A.V. Schally, *Peptides*, 1992, **13**, 267.
253. R. De Castiglione, L. Gozzini, M. Galantino, F. Corradi, M. Ciomei, F. Roletto and F. Bertolero, *Farmaco*, 1992, **47**, 855.
254. J. Jurayj and M. Cushman, *Tetrahedron*, 1992, **48**, 8601.
255. K. Maruyama, H. Nagasawa, A. Isogai, H. Ishizaki and A. Suzuki, *J. Protein Chem.*, 1992, **11**, 13.
256. K. Nagata, K. Maruyama, H. Nagasawa, I. Urushibata, I. Isogai, H. Ishizaki and A. Suzuki, *Peptides*, 1992, **13**, 653.
257. J.C. Cheronis, E.T. Whalley, K.T. Nguyen, S.R. Eubanks, L.G. Allen, M.J. Duggan, S.D. Loy, K.A. Bonham and J.K. Blodgett, *J.Med.Chem.*, 1992, **35**, 1563.
258. B. Lammek, Y. Ito, I. Gavras and H. Gavras, *Coll.Czech.Chem.Commun.*, 1992, **57**, 1960.
259. C. Choi, *Han'guk Nonghwa Hakhoechi*, 1991, **34**, 334.
260. J.C. Hendrix and P.T. Lansbury, *J.Org.Chem.*, 1992, **57**, 3421.
261. K. Higashiura and K. Ienaga, *Synthesis*, 1992, 353.
262. D.J.M. Stone, R.J. Waugh, J.H. Bowie, J.C. Wallace and M.J. Tyler, *J. Chem.Soc., Perkin Trans. 1*, 1992, 3173.
263. H. Kuroda, Y.N. Chen, T.X. Watanabe, T. Kimura and S. Sakakibara, *Pept. Res.*, 1992, **5**, 265.
264. J.W. Perich and R.B. Johns, *Austr.J.Chem.*, 1992, **45**, 1857.
265. V.V. Ryakhovskii, S.V. Agafonov, O.B. Vinogradova, Yu.M. Kosyrev, M.V. Kiselevskii and V.S. Dobryanskii, *Khim.-Farm. Zh.*, 1992, **26**, 46.
266. Yu.M. Taskaeva, G.A. Korshunova and Yu. P. Svachkin, *Zh.Obshch.Khim.*, 1992, **62**, 717.

267. K. Shiosaki, C.W. Lin, H. Kopecka, R.A. Craig, B.R. Bianchi, T.R. Miller, D.G. Witte, M. Stashko and A.M. Nadzan, *J.Med.Chem.*, 1992, **35**, 2007.
268. D. Ron, C. Gilon, M. Hanani, A. Vromen, Z. Selinger and M. Chorev, *J. Med.Chem.*, 1992, **35**, 2806.
269. I. McCort-Tranchepain, D. Ficheux, C. Durieux and B.P. Roques, *Int.J. Peptide Protein Res.*, 1992, **39**, 48.
270. W. Danho, J.W. Tilley, S.-J. Shiuey, I. Kulesha, J. Swistok, R. Makofske, J. Michalewsky, R. Wagner, J. Triscari, D. Nelson, F.Y. Chiruzzo and S. Weatherford, *Int.J.Peptide Protein Res.*, 1992, **39**, 337.
271. J.W. Tilley, W. Danho, S.-J. Shiuey, I. Kulesha, J. Swistok, R. Makofske, J. Michalewsky, J. Triscari, D. Nelson, S. Weatherford, V. Madison, D. Fry and C. Cook, *J.Med.Chem.*, 1992, **35**, 3774.
272. J.W. Tilley, W. Danho, V. Madison, D. Fry, J. Swistok, R. Makofske, J. Michalewsky, A. Schwartz, S. Weatherford, J. Triscari and D. Nelson, *J. Med.Chem.*, 1992, **35**, 4249.
273. M.W. Holladay, M.J. Bennett, M.D. Tufano, C.W. Lin, K.E. Asin, D.G. Witte, T.R. Miller, B.R. Bianchi, A.L. Nikkel, L. Bednarz and A.M. Nadzan, *J.Med.Chem.*, 1992, **35**, 2919.
274. J.W. Tilley, W. Danho, S.-J. Shiuey, I. Kulesha, R. Sarabu, J. Swistok, R. Makofske, G.L. Olson, E. Chiang, V.K. Rusiecki, R. Wagner, J. Michalewsky, J. Triscari, D. Nelson, F.V. Chiruzzo and S. Weatherford, *Int.J.Peptide Protein Res.*, 1992, **39**, 322.
275. P.J. Corringer, C. Durieux, M. Ruiz-Gayo and B.P. Roques, *J.Labelled Compd.Radiopharm.*, 1992, **31**, 459.
276. W.D. Kornreich, R. Galyean, J.F. Hernandez, A.G. Craig, C.J. Donaldson, G. Yamamoto, C. Rivier, W. Vale and J. Rivier, *J.Med.Chem.*, 1992, **35**, 1870.
277. H. Rao, S.C. Mohr, H. Fairhead and P. Setlow, *FEBS Lett.*, 1992, **305**, 115.
278. H. Hojo and S. Aimoto, *Bull.Chem.Soc.Jpn.*, 1992, **65**, 3055.
279. A.M. Tamburro, V. Guantieri and D. Daga Gordini, *J.Biomol.Struct.Dyn.*, 1992, **10**, 441.
280. K. Ishikawa, T. Fukami, T. Nagase, K. Fujita, T. Hayama, K. Niiyama, T. Mase, M. Ihara and M. Yano, *J.Med.Chem.*, 1992, **35**, 2139.
281. C. Valembois, C. Mendre, J.C. Cavadore and B. Calas, *Tetrahedron Lett.*, 1992, **33**, 4005.
282. W.L. Cody, A.M. Doherty, J.X. He, P.L. DePue, S.T. Rapundalo, G.A. Hingorani, T.C. Major, R.L. Panek, D.T. Dudley, S.J. Haleen, D. LaDouceur, K.E. Hill, M.A. Flynn and E.E. Reynolds, *J.Med.Chem.*, 1992, **35**, 3301.
283. G. Fassina, R. Consonni, L. Zetta and G. Cassani, *Int.J.Peptide Protein Res.* 1992, **39**, 540.
284. G. Fassina, A. Corti and G. Cassani, *Int.J.Peptide Protein Res.*, 1992, **39**, 549.
285. N. Ueyama, S. Ueno, A. Nakamura, K. Wada, H. Matsubara, S. Kumagai, S. Sakakibara, T. Tsukihara, *Biopolymers*, 1992, **32**, 1535.
286. M. Hage-van Noort, W.C. Puijk, H.H. Plasman, D. Kuperus, W.M.M. Schaaper, N.J.C.M. Beekman, J.A. Grootegoed and R.H. Meloen, *Proc.Natl. Acad. Sci., U.S.A.*, 1992, **89**, 3922.
287. Ü. Langel, T. Land and T. Bartfai, *Int.J.Peptide Protein Res.*, 1992, **39**, 516.
288. M. Mokotoff, K. Ren, L.K. Wong, A.V. LeFever and P.C. Lee, *J.Med.Chem.*, 1992, **35**, 4696.
289. S. Mojsov, *Int.J.Peptide Protein Res.*, 1992, **40**, 333.
290. M.H. Lyttle, D.T. Aaron, M.D. Hocker and B.R. Hughes, *Pept.Res.*, 1992, **5**, 336.

291. C. Celma, F. Albericio, E. Pedroso and E. Giralt, *Pept.Res.*, 1992, **5**, 62.
292. G. Flouret, K. Mahan and T. Majewski, *J.Med.Chem.*, 1992, **35**, 636.
293. E. Masiukiewicz, B. Rzeszotarska, G. Fortuna and K. Kochman, *J.Prakt. Chem.*, 1991, **333**, 573.
294. J. Rivier, J. Porter, C. Hoeger, P. Theobald, A.G. Craig, J. Dykert, A. Corrigan, M. Perrin, W.A. Hook, R.P. Siraganian, W. Vale and C. Rivier, *J.Med.Chem.*, 1992, **35**, 4270.
295. F. Haviv, T.D. Fitzpatrick, C.J. Nichols, R.E. Swenson, E.N. Bush, G. Diaz, A. Nguyen, H.N. Nellans, D.J. Hoffman, H. Ghanbari, E.S. Johnson, S. Love, V. Cybulski and J. Greer, *J.Med.Chem.*, 1992, **35**, 3890.
296. C. Somlai, W. König and J. Knolle, *Liebig's Ann.Chem.*, 1992, 1055.
297. J.J. Nestor, R. Tahilramani, T.L. Ho, J.C. Goodpasture, B.H. Vickery and P. Ferrandon, *J.Med.Chem.*, 1992, **35**, 3942.
298. J. Bongers, A.M. Felix, R.M. Campbell, Y. Lee, D.J. Merkler and E.P. Heimer, *Pept.Res.*, 1992, **5**, 183.
299. P. Gaudreau, L. Boulanger and T. Abribat, *J.Med.Chem.*, 1992, **35**, 1864.
300. A.R. Friedman, A.K. Ichhpurani, W.M. Moseley, G.R. Alaniz, W.H. Claflin, D.L. Cleary, M.D. Prairie, W.C. Kreuger, L.A. Frohman, T.R. Downs and R.M. Epand, *J.Med.Chem.*, 1992, **35**, 3928.
301. M. Zarandi, P. Serfozo, J. Zsigo, A.H. Deutch, T. Janaky, D.B. Olsen, S. Bajusz and A.V. Schally, *Pept.Res.*, 1992, **5**, 190.
302. M. Zarandi, P. Serfozo, J. Zsigo, L. Bokser, T. Janaky, D.B. Olsen, S. Bajusz and A.V. Schally, *Int.J.Peptide Protein Res.*, 1992, **39**, 211.
303. D.P. Smith, M.L. Heiman, J.F. Wagner, R.L. Jackson, R.A. Bimm and H.M. Hsiung, *Bio/Technology*, 1992, **10**, 315.
304. R.M. Campbell, Y. Lee, T.F. Mowles, K.W. McIntyre, M. Ahmad, A.M. Felix and E.P. Heimer, *Peptides (Pergamon)*, 1992, **13**, 787.
305. Y. Xu, D. Cui, H. Yang, S. Zhu, *Shengwu Huaxue Zazhi*, 1992, **8**, 230.
306. J. Fan, Y. Tang, Y. Feng and Y. Zhang, *Shengwu Huaxue Yu Shengwu Wuli Xuebao*, 1992, **24**, 7.
307. M. Xu, Y. Ye, X. Zhang, and S. Zhu, *Shengwu Huaxue Zazhi*, 1992, **8**, 410.
308. Y. Hirano, M. Okuno, T. Hayashi and A. Nakajima, *Polym.J. (Tokyo)*, 1992, **24**, 465.
309. M. Nomizu, A. Utani, N. Shiraishi, Y. Yamada and P.P. Roller, *Int.J. Peptide Protein Res.*, 1992, **40**, 72.
310. K.P. Soteriadou, M.S. Remoundos, M.C. Katsikas, A.K. Tzinia, V. Tsikaris, C. Sakarellos and S.J. Tzartos, *J.Biol.Chem.*, 1992, **267**, 13980.
311. S. Kuwata, A. Nakanishi, T. Yamada and T. Miyazawa, *Tetrahedron Lett.*, 1992, **33**, 6995.
312. Y. Okada, S. Nakayama, S. Iguchi, Y. Kikuchi, M. Irie, J.-i. Sawada, H. Ikebuchi and T. Terao, *Chem.Pharm.Bull.*, 1992, **40**, 1029.
313. E. Eriotou-Bargiota, C.-B. Xue, F. Naider and J.M. Becker, *Biochemistry*, 1992, **31**, 551.
314. M. Kohmura, N. Nio and Y. Ariyoshi, *Biosci.Biotechnol.Biochem.*, 1992, **56**, 472.
315. C. Hashimoto, *Bull.Chem.Soc.Jpn.*, 1992, **65**, 1268.
316. A.B. McElroy, S.P. Clegg, M.J. Deal, G.B. Ewan, R.M. Hagan, S.J. Ireland, C.C. Jordan, B. Porter, B.C. Ross, P. Ward and A.R. Whittington, *J.Med.Chem.*, 1992, **35**, 2582.
317. S.L. Harbeson, S.A. Shatzer, T.-B. Le and S.H. Buck, *J.Med.Chem.*, 1992, **35**, 3949.

318. M.J. Deal, R.M. Hagan, S.J. Ireland, C.C. Jordan, A.B. McElroy, B. Porter, B.C. Ross, M. Stephens-Smith and P. Ward, *J.Med.Chem.*, 1992, **35**, 4195.
319. R.D. Feinstein, J.H. Boublik, D. Kirby, M.A. Spicer, A.C. Craig, K. Malewicz, N.A. Scott, M.R. Brown and J.E. Rivier, *J.Med.Chem.*, 1992, **35**, 2836.
320. D.F. Mierke, H. Dürr, H. Kessler and G. Jung, *Eur.J.Biochem.*, 1992, **206**, 39.
321. M.T. Reymond, L. Delmas, S.C. Koerber, M.R. Brown and J.E. Rivier, *J. Med.Chem.*, 1992, **35**, 3653.
322. S. Doulut, M. Rodriguez, D. Lugrin, F. Vecchini, P. Kitabgi, A. Aumelas and J. Martinez, *Pept.Res.*, 1992, **5**, 30.
323. R.J. Nachman, G.M. Coast, G.M. Holman and W.F. Haddon, *Int.J.Peptide Protein Res.*, 1992, **40**, 423.
324. H. Ito, S. Tsubuki, Y. Saito and S. Kawashima, *Nippon Kagaku Kaishi*, 1992, 1363.
325. L. Varga-Defterdarovic, S. Horvat, N.N. Chung and P.W. Schiller, *Int.J. Peptide Protein Res.*, 1992, **39**, 12.
326. D.W. Hansen, A. Stapelfeld, M.A. Savage, M. Reichman, D.L. Hammond, R.C. Haaseth and H.I. Mosberg, *J.Med.Chem.*, 1992, **35**, 684.
327. G. Toth, K.C. Russell, G. Landis, T.H. Kramer, L. Fang, R. Knapp, P. Davis, T.F. Burks, H.I. Yamamura and V.J. Hruby, *J.Med.Chem.*, 1992, **35**, 2384.
328. N.S. Chandrakumar, A. Stapelfeld, P.M. Beardsley, O.T. Lopez, B. Drury, E. Anthony, M.A. Savage, L.N. Williamson and M. Reichman, *J.Med.Chem.*, 1992, **35**, 2928.
329. I. Ojima, K. Kato, F.A. Jameison, J. Conway, K. Nakahashi, M. Hagiwara, T. Miyamae and H.E. Radunz, *Bioorg.Med.Chem.Lett.*, 1992, **2**, 219.
330. R. Paruszewski, R. Matusiak, G. Rostafinska-Suchar, S.W.G. Gumulka, K. Misterek and A. Dorociak, *Pol.J.Pharmacol.Pharm.*, 1991, **43**, 381.
331. R. Matsueda, T. Yasunaga, H. Kodama, M. Kondo, T. Costa, Y. Shimohigashi, *Chem.Lett.*, 1992, 1259.
332. A. Polinsky, M.G. Cooney, A. Toy-Palmer, G. Ösapay and M. Goodman, *J.Med.Chem.*, 1992, **35**, 4185.
333. W.M. Kazmierski, R.D. Ferguson, R.J. Knapp, G.K. Lui, H.I. Yamamura and V.J. Hruby, *Int.J.Peptide Protein Res.*, 1992, **39**, 401.
334. D. Tourwé, J. Couder, M. Ceusters, D. Meert, T.F. Burks, T.H. Kramer, P. Davis, R. Knapp, H.I. Yamamura, J.E. Leysen and G. van Binst, *Int.J. Peptide Protein Res.*, 1992, **39**, 131.
335. S. Salvadori, R. Guerrini, P.A. Borea and R. Tomatis, *Int.J. Peptide Protein Res.*, 1992, **40**, 437.
336. D.L. Heyl and H.I. Mosberg, *Int.J.Peptide Protein Res.*, 1992, **39**, 450.
337. L. Fang, R.J. Knapp, T. Matsunaga, S.J. Weber, T. Davis, V.J. Hruby and H.I. Yamamura, *Life Sci.*, 1992, **51**, PL189.
338. L.H. Lazarus, S. Salvadori, P. Grieco, W.E. Wilson and R. Tomatis, *Eur. J.Med.Chem.*, 1992, **27**, 791.
339. L.H. Lazarus, S. Salvadori, G. Balboni, R. Tomatis and W.E. Wilson, *J. Med.Chem.*, 1992, **35**, 1222.
340. D.L. Heyl and H.I. Mosberg, *J.Med.Chem.*, 1992, **35**, 1535.
341. S. Salvadori, C. Bianchi, L.H. Lazarus, V. Scaranari, M. Attila and R. Tomatis, *J.Med.Chem.*, 1992, **35**, 4651.
342. P.W. Schiller, G. Weltrowska, T.M.-D. Nguyen, B.C. Wilkes, N.N. Chung and C. Lemieux, *J.Med.Chem.*, 1992, **35**, 3956.
343. B. Buzas, G. Toth, S. Cavagnero, V.J. Hruby and A. Borsodi, *Life Sci.*, 1992, **50**, PL75.

344. H. Choi, T.F. Murray, G.E. Delander, V. Caldwell and J.V. Aldrich, *J. Med.Chem.*, 1992, **35**, 4638.
345. S.C. Story, T.F. Murray, G.E. Delander and J.V. Aldrich, *Int.J.Peptide Protein Res.*, 1992, **40**, 89.
346. J. Erchegyi, A.J. Kastin, J.E. Zadina and X.-D. Qiu, *Int.J.Peptide Protein Res.*, 1992, **39**, 477.
347. N.G.J. Delaet, P.M.F. Verheyden, D. Tourwe, G. van Binst, P. Davis and T.F. Burks, *Biopolymers*, 1992, **32**, 957.
348. K.R. Snyder, S.C. Story, M.E. Heidt, T.F. Murray, G.E. DeLander and J.V. Aldrich, *J.Med.Chem.*, 1992, **35**, 4330.
349. H. Mickos, K. Sundberg and B. Lüning, *Acta Chem.Scand.*, 1992, **46**, 989.
350. W. Zeng and B. Hemmasi, *Liebig's Ann.Chem.*, 1992, 311.
351. H. Nakashima, M. Masuda, T. Murakami, Y. Koyanagi, A. Matsumoto, N. Fujii and N. Yamamoto, *Antimicrob.Agents Chemother.*, 1992, **32**, 1249.
352. R. Jezek, J. Franc, J. Slaninová and M. Lebl, *Coll.Czech.Chem.Commun.*, 1992, **57**, 621.
353. M. Lebl, G. Toth, J. Slaninová and V.J. Hruby, *Int.J.Peptide Protein Res.*, 1992, **40**, 148.
354. Z. Procházka. J. Slaninová, T. Barth, A. Stierandová, J. Trojnar, P. Melin and M. Lebl, *Coll.Czech.Chem.Commun.*, 1992, **57**, 1335.
355. B.C. Pal, J. Slaninová, T. Barth, J. Trojnar and M. Lebl, *Coll.Czech. Chem.Commun.*, 1992, **57**, 1345.
356. P.D. Williams, M.G. Bock, R.D. Tung, V.M. Garsky, D.S. Perlow, J.M. Erb, G.F. Lundell, N.P. Gould, W.L. Whitter, J.B. Hoffman, M.J. Kaufman, B.V. Clinesch-midt, D.J. Pettibone, R.M. Freidinger and D.F. Veber, *J.Med. Chem.*, 1992, **35**, 3905.
357. D.D. Smith, J. Slaninová and V.J. Hruby, *J.Med.Chem.*, 1992, **35**, 1558.
358. M. Manning, J. Przybylski, Z. Grzonka, E. Nawrocka, B. Lammek, A. Misicka, L.L. Cheng, W.Y. Chan, N.C. Wo and W.H. Sawyer, *J.Med.Chem.*, 1992, **35**, 3895.
359. M. Manning, K. Bankowski, C. Barberis, S. Jard, J. Elands and W.Y. Chan, *Int.J.Peptide Protein Res.*, 1992, **40**, 261.
360. M. Manning, S. Stoev, K. Bankowski, A. Misicka, B. Lammek, N.C. Wo and W.H. Sawyer, *J.Med.Chem.*, 1992, **35**, 382.
361. J. Howl, D.C. New and M. Wheatley, *J.Mol.Endocrinol.*, 1992, **9**, 123.
362. M. Zertová, Z. Procházka, J. Slaninová, J. Skopková, T. Barth and M. Lebl, *Coll.Czech.Chem.Commun.*, 1992, **57**, 604.
363. M. Zertová, Z. Procházka, J. Slaninová, T. Barth, P. Majer and M. Lebl, *Coll.Czech.Chem.Commun.*, 1992, **57**, 1103.
364. D. Barbeau, S. Guay, W. Neugebauer and E. Escher, *J.Med.Chem.*, 1992, **35**, 151.
365. G. Guillon, D. Barbeau, W. Neugebauer, S. Guay, L. Bilodeau, M.-N. Balestre, N. Gallo-Payet and E. Escher, *Peptides*, 1992, **13**, 7.
366. S.G. Nadler, J.L. Kapouch, J.I. Elliott and K.R. Williams, *J.Biol.Chem.*, 1992, **267**, 3750.
367. Y. Tor, J. Libman, A. Shanzer, C.E. Felder and S. Lifson, *J.Amer.Chem. Soc.*, 1992, **114**, 6653.
368. J. Xu, J. Wu, L. Chen, W. Shen, W. Ma, S. Liu and W. Zhang, *Shengwu Huaxue Yu Shengwu Wuli Xuebao*, 1992, **24**, 109.
369. Z. Huang, Y.-B. He, K. Raynor, M. Tallent, T. Reisine and M. Goodman, *J.Amer.Chem.Soc.*, 1992, **114**, 9390.

370. S.V. Egorova, E.B. Gurina, V.P. Golubovich and A.A. Akhrem, *Vestsi Akad Navuk BSSR, Ser.Khim.Navuk*, 1991, 61.
371. D. Hagiwara, H. Miyake, H. Morimoto, M. Murai, T. Fujii and M. Matsuo, *J.Med.Chem.*, 1992, **35**, 2015.
372. D. Hagiwara, H. Miyake, H. Morimoto, M. Murai, T. Fujii and M. Matsuo, *J.Med.Chem.*, 1992, **35**, 3184.
373. M. Antoniou, C. Poulos and T. Tsegenidis, *Int.J.Peptide Protein Res.*, 1992, **40**, 395.
374. P. Kaur, G.K. Patnaik, R. Raghubir and V.S. Chauhan, *Bull.Chem.Soc.Jpn.*, 1992, **65**, 3412.
375. A. Jenmalm, K. Luthman, G. Lindeberg, F. Nyberg, L. Terenius and U. Hacksell, *Bioorg.Med.Chem.Lett.*, 1992, **2**, 1693.
376. N.-h. Guo, H.C. Krutzsch, E. Nègre, V.S. Zabrenetzky and D.D. Roberts, *J.Biol.Chem.*, 1992, **267**, 19349.
377. S. Hoerger, B. Gallert, H. Echner and W. Voelter, *Z.Naturforsch.,B:Chem. Sci.*, 1992, **47**, 1170.
378. A. Kapurniotu and W. Voelter, *Liebig's Ann.Chem.*, 1992, 361.
379. T. Abiko and H. Sekino, *Amino Acids*, 1991, **1**, 215.
380. J. Freund, A. Kapurniotu, T.A. Holak, M. Lenfant and W. Voelter, *Z. Naturforsch.,B:Chem.Sci.*, 1992, **47**, 1324.
381. A.A. Mazurov, S.A. Andronati, B.A. Labasyuk, V.M. Kabanov and A.N. Mokhovikov, *Bioorg.Med.Chem.Lett.*, 1992, **2**, 649.
382. L. Lankiewicz, C.Y. Bowers, G.A. Reynolds, V. Labroo, L.A. Cohen, S. Vonhof, A.L. Siren and A.F. Spatola, *Biochem.Biophys.Res.Commun.*, 1992, **184**, 359.
383. A. Tromelin, M.-H. Fulachier, G. Mourier and A. Ménez, *Tetrahedron Lett.*, 1992, **33**, 5197.
384. L.D. Heerze, P.C.S. Chong and G.D. Armstrong, *J.Biol.Chem.*, 1992, **267**, 25810.
385. L. Kovács and M. Hesse, *Helv.Chim.Acta*, 1992, **75**, 1909.
386. J. Bahr and B. Lüning, *Acta Chem.Scand.*, 1992, **46**, 266.
387. J.P. Tam and Z.-Y. Shen, *Int.J.Peptide Protein Res.*, 1992, **39**, 464.
388. S.-M. Ngai and R.S. Hodges, *J.Biol.Chem.*, 1992, **267**, 15715.
389. G.S. Shaw, W.A. Findlay, P.D. Semchuk, R.S. Hodges and B.D. Sykes, *J. Amer.Chem.Soc.*, 1992, **114**, 6258.
390. E. Nawrocka-Bolewska, A. Kubik, A. Szewczuk, I.Z. Siemion, E. Obuchowicz, K. Golba and Z.S. Herman, *Pol.J.Pharmacol.Pharm.*, 1991, **43**, 281.
391. M. Gobbo, L. Biondi, F. Filira, B. Scolaro, R. Rocchi and T. Piek, *Int. J.Peptide Protein Res.*, 1992, **40**, 54.
392. S. Chen and J. Xu, *Youji Huaxue*, 1992, **12**, 418.
393. C. Carreño, X. Roig, J. Cairo, J. Camamero, M.G. Mateu, E. Domingo, E. Giralt and D. Andreu, *Int.J.Peptide Protein Res.*, 1992, **39**, 41.
394. M. Sobel, D.F. Soler, J.C. Kermode and R.B. Harris, *J.Biol.Chem.*, 1992, **267**, 8857.
395. J. Takagi, H. Asai and Y. Saito, *Biochemistry*, 1992, **31**, 8530.
396. S. Ranganathan, N. Jayaraman and R. Roy, *Tetrahedron*, 1992, **48**, 931.
397. S.F. Michael, V.J. Kilfoil, M.H. Schmidt, B.T. Amann and J.M. Berg, *Proc.Natl.Acad.Sci.,U.S.A.*, 1992, **89**, 4796.
398. F.H. Tsai, C.G. Overberger and R. Zand, *J.Polym.Sci.,Part A:Polym.Chem.*, 1992, **30**, 551.
399. T. Munegumi, N. Suzuki, N. Tanikawa and K. Harada. *Chem.Lett.*, 1992, 1679.
400. Y. Nosho, T. Ikehara, Y. Sasatani, H. Yamauchi and S. Hashimoto, *Chem. Express*, 1992, **7**, 753.
401. M.E. Gelbin and J. Kohn, *J.Amer.Chem.Soc.*, 1992, **114**, 3962.

402. J.N. Zeng, D.J. Magiera and I.S. Krull, *J.Polym.Sci.,Part A:Polym. Chem.*, 1992, **30**, 1809.
403. N. Nishino, Y. Makinose and T. Fujimoto, *Chem.Lett.*, 1992, 77.
404. H. Grun, M. Meldal and K. Breddam, *Biochemistry*, 1992, **31**, 6011.
405. L.L. Maggiora, C.W. Smith and Z.-Y. Zhang, *J.Med.Chem.*, 1992, **35**, 3727.
406. W. Hong, L. Dong, Z. Cai and R. Titmas, *Tetrahedron Lett.*, 1992, **33**, 741.
407. J.W. Skiles, V. Fuchs, C. Miao, R. Sorcek, K.G. Grozinger, S.C. Mauldin, J. Vitous, P.W. Mui, S. Jacober, G. Chow, M. Matteo, M. Skoog, S.M. Weldon, G. Possanza, J. Kierns, G. Letts and A.S. Rosenthal, *J.Med. Chem.*, 1992, **35**, 641.
408. M.R. Angelastro, J.P. Burkhart, P. Bey, and N.P. Peet, *Tetrahedron Lett.*, 1992, **33**, 3265.
409. J.W. Skiles, C. Miao, R. Sorcek, S. Jacober, P.W. Mui, G. Chow, S.M. Weldon, G. Possanza, M. Skoog, J. Keirns, G. Letts and A.S. Rosenthal, *J.Med.Chem.*, 1992, **35**, 4795.
410. P.D. Edwards, E.F. Meyer, J. Vijayalakshmi, P.A. Tuthill, D.A. Andisik, B. Gomes and A. Strimpler, *J.Amer.Chem.Soc.*, 1992, **114**, 1854.
411. H.W. Pauls, B. Cheng and L.S. Reid, *Bioorg.Chem.*, 1992, **20**, 124.
412. K. Hayashi, Y. Hamada and T. Shioiri, *Tetrahedron Lett.*, 1992, **33**, 5075.
413. M. Ferrer, C. Woodward and G. Barany, *Int.J.Peptide Protein Res.*, 1992, **40**, 194.
414. D.L. Maeder, M. Sunde and D.P. Botes, *Int.J.Peptide Protein Res.*, 1992, **40**, 97.
415. K. Kawasaki, K. Hirase, M. Miyano, T. Tsuji and M. Iwamoto, *Chem.Pharm. Bull.*, 1992, **40**, 3253.
416. H. Nakanishi, R.A. Chrusciel, R. Shen, S. Burtenshaw, M.E. Johnson, T.J. Rydel, A. Tulinsky and M. Kahn, *Proc.Natl.Acad.Sci., U.S.A.*, 1992, **89**, 1705.
417. Z. Zheng, R.W. Ashton, F. Ni and H.A. Scheraga, *Biochemistry*, 1992, **31**, 4426.
418. S.A. Poyarkova, V.P. Kukhar, M.J. Kolicheva, S.N. Khrapunov and A.I. Dragan, *Biopolim.Kletka*, 1992, **8**, 20.
419. Z. Szewczuk, B.F. Gibbs, S.Y. Yue, E.O. Purisima and Y. Konishi, *Biochemistry*, 1992, **31**, 9132.
420. J. DiMaio, B. Gibbs, J. Lefebvre, Y. Konishi, D. Munn, S.Y. Yue and W. Hornberger, *J.Med.Chem.*, 1992, **35**, 3331.
421. C.-L.J. Wang, T.L. Taylor, A.J. Mical, S. Spitz and T.M. Reilly, *Tetrahedron Lett.*, 1992, **33**, 7667.
422. A.C. Lellouch and P.T. Lansbury, *Biochemistry*, 1992, **31**, 2279.
423. J.A. Zablocki, M. Miyano, S.N. Rao, S. Panzer-Knodle, N. Nicholson and L. Feigen, *J. Med. Chem.*, 1992, **35**, 4914.
424. P.L. Barker, S. Bullens, S. Bunting, D.J. Burdick, K.S. Chan, T. Deisher, C. Eigenbrot, T.R. Gadek, R. Gantzos, M.T. Lipari, C.D. Muir, M.A. Napier, R.M. Pitti, A. Padua, C. Quan, M. Stanley, M. Struble, J.Y.K. Tom and J.P. Burnier, *J.Med.Chem.*, 1992, **35**, 2040.
425. F. Ni, D.R. Ripoll, P.D. Martin and B.F.P. Edwards, *Biochemistry*, 1992, **31**, 11551.
426. S. Elgendy, J. Deadman, G. Patel, D. Green, N. Chino, C.A. Goodwin, M.F. Scully, V.V. Kakkar and G. Claeson, *Tetrahedron Lett.*, 1992, **33**, 4209.
427. L. Cheng, C.A. Goodwin, M.F. Schully, V.V. Kakkar and G. Claeson, *J.Med. Chem.*, 1992, **35**, 3364.
428. E.J. Iwanowicz, J. Lin, D.G.M. Roberts, I.M. Michel and S.M. Seiler, *Bioorg.Med.Chem.Lett.*, 1992, **2**, 1607.
429. A. Chattopadhyay, H.L. James and D.S. Fair, *J.Biol.Chem.*, 1992, **267**, 12323.
430. R.M. Scarborough, M.A. Naughton, W. Teng, D.T. Hung, J. Rose, T.-K.H. Vu, V.I. Wheaton, C.W. Turck and S.R. Coughlin, *J.Biol.Chem.*, 1992, **267**, 13146.

431. B.H. Chao, S. Kalkunte, J.M. Maraganore and S.R. Stone, *Biochemistry*, 1992, **31**, 6175.
432. M.S. Deshpande and J. Burton, *J.Med.Chem.*, 1992, **35**, 3094.
433. K. Midura-Nowaczek, I. Bruzgo, J. Siemieniuk, W. Roszkowska-Jakimiec and K. Worowski, *Acta Pol.Pharm.*, 1990, **47**, 31.
434. B.K. Handa and C.Kay, *Int.J.Peptide Protein Res.*, 1992, **40**, 363.
435. N.P. Camp, P.C.D. Hawkins, P.B. Hitchcock and D. Gani, *Bioorg.Med.Chem. Lett.*, 1992, **2**, 1047.
436. T. Mimoto, J. Imai, S. Kisanuki, H. Enomoto, N. Hattori, K. Akaji and Y. Kiso, *Chem.Pharm.Bull.*, 1992, **40**, 2251.
437. S. Ikeda, J.A. Ashley, P. Wirsching and K.D. Janda, *J.Amer.Chem.Soc*,. 1992, **114**, 7604.
438. S.K. Grant, M.L. Moore, S.A. Fakhoury, T.A. Tomaszek and T.D. Meek, *Bioorg.Med.Chem.Lett.*, 1992, **2**, 1441.
439. T.J. Tucker, W.C. Lumma, L.S. Payne, J.M. Wai, S.J. de Solms, E.A. Giuliani, P.L. Darke, J.C. Heimbach, J.A. Zugay, W.A. Schleif, J.C. Quintero, E.A. Emini, J.R. Huff and P.S. Anderson, *J.Med.Chem.*, 1992, **35**, 2525.
440. G.B. Dreyer, D.M. Lambert, T.D. Meek, T.J. Carr, T.A. Tomaszek, A.V. Fernandez, H. Bartus, E. Cacciavillani, A.M. Hassell, M. Minnich, S.R. Petteway, B.W. Metcalf and M. Lewis, *Biochemistry*, 1992, **31**, 6646.
441. S.D. Young, L.S. Payne, W.J. Thompson, N. Gaffin, T.A. Lyle, S.F. Britcher, S.L. Graham, T.H. Schultz, A.A. Deana, P.L. Darke, J. Zugay, W.A. Schleif, J.C. Quintero, E.A. Emini, P.S. Anderson, and J.R. Huff, *J.Med.Chem.*, 1992, **35**, 1702.
442. T.K. Sawyer, D.J. Staples, L.Liu, A.G. Tomaselli, J.O. Hui, K. O'Connell, H. Schostarez, J.B. Hester, J. Moon, W.J. Howe, C.W. Smith, D.L. Decamp, C.S. Craik, B.M. Dunn, W.T. Lowther, J. Harris, R.A. Poorman, A. Wlodawer, M. Jaskolski and R.L. Heinrikson, *Int.J.Peptide Protein Res.*, 1992, **40**, 274.
443. D.H. Rich, J.V.N.V. Prasad, C.-Q. Sun, J. Green, R. Mueller, K. Houseman, D. MacKenzie and M. Malkovsky, *J.Med.Chem.*, 1992, **35**, 3803.
444. J. Tözsér, I.T. Weber, A. Gustchina, I. Bláha, T.D. Copeland, J.M. Louis and S. Oroszlan, *Biochemistry*, 1992, **31**, 4793.
445. A. Peyman, K.-H. Budt, J. Spanig, B. Stowasser and D. Ruppert, *Tetrahedron Lett.*, 1992, **33**, 4549.
446. Z. Szewczuk, K.L. Rebholz and D.H. Rich, *Int.J.Peptide Protein Res.*, 1992, **40**, 233.
447. A.M. Doherty, I. Sircar, B.E. Koenberg, J. Quin, R.T. Winters, J.S. Kaltenbronn, M.D. Taylor, B.L. Batley, S.R. Rapundalo, M.J. Ryan and C.A. Painchaud, *J.Med.Chem.*, 1992, **35**, 2.
448. S.E. de Laszlo, B.L. Bush, J.J. Doyle, W.J. Greenlee, D.G. Hangauer, T.A. Halgren, R.J. Lynch, T.W. Schorn and P.K.S. Siegl, *J.Med.Chem.*, 1992, **35**, 833.
449. W.R. Baker, H.-S. Jae, S.F. Martin, S.L. Condon, H.H. Stein, J. Cohen and H.D. Kleinert, *Bioorg.Med.Chem.Lett.*, 1992, **2**, 1405.
450. T.D. Ocain, D.D. Deininger, R. Russo, N.A. Senko, A. Katz, J.M. Kitzen, R. Mitchell, G. Oshiro, A. Russo, R. Stupienski and R.J. McCaully, *J. Med.Chem.*, 1992, **35**, 823.
451. W.C. Patt, H.W. Hamilton, M.D. Taylor, M.J. Ryan, D.G. Taylor, C.J.C. Connolly, A.M. Doherty, S.R. Klutchko, I. Sircar, B.A. Steinbaugh, B.L. Batley, C.A. Painchaud, S.T. Rapundalo, B.M. Michniewicz and S.C. Olson, *J.Med.Chem.*, 1992, **35**, 2562.
452. K.Y. Hui, H.M. Siragy and E. Haber, *Int.J.Peptide Protein Res.*, 1992, **40**, 152.

453. W.T. Ashton, C.L. Cantone, R.L. Tolman, W.J. Greenlee, R.J. Lynch, T.W. Schorn, J.F. Strouse and P.K.S. Siegl, *J.Med.Chem.*, 1992, **35**, 2772.
454. S. Atsuumi, M. Nakano, Y. Koike, S. Tanaka, H. Funabashi, K. Matsuyama, M. Nakano, Y. Sawasaki, K. Funabashi and H. Morishima, *Chem.Pharm.Bull.*, 1992, **40**, 3214.
455. S. Atsuumi, M. Nakano, Y. Koike, S. Tanaka, K. Matsuyama, M. Nakano and H. Morishima, *Chem.Pharm.Bull.*, 1992, **40**, 364.
456. D.V. Patel and D.E. Ryono, *Bioorg.Med.Chem.Lett.*, 1992, **2**, 1089.
457. M.N.G. James, A.R. Sielecki, K. Hayakawa and M.H. Gelb, *Biochemistry*, 1992, **31**, 3872,
458. K.D. Parris, D.J. Hoover, D.B. Damon, and D.R. Davies, *Biochemistry*, 1992, **31**, 8125.
459. M.-C. Fournié-Zaluski, P. Coric, S. Turcaud, E. Lucas, F. Noble, R. Maldonado and B.P. Roques, *J.Med.Chem.*, 1992, **35**, 2473.
460. R. Herranz, S. Vinuesa, C. Pérez, M.T. García-Lopez, M.L. De Ceballos, F.M. Murillo and J. Del Rio, *Arch.Pharm.(Wenheim, Ger.)*, 1992, **325**, 515.
461. R. Gonzalez-Muniz, J.R. Harto, M.L. De Ceballos, J. Del Rio and M.T. García-Lopez, *Arch.Pharm.(Weinheim, Ger.)*, 1992, **325**, 743.
462. M. Thierry, K. Mitsuharu, D. Lucette, D. Pierre, G. Claude, N. Nadine, S.J. Charles and L.J. Marie, *Bioorg.Med.Chem.Lett.*, 1992, **2**, 949.
463. S.S. Ghosh, O. Said-Nejad, J. Roestamadji and S. Mobashery, *J.Med.Chem.*, 1992, **35**, 4175.
464. M. Vincent, C. Pascard, M. Cesario, G. Rémond, J.-P. Bouchet, Y. Charton and M. Laubie, *Tetrahedron Lett.*, 1992, **33**, 7369.
465. D. Grobelny, L. Poncz and R.E. Galardy, *Biochemistry*, 1992, **31**, 7152.
466. V. Dive, A. Yiotakis, C. Roumestand, B. Gilquin, J. Labadie and F. Toma, *Int.J.Peptide Protein Res.*, 1992, **39**, 506.
467. C. Giordano, R. Calabretta, C. Gallina, V. Consalvi and R. Scandurra, *Farmaco*, 1991, **46**, 1497.
468. T.L. Graybill, M.J. Ross, B.R. Gauvin, J.S. Gregory, A.L. Harris, M.A. Ator, J.M. Rinker and R.E. Dolle, *Bioorg.Med.Chem.Lett.*, 1992, **2**, 1375.
469. V.J. Robinson, H.W. Pauls, P.J. Coles, R.A. Smith and A. Krantz, *Bioorg. Chem.*, 1992, **20**, 42.
470. N. Marks, M.J. Berg, R.C. Makofske, J. Swistok, E.J. Simon, D. Ofri, K. Del Compare and W. Danho, *Pept.Res.*, 1992, **5**, 194.
471. T. Fox, E. de Miguel, J.S. Mort and A.C. Storer, *Biochemistry*, 1992, **31**, 12571.
472. S.-T. Chen, S.-L. Lin, S.-C. Hsiao and K.-T. Wang, *Bioorg.Med.Chem. Lett.*, 1992, **2**, 1685.
473. H. Angliker, J. Anagli and E. Shaw, *J.Med.Chem.*, 1992, **35**, 216.
474. M. Hatsu, M. Tuda, Y. Muraoka, T. Aoyagi and T. Takeuchi, *J.Antibiot.*, 1992, **45**, 1088.
475. G.W. Huffman, P.D. Gesellchen, J.R. Turner, R.B. Rothenberger, H.E. Osborne, F.D. Miller, J.L. Chapman and S.W. Queener, *J.Med.Chem.*, 1992, **35**, 1897.
476. P.D. Edwards, *Tetrahedron Lett.*, 1992, **33**, 4279.
477. B.K. Handa and E. Keech, *Int.J.Peptide Protein Res.*, 1992, **40**, 66.
478. G.T. Wang and G.A. Krafft, *Bioorg.Med.Chem.Lett.*, 1992, **2**, 1665.
479. E. Kasafirek, A. Sturc and A. Roubalová, *Coll.Czech.Chem.Commun.*, 1992, **57**, 179.
480. W. Sommergruber, H. Ahorn, A. Zöphel, I. Maurer-Fogy, F. Fessl, G. Schnorrenberg, H.-D. Liebig, D. Blaas, E. Kuechler and T. Skern, *J.Biol. Chem.*, 1992, **267**, 22639.

481. B.A. Malcolm, S.M. Chin, D.A. Jewell, J.R. Stratton-Thomas, K.B. Thudium, R. Ralston and S. Rosenberg, *Biochemistry*, 1992, **31**, 3358.
482. A. Ewenson, R. Laufer, J. Frey, M. Chorev, Z. Selinger and C. Gilon, *Eur.J.Med.Chem.*, 1992, **27**, 179.
483. R.P. Robinson and K.M. Donahue, *J.Org.Chem.*, 1992, **57**, 7309.
484. J.L. Colbran, S.H. Francis, A.B. Leach, M.K. Thomas, H. Jiang, L.M. McAllister and J.D. Corbin, *J.Biol.Chem.*, 1992, **267**, 9589.
485. H.-C. Cheng, H. Nishio, O. Hatase, S. Ralph and J.H. Wang, *J.Biol.Chem.*, 1992, **267**, 9248.
486. A. Kuliopulos, C.E. Cieurzo, B. Furie, B.C. Furie and C.T. Walsh, *Biochemistry*, 1992, **31**, 9436.
487. B. Imperiali, K.L. Shannon, M. Unno and K.W. Rickert, *J.Amer.Chem.Soc.*, 1992, **114**, 7944.
488. P. Gaudreau, P. Brazeau, M. Richer, J. Cormier, D. Langlois and Y. Langelier, *J.Med.Chem.*, 1992, **35**, 346.
489. L.L. Chang, J. Hannah, W.T. Ashton, G.H. Rasmusson, T.J. Ikeler, G.F. Patel, V. Garsky, C. Uncapher, G. Yamanaka, W.L. McClements and R.L. Tolman, *Bioorg.Med.Chem.Lett.*, 1992, **2**, 1207.
490. T.W.C. Lo and P.J. Thornalley, *J.Chem.Soc., Perkin Trans. 1*, 1992, 639.
491. Ösapay and J.W. Taylor, *J.Amer.Chem.Soc.*, 1992, **114**, 6966.
492. P.C. Lyu, H.X. Zhou, N. Jelveh, D.E. Wemmer and N.R. Kallenbach, *J.Amer. Chem.Soc.*, 1992, **114**, 6560.
493. N.E. Zhou, C.M. Kay and R.S. Hodges, *J.Biol.Chem.*, 1992, **267**, 2664.
494. M.R. Ghadiri, C. Soares and C. Choi, *J.Amer.Chem.Soc.*, 1992, **114**, 825.
495. B. Imperiali, S.L. Fisher, R.A. Moats and T.J. Prins, *J.Amer.Chem.Soc.*, 1992, **114**, 3182.
496. C.M. Falcomer, Y.C. Meinwald, I. Choudhary, S. Talluri, P.J. Milburn, J. Clardy and H.A. Scheraga, *J.Amer.Chem.Soc.*, 1992, **114**, 4036.
497. A. Bharadwaj, A. Jaswal and V.S. Chauhan, *Tetrahedron*, 1992, **48**, 2691.
498. K.-H. Altmann, E. Altmann and M. Mutter, *Helv.Chim.Acta*, 1992, **75**, 1198.
499. S. Miick, G.V. Martinez, W.R. Fiori, A.P. Todd and G.L. Millhauser, *Nature*, 1992, **359**, 653.
500. H. Mihara, N. Nishino, R. Hasegawa, T. Fujimoto, S. Usui, H. Ishida and K. Ohkubo, *Chem.Lett.*, 1992, 1813.
501. H. Mihara, N. Nishino and T. Fujimoto, *Chem.Lett.*, 1992, 1809.
502. G. Valle, W.M. Zazmierski, M. Crisma, G.M. Bonora, C. Toniolo and V.J. Hruby, *Int.J.Peptide Protein Res.*, 1992, **40**, 222.
503. B.-Y. Zhu, N.E. Zhou, P.D. Semchuk, C.M. Kay and R.S. Hodges, *Int.J. Peptide Protein Res.*, 1992, **40**, 171.
504. N. Nishino, H. Mihara, Y. Tanaka and T. Fujimoto, *Tetrahedron Lett.*, 1992, **33**, 5767.
505. S.-C. Li and C.M. Deber, *Int.J. Peptide Protein Res*, 1992, **40**, 243.
506. R. Moser, *Protein Eng.*, 1992, **5**, 323.
507. C. Toniolo, F. Formaggio, M. Crisma, G.M. Bonora, S. Pegoraro, S. Polinelli, W.H.J. Boesten, H.E. Shoemaker, Q.B. Broxtermann and J. Kamphius, *Pept.Res.*, 1992, **5**, 56.
508. D.E. Palmer, C. Pattaroni, K. Nunami, R.K. Chadha, M. Goodman, T. Wakamiya, K. Fukase, S. Horimoto, M. Kitazawa, H. Fujita, A. Kubo and T. Shiba, *J.Amer.Chem.Soc.*, 1992, **114**, 5634.
509. A. Lecoq, G. Boussard, M. Marraud and G. Aubry, *Tetrahedron Lett.*, 1992, **33**, 5209.

510. Y. Tor, J. Libman, A. Shanzer, C.E. Felder and S. Lifson, *J.Amer.Chem. Soc.*, 1992, **114**, 6653.
511. C.R. Bertozzi, P.D. Hoeprich and M.D. Bednarski, *J.Org.Chem.*, 1992, **57**, 6092.
512. B. Luening, T. Norberg and J. Tejbrant, *J.Carbohydr.Chem.*, 1992, **11**, 933.
513. R. Polt, L.Szabó, J. Treiberg, Y. Li and V.J. Hruby, *J.Amer.Chem.Soc.*, 1992, **114**, 10249.
514. S. Peters, T. Bielfeldt, M. Meldal, K. Bock and H. Paulsen, *Tetrahedron Lett.*, 1992, **33**, 6445.
515. K. Kojima, T. Takeda and Y. Ogihara, *Chem.Pharm.Bull.*, 1992, **40**, 296.
516. V.O. Kur'yanov, A.E. Zemlyakov and V.Ya. Chirva, *Khim.Prir.Soedin*, 1991, 553.
517. A.M. Jansson, M. Meldal and K. Bock, *J.Chem.Soc., Perkin Trans. 1*, 1992, 1699.
518. H. Iijima, Y. Nakahara and T. Ogawa, *Tetrahedron Lett.*, 1992, **33**, 7907.
519. B. Luening, T. Norberg, C. Rivera-Baeza and J. Tejbrant, *Glycoconjugate J.*, 1991, **8**, 450.
520. L. Urge, L. Otvos, E. Lang, K. Wroblewski, I. Laczko and M. Hollosi, *Carbohydr.Res.*, 1992, **235**, 83.
521. S. Peters, T. Bielfeldt, M. Meldal, K. Bock and H. Paulsen, *J.Chem.Soc., Perkin Trans. 1*, 1992, 1163.
522. M. Schultz and H. Kunz, *Tetrahedron Lett.*, 1992, **33**, 5319.
523. S. Attal, S. Bay and D. Cantacuzene, *Tetrahedron*, 1992, **48**, 9251.
524. H. Kunz and J. März, *Synlett.*, 1992, 591.
525. F. Siedler, S. Rudolph, H.J. Musiol and L. Moroder, *Pept.Res.*, 1992, **5**, 39.
526. M.K. Gurjar and U.K. Saha, *Tetrahedron Lett.*, 1992, **33**, 4979.
527. T. Wakamiya, *Chem.Express*, 1992, **7**, 577.
528. J.W. Perich, R.B. Johns and E.C. Reynolds, *Austr.J.Chem.*, 1992, **45**, 385.
529. W. Bannwarth and E.A. Kitas, *Helv.Chim.Acta*, 1992, **75**, 707.
530. S.M. Domchek, K.R. Auger, S. Chatterjee, T.R. Burke and S.E. Shoelson, *Biochemistry*, 1992, **31**, 9865.
531. C. Garbay-Jaureguiberry, D. Ficheux and B.P. Roques, *Int.J.Peptide Protein Res.*, 1992, **39**, 523.
532. G. Tong, J.W. Perich and R.B. Johns, *Austr.J.Chem.*, 1992, **45**, 1225.
533. J.W. Perich and E.C. Reynolds, *Austr.J.Chem.*, 1992, **45**, 1765.
534. J.W. Perich, R.M. Valerio and R.B. Johns, *Austr.J.Chem.*, 1992, **45**, 919.
535. J.W. Perich, *Int.J.Peptide Protein Res.*, 1992, **40**, 134.
536. P. Coutrot, C. Grison and C. Charbonnier-Gérardin, *Tetrahedron*, 1992, **48**, 9841.
537. J.W. Perich and E.C. Reynolds, *Bioorg.Med.Chem.Lett.*, 1992, **2**, 1153.
538. J.W. Perich, D.P. Kelly and E.C. Reynolds, *Int.J.Peptide Protein Res.*, 1992, **40**, 81.
539. R. Chen, Y. Zhang and M. Cheng, *Gaodeng Xuexiao Huaxue Xuebao*, 1992, **13**, 1075.
540. P. Wohlfart, W. Haase, R.S. Molday and N.J. Cook, *J.Biol.Chem.*, 1992, **267**, 644.
541. J.-P. Defoort, B. Nardelli, W. Huang and J.P. Jam, *Int.J.Peptide Protein Res.*, 1992, **40**, 214.
542. F. Baleux and P. Dubois, *Int.J.Peptide Protein Res.*, 1992, **40**, 7.
543. R.J. Epstein, B.J. Druker, T.M. Roberts and C.D. Stiles, *Proc.Natl.Acad. Sci., U.S.A.*, 1992, **89**, 10435.
544. H.J. Dyson, E. Norrby, K. Hoey, D.E. Parks, R.A. Lerner and P.E. Wright, *Biochemistry*, 1992, **31**, 1458.
545. G. Tuchscherer, C. Servis, G. Corradin, U. Blum, J. Rivier and M. Mutter, *Protein Sci.*, 1992, **1**, 1377.

546. K.S.N. Iyer, S. Upadhye, L.R. Kadam, S.D. Mahale, S. Dhanasekharan, U. Natraj and T.D. Nandedkar, *Int.J.Peptide Protein Res.*, 1992, **39**, 137.
547. Z. Mackiewicz, H. Swiderska, M. Kalmanowa, Z. Smiatacz, A. Nowoslawski and G. Kupryszewski, *Coll.Czech.Chem.Commun.*, 1992, **57**, 204.
548. A.M. Kolodziejczyk, A.S. Kolodziejczyk and S. Stoev, *Int.J.Peptide Protein Res.*, 1992, **39**, 382.
549. Z. Balajthy, J. Aradi, I. Kiss and P. Elödi, *J.Med.Chem.*, 1992, **35**, 3344.
550. M. Egholm, O. Buchardt, P.E. Nielsen and R.H. Berg, *J.Amer.Chem.Soc.*, 1992, **114**, 1895.
551. M. Hagihara, N.J. Anthony, T.J. Stout, J. Clardy and S.L. Schreiber, *J. Amer.Chem.Soc.*, 1992, **114**, 6568.
552. D. Ranganathan, K. Shah and N. Vaish, *J.Chem.Soc., Chem.Commun.*, 1992, 1145.
553. G. Angelini, C. Sparapani and A. Margonelli, *J.Labelled Compd. Radiopharm.*, 1992, **31**, 739.
554. F. Berti, C. Ebert and L. Gardossi, *Tetrahedron Lett.*, 1992, **33**, 8145.
555. B. Padmanabhan, S. Dey, D. Khandelwal, G.S. Rao and T.P. Singh, *Biopolymers*, 1992, **32**, 1271.
556. A. Hammadi, J.M. Nuzillard, J.C. Poulin and H.B. Kagan, *Tetrahedron Assym.*, 1992, **3**, 1247.
557. N. Voyer and B. Guérin, *J.Chem.Soc., Chem.Commun.*, 1992, 1253.
558. N.G. Luk'yanenko, S.S. Basok, N.V. Kulikov, T.L. Karaseva and Zh.N. Tsapenko, *Khim.-Farm.Zh.*, 1992, **26**, 63.
559. N. Bodor, L. Prokai, W.-M. Wu, H. Farag, S. Jonalagadda, M. Kawamura and J. Simpkins, *Science*, 1992, **257**, 1698.
560. M. Zinic, L. Frkanec, V. Skaric, J. Trafton and G.W. Gokel, *Supramol. Chem.*, 1992, **1**, 47.
561. P. Hermann, H. Baumann, C. Herrnstadt and D. Glanz, *Amino Acids*, 1992, **3**, 105.
562. V. Magafa, G. Stavropoulos, T. Zafiropoulos, J.P. Laussac and N. Hadjiliadis, *C.R.Acad.Sci., Ser II*, 1992, **315**, 169.
563. S. Ranganathan and N. Jayaraman, *Tetrahedron Lett.*, 1992, **33**, 6681.
564. G. Byk and C. Gilon, *J.Org.Chem.*, 1992, **57**, 5687.
565. B. Di Blasio, F. Rossi, E. Benedetti, V. Pavone, M. Saviano, C. Pedone, G. Zanotti and T. Tancredi, *J.Amer.Chem.Soc.*, 1992, **114**, 8277.
566. I. Toth, G.J. Anderson, R. Hussain, I.P. Wood, E. del Olmo Fernandez, P. Ward and W.A. Gibbons, *Tetrahedron*, 1992, **48**, 923.
567. R. Hussain, I. Toth and W.A. Gibbons, *Liebig's Ann. Chem.*, 1992, 169.
568. H.B.A. De Bont, J.H. Van Boom and R.M.J. Liskamp, *Rec.Trav.Chim.Pays- Bas*, 1992, **111**, 222.
569. S. Natarajan, S.M. Festin, A. Hedberg, E.C.-K. Liu, D.M. Floyd and J.T. Hunt, *Int.J.Peptide Protein Res.*, 1992, **40**, 567.
570. C.H. Tung, T. Zhu, H. Lackland and S. Stein, *Peptide Res.*, 1992, **5**, 115.
571. H. Tamiaki and K. Maruyama, *J.Chem.Soc., Perkin Trans. 1*, 1992, 2431.
572. N. Galéotti, C. Montagne, J. Poncet and P. Jouin, *Tetrahedron Lett.*, 1992, **33**, 2807.
573. Y. Yuasa, T. Watanabe, A. Nagakura, H. Tsuruta, G.A. King, J.G. Sweeny and G.A. Iacobucci, *Tetrahedron*, 1992, **48**, 3473.
574. F. Formaggio, M. Crisma, G. Valle, C. Toniolo, W.H.J. Boesten, H.E. Schoemaker, J. Kamphuis and P.A. Temussi, *J.Chem.Soc., Perkin Trans. 2*, 1992, 1945.
575. S. Polinelli, Q.B. Broxterman, H.E. Schoemaker, W.H.J. Boesten, M. Crisma, G. Valle, C. Toniolo and J. Kamphuis, *Bioorg.Med.Chem.Lett.*, 1992, **2**, 453.
576. B. Imperiali and S.L. Fisher, *J.Org.Chem.*, 1992, **57**, 757.

577. D. Permentier, S. Vansteenkiste, E. Schacht, H. Vermeersch and J.P. Remon, *Bull.Soc.Chim.Belg.*, 1992, **101**, 701.
578. Y. Hirano, Y. Kando, T. Hayashi, K. Goto and A. Nakajima, *J.Biomed. Mater.Res.*, 1991, **25**, 1523.
579. M.J. Bogusky, A.M. Naylor, S.M. Pitzenberger, R.F. Nutt, S.F. Brady, C.D. Colton, J.T. Sisko, P.S. Anderson and D.F. Veber, *Int.J.Peptide Protein Res.*, 1992, **39**, 63.
580. F.A. Robey, T.A. Harris, N.H.H. Heegaard, A.K. Nguyen and D. Batinic, *Chim.Oggi*, 1992, **10**, 27.
581. T. Kolter, A. Klein and A. Giannis, *Angew.Chem., Int.Ed.*, 1992, **31**, 1391.
582. C.-B. Xue, J.M. Becker and F. Naider, *Tetrahedron Lett.*, 1992, **33**, 1435.
583. R. Herranz, S. Vinuesa, C. Pérez, M.T. García-López, E. López, M.L. de Caballos and J. Del Rio, *J.Med.Chem.*, 1992, **35**, 889.
584. T.C. Bruice, H.-Y. Mei, G.-X. He and V. Lopez, *Proc.Natl.Acad.Sci., U.S.A.*, 1992, **89**, 1700.
585. W.S. Wade, M. Mrksich and P.B. Dervan, *J.Amer.Chem.Soc.*, 1992, **114**, 8783.
586. C. García-Echeverría, M. Pons, E. Giralt and F. Albericio, *Bioorg.Med. Chem.Lett.*, 1992, **2**, 281.
587. A.V. Reddy and B. Ravindranath, *Int.J.Peptide Protein Res.*, 1992, **40**, 472.
588. I. Marle, A. Karlsson and C. Petterson, *J.Chromatogr.*, 1992, **604**, 185.
589. K. Oyama, M. Itoh and Y. Nonaka, *Bull.Chem.Soc.Jpn.*, 1992, **65**, 1751.
590. M. Pugnière. B. Castro, N. Domergue and A. Previero, *Tetrahedron Assym.*, 1992, **3**, 1015.
591. T. Miyazawa, H. Iwanaga, T. Yamada and S. Kuwata, *Chirality*, 1992, **4**, 427.
592. A.J. Chiou, S.H. Wu and K.T. Wang, *Biotechnol.Lett.*, 1992, **14**, 461.
593. F.M. Bautista, J.M. Campelo, A. Garcia, D. Luna and J.M. Marinas, *Amino Acids*, 1992, **2**, 87.
594. C. Chabanet and M. Yvon, *J.Chromatogr.*, 1992, **599**, 211.
595. A.E. Zemlyakov, D.A. Poteev and V.Ya. Chirva, *Khim.Prir.Soedin*, 1991, 865.
596. M. Knight and K. Takahashi, *Liq.Chromatogr.*, 1992, **15**, 2819.
597. M.H. Hyun, J.J. Ryoo, C.S. Min and W.H. Pirkle, *Bull.Korean Chem.Soc.*, 1992, **13**, 407.
598. O. Boniface, P. Fourquet, E. Boschetti and L. Guerrier, *J.Liq. Chromatogr.*, 1992, **15**, 2183.
599. S.E. Blondelle, K. Büttner and R.A. Houghten, *J. Chromatogr.*, 1992, **625**, 199.

3
Analogue and Conformational Studies on Peptide Hormones and Other Biologically Active Peptides

By C. M. BLADON

1 Introduction

Material for this review, the format of which is the same as last year's report, was obtained by scanning the major organic and bioorganic journals and Chem. Abs. Selects: Amino Acids, Peptides, and Proteins up to May 17th 1993 (issue 10). The numbers of papers relevant to this chapter continues to increase and 10% more articles are reviewed this year compared to last year. As is common practice no attempt has been made to review patent or symposium proceedings. The ratio observed over recent years for the numbers of papers included in this review which originate from academic institutes compared with those from industrial organisations remains approximately the same at 3 : 1.

2 Peptide-backbone Modifications

2.1 *ψ[CSNH]-Thioamide Analogues.*

Elongation of the C-terminus of thiodipeptides has been achieved using proteases.[1] Reaction of Boc-D-Alaψ[CSNH]Phe-OMe with valine allyl ester in the presence of chymotrypsin gave a 70% yield of the thiopeptide Boc-D-Alaψ[CSNH]Phe-Val-OAll which was subsequently incorporated into larger fragments. Chain extension at the N-terminus of H_2N-Leuψ[CSNH]Met-OMe, first by addition of a single histidine residue and then by coupling a hexapeptide, was the route adopted for the synthesis of the bombesin analogue H-Asn-Gln-Trp-Ala-Val-Gly-His-Leuψ[CSNH]Met-NH_2.[2] Replacing the amide bonds of the pyroglutamic acid and/or the prolinamide moieties of thyrotropin-releasing hormone (TRH) with the thioamide isostere resulted in compounds (1a,b) which were more selective (higher affinity to pituitary rather than cortical receptors) than the parent hormone.[3] The analogue (1c) containing a thioamide replacement in the pyroglutamyl ring was not selective. The syntheses of the bombesin and TRH analogues both used Lawesson's reagent for the thionation steps. Lawesson's reagent proved

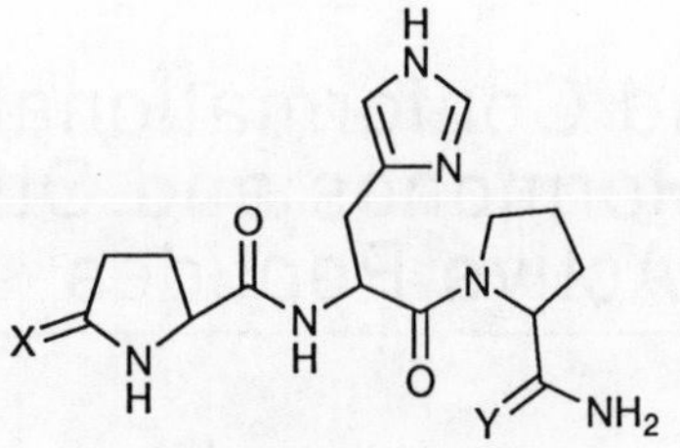

(1) a, X = Y = S;
b, X = O, Y = S;
c, X = S, Y = O

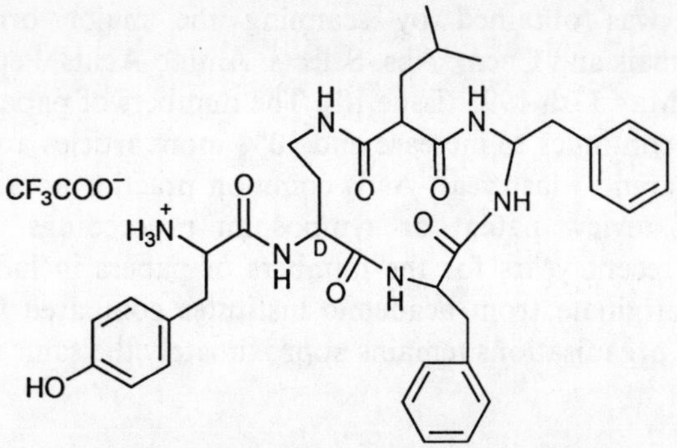

(2) a, TFA, H-Tyr-C[D-A_2bu-Phe-gPhe-*S*-mLeu]
b, TFA, H-Tyr-C[D-A_2bu-Phe-gPhe-*R*-mLeu]

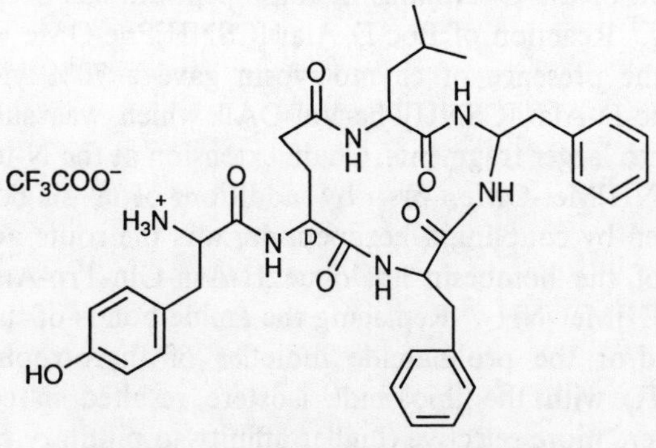

(3) a, TFA, H-Tyr-C[D-Glu-Phe-gPhe-L-rLeu]
b, TFA, H-Tyr-C[D-Glu-Phe-gPhe-D-rLeu]

unsuitable for the thionation of cyclo(Gly-Pro-Phe-Val-Phe-Phe) but a variant, Yokoyama's reagent, effected O/S exchange[4] at the Phe^5 C=O. Although substitution at this position was unexpected, NMR conformational studies indicated that this carbonyl oxygen was more accessible to the bulky reagent than that of glycine.

2.2 *ψ[NHCO]-Retro-inverso Analogues.*

The retro-inverso moiety of H-Thr-Tyr-gGln-m(*R,S*)Arg-Thr-Arg-Ala-Leu-Val-OH was generated by coupling H_2N-D-Glu(Bzl)←m(*R,S*)Arg(Mtr)-OH to the resin bound C-terminal pentapeptide followed by a Hofmann rearrangement with [bis(trifluoroacetoxy)iodo]benzene to convert the *N*-terminal amide to the amine.[5] Hofmann rearrangement of Boc-Phe-Phe-NH_2 to the *gem*-diaminoalkyl dipeptide, chain extension, and cyclisation of the resulting linear pentapeptide was the route adopted for the preparation[6] of the 14-membered cyclic dermorphin analogues (2, 3) containing either one or two reversed amide bonds. NMR studies, supported by molecular dynamics simulations, suggested that H-Tyr-cyclo(D-A_2bu-Phe-gPhe-*S*-mLeu) (2a), which is superactive at both the μ- and δ-opioid receptors, was conformationally flexible and adopted both the extended and folded conformations required for bioactivity at the μ- and δ-receptors respectively.[7] The *R*-mLeu analogue (2b) adopted only extended conformations and showed higher affinity for the μ-receptor then the δ-receptor whereas Tyr-cyclo(D-Glu-Phe-gPhe-L-rLeu) (3a), the only analogue which displayed selectivity for the δ-receptor, assumed a compact conformation with the Tyr side chain folded back over the 14-membered ring. The unambiguous assignment of the absolute configuration for the peptidomimetic residues in these analogues was based on the use of nOes and vicinal $J_{NH\text{-}CH}$ coupling constants. This technique has also been applied to the assignment of the mPhe residue in the cyclic hexapeptide somatostatin analogues cyclo[$gSar^6$-(*S* and *R*)-$mPhe^7$-D-Trp^8-Lys^9-Thr^{10}-Phe^{11}] and to the stereoisomers of 2-Ac^5c in the somatostatin and morphiceptin analogues cyclo[$(2\text{-}Ac^5c)^6$-Phe^7-D-Trp^8-Lys^9-Thr^{10}-Phe^{11}] and H-Tyr-*cis*-2-Ac^5c-Phe-Val-NH_2 respectively.[8]

Application of the retro-inverso modification to the tripeptide alcohol renin inhibitor Boc-Phe-His-Leu-ol is described in section 5.2. The mass spectrometry of retro-inverso peptides has been reviewed.[9]

2.3 *ψ[CH_2NH]-Amino Methylene Analogues.*

Sequential replacement of each peptide bond with the reduced isostere in the C-terminal hexapeptide of neurotensin combined with amino acid substitutions yielded H-Lysψ[CH_2NH]Lys-Pro-Tyr-Ile-Leu-

OH. This compound exhibited the same affinity as the parent peptide for neurotensin receptors but was approximately 10 times more potent in stimulating guinea pig ileum contraction.[10] A similar type of synthetic approach with neurokinin A fragments led to H-Asp-Ser-Phe-Val-βAla-Leuψ[$CH_2N(CH_2)_2CH_3$]Phe-NH_2 which was the most potent member of a new class of neurokinin A partial agonists/antagonists.[11] Enhanced enzymatic resistance and improved selectivity for the μ-opioid receptor were observed in β-casomorphin-5 and morphiceptin analogues with Pheψ[CH_2N]Pro or Proψ[CH_2NH]Phe moieties.[12] Furthermore, reduction of the Pro^2-Phe^3 amide bond in β-casomorphin-5 resulted in the conversion of an agonist into an antagonist. μ-Opioid receptor selectivity was increased 5-fold by substituting the D-Ala-Phe amide bond in the dermorphin tetrapeptide H-Tyr-D-Ala-Phe-Gly-NH_2 with the reduced isostere.[13] This increase in selectivity was largely due to a reduction in binding affinity for the δ-receptor. All the analogues of the μ-receptor antagonist H-D-Phe-Cys-Tyr-D-Trp-Orn-Thr-Pen-Thr-NH_2 with a ψ[CH_2NH] bond at positions 2-3 or 3-4 and/or a pipecolic acid derived amino acid at position 3 or 4 exhibited lower affinity for both the δ- and μ-receptors than the parent peptide.[14]

2.4 *ψ[CH=CH] and ψ[CH_2CH_2]-Ethylenic and Carba Analogues.*

Allylic rearrangement of γ-mesyloxy-α,β-unsaturated esters using mixed metal cyanocuprates $RCu(CN)M.BF_3$ to produce (E)-alkene dipeptide isosteres has been reviewed.[15] Side reactions leading to reductive elimination products were effectively suppressed by using organozinc cuprates $R_2Cu(CN)(ZnCl)_2.2Mg(X)Cl.\ nLiCl$ instead of the organocyanocopper reagents.[16] The versatility of an alternative route to alkene dipeptide isosteres using the [2,3]-Wittig-Still rearrangement was demonstrated by the synthesis of Glyψ[CH=CH]Ala and Pheψ[CH=CH]Gly (Scheme 1).[17] By switching from THF to hexane for the rearrangement reaction the yields of the required [2,3]-products were increased and only trace amounts of the [1,2]-byproduct were isolated. Investigations into the role of the peptide backbone on receptor interactions prompted the synthesis of Boc-Glyψ[E,CH=CH]D,L-Phe-OH, its saturated analogue, and their incorporation into enkephalin analogues. Routes to the dipeptide units include alkylation and decarboxylation of 3-carboethoxy-2-piperidone[18] or asymmetric alkylation of Glyψ[E,CH=CH]Gly using Evans-type methodology[19] (see Scheme 2). In general the enkephalin analogues incorporating this modification all showed greatly diminished potencies. These results supported the proposals that the Gly^3 carbonyl is essential for interaction with the opioid receptors. Reversal of the Phe^3-Gly^4 sequence

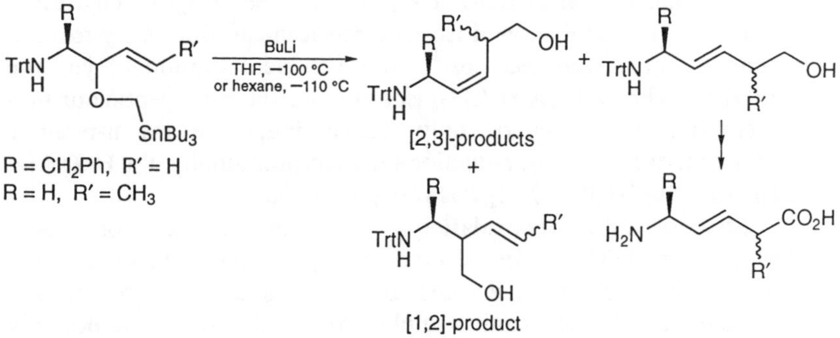

R = CH_2Ph, R′ = H
R = H, R′ = CH_3

BuLi
THF, −100 °C
or hexane, −110 °C

[2,3]-products
+
[1,2]-product

Scheme 1

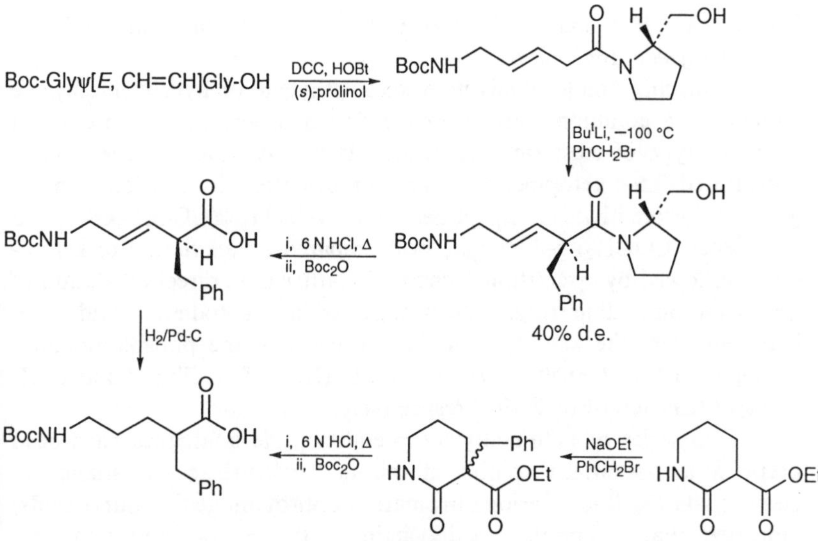

Boc-Glyψ[*E*, CH=CH]Gly-OH

DCC, HOBt
(*s*)-prolinol

Bu^tLi, −100 °C
$PhCH_2Br$

40% d.e.

i, 6 N HCl, Δ
ii, Boc_2O

H_2/Pd-C

NaOEt
$PhCH_2Br$

i, 6 N HCl, Δ
ii, Boc_2O

Scheme 2

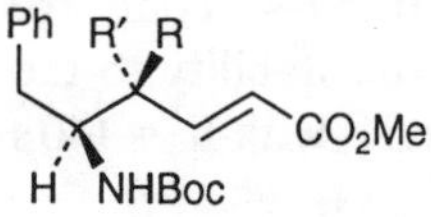

(4)

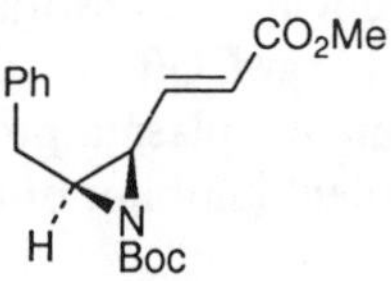

(5)

in dermorphin (H-Tyr-D-Ala-Phe-Gly-Tyr-Pro-Ser-NH_2) to Gly^3-Phe^4, combined with a double bond isosteric replacement, drastically reduced binding to the opioid receptor.[20] However incorporation of just the ψ[E,CH=CH] or the ψ[CH_2CH_2] isostere into the heptapeptide or into the N-terminal tetrapeptide only slightly influenced the μ-receptor affinity and, due to a large reduction on δ-receptor affinity, the H-Tyr-D-Ala-Pheψ[CH_2CH_2]Gly-NH_2 was very μ-selective.

Stereoselective epoxidation on the β-face of Boc-Pheψ[E,CH=CH]Gly-OMe followed by ring opening with fluoride ion yielded the (*S*)-alcohol (4, R=OH, R^1=H).[21] Reaction of the unsaturated alcohol with diethylamino sulphur trifluoride gave rise to not only the expected fluoro compound (4, R=H, R^1=F) but also the aziridine (5).

2.5 *ψ[$COCH_2$] and ψ[$CH(OH)CH_2$]-Ketomethylene and Hydroxymethylene Analogues.*

Replacing amide bonds with ketomethylene or hydroxymethylene moieties is a common strategy in the design of enzyme inhibitors and thus this type of isosteric modification is largely covered in section 5. Substituted 2,5-diketopiperidines are conformationally restricted analogues of ketomethylene dipeptides and cyclo[Trpψ($COCH_2$)Gly] and cyclo[Pheψ($COCH_2$)-ξ-Phe] were best prepared by alkylation of ketodiesters followed by cyclisation (Scheme 3) rather than direct alkylation of the diketopiperidine ring.[22] Alkylation of a 4-ketodiester and a β-ketoester were the key steps in the syntheses of the phosphoramidon analogue H_2O_3P-Leuψ[$COCH_2$]Xaa-OMe (Xaa=Phe, Trp)[23] and of Z-Pheψ[$COCH_2$]Gly-Pro-OMe[24] respectively.

Two series of ketomethylene pseudopeptide analogues have been prepared as potential thrombin inhibitors. Structure-activity studies on compounds (6, Xaa=various aromatic nonproteinogenic amino acids) indicated that a lipophilic side chain at the P_3 position was very important for binding to the apolar site of thrombin; for example (6, Xaa=D-Dpa) (Dpa=β,β-diphenylalanine) was 7 times more active than the D-Phe analogue.[25] Substitution of the scissile amide function in hirutonins by [ArgψCO(CH_2)$_n$CO] (n=1-4) conferred subnanomolar affinity and complete plasma proteolytic stability to the peptide and (7) was the most potent inhibitor of the series with K_i=140+/-20pM.[26]

2.6 *Phosphono-Peptides.*

Methods for the synthesis and incorporation into peptide sequences of the α-aminophosphonic acid derivatives (8) have been described by several groups of workers.[27-30] Phosphono and phosphino analogues of

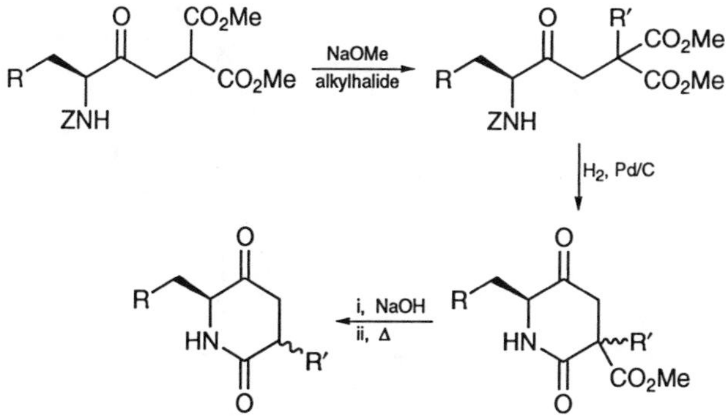

NaOMe alkylhalide

H_2, Pd/C

i, NaOH ii, Δ

Scheme 3

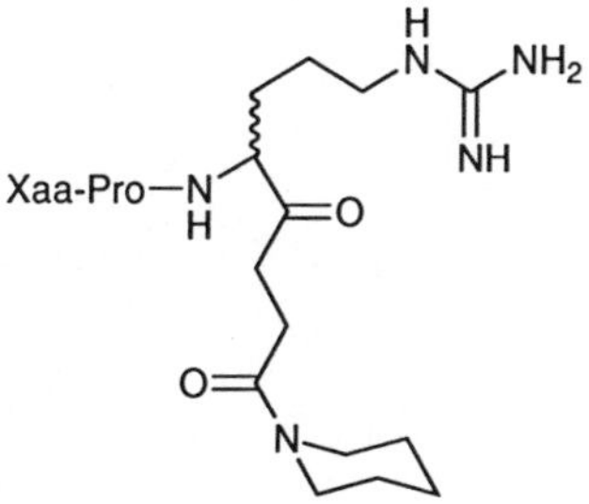

(6)

Ac—D-Phe-Pro-Argψ[CO(CH$_2$)$_2$CO]Gln-Ser-His-Asn-Asp-Gly-Asp-Phe-Glu-Glu-Ile-Pro-Glu
HO—Gln-Leu-Tyr-Glu

(7)

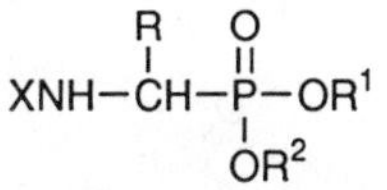

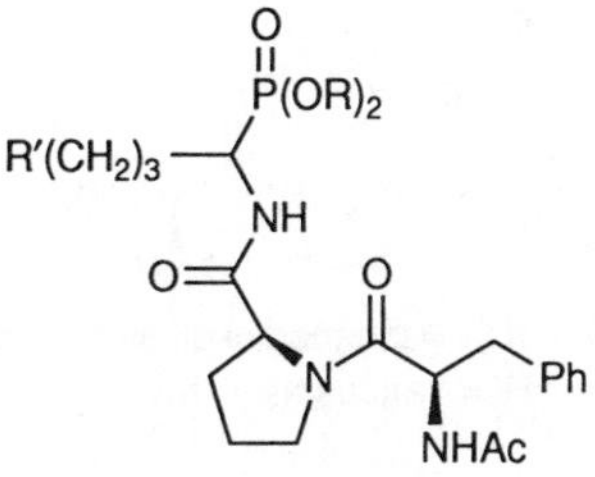

(8) X = PhthNCH$_2$CO, Z; $R^1 = R^2$ = Et
X = Fmoc; R^1 = H; R^2 = Bn
X = H; $R^1 = R^2$ = Me, Et

(9) R = Ph, Me; R′ = NHC(=NH)NH$_2$, CH$_2$NH$_2$

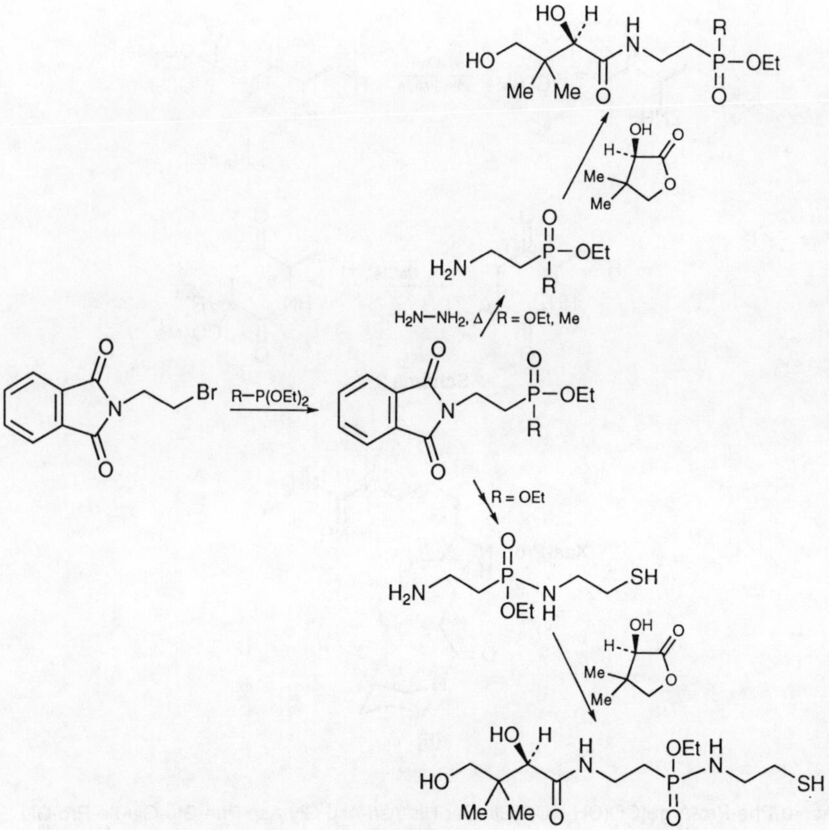

HO H H R
HO N P–OEt
Me Me O O

OH
H O
Me
Me O

O
H_2N P–OEt
R

$H_2N{-}NH_2$, Δ R = OEt, Me

O O
N Br R–P(OEt)$_2$ N P–OEt
O O R

R = OEt

O
H_2N P N SH
OEt H

OH
H O
Me
Me O

HO H H OEt H
HO N P N SH
Me Me O O

Scheme 4

R′ B(OR)$_2$
NH
O O
N Ph
ZNH

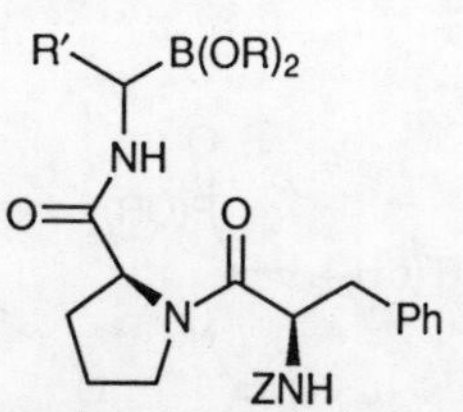

(10) $(OR)_2$ = pinanediol diester, pinacol diester
R′ = various alkyl, aryl

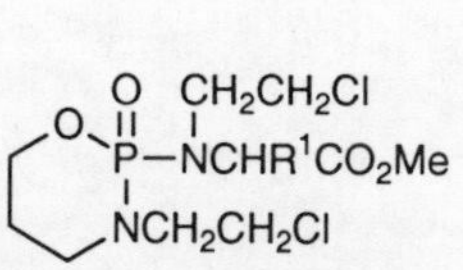

O O CH_2CH_2Cl
P–$NCHR^1CO_2Me$
NCH_2CH_2Cl

(11)

O Ph
R^2 P
NH
HN
R^1
O

O
[N]$_2$–$PNRCHR^1CO_2Me$

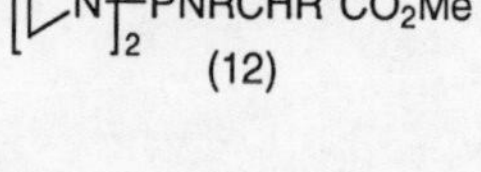

(12)

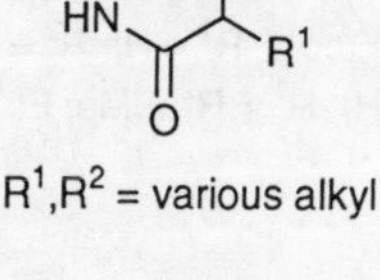

(15) R^1,R^2 = various alkyl

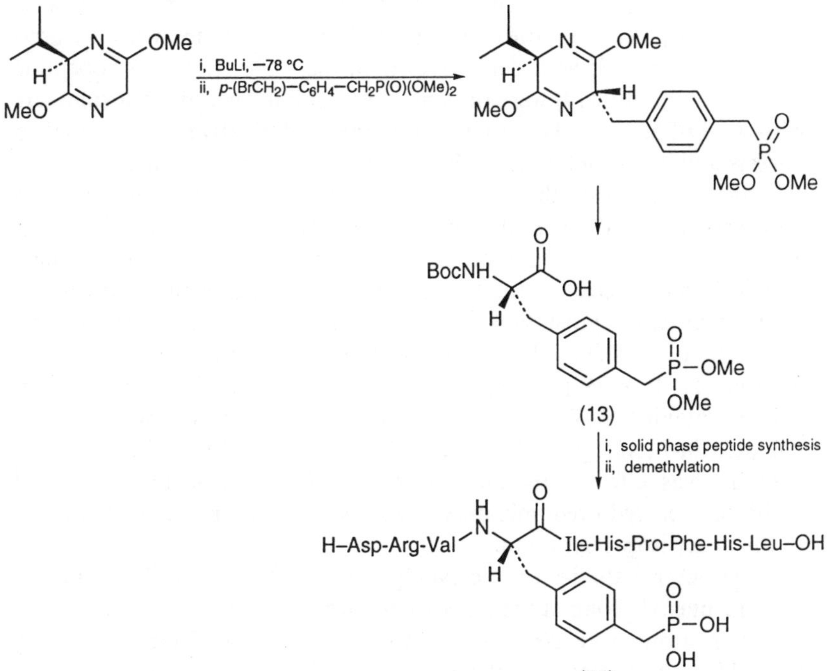

i, BuLi, −78 °C
ii, p-($BrCH_2$)−C_6H_4−$CH_2P(O)(OMe)_2$

(13)

i, solid phase peptide synthesis
ii, demethylation

H–Asp-Arg-Val Ile-His-Pro-Phe-His-Leu–OH

(14)

Scheme 5

EDC

BPCONH CO_2H
(BP = biphenyl)

Na/Hg

$NaBH_4$
$R^1 = R^2 = O$
$R^1 = H,\ R^2 = OH$

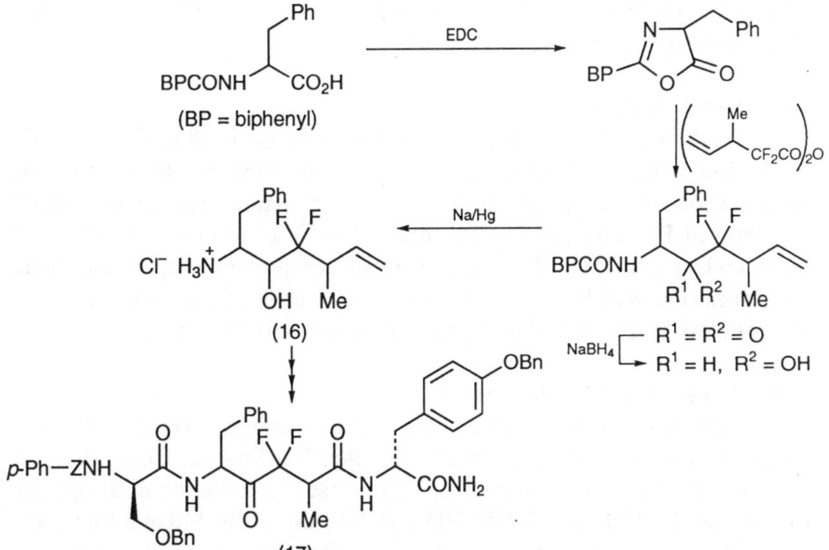

(16)

(17)

Scheme 6

members of the pantothenic acid family of natural products were prepared by condensation of (−)-(*R*)-pantolactone with (2-aminoethyl)phosphonic acid derivatives (Scheme 4).[31] The phosphopeptide thrombin inhibitors (9) were obtained by coupling an aminophosphonate with Ac-D-Phe-Pro-OH.[32] Only the diphenyl derivatives were found to possess any inhibitory activity (IC_{50}s in μmolar range) but these were less effective as thrombin inhibitors than the boronic acid inhibitors on which they were based. The peptide boronic acids (10) with a neutral side chain at the P_1 position showed good thrombin inhibition (K_i in nmolar range) and were more selective for this enzyme than other serine proteases.[33] Various levels of cytostatic activity were exhibited by the *N*-phosphorylated amino acids (11, 12) and the most potent was the 1,3,2-oxazaphosphorine (11, R^1 = Me).[34] Preparation of Boc-*p*-dimethyl phosphomethyl-L-phenylalanine (13) was a key element in the synthesis of the angiotensin-I analogue (14) (Scheme 5).[35] Stereocontrol in the alkylation reaction was effected by the side chain of the bis-lactim ether and addition occurred predominantly to the side of the ring opposite to that of the bulky isopropyl group.

Structural studies on the cyclophosphonodipeptides (15) indicated that the phenyl group at the phosphorus atom and the substituents at the α-carbon atom were *cis* to each other and that the 6-membered ring adopted a twist-boat conformation.[36] The linear phosphonodipeptides $R_2O_3PCH_2NHCH_2CONHCHR^1PO_3Ph_2$ (R = Et, Ph; R^1 = H, Ph) and $Et_2O_3PCH_2NHCH_2CONHCHR^1CO_2R^2$ (R^1 = H, Me, CH_2Ph; R^2 = Me, Et) showed poor herbicidal activities.[37]

2.7 *ψ[SO₂NH] Analogues.*

Acylation of phenylalanine methyl ester or proline methylamide with $BocNHCH(CH_2Ph)CH_2SOCl$ was an effective route for the preparation of the ψ[SONH], and by further oxidation, the ψ[SO_2NH] isosteres of Phe-Phe and Phe-Pro dipeptides.[38] In the crystal structure of Bz-Phe-Tau-Leu-OMe the central SO_2NH component adopted a cisoidal conformation which maintained the two C^α atoms in a spatial orientation similar to that found in peptides with a *cis* amide bond.[39]

2.8 *C-Terminal Modifications.*

The synthetic utility of α-amino pentafluoroethyl ketones has been demonstrated by the preparation of Boc-Val-Pro-Val-$COCF_2CF_3$, a potent inhibitor of human neutrophil elastase.[40] Alkylation of azaglycine dipeptides XNH-Phe-CO-NH-NH-$CONH_2$ using the Mitsunobu procedure (PPh_3, diethyl azodicarboxylate) was found to be regioselective with reaction only occurring at the more acidic $N_\beta H$ proton.[41]

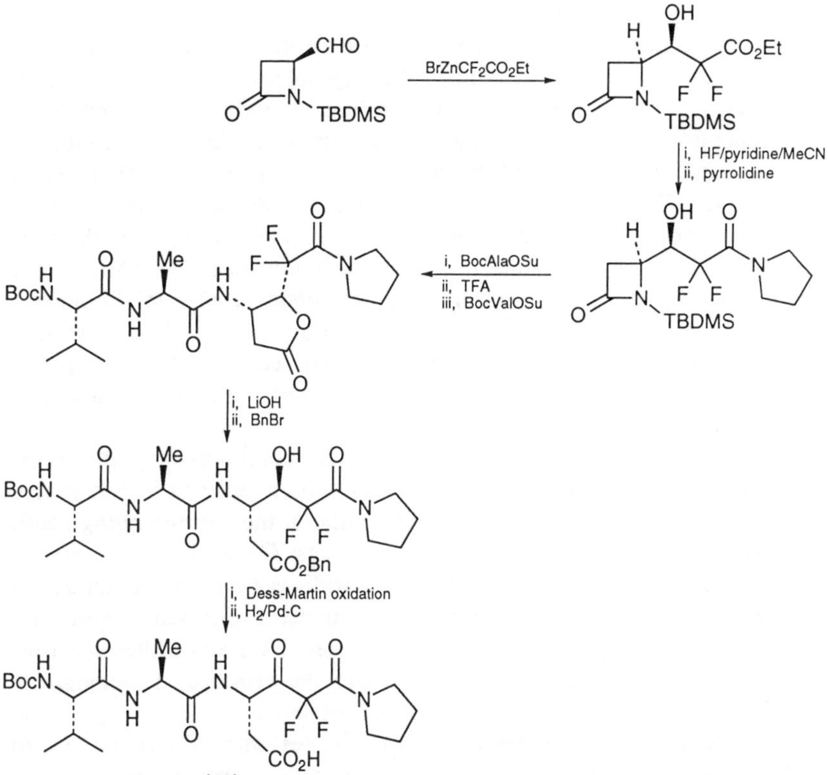

(18)

Scheme 7

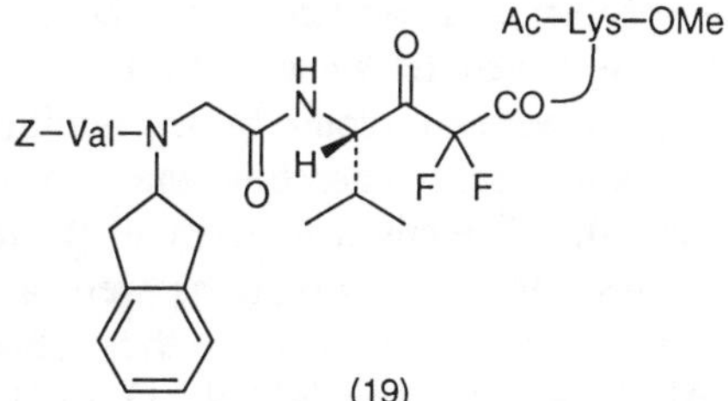

(19)

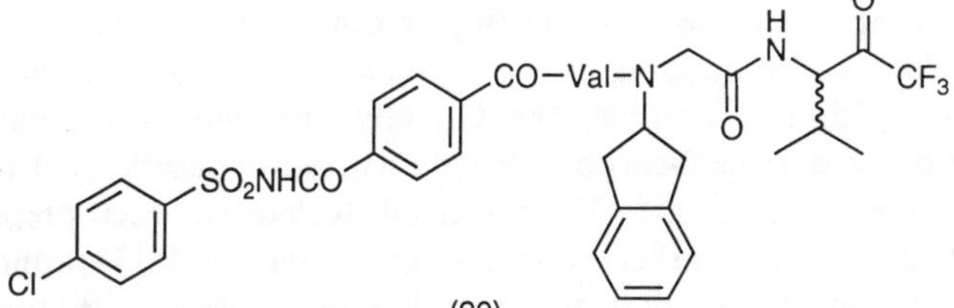

(20)

2.9 *Miscellaneous Modifications.*

A versatile route to peptidyl α,α-difluoroalkyl ketones has been developed (Scheme 6) in which the key intermediate (16) could be extended at either the N- or C-terminus to yield, for example, the tetrapeptide species (17).[42] In a different approach the peptidyl difluoro ketone inhibitor (18) of interleukin-1β converting enzyme was prepared by a route (Scheme 7) involving chain extension at the N-terminus of the difluoro moiety.[43] In a series of α,α-difluoromethylene ketone inhibitors of human leukocyte elastase the preferred residue at the P_1 position was α,α-difluorostatone and one of the most potent *in vitro* inhibitors was (19) with an IC_{50} of 0.057μM.[44] The related inhibitor (20) (IC_{50} = 0.084μM) was also evaluated for *in vivo* activity and was found to inhibit elastase-induced pulmonary haemorrhage in hamsters in a dose dependent manner with an ED_{50} of 4.8μg.[45]

Cyclisation of serine and threonine containing thiopeptides to the thiazolines under Mitsunobu conditions has been reported by two groups of workers (Scheme 8). However, cyclisation of the corresponding amide derivatives led to different reaction products; Galeotti and co-workers isolated the *cis*-oxazolines[46] whereas aziridines (21) were obtained by Wipf and Miller.[47] On the basis of further experiments with *allo*-threonine, which yielded the *trans*-oxazolines, Wipf and Miller proposed that destabilising *gauche* interactions and the presence of moderate to strong base determined the course of the reactions.

Cyclisation of a series of nitrile oxides with *N*-acryloyl-(2*R* or 2*S*)-borane-10,2-sultam gave a series of pseudodipeptides incorporating the 2-isoxazoline moiety as the amide bond surrogate (Scheme 9).[48] Racemisation problems associated with the conversion of protected dipeptides into protected tetrazole dipeptides (conformational mimics of *cis*-amide bond) were overcome by the addition of quinoline to the reaction with PCl_5 when generating the imidoyl chloride from the amide (Scheme 10).[49] The chiral integrity of the tetrazole dipeptide was maintained if basic reaction conditions were avoided during peptide synthesis, and for example, the bradykinin analogues [Pro^{2}ψ[CN_4]Ala^3]-BK and [Ala^{6}ψ[CN_4]-D or L-Ala^7]-BK were prepared using acid labile protecting groups. Several Glyψ[C_4N]Xaa dipeptides (22) containing the *cis* peptide bond isostere have been prepared and, for example, the Gly-Gly analogue was obtained by alkylation of pyrrole-2-carboxaldehyde with bromoacetic acid followed by reductive amination.[50] The ψ[CH_2O] isostere has been prepared by a method involving carbene insertion into O-H bonds. Treatment of a series of *N*-protected derivatives of amino alcohols with methyl or ethyl diazoacetates in the presence of rhodium (II) acetate gave the

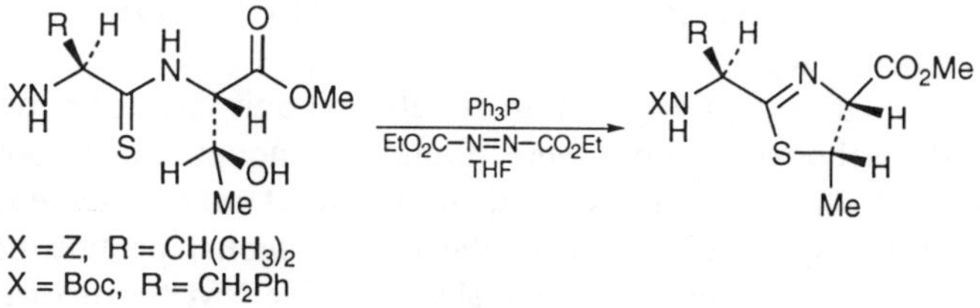

$$\xrightarrow[\text{EtO}_2\text{C–N=N–CO}_2\text{Et, THF}]{\text{Ph}_3\text{P}}$$

X = Z, R = $CH(CH_3)_2$
X = Boc, R = CH_2Ph

Scheme 8

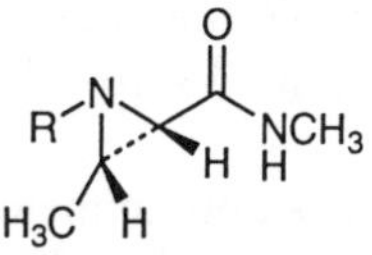

(21) R = Z-Pro, Z-Gly

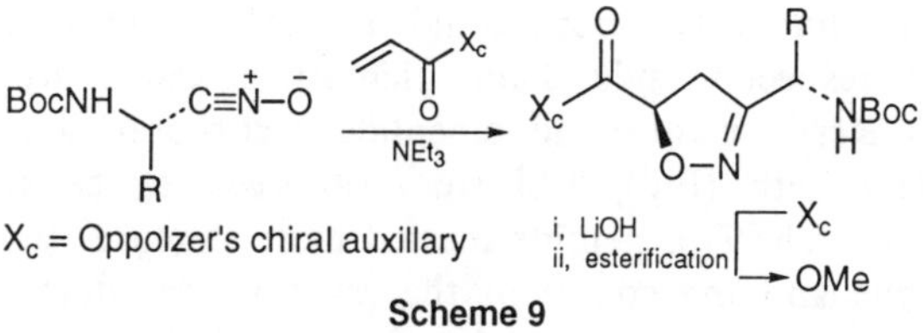

BocNH, $C{\equiv}\overset{+}{N}{-}\overset{-}{O}$, R

NEt₃

X_c = Oppolzer's chiral auxillary

i, LiOH
ii, esterification

$X_c \rightarrow$ OMe

Scheme 9

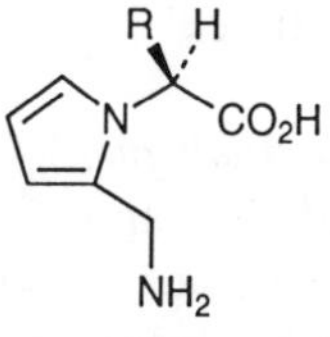

(22)

$$\text{ZNH–CH(R}^1\text{)–C(=O)–NH–CH(R}^2\text{)–CO}_2\text{Bzl} \xrightarrow{\text{PCl}_5\text{/quinoline}} \text{ZNH–CH(R}^1\text{)–C(Cl)=N–CH(R}^2\text{)–CO}_2\text{Bzl}$$

$$\downarrow \text{HN}_3$$

$$\text{ZNH–CH(R}^1\text{)–C(tetrazole)–N–CH(R}^2\text{)–CO}_2\text{Bzl}$$

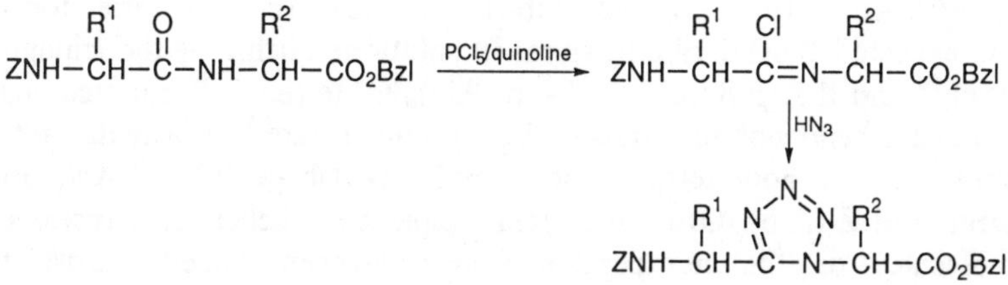

Scheme 10

fully protected pseudodipeptides Z-Xaaψ[CH_2O]Gly-OR^1 (Xaa = Ala, Val, Ile, Leu, Phe).[51]

Vinylogous peptides with an (*E*)-ethenyl unit inserted between the C^α and carbonyl carbon were found to adopt novel secondary structures.[52] A two-stranded antiparallel sheet was observed in the crystal packing of (23) whereas the di(vinylogous) peptide (24) was organised into long stacks of parallel sheets. The NMR structure of (25, R = NMe_2) suggested an antiparallel sheet conformation with a tight turn at the prolyl-glycyl unit. Compound (26), a minor byproduct isolated during the synthesis of (25, R = OMe), adopted a novel helical conformation in both solution and the solid state which is related to the 3_{10} helix of normal peptides.

The pyrrolinone-based peptidomimetic (27) was synthesised in an iterative two-step procedure outlined in Scheme 11.[53] X-Ray analysis of (27) indicated a β-strand conformation similar to that found for the tetrapeptide H-Leu-Leu-Val-Tyr-OMe and this confirmed the prediction that the pyrrolinone backbone could mimic the conformation of a β-strand with respect to side chain orientations and hydrogen bonding capabilities. Reactions of pseudodipeptides incorporating an α,β-unsaturated moiety with $(Bu^t)_2CuLi$ were observed to be stereoselective (Scheme 12).[54] The D,L peptide reacted more stereoselectively than the L,L analogue and the course of the reaction was dominated by the directly adjacent stereocentre (1,2-asymmetric induction) although the more distant stereocentre (1,5-asymmetric induction) also contributed to the direction of nucleophilic attack.

2.10 *α,α-Dialkylated Glycine Analogues.*

Structural analysis of N- and C-terminally blocked peptides containing the monomer units of a potential β-bend ribbon spiral (approximate 3_{10}-helix) concluded that the sequences -MeAib-Aib-, -MeAib-L-Ala-, and Aib-MeAib- induced stable β-bend structures,[55] the -(L-Pro-Aib)$_n$- (n = 1-5), and -Aib-(L-Pro-Aib)$_n$- (n = 1-4) series formed the expected β-bend ribbon spiral in solution starting at the tripeptide level,[56] and that *p*-BrBz-Aib-(L-Pro-Aib)$_n$-OMe (n = 3,4) adopted right-handed β-bend ribbon spirals in the crystalline state.[57] X-Ray diffraction studies of the homotetra-[58] and homohexa- Aib peptides,[59] Ac-(Aib)$_4$-OBu^t and Z-(Aib)$_6$-OBu^t, indicated incipient 3_{10} helical structures. The molecules of the hexapeptide were hydrogen-bonded head-to-tail forming infinitely long, parallel helical columns while a regular alternation of right- and left-handed 3_{10} helices hydrogen-bonded head-to-tail was observed in the crystal structure of Z-D-Val-(Aib)$_2$-L-Phe-OMe.[60] A C_{11} hydrogen-bonded ring characterised the second of the two β-bends in

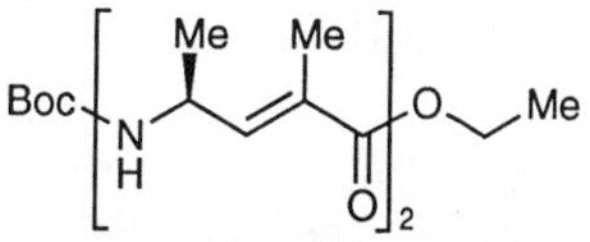

(23)

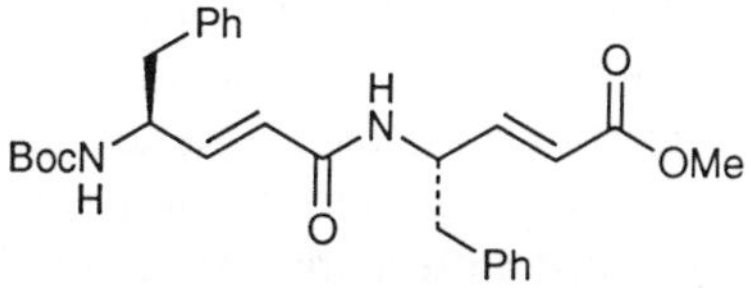

(24)

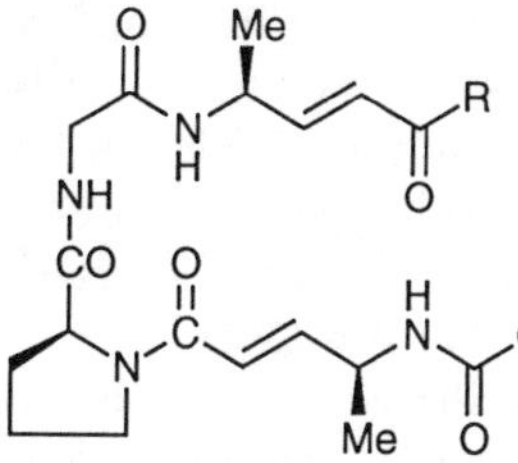

(25)

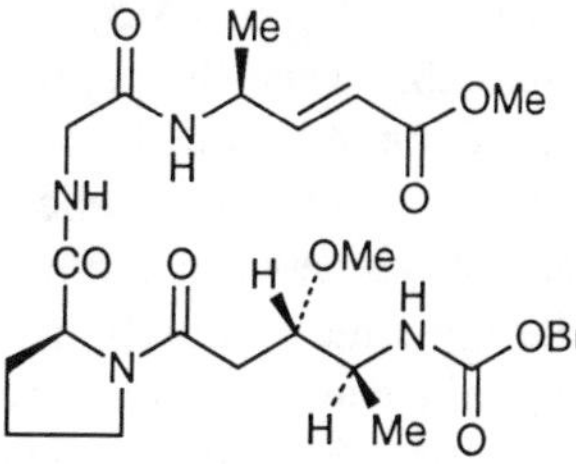

(26)

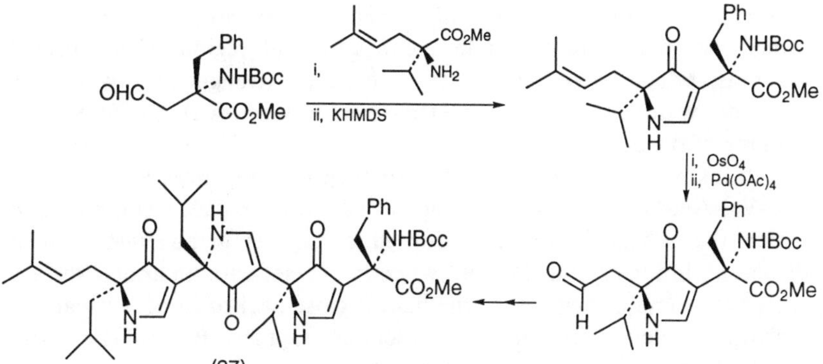

i, ii, KHMDS

i, OsO_4
ii, $Pd(OAc)_4$

(27)

Scheme 11

$(Bu^t)_2CuLi$ | Me_3SiCl, THF, −78 °C

(D,L) > 95% | < 5%
(L,L) 80% | 20%

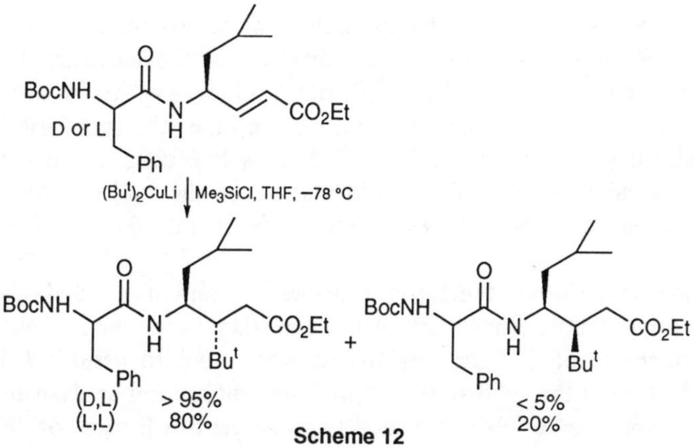

Scheme 12

the crystal structure of Boc-Aib-Aib-β-Ala-NHMe[61] and indicated that a β-amino acid could be incorporated into a β-bend without causing a major perturbation of the overall geometry.

Factors that control the helical folding of Aib-rich peptides were incorporated into a theoretical model to describe the 3_{10}-helix/α-helix equilibrium constant for a given peptide.[62] A molecular mechanics study of the structure of poly(Aib) revealed that the 3_{10}-helical conformation was more stable when the peptide formed part of a dimer but in isolated molecules, depending on the length of the chain and dielectric constant of the environment, the α-helical conformation appeared more stable.[63] The interchange between a type-II β-turn and a γ-turn was seen during molecular dynamics simulations of Boc-Ala-Aib-Ala-OMe.[64]

Crystallographic studies revealed an almost completely α-helical structure for Boc-Val-Ala-Leu-Aib-Val-Ala-Leu-(Val-Ala-Leu-Aib)$_2$-OMe but the hydration and packing of the helices was influenced by the crystallisation solvent.[65] The polymorph from isopropanol-water contained two water molecules in the head-to-tail region of the helical columns, which were packed in a parallel arrangement, while the methanol-water form was more extensively solvated with an antiparallel packing of the helices.

Linking two of the α-helical heptapeptide sequences Val-Ala-Leu-Aib-Val-Ala-Leu *via* an ε-aminocaproic acid residue gave a compound in which the α-helical modules adopted a parallel arrangement, albeit displaced, in the crystal state but when in contact with an apolar surface there was some evidence that the helices close-packed in an antiparallel fashion.[66] Calculations on the tetrapeptide fragment Boc-Aib-Leu-Leu-Aib-OMe of leucinostatin (a nonapeptide antibiotic) indicated that the peptide backbone folded into a right-handed 3_{10} helical conformation and that the pore enclosed by the helix was approximately 3Å.[67] The solution phase synthesis of another member of the leucinostatins has been reported.[68] Although [Aib^{5,6},D-Ala^{8}]cyclolinopeptide A adopted a more rigid structure than the native peptide the conformational constraint imposed by the Aib and D-Ala residues caused a lowering of the biological activity.[69] The conformation of the cyclolinopeptide A analogue has also been investigated by molecular dynamics simulations.[70]

Conformational studies on peptides containing α,α-dialkylsubstituted amino acids other than Aib have also been carried out. The homotripeptide of $C^{\alpha,\alpha}$-dipropylglycine was found to assume a helical conformation in the solid state[71] while X-ray diffraction analysis of Z-D-Val-Ac^{6}c-Gly-L-Phe-OMe indicated two consecutive β-turns of different types (II′ and I), giving rise to an unusual three-dimensional zig-zag structure.[72] Analysis of the X-ray crystal structures of $AcNHC(Et)_2CONHMe$[73] and $BocNHC(Me)_2CONHC(Bu)_2CO_2Me$[74] revealed that the α,α-dibutylglycyl residue of the latter compound adopted a fully extended conformation while the geometry of the $C^{\alpha,\alpha}$-diethylglycyl residue in the former compound was closer to that of a 3_{10}- rather than an α-helix. Investigations into the conformational properties of model peptides containing α-methyl-α-amino acids concluded that the (*R*)-α-MeAsp has a higher helix-inducing potential than the (*S*)-enantiomer[75] and that the αMePhe residue is a stronger β-turn and helix promoter than phenylalanine.[76]

Finally, several racemic α,α-disubstituted α-amino acid amides $H_2NCR^1R^2CONH_2$ (R^1, R^2 = various alkyl, aryl) have been prepared[77] by phase-transfer catalysed alkylation of *N*-benzylidene amino acid derivatives $PhCH{=}NCHR^1CONH_2$ followed by weak acidic hydrolysis of the Schiff bases. Optically pure L-(+)-α-methylarginine was obtained by chemical resolution of D,L-C^{α}-methylornithine followed by guanidination.[78]

3 Conformationally Restricted Cyclic and Bridged Analogues

3.1 *Rings and Bridges formed via Amide Bonds.*

Lactamisation of the weak oxytocin antagonist dPen-Tyr-Ile-Glu-Asn-Cys-Pro-Lys-Gly-NH_2 produced the bicyclic analogue (28) which proved to be more potent than the monocyclic compound in the uterotonic assay [pA_2 of 8.4 for (28) compared to a pA_2 of 5.8 for the monocyclic compound].[79] By comparing the activity and potency of the bicyclic compound with those of analogues with neutral and/or charged side chains in positions 4 and 8 the authors concluded that the potent antagonistic properties of (28) was a consequence of the bicyclic nature of the molecule and not merely the result of obtaining an optimal degree of lipophilicity. The cyclic β-casomorphin analogues H-Tyr-D-Lys-Phe-Pro-Gly and H-Tyr-D-Orn-Phe-Pro-Gly were approximately 50 times and 150 times more effective analgesics in *in vivo* assays than their linear counterparts.[80] The conformation of the bridged, potent, LHRH antagonist Ac-Δ^3-Pro^1-D-*p*$FPhe^2$-D-Trp^3-cyclo(Asp^4-Tyr^5-D-$2Nal^6$-Leu^7-Arg^8-Pro^9-Dpr^{10})-NH_2 (Dpr = 2,4-diamino propionic acid) has been studied using 2D NMR experiments[81] and molecular dynamics simulations.[82] The data indicated that residues 5 to 8 adopted a β-hairpin conformation stabilised by two transannular hydrogen bonds and that the linear part of the molecule (residues 1-3) was more flexible and was orientated above the ring.

The effects of introducing an i-(i+4) lactam bridge into a number dPen-Tyr-Ile-Glu-Asn-Cys-Pro-Lys-Gly—NH_2

(28)

H—(Lys-Leu-Lys-Glu-Leu-Lys-Xaa)$_3$—OH

(29)

(30)

CO—D-Ser-Lys-Gly-D,L-OHOrn-Lys-D,L-OHOrn-Ser

H—Glu—NH—CH—CH_2—OH

(31)

H—Tyr—NH ... CH_2—S—CH_2 ... CO—Gly—Phe—NH ... $CONH_2$

(32)

H_2N—$(CH_2)_n$—NH—CH(R)—CO_2H

(33) $n = 2, 3, 6$

C—Arg-Phe-Phe—N—CH_2—CO—Leu-Met—NH_2

$(CH_2)_n$—CONH—$(CH_2)_3$

(34)

H—Lys-Asp-Ser-Phe-Val—NH— ... Met—NH_2

(35)

of peptides has been investigated. CD studies of the model peptides (29, Xaa = Asp, Glu) and the acyclic analogue H-[Lys-Leu-Lys(Ac)-Glu-Leu-Lys-Gln]$_3$-OH concluded that incorporation of multiple lactam bridges stabilised the α-helical conformation in aqueous solution in the order Lys^{i},Asp^{i+4} bridges >> Lys^{i},Glu^{i+4} bridges > no bridges.[83] The α-helical content of cyclo^{18,22}-[Lys^{18},Asp^{22}]NPY(10-36) (NPY = neuropeptide Y) was significantly higher than the linear analogue N$^{\alpha}$-Ac-NPY(10-36). These compounds were equipotent in rat brain receptor binding assays which suggested that the α-helical conformation was probably the correct conformation for the lactam-bridged region.[84] However the same effect was not observed when a similar constraint was introduced into the full length 1–36mer. Both the α-helical content and the receptor binding activity of cyclo^{18,22}-[Lys^{18},Asp^{22}]NPY were significantly lower than those of NPY which indicated, in this case, that the lactam bridge did interfere with receptor recognition. Stabilisation of the putative C-terminal α-helix of NPY(18-36) by bridging the side chains of amino acids at positions 30/34 and 33/36 yielded analogues with poor affinities for both the Y_1 and Y_2 receptors.[85] A combination of NMR, CD, and constrained molecular dynamics established that in GRF(1-29)NH_2 (GRF = growth-hormone releasing factor) the helical regions between residues 7–14 and 21–28 were stabilised by incorporating i-(i + 4) lactam bridges at positions 8/12 and/or 21/25.[86] Furthermore all these GRF constrained analogues exhibited substantial biological activity and were more resistant than the linear compound to degradation during incubation with human plasma.

Resistance to enzymatic degradation and a higher activity than the linear counterpart were two properties also exhibited by the cyclic peptide T analogue cyclo(Thr-Thr-Asn-Tyr-Thr-Asp).[87] Six diastereoisomeric cyclic hexapeptides of the human thymopoietin (III) partial sequence -Leu^{39}-Tyr^{40}-Leu^{41}-Gln^{42}-Ser^{43}-Leu^{44}- were prepared by separately replacing each residue with the D-isomer and forming an amide bond between the terminal amino and carboxyl groups.[88] The endothelin antagonist cyclo(D-Trp-D-Asp-Pro-D-Val-Leu) was developed from a cyclic pentapeptide lead compound isolated from *Streptomyces misakiensis*.[89]

A detailed conformational study of cyclo(*S*-Phe-*S*-His) (30) revealed that the dipeptide existed as a mixture of two slowly interconverting isomers in solution.[90] In the major conformer the phenyl group was bent over the diketopiperazine ring and the imidazole ring pointed outwards while the minor conformer was U-shaped with the two aromatic rings facing each other. The presence of a 13-membered ring in desferriferribactin (31), (a peptide produced by a strain of *Pseudomonas fluorescens*), was confirmed by 2D NMR experiments.[91]

3.2 *Bridges formed by Disulphide Bonds.*

Conformational analysis of the lanthionine (*i.e.* monosulphide) bridged opioid peptide (32) using NMR and molecular dynamics techniques revealed that there were three principal interconverting families of backbone ring conformations.[92] Although (32) was superactive in both the guinea pig ileum and mouse vas deferens *in vitro* tests it showed no preference for either of the opioid receptor subtypes. From these observations the authors concluded that factors other than backbone ring flexibility may determine δ vs μ selectivity in this class of molecule. The sterically allowed backbone conformations of the disulphide bridged tripeptides Xaa-Ala-Yaa [Xaa, Yaa = Cys, Hcy (homocysteine), MPc (*cis*-4-mercaptoproline), and MPt (*trans*-4-mercaptoproline)] were elucidated using a systematic conformational search procedure.[93] Using these results Marshall and co-workers predicted that in the receptor-bound conformation of highly active angiotensin II analogues containing the Xaa^3-Tyr^4-Yaa^5 sequence the Tyr^4 assumed the torsional values associated with a left-handed α-helix. A CD study of MCH (melanin concentrating hormone, DTMRCMVGRVYRPCWEV) analogues differing in length and/or in the size of the disulphide bridge has been carried out.[94] Analogues containing a 17-membered ring, *e.g.* [Ala^5,Cys^{10}]MCH$_{1\text{-}17}$ exhibited a greater tendency to form an ordered conformation than those containing a 26-membered ring such as [Ala^5,Cys^7]MCH$_{1\text{-}17}$.

3.3 *Miscellaneous Bridges and β-Turn Mimetics.*

Several *N*-(ω-aminoalkylene) amino acids (33) have been prepared by alkylation of alkylenediamines with α-halogeno acids.[95] Protected derivatives were coupled with several protected amino acids to yield dipeptide building blocks such as Boc-Phe-N[$(CH_2)_n$NHZ]CHRCO_2Me (R = H, Me, CH_2CHMe, n = 2,3,6), useful for the synthesis of conformationally constrained N-backbone cyclic peptides. The synthesis of two N-backbone cyclic analogues (34, n = 2, 3) of substance P was included in last year's review and that paper has now been followed by a second report in which a conformational analysis of the two compounds is described.[96] NMR experiments and molecular dynamics simulations established that the cyclic portion of the molecule was conformationally well defined although the CH_2-CH_2 bridging fragment conferred some flexibility to the ring structure. Both analogues were found to exist in DMSO as an equilibrium mixture of *cis* and *trans* isomers about the substituted amide bond of glycine. However these studies did not indicate which of the ring conformations related to the bioactive conformation that selectively activates the NK-1 receptor.

Introduction of a γ-lactam constraint between the Gly^9 and Leu^{10} residues of [Lys^3]NKA$_{3\text{-}10}$ (H-Lys-Asp-Ser-Phe-Val-Gly-Leu-Met-NH$_2$) gave a compound (35) which was a much more selective NK-2 agonist than the linear analogue. Furthermore, in addition to its high selectivity (> 1,000-fold and > 300-fold with respect to NK-1 and NK-3 receptors respectively), (35) exhibited increased resistance to peptidases in biological preparations when compared with NKA.[97] Four diastereoisomers of the more rigid constraint (36) were also prepared by, in effect, fusing the γ-lactam with a D-Pro moiety (replacing Gly at position 9 with D-Pro is known to enhance NK-2 receptor selectivity) (Scheme 13). Not surprisingly, only compound (37), obtained by incorporation of the *RRS* stereoisomer into the corresponding region of [Ava^6]SP$_{6\text{-}11}$ (Ava = δ-aminovaleric acid), showed full intrinsic NK-2 agonist activity relative to NKA in the rat colon.

The cyclic N-terminal fragment (38) of elcatonin was prepared by cyclising a linear precursor and the best yields were obtained if either the Leu-Ser or the Thr-NHCH(R) linkage was formed in the cyclisation step.[98] In a series of papers investigating the synthesis of model peptides with multiple conformational constraints the cyclic hexapeptide cyclo(Lys-Lys-Gly)$_2$ and derivatives in which pairs of ε-nitrogens of the lysine residues were cross-linked have been prepared.[99-101] The efficiency of the photocyclisation of (*N*-acetylglycyl)oligopeptide-linked anthraquinones (Scheme 14) was found to depend on the size and sequence of the oligospacers.[102] The highest conversions were achieved with a dipeptide spacer unit.

Tetrapeptides[103] Ac-Cys-Pro-Xaa-Cys-NHMe (Xaa = Asn, Gly, Ser, Phe, Val, Aib) and cyclic hexapeptides[104] cyclo(D-Pro-Pro-Gly-Arg-Gly-Asp) and cyclo(D-Pro-Pro-Arg-Gly-Asp-Gly) were constructed to study β-turn conformations. The chemical equilibrium between the cyclic and linear forms of the cystine containing peptides indicated that the order of preference of the amino acid at position Xaa to form a β-turn at the Pro-Xaa fragment was Asn > Aib > Gly > Ser > Phe > Val. The D-Pro-Pro sequence in the cyclic hexapeptides favoured a type II β-turn and this stabilised a second turn in the ring. The most populated conformation of the pentapeptide cyclo(Gly-Pro-D-Phe-Gly-Xaa) (Xaa = Ala, Val) in DMSO-d$_6$ consisted of a γ-turn around D-Phe^3 fused with an inverse γ-turn around residue 5.[105] Molecular dynamics simulations[106] revealed that conformational flexibility around the D-Phe^3-Gly^4 peptide bond enabled the formation of the γ-turn around D-

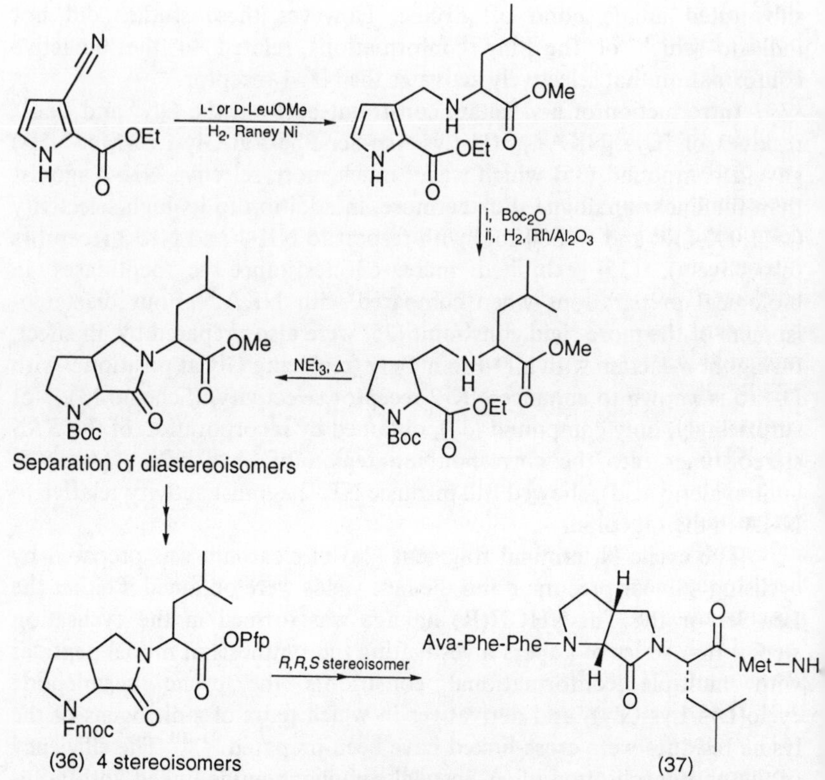

L- or D-LeuOMe
H_2, Raney Ni

i, Boc_2O
ii, H_2, Rh/Al_2O_3

NEt_3, Δ

Separation of diastereoisomers

R,R,S stereoisomer

(36) 4 stereoisomers

Ava-Phe-Phe—N ... Met—NH

(37)

Scheme 13

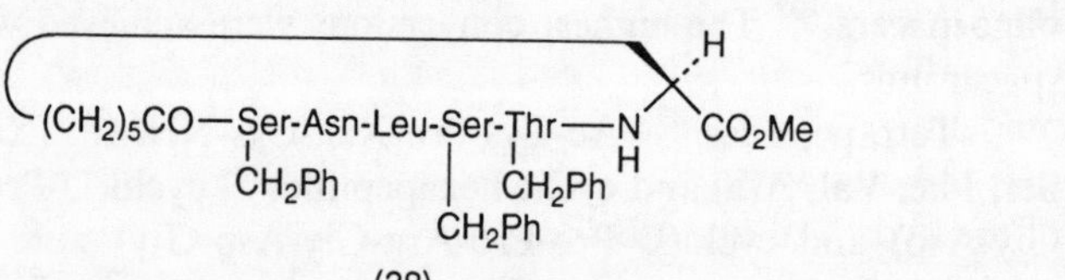

$(CH_2)_5CO$—Ser-Asn-Leu-Ser-Thr—N(H)—CO_2Me with CH_2Ph groups

(38)

Ac—Gly—(spacer)—NHCHCO$_2$Me, CO_2CH_2

hν, CH_3CN

AcNH—CO—(spacer)—NHCHCO$_2$Me, CO_2CH_2, OH

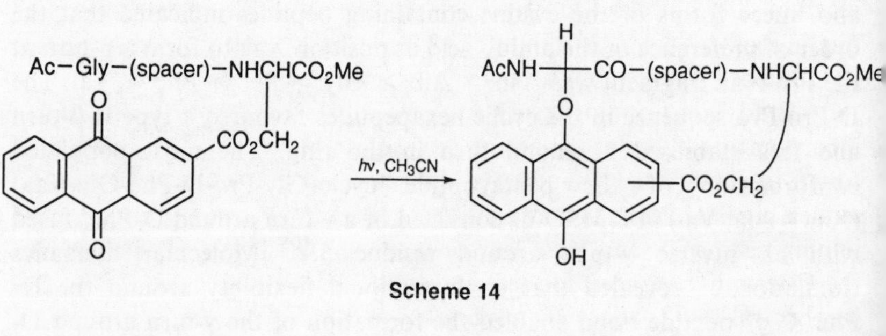

Scheme 14

Phe^3. A minor conformer (< 5%) in DMSO contained a *cis* Gly-Pro bond. Energy minimisation studies have also been carried out on tripeptides containing the Xaa-*cis*-Pro system (Xaa = Gly, Ala, D-Ala, L-Pro).[107]

In the crystal structure of Boc-glycylglycine-cyclohexylamide (39) two of the amide protons were within hydrogen bonding distance of the preceding amidic nitrogen and the β-turn conformation of this compound was stabilised by two successive five-membered ring N-H···N interactions rather than the 10-membered ring C=O···H-N interaction characteristic of the idealised type I β-turn.[108] An intramolecular hydrogen bond [(^{i}Pr)NH···O=C(urethane)] stabilising a β-turn was indicated in both the solution and crystal structures of Boc-Ala-AzaPro-NH^{i}Pr but no analogous interaction was detected for the related compound Z-AzaPro-Ala-NH^{i}Pr.[109] Thermodynamic data for Ac-Pro-Ala-NHMe suggested that a β-turn was favoured at low temperature (205K) but at higher temperatures the 7-membered γ-turn became increasingly competitive with the β-turn.[110] The solution conformation of cyclo(Pro-β-Ala-Pro-β-Ala) was characterised by two γ-turns stabilised by hydrogen bonds between the CO and NH groups of the two β-alanine residues.[111] Evidence of a β-turn motif was detected in the IR vibrational CD spectrum of cyclo(Gly-Pro-Gly-D-Ala-Pro)[112] and in the Raman and IR spectra of Boc-(D-alle-L-Ile)$_3$-OMe.[113]

The spiro-bicyclic system (40),[114] the 10-membered ring compound (41),[115] the substituted 3-oxoindolizidines (42),[116] and the *syn* and *anti* isomers of (43)[117] were all designed and synthesised to mimic the β-turn motif in peptides. Both the spiro-bicyclic compound (40, X = Boc, R = OMe) and the oxoindolizines had N- and C-protecting groups compatible with standard methods of peptide synthesis. Molecular modelling and NMR studies on (40, X = COMe, R = NHMe) showed that the conformation of the molecule was in close agreement with that of the hydrogen-bonded type II β-turn with respect to torsion angles and interatomic distances. The 10-membered ring peptidomimetic (41) of residues 40-45 of the CD4 glycoprotein was found to effectively inhibit binding of the human immunodeficiency virus envelope glycoprotein gp120 to cell surface CD4 at low micromolar levels.

Bradykinin analogues D-Arg-Arg-Pro-Hyp-Gly-Thr-Ser-X-Y-Arg incorporating either of the amino acid pairs (44, 45) at position 7/8 were found to exhibit high (nanomolar) affinity to the bradykinin receptor in guinea pig ileal tissue as long as the ether group was *trans* to the carboxyl joining residues X and Y.[118] Furthermore, NMR studies established that the C-terminal tetrapeptide sequence Ser-D-Hype(*trans* propyl)-Oic-Arg [i.e. central two residues corresponding to (44) with a *trans* propyl ether

(39) (40) (41)

(42) R′ = side chain of Phe, Asp, Trp

(43)

(44) OR = *cis* or *trans* alkyl or aryl ether

(45) OR as (44)

Arg-Pro-Pro-Gly-Phe—N

(46)

group] adopted a β-turn conformation in water. Replacing the turn residues Ser^6 - Phe^8 of bradykinin with a substituted lactam moiety gave (46) which was found to possess micromolar binding affinity for the bradykinin receptor in NG108-15 nerve cells.[119]

The ring system of the bicyclic analogue (47) of the Phe-Pro dipeptide was obtained by condensing an (α-CH_2CHO)phenylalanine derivative with L-cysteine but the key step in the synthesis was bis-alkylation of the diphenyloxazinone (Scheme 15).[120] The key steps in the syntheses of the two constrained Pro-Phe building blocks (48, 49) were regioselective reduction of an *N*-acylpyrrolidinone (Scheme 16)[121] and anodic oxidation of a proline derivative (Scheme 17).[122] Cyclisation of the aminal (50) and the oxazolidinone (51), both prepared from *N*-phthaloyl-L-phenylalanine, gave respectively, the *cis* and *trans* isomeric forms of the conformationally constrained '*anti*'-Phe-Gly-dipeptide. Each benzolactam was incorporated into a short peptide sequence.[123]

A series of disubstituted indanes (52) have been prepared in which the groups attached at positions 1 and 6 of the ring system were in an orientation which mimicked that of the side chains of two adjacent amino acids of an α-helix.[124]

4 Dehydroamino Acid Analogues

The properties and synthesis of α,β-dehydropeptides has been reviewed.[125] Most of the X-ray diffraction studies on α,β-dehydropeptides have been carried out on peptides containing the α,β-dehydrophenylalanine residue (ΔPhe). The tripeptide amide[126] Z-D-Ala-Δ^zPhe-Gly-NH_2 and tetrapeptide[127] Boc-Leu-Δ^zPhe-Ala-Leu-OMe exhibited two consecutive β-turns (type II′-1 and III-1 respectively) in the solid state with the Δ^zPhe as the common residue in the double bend. Ac-ΔPhe-Ala-ΔPhe-NHMe also assumed an incipient 3_{10} helical conformation[128] while three full turns of a 3_{10} helix were observed for Boc-Val-ΔPhe-Phe-Ala-Phe-ΔPhe-Val-ΔPhe-Gly-OMe.[129] Both the doubly and triply unsaturated molecules were linked head-to-tail in the crystal by intermolecular hydrogen bonds forming long helical chains. In the tripeptide the neighbouring helices ran in an antiparallel direction while adjacent helices were parallel in the nonapeptide. The structure of Boc-Gly-Δ^zPhe-Leu-Δ^zPhe-Ala-NHMe and Boc-Val-Δ^zPhe-Phe-Ala-Leu-Ala-Δ^zPhe-Leu-OMe were investigated by NMR techniques.[130] Evidence pointed to 3_{10} and α-helical conformations for the pentapeptide and octapeptide respectively in $CDCl_3$ but in a more polar solvent (DMSO) the folded conformations were not as stable and some conformational heteroge-

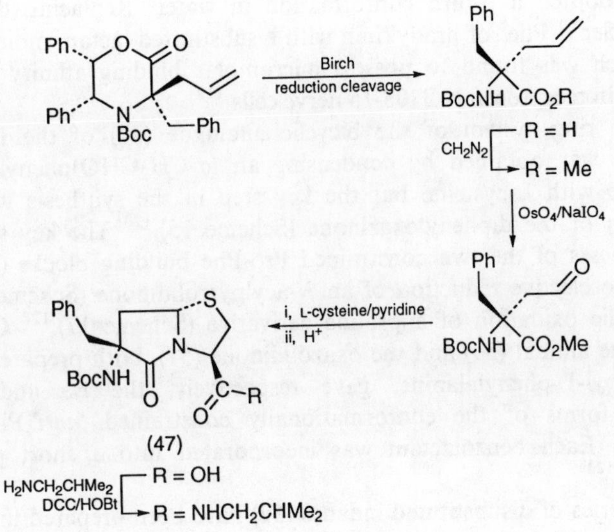

Birch reduction cleavage

BocNH CO_2R

CH_2N_2: R = H → R = Me

$OsO_4/NaIO_4$

BocNH CO_2Me

i, L-cysteine/pyridine
ii, H^+

(47)

$H_2NCH_2CHMe_2$, DCC/HOBt: R = OH → R = $NHCH_2CHMe_2$

Scheme 15

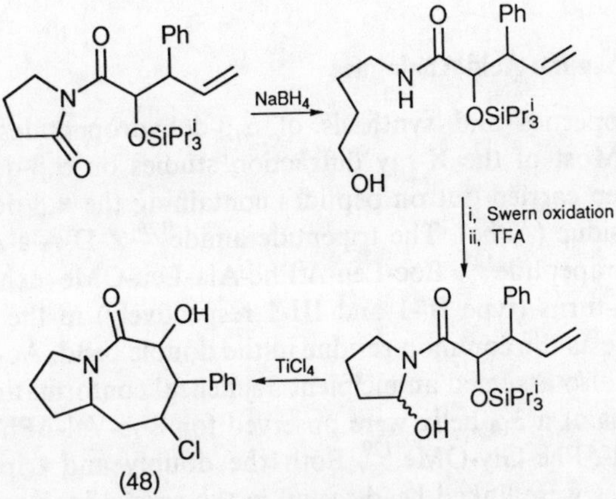

$NaBH_4$

i, Swern oxidation
ii, TFA

$TiCl_4$

(48)

Scheme 16

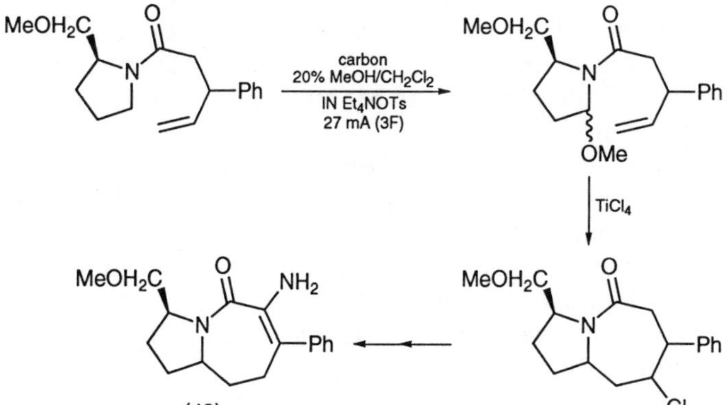

MeOH$_2$C O N Ph

carbon
20% MeOH/CH_2Cl_2
IN Et_4NOTs
27 mA (3F)

MeOH$_2$C O N Ph OMe

$TiCl_4$

MeOH$_2$C O N Ph Cl

MeOH$_2$C O N NH_2 Ph

(49)

Scheme 17

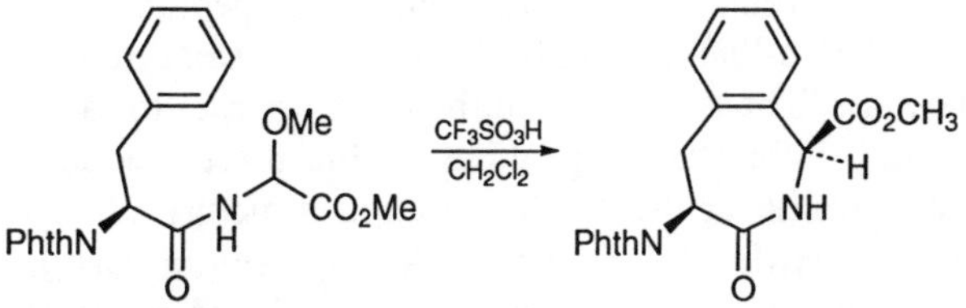

PhthN O N H OMe CO_2Me

(50)

CF_3SO_3H
CH_2Cl_2

PhthN O NH CO_2CH_3 H

"*cis*"

PhthN O N O O

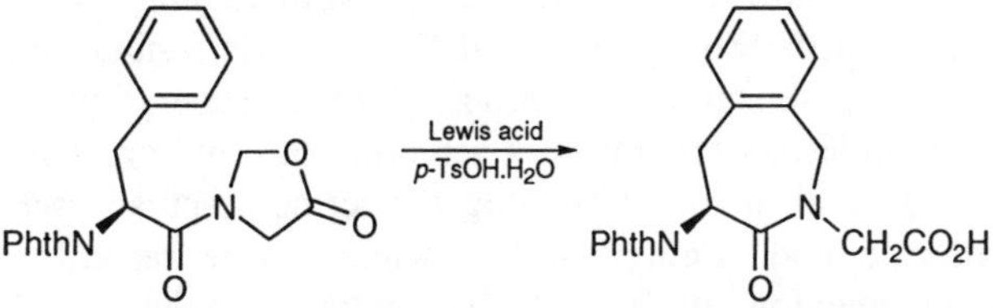

(51)

Lewis acid
p-TsOH.H_2O

PhthN O N CH_2CO_2H

"*trans*"
(mixture of enantiomers)

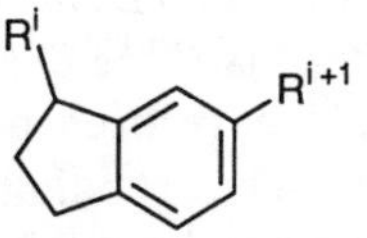

R^i R^{i+1}

(52) R^i = amino acid side chain of Ala, Phe, Asp
R^{i+1} = amino acid side chain of Ala, Glu, Leu, Ser

neity was observed. Crystal and molecular structures of Ac-Δ^{z}Phe-β-Ala-OH and Ac-Δ^{z}Tyr(Me)-β-Ala-OH have also been reported.[131] The high resolution electron density map of Ac-Δ^{z}Phe-NHMe obtained from X-ray diffraction data was found to be very similar to that derived from *ab initio* calculations.[132]

Type II β-turns dominated the solution conformations of the model peptides Ac-Pro-ΔXaa-NHMe (ΔXaa = ΔVal, Δ^{z}Phe, Δ^{z}Leu, Δ^{z}Abu).[133] In the corresponding saturated series, Ac-Pro-L-Xaa-NHMe, there was conformational flexibility with an inverse γ-bend, a β-turn, and open forms in equilibrium depending on the nature of the Xaa side chain. A β-bend between the type I and type III conformation was observed in the crystal structure of Ac-Pro-ΔVal-NHMe.[134] Both Ac-ΔLeu-NMe_2 and Ac-DL-Leu-NMe_2 are associated as cyclic dimers in the solid state and the major difference between the two structures was in the orientation of the side chains.[135]

Structural studies on ΔAla containing peptides Ac-Pro-ΔAla-NHMe[133] and Boc-Phe-ΔAla-OMe[136] revealed extended conformations in the solid state. The ΔAla residue therefore appears to exert a different conformational influence on peptides to the β-turn inducing effect of most dehydroamino acids. In the solution structure of Ac-D-Ala-Ile-ΔAla-Leu-Ala-NHMe (53) (the ring A fragment of nisin), the ΔAla assumed a roughly planar conformation and induced an inverse γ-turn in the preceding residue.[137] Replacing the ΔAla in (53) with either L- or D-Ala affected the conformation and these substituted analogues adopted a γ- or inverse γ-turn about the central D- or L-Ala residue respectively. Two one-pot reactions have been reported for the generation of the ΔAla residue in peptides. In one report a serendipitous finding, which proved to be quite general, involved treating the serine precursor peptide with oxalyl chloride and triethylamine[138] while in the second paper the dehydration reaction utilised diethyl chlorophosphate and sodium hydride.[139] The latter reaction was also stereospecific giving exclusively the Z-isomer from the threonine precursors.

Asymmetric hydrogenation of various protected mono (Z)-dehydro enkephalins[140] and Ac-Δ^{z}Phe-Xaa-OMe (Xaa = Gly, D- or L-Ala, D- or L-phenylglycine) dipeptides[141] have been performed in the presence of chiral rhodium complexes. The rhodium complex containing the (*R,R*)dipamp ligand was found to be the best catalyst for formation of the (*S*)-configuration at the new chiral centre in the enkephalins. In the dipeptide series the best diastereoselectivities were observed using the (*S*)-dipeptide substrates and the tetrasulphonated (*S,S*)-BDPP ligand. Dehydro residues in substance P hexa- and octa-peptide analogues were incorporated as dipeptide sequences, which in turn were obtained by β-elimination of water from the C-terminal D,L-phenylserine moiety of Boc-Phe-D,L-Phe(β-OH)-OH and Boc-Pro-D,L-Phe(β-OH)-OH.[142] Most of the compounds prepared were substantially less active in *in vitro* and *in vivo* assays than substance P and only [ΔPhe^{5,8} Sar^{9}]SP$_{4\text{-}11}$ demonstrated significant bioactivity (60% that of substance P in guinea pig ileum assay).

Horner-Emmons condensation of XNHCH[PO(OMe)$_2$]CO$_2$R with various aldehydes and ketones in the presence of DBU gave predominantly the (Z)-dehydroamino acid derivatives.[143] Furthermore these products did not contain any poisons for the rhodium catalysts used in subsequent hydrogenation reactions. Mixtures of (Z)- and (E)-*N*-trifluoroacetyl-α,β-dehydro amino acids have been prepared by a route involving a zinc bromide catalysed reaction between a silylated derivative of *N*-trifluoroacetylglycine with various aliphatic and aromatic aldehydes.[144]

Thioamidation of *N,O*-diprotected-Thr-ΔAbu-NH$_2$ (54) with Lawesson's reagent followed by cyclisation with ethyl bromopyruvate afforded the thiazole a-dehydroamino acid (55) (Scheme 18).[145] This sequence of reactions appeared to be quite general and several other thiazole amino acids were prepared including H-D-(Val)Thz-OEt (56). Coupling of (56) to (55, R = H) gave the thiazole moiety (57) of the Micrococcin P$_1$ antibiotic. The (E)-1-azabicyclo[3.1.0]hex-2-ylidene ring system (58) present in the azinomycin series of antitumor antibiotics was prepared by formation of β-bromodehydroamino acid derivatives followed by intramolecular Michael addition-elimination. This reaction proceeded with complete retention of the starting olefin geometry (Scheme 19).[146]

Coupling of R-Δ^{z}Asp(OBu^{t})-OH (R = Ac or Z) to other amino acids was best achieved using either the benzyloxycarbonyl derivative or by first converting the α-carboxylic acid to an active ester.[147] If the *N*-acetyl derivative was used low yields of the dipeptides were obtained due to formation and breakdown of an oxazolone intermediate. Ring opening of the pyranone ring in (59) with KOH afforded the β-naphthyl-α,β-dehydroamino acid derivative.[148]

5 Enzyme Inhibitors

5.1 *Angiotensin Converting Enzyme (ACE) Inhibitors.*

The chiral synthesis of the ACE inhibitors enalapril and lisinopril and a cephem-based elastase inhibitor have been reviewed.[149] The methyl ester analogue (60) of enalapril has been prepared in a 7-step route starting from a tetrahydro-1,4-oxazine-2-one derivative (Scheme 20).[150]

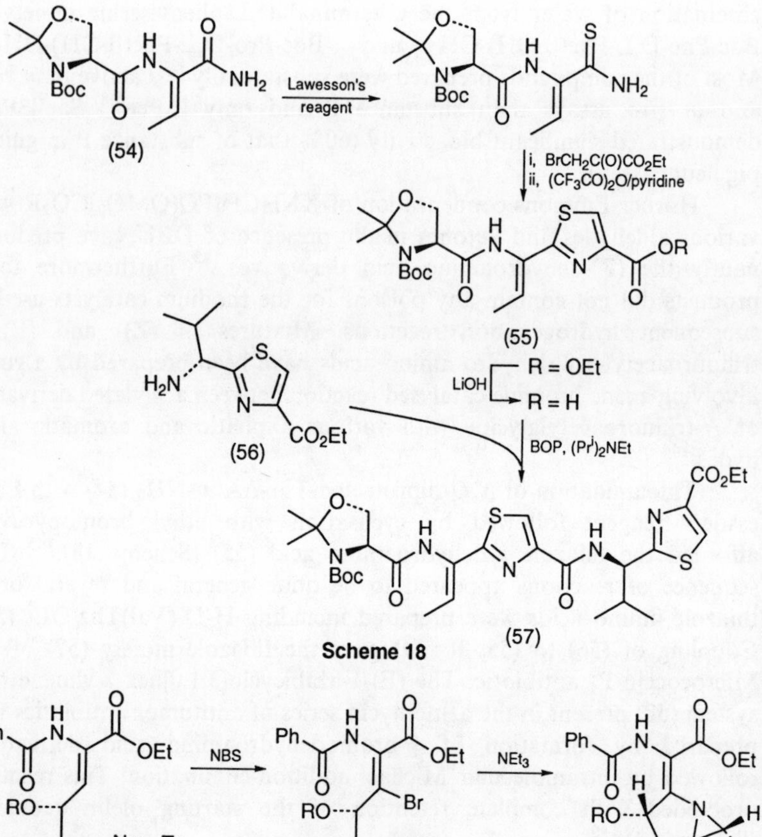

(54) Lawesson's reagent

i, $BrCH_2C(O)CO_2Et$
ii, $(CF_3CO)_2O$/pyridine

(55) LiOH R = OEt → R = H

(56) BOP, $(Pr^i)_2NEt$

(57)

Scheme 18

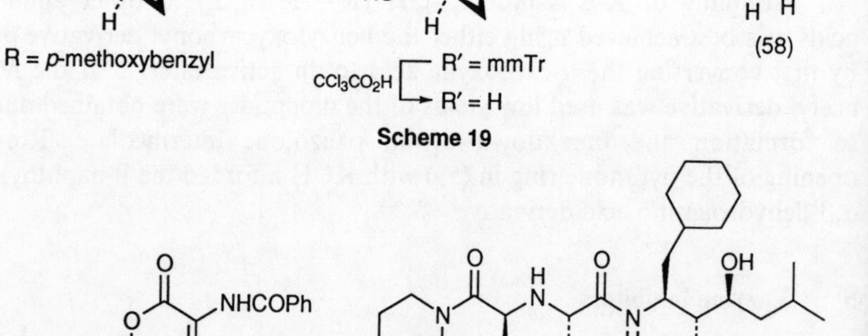

R = *p*-methoxybenzyl

NBS

NEt_3

CCl_3CO_2H R′ = mmTr → R′ = H

(58)

Scheme 19

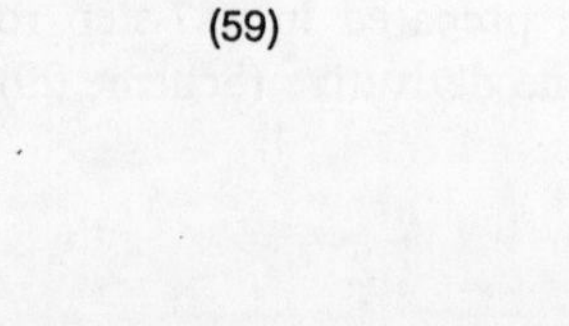

(59)

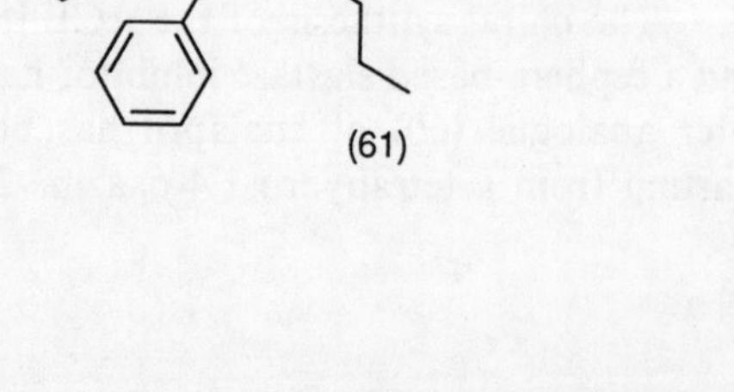

(61)

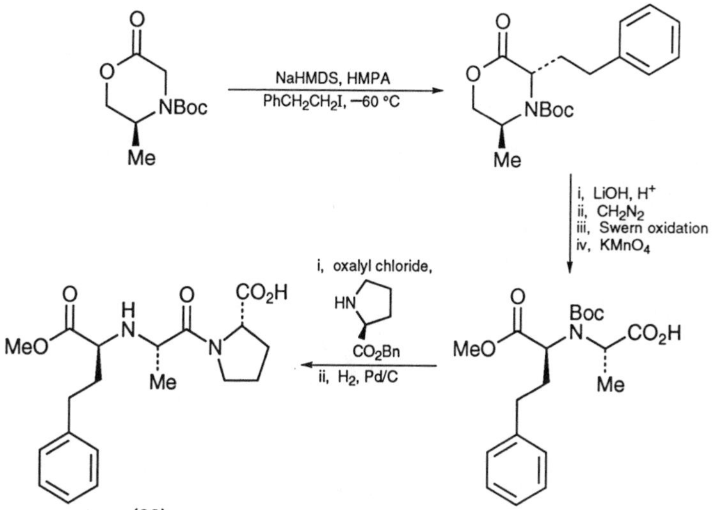

NaHMDS, HMPA
$PhCH_2CH_2I$, −60 °C
i, LiOH, H^+
ii, CH_2N_2
iii, Swern oxidation
iv, $KMnO_4$
i, oxalyl chloride,
CO_2Bn
ii, H_2, Pd/C

(60)

Scheme 20

NaOH
i, resolution
ii, $PhCH_2OH$, H^+(cat.)
DCC, HOBt
H_2, Pd/C

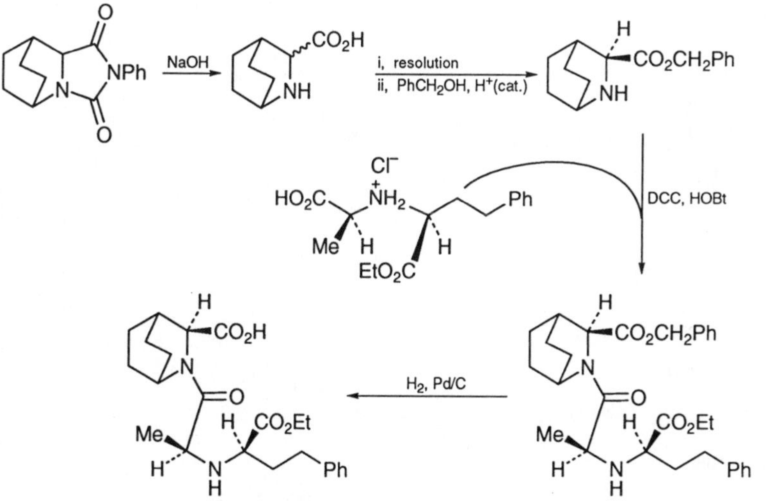

isolated as the hydrochloride

Scheme 21

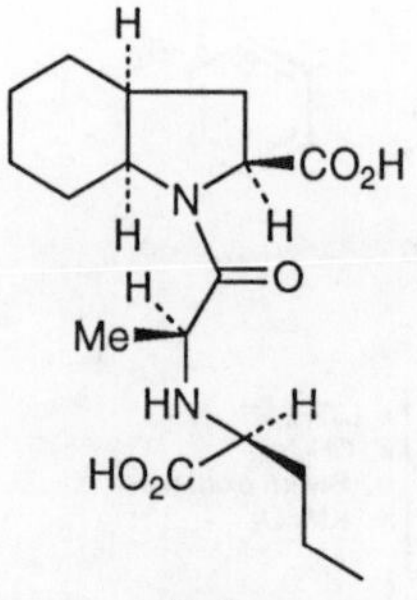

(62) 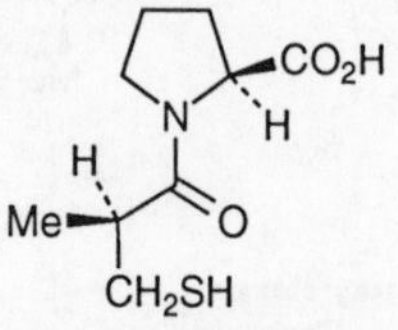 (63) 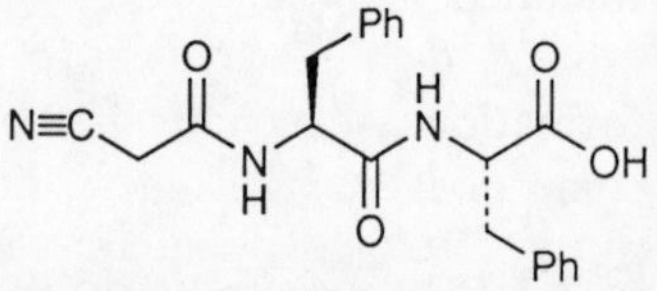 (64)

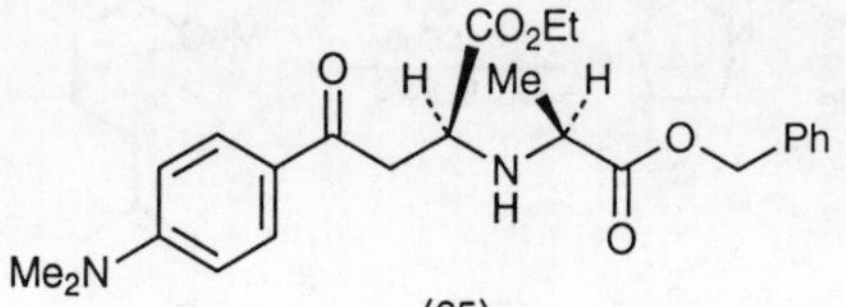

(65)

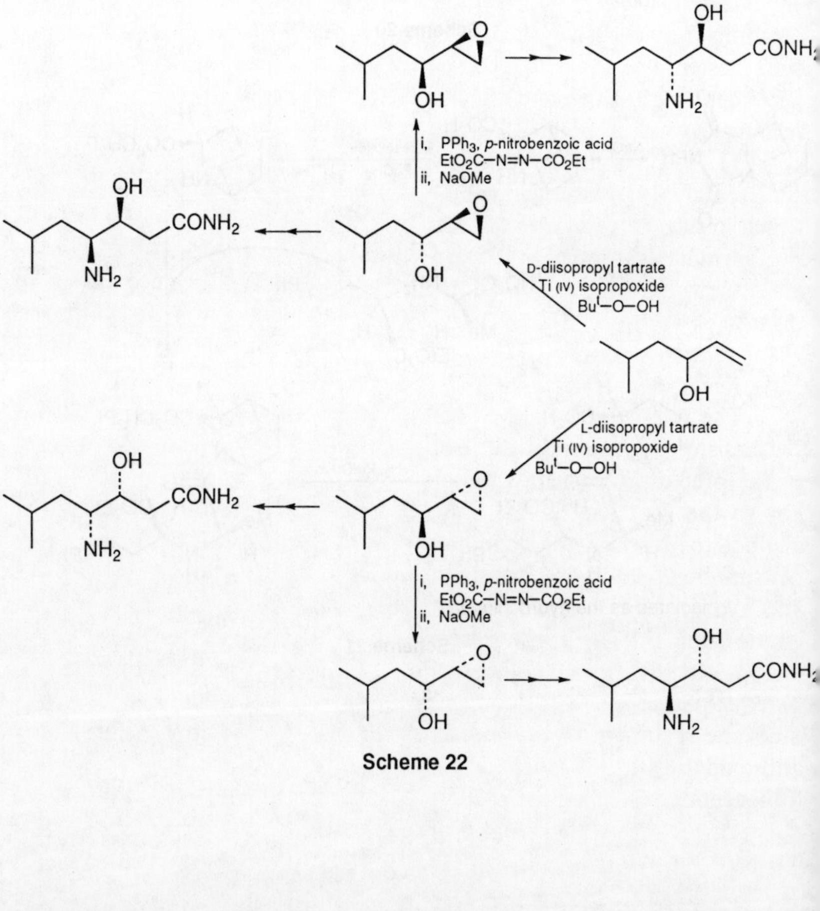

i, PPh_3, *p*-nitrobenzoic acid
$EtO_2C-N{=}N-CO_2Et$
ii, NaOMe

D-diisopropyl tartrate
Ti (IV) isopropoxide
Bu^t-O-OH

L-diisopropyl tartrate
Ti (IV) isopropoxide
Bu^t-O-OH

i, PPh_3, *p*-nitrobenzoic acid
$EtO_2C-N{=}N-CO_2Et$
ii, NaOMe

Scheme 22

The key step in the synthesis was stereospecific alkylation of the lactone starting material and a similar reaction was also employed in the preparation of the renin inhibitor (61). A convergent route (Scheme 21) was adopted for the synthesis of zabicipril, an orally active, well tolerated, and long lasting inhibitor of ACE.[151]

A detailed analysis of the physicochemical properties and conformational behaviour of the dicarboxylic acid form of perindopril (62), some of its stereoisomers, and a desmethyl analogue has been carried out.[152] Results of kinetic and thermodynamic studies on the *cis-trans* isomerisation of captopril (63) indicated that the dependence of the equilibrium constant $K_{t/c}$ on the protonation state of the carboxylic acid and thiol groups was due mainly to differences in rate constants for *cis* to *trans* interconversion for the (CO_2H, SH), (CO_2^-, SH), and (CO_2^-, S^-) forms.[153]

Other potent ACE inhibitors have also been prepared,[154-156] including (64), formed by coupling *N*-cyanoacetylphenylalanine with phenylalanine *tert*-butyl ester, and (65), one of the diastereoisomers resulting from reaction of an alkyl 4-hydroxy-2-butynoate with alanine benzyl ester. The *N*-cyanoacetyl dipeptide (64) inactivated ACE in a time dependent manner[154] and the kinetic parameters for inactivation and turnover were reported as $k_{inact} = 0.08 min^{-1}$, $K_m = 10.5 mM$ and $k_{cat} = 11.1 s^{-1}$. The mechanism of inactivation was envisaged as involving ACE catalysed rearrangement of the nitrile group to the ketenimine which in turn trapped an active-site nucleophile. The *S,S*-chirality at the two asymmetric centres in (65) was confirmed by X-ray analysis and the molecule was also found to adopt an extended conformation in the solid state.[155]

5.2 *Renin Inhibitors.*

The methodology developed by Bessodes and co-workers for the synthesis of statine, and described in last year's review, has been applied to the preparation of statine amide and its three stereoisomers.[157] The key to the route lay in the Sharpless epoxidation of a racemic allylic alcohol to generate the stereoisomeric epoxy alcohols (Scheme 22). A Mitsunobu reaction to generate the epoxy azide, ring opening with KCN and hydrolysis and hydrogenolysis converted each epoxy alcohol to the amino amides. The eight stereoisomers of isostatine (3*S*, 4*R*, 5*S*)-4-amino-3-hydroxy-5-methylheptanoic acid (66) have also been prepared.[158] Starting from Boc-L- and D-Ile and Boc-L- and D-*allo*-Ile, a diborane reduction of the carboxylic acid to the corresponding alcohol and oxidation to the aldehyde was followed by condensation with ethyl lithioacetate. Each of the 4 diastereoisomeric mixtures of Boc-Ist-OEt

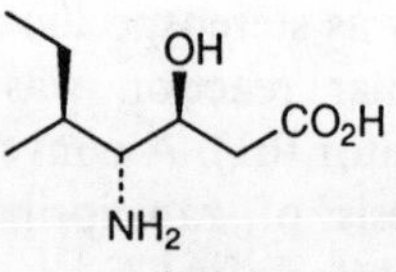

OH

CO_2H

NH_2

(66)

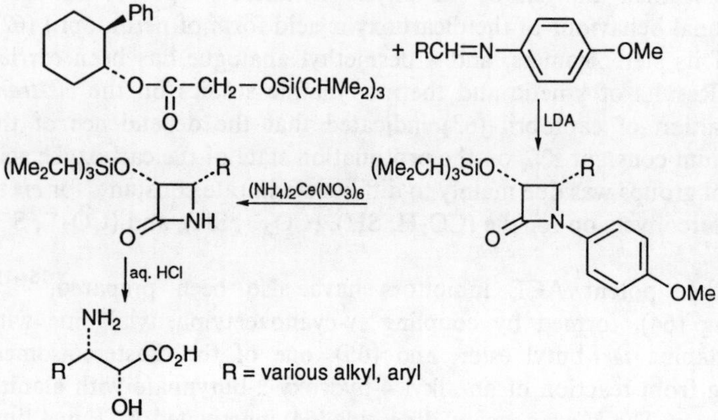

Ph

+ RCH=N–⟨⟩–OMe

O–C(=O)–CH_2–OSi$(CHMe_2)_3$

LDA

$(Me_2CH)_3SiO$ ... R

$(NH_4)_2Ce(NO_3)_6$

$(Me_2CH)_3SiO$ ... R

NH

N

OMe

aq. HCl

NH_2

CO_2H

R

OH

R = various alkyl, aryl

Scheme 23

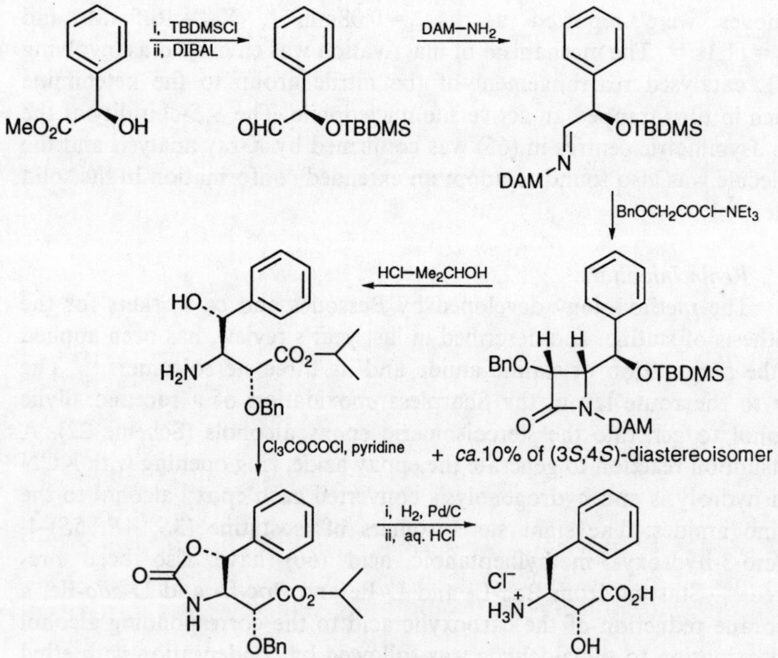

i, TBDMSCl

ii, DIBAL

DAM–NH_2

MeO_2C OH

OHC OTBDMS

OTBDMS

N

DAM

$BnOCH_2COCl–NEt_3$

HCl–Me_2CHOH

HO

CO_2

H_2N

OBn

H H

BnO

OTBDMS

N DAM

+ *ca.*10% of (3*S*,4*S*)-diastereoisomer

$Cl_3COCOCl$, pyridine

i, H_2, Pd/C

ii, aq. HCl

O

CO_2

N

H

OBn

Cl^-

$H_3\overset{+}{N}$

CO_2H

OH

Scheme 24

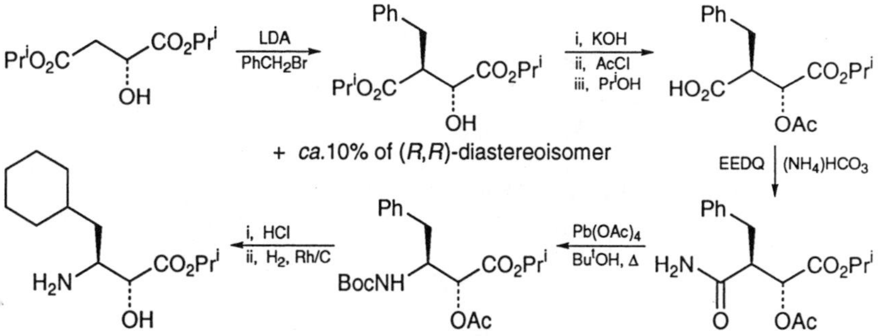

Pr^iO_2C CO_2Pr^i OH → (LDA, $PhCH_2Br$) → Ph Pr^iO_2C CO_2Pr^i OH

\+ *ca.*10% of (*R*,*R*)-diastereoisomer

→ (i, KOH; ii, AcCl; iii, Pr^iOH) → Ph HO_2C CO_2Pr^i OAc

→ (EEDQ, $(NH_4)HCO_3$) → Ph H_2N CO_2Pr^i O OAc

→ ($Pb(OAc)_4$, Bu^tOH, Δ) → Ph BocNH CO_2Pr^i OAc

→ (i, HCl; ii, H_2, Rh/C) → H_2N CO_2Pr^i OH

Scheme 25

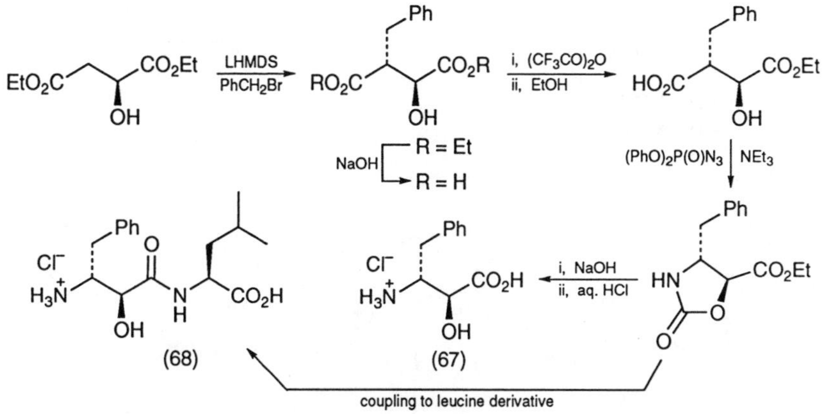

EtO_2C CO_2Et OH → (LHMDS, $PhCH_2Br$) → Ph RO_2C CO_2R OH

NaOH: R = Et → R = H

→ (i, $(CF_3CO)_2O$; ii, EtOH) → Ph HO_2C CO_2Et OH

→ ($(PhO)_2P(O)N_3$, NEt_3) → Ph HN CO_2Et O O

→ (i, NaOH; ii, aq. HCl) → Ph Cl^- $H_3\overset{+}{N}$ CO_2H OH (67)

Ph Cl^- $H_3\overset{+}{N}$ O N H CO_2H OH (68)

coupling to leucine derivative

Scheme 26

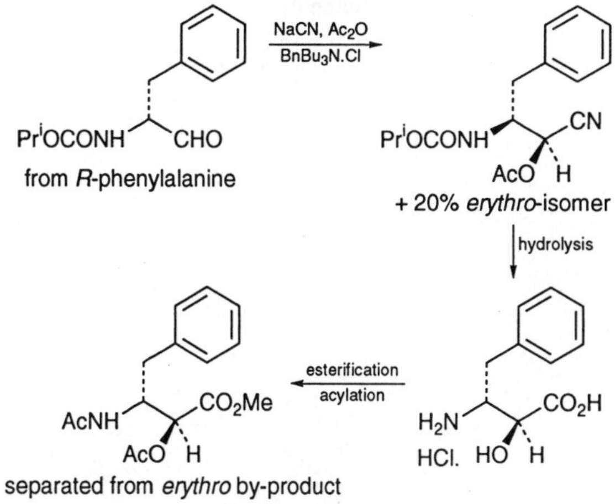

Pr^iOCONH CHO from *R*-phenylalanine → (NaCN, Ac_2O, $BnBu_3N.Cl$) → Pr^iOCONH CN AcO H

\+ 20% *erythro*-isomer

→ (hydrolysis) → H_2N CO_2H HCl. HO H

→ (esterification, acylation) → AcNH CO_2Me AcO H

separated from *erythro* by-product

Scheme 27

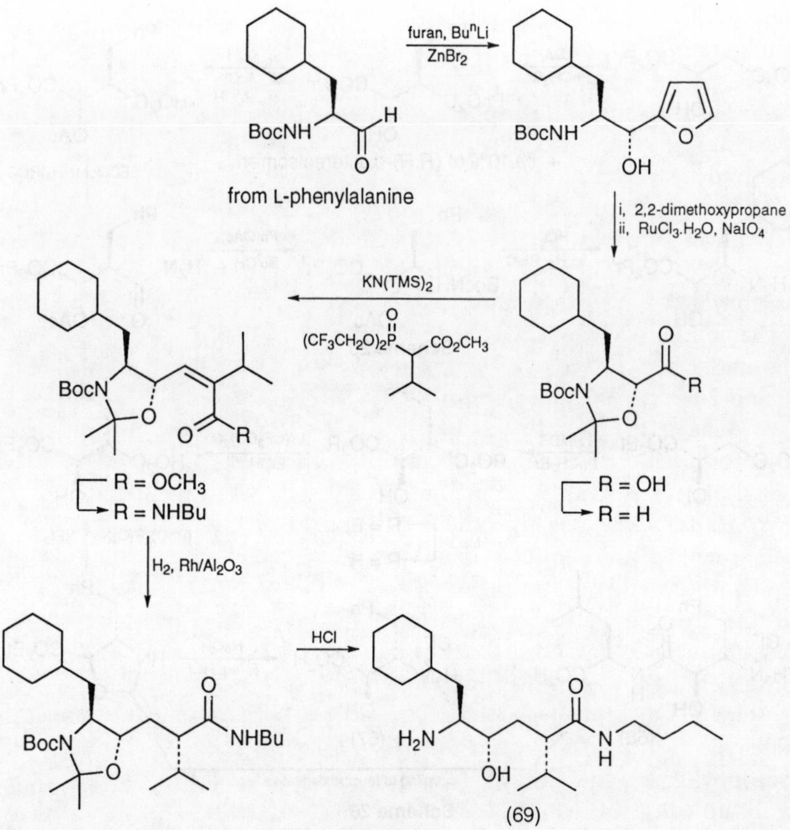

furan, Bu^nLi
$ZnBr_2$

BocNH

from L-phenylalanine

i, 2,2-dimethoxypropane
ii, $RuCl_3.H_2O$, $NaIO_4$

$KN(TMS)_2$

$(CF_3CH_2O)_2P(O)CH(iPr)CO_2CH_3$

R = OH
R = H

R = OCH_3
R = NHBu

H_2, Rh/Al_2O_3

HCl

NHBu

(69)

Scheme 28

were purified by HPLC and the stereochemistry at C-3 and the absolute configuration of each isomer was established on the basis of the ^{1}H NMR spectra of the acetyl-γ-lactam derivatives.

Three syntheses of (2*R*, 3*S*)-3-amino-2-hydroxybutyric acid derivatives (*i.e.* norstatine analogues) have been described. Two of the routes involved stereoselective [2+2] cycloadditon of an imine with an enolate[159] or a ketene[160] to produce 3,4-*cis*-disubstituted β-lactam derivatives (Schemes 23 and 24 respectively). In the third route the key steps were benzylation of diisopropyl malate, differentiation of the two carboxylates, and a Hofmann degradation of one of the carboxylates (Scheme 25).[161] Stereocontrolled benzylation of diethyl (*S*)-malate was the initial step in the synthesis of the isomeric compound (2*S*, 3*R*)-3-amino-2-hydroxy-4-phenylbutyric acid (AHPBA) (67) and of (-) bestatin (68) (Scheme 26).[162] An alternative route to (67) involved diastereoselective formation of cyanohydrin acetates from (*R*)-*N*-(isopropoxycarbonyl)phenylalaninal (Scheme 27).[163]

Hydroxyethylene dipeptide isosteres Xaaψ[CH(OH)CH_2]Yaa have been prepared utilising both chiral and non-chiral starting materials. In the synthesis of (69), starting from L-phenylalanine, the stereochemistry of the hydroxyl group was set by a highly diastereoselective aldol reaction with furan and that of the isopropyl group was controlled by a facial-selective hydrogenation reaction (Scheme 28).[164] The Leu-Val isostere (71), isolated as the lactone, was prepared from isovaleryl aldehyde (Scheme 29)[165] and the amide intermediate (70), with the *S*-isopropyl group at C-2, crystallised in high diastereoisomeric excess after work-up of the one-pot Claisen rearrangement/amidation reaction. The *S*-isopropyl group directed epoxidation to the *syn* face of the acyclic system which effectively transferred the asymmetry at C-2 to the C-4 and C-5 centres. Two other approaches to the hydroxyethylene lactone precursor have also been reported and both routes involved stereoselective reduction of a carbonyl group to generate a *syn*-alcohol and stereoselective alkylation of a lithium enolate to introduce side chain functionality (Schemes 30[166] and 31[167]) . Allylsilane chemistry was utilised in the preparation of protected Ileψ[CH(OH)CH_2]Glu and Ileψ[CH(OH)CH_2]Ala isosteres.[168] Reaction of protected amino aldehydes with $Me_3SiCH_2C(CH_2Cl)=CH_2$ in the presence of BF_3OEt_2 gave selectively the corresponding *S*-homoallylic alcohols which were further elaborated into protected derivatives of the dipeptide isosteres (Scheme 32).

A general method for the stereocontrolled construction of the four diastereoisomers of the ψ[CH(OH)CH(OH)] moiety has been developed.[169,170] The key steps, outlined in Scheme 33 for one of the Leu-Ala

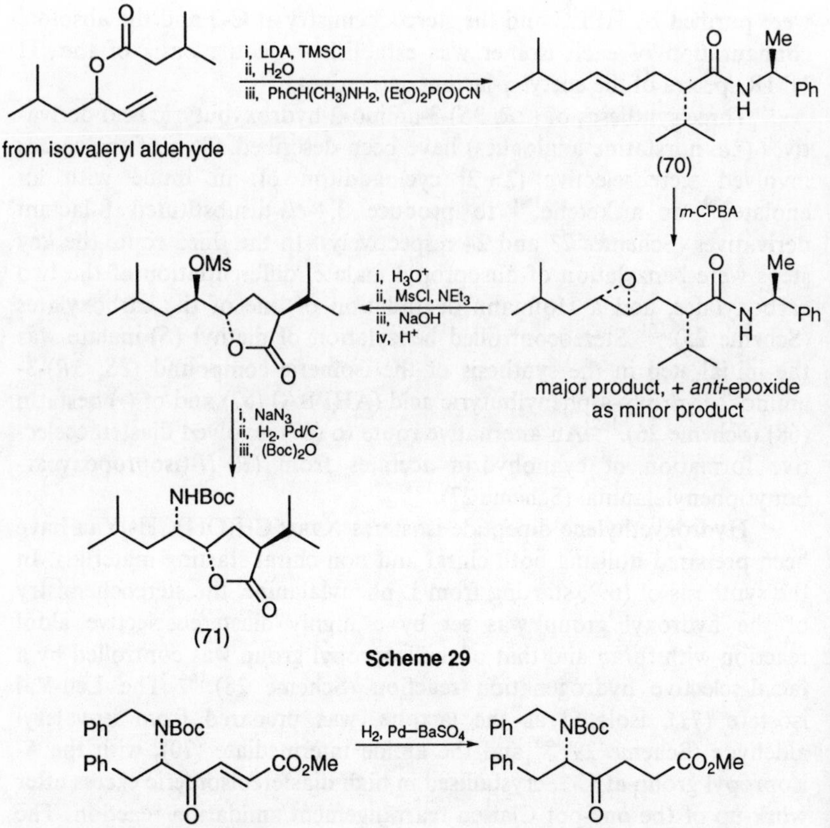

from isovaleryl aldehyde

i, LDA, TMSCl
ii, H_2O
iii, $PhCH(CH_3)NH_2$, $(EtO)_2P(O)CN$

(70)

m-CPBA

major product, + *anti*-epoxide as minor product

i, H_3O^+
ii, MsCl, NEt_3
iii, NaOH
iv, H^+

OMs

i, NaN_3
ii, H_2, Pd/C
iii, $(Boc)_2O$

NHBoc

(71)

Scheme 29

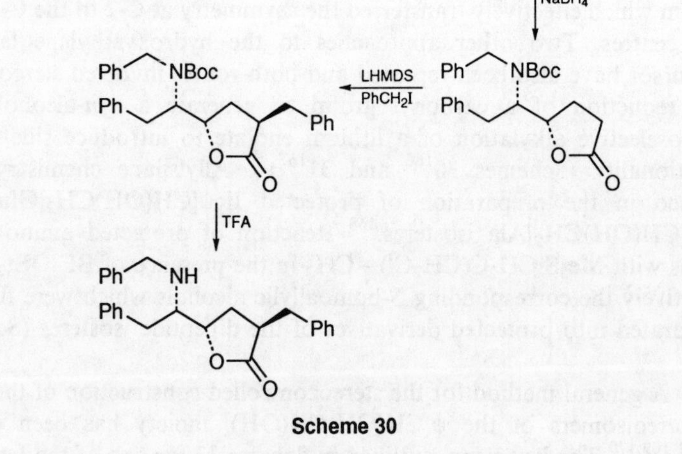

from L-phenylalanine

H_2, Pd–$BaSO_4$

$NaBH_4$

LHMDS
$PhCH_2I$

TFA

Scheme 30

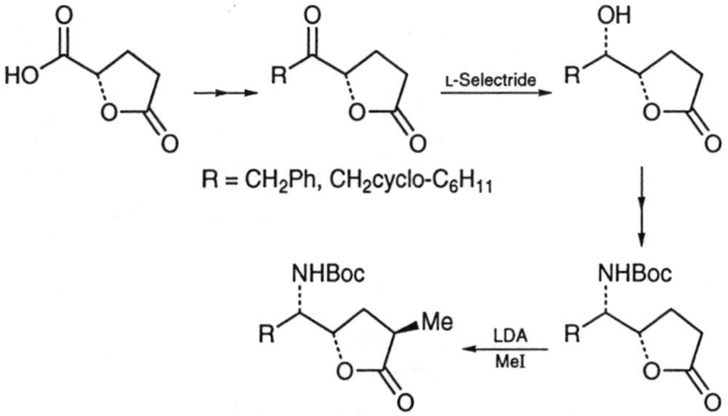
HO
R
L-Selectride
OH
R = CH2Ph, CH2cyclo-C6H11
NHBoc
Me
LDA
MeI

Scheme 31

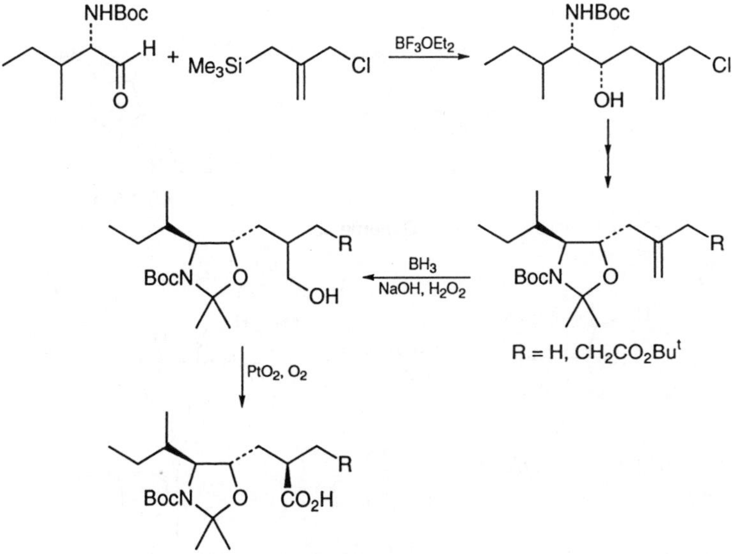
NHBoc
H
Me3Si
Cl
BF3OEt2
OH
BocN
R
BH3
NaOH, H2O2
R = H, CH2CO2Bu^t
PtO2, O2
CO2H

Scheme 32

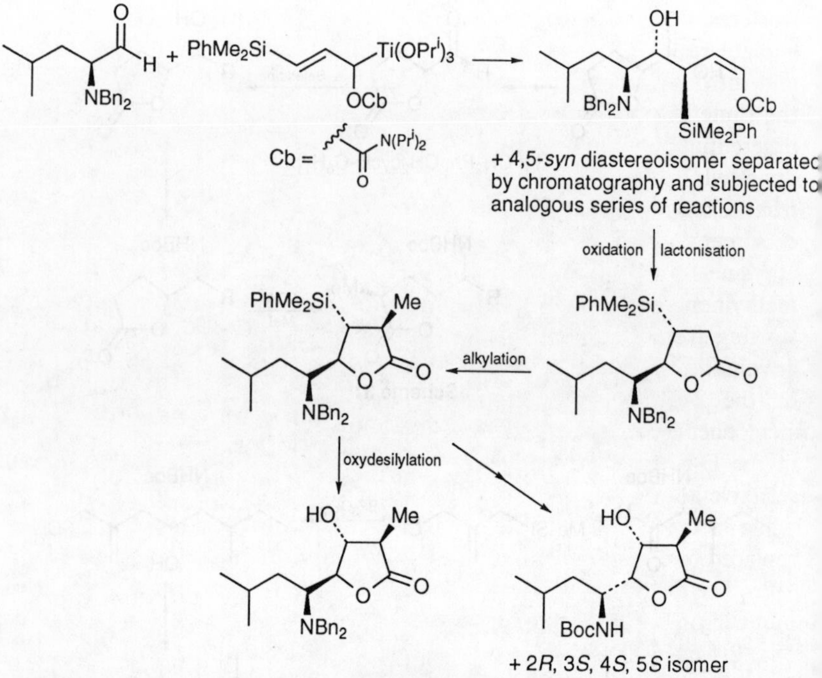

O H NBn$_2$ + PhMe$_2$Si Ti(OPr^i)$_3$ OCb → OH Bn$_2$N OCb SiMe$_2$Ph

Cb = N(Pr^i)$_2$ O

+ 4,5-*syn* diastereoisomer separated by chromatography and subjected to analogous series of reactions

oxidation | lactonisation

PhMe$_2$Si O O NBn$_2$

alkylation

PhMe$_2$Si Me O O NBn$_2$

oxydesilylation

HO Me O O NBn$_2$

HO Me O O BocNH

+ 2*R*, 3*S*, 4*S*, 5*S* isomer

Scheme 33

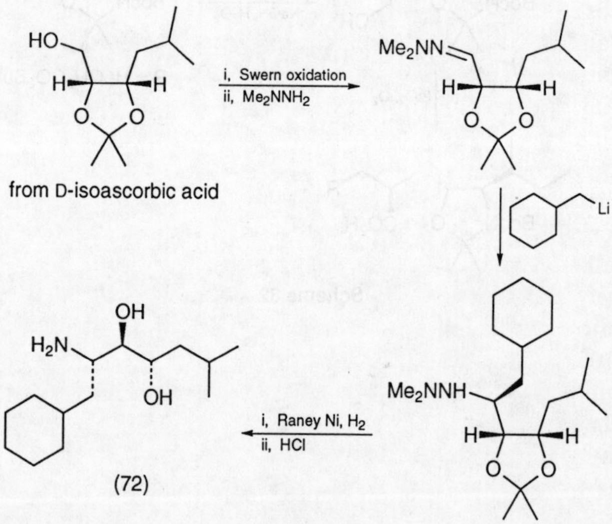

HO H H O O

from D-isoascorbic acid

i, Swern oxidation
ii, Me$_2$NNH$_2$

Me$_2$NN= H H O O

Li

Me$_2$NNH H H O O

i, Raney Ni, H$_2$
ii, HCl

H$_2$N OH OH

(72)

Scheme 34

isosteres, are the C_3-extension of (*S*)-*N*,*N'*-dibenzylleucinal with a homoenolate reagent, conversion of the resulting adduct into the γ-lactone, and introduction of the methyl group with enolate methodology using the dimethylphenylsilyl group as a directing device for diastereofacial differentiation. Finally, either stereospecific oxydesilylation of the 3-dimethylphenylsilyl function or *anti*-selective Peterson elimination to the *trans*-alkenoic acid and *syn*-bishydroxylation yielded the various diastereoisomers of the dihydroxyethylene dipeptide isosteres. The dihydroxyethylene dipeptide isostere (2*S*,3*R*,4*S*)-2-amino-1-cyclohexyl-6-methylheptane-3,4-diol (72) has been prepared by two different routes starting from either D-isoascorbic acid or D-ribose. Both routes took advantage of the fact that the 4- and 5-OH groups in D-isoascorbic acid or the 2- and 3-OH groups in D-ribose have the same absolute stereochemistry as the dihydroxyethylene moiety. The C-2 centre of (72) in the D-isoascorbic acid route (Scheme 34)[171] was formed by stereoselective addition of cyclohexylmethyllithium to the dimethyl hydrazone while in the D-ribose route[172] the 4-OH group (ribose numbering) was replaced by an amine with inversion of configuration (Scheme 35).

The dihydroxyethylene isostere (72) was the P_1/P_1' component in a number of renin inhibitors. In the series (73) with a (2*S*,4*S*)-3-(aza or oxa)-2,4-dialkyl glutaric acid amide moiety as the P_2/P_3 amide bond replacement the optimum potency was achieved with (73, R=OCH_2OMe, R^1=CH_2Ph, R^2=Bu, X=N or O).[173] Introducing a basic substituent into the hydroxyethylene component improved the potency and solubility and the morphinolylpropyl amide derivative (74) is a promising orally active renin inhibitor.[174] Parke-Davis has selected (75, R=morpholino, n=1) from a series of 2-amino-4-thiazole-containing inhibitors (R=morpholino, piperazino, n=0-2) for *in vivo* evaluation on the basis of its superior efficacy and duration of action in *in vitro* assays against monkey renin.[175] Compound (76) from Abbott Laboratories, structurally very similar to Parke-Davis' (75), had oral bioavailabilities of 8, 24, 32, and 53% in the monkey, rat, ferret, and dog respectively and is now under clinical evaluation.[176] In another Parke-Davis series, containing a α-heteroatom amino acid as the P_2 residues, many of the compounds exhibited subnanomolar potency when tested *in vitro* against monkey renin but only (77) showed moderate hypotensive effects when given orally.[177]

A hypotensive effect was observed when the pentahydroxy inhibitor (78) was administered intraduodenally to sodium-depleted rhesus monkeys.[178] The pentol moiety was derived from D-(+)-mannose.

Potent renin inhibitors containing transition-state mimetics other than the hydroxyethylene isostere at the P_1-P_1' locus have also been

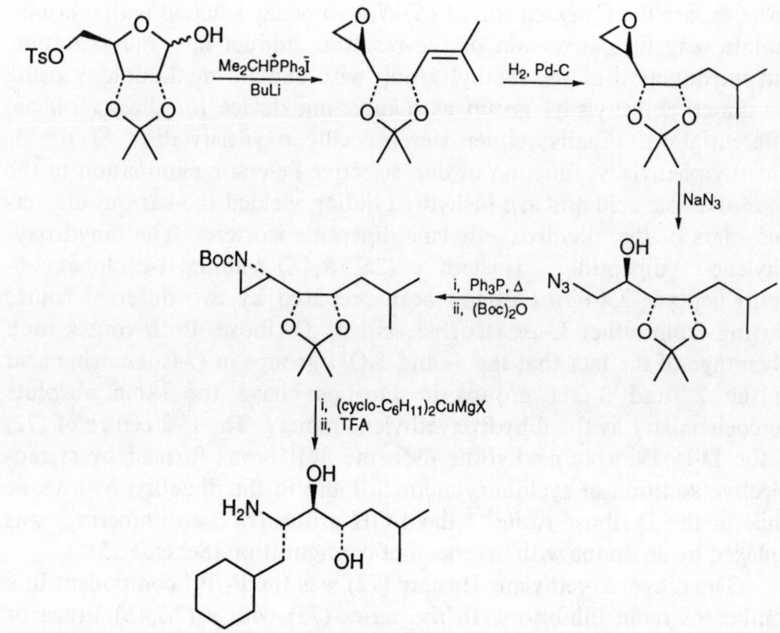

TsO OH Me$_2$CH$\overset{+}{P}$Ph$_3\bar{I}$ BuLi H$_2$, Pd-C NaN$_3$ OH N$_3$ BocN i, Ph$_3$P, Δ ii, (Boc)$_2$O i, (cyclo-C$_6$H$_{11}$)$_2$CuMgX ii, TFA OH H$_2$N OH

(72)

Scheme 35

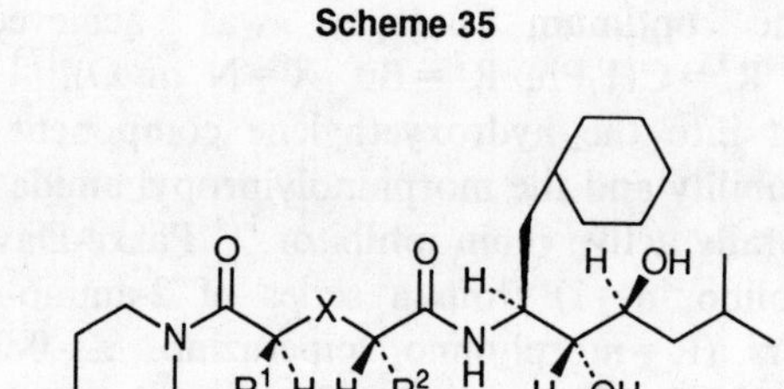

(73) R = various alkoxy, alkyl
R^1,R^2 = various alkyl
X = O, NH

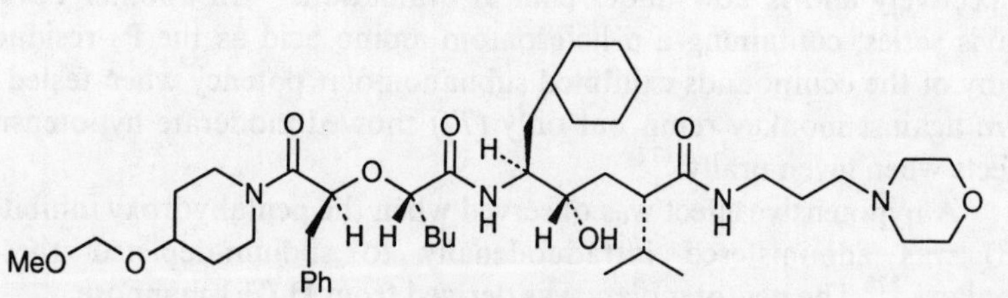

(74) (A-74273)

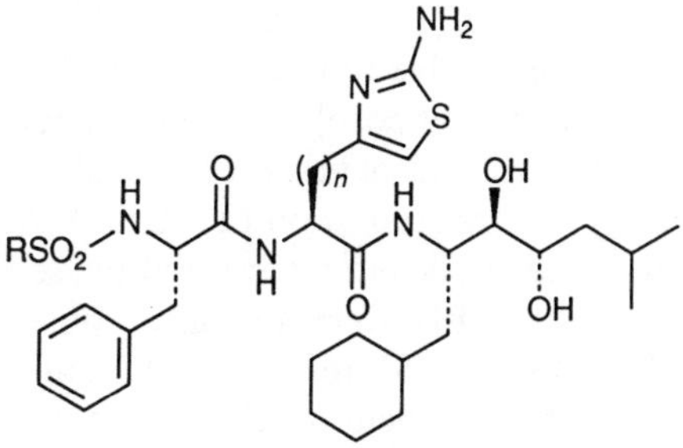

(75) (PD 134672)

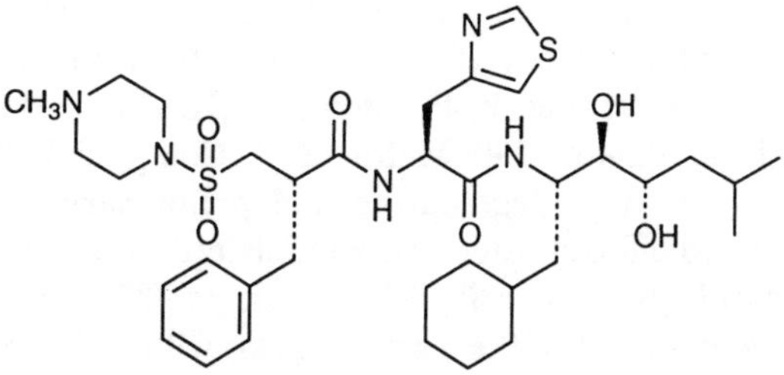

(76) (A-72517)

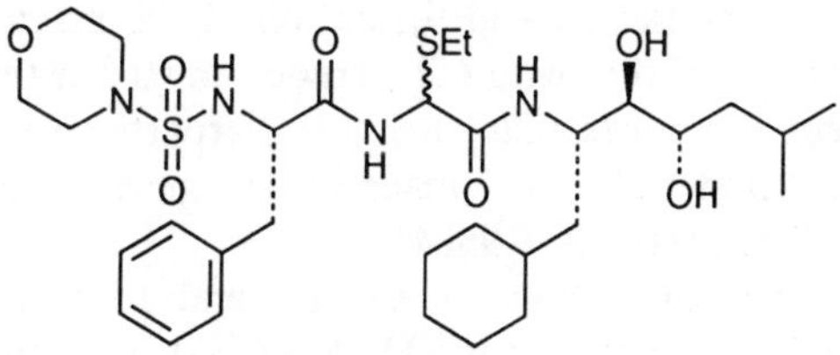

(77)

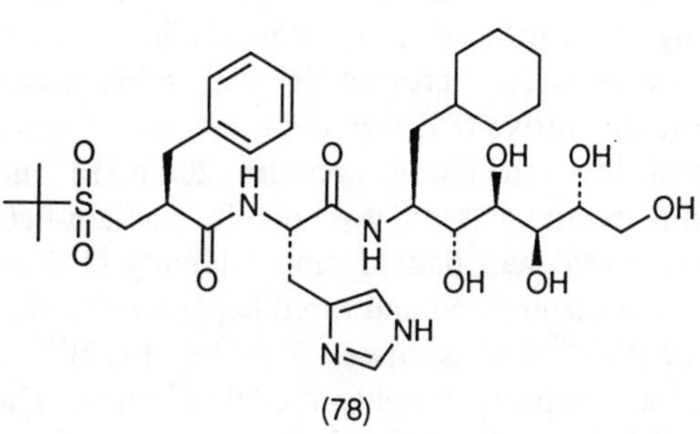

(78)

reported. The difluoroketone (79) showed oral efficacy in salt-depleted cynomolgus monkeys comparable to that of the Parke-Davis compound (75).[179] Comparison of the difluoroketone analogues with the corresponding alcohols indicated that oxidation markedly decreased renin *vs* cathepsin D selectivity but incorporation of a basic moiety such as 4-(2-aminoethyl)morpholine at P_2' position increased the selectivity in favour of renin. Compounds (80, 81) were the most potent *in vitro* renin inhibitors (IC_{50} of 31 and 4.2nM respectively) from series bearing C-terminii derived from mercaptoheterocycles[180] and α-mercaptoalkanoic acids, esters, and amides.[181] Although these compounds substantially reduced plasma renin activity when administered i.v. to sodium-depleted rhesus monkeys the accompanying drop in blood pressure was of short duration. Inhibitor (82), from a series bearing a heterocyclic moiety at P_1', exhibited *in vitro* potency 20 times that of the lead compound (83) containing ACHPA as the P_1-P_1' mimetic [IC_{50} of 9.3nM for (82) compared with an IC_{50} of 190nM for (83)]. Compound (82) was also resistant to chymotryptic degradation and preliminary *in vivo* studies demonstrated good efficacy after oral administration to sodium-depleted cynomolgus monkeys.[182] The P_1-P_1' mimetic in (82) was obtained by nucleophilic displacement of the mesylate (84a) with 2-mercaptopyridine. Removal of the protecting groups from (84b) and stepwise chain extension gave (82) whereas removal of the protecting groups from (85) followed by treatment, in high dilution, with EDC/HOBt/*N*-methylmorpholine resulted in cyclisation and formation of the macrocycle (86).[183] The diastereoisomers of (86) with (*S*) stereochemistry at the P_2' position were more potent inhibitors than their counterparts with (*R*) chirality. Further modification of the macrocycle gave inhibitors (87a) (IC_{50} = 5nM) and (87b) (IC_{50} = 6.9nM).

Replacing one amide bond in the 13- and 14-membered macrocycles (88, X = CH_2, Y = NH; X = NH, Y = CH_2) with an ester linkage gave glutamate-derived inhibitors (88, X = CH_2, Y = O) and serine derived inhibitors (88, X = O, Y = CH_2).[184] While the oxygen-for-nitrogen exchange had little effect on potency in the glutamate series, potency was dramatically increased in the serine series. The 3(*S*)-quinuclidinyl-Phe derivative (89) was the most potent compound of the serine series and lowered blood pressure 20mmHg and completely inhibited plasma renin activity for 6h in sodium-depleted rhesus monkeys. However (89) had limited bioavailability (1% in rats) due to cleavage of the serine ester bond and rapid hepatic extraction.

Macrocycle (90)[185] and compounds (91) and (92)[186] were designed with the aid of a computer graphic model of renin. The macrocycle displayed poor binding affinity and the most potent of the linear

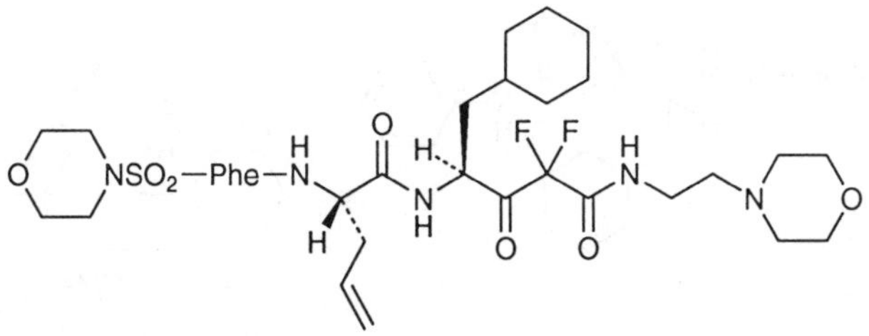
O
NSO2—Phe—N
H
O
H
N
H
F F
O
O
H
N
N
O

(79)

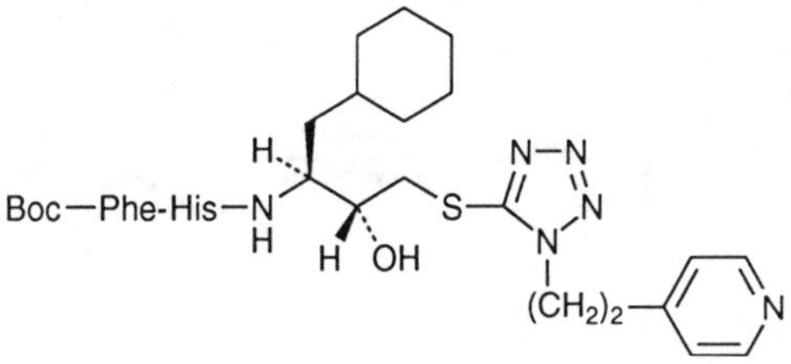
Boc—Phe-His—N
H
H
OH
S
N—N
N
N
(CH2)2
N

(80)

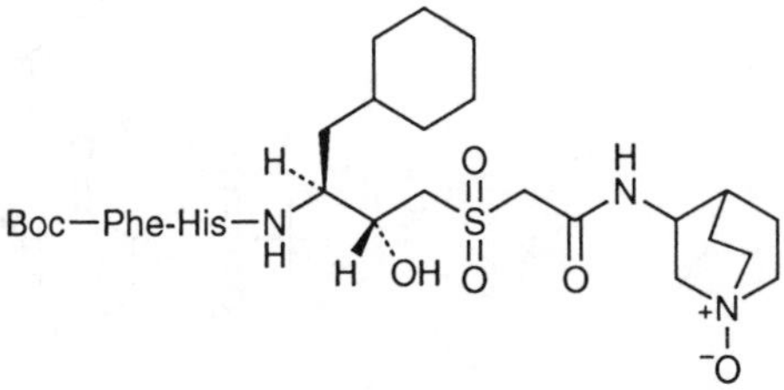
Boc—Phe-His—N
H
H
OH
O
S
O
H
N
O
+N
-O

(81)

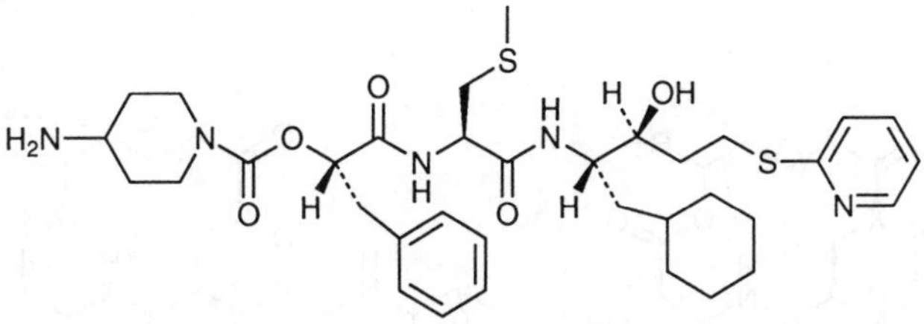
H2N
N
O
O
H
N
H
O
S
H
N
H OH
H
S
N

(82)

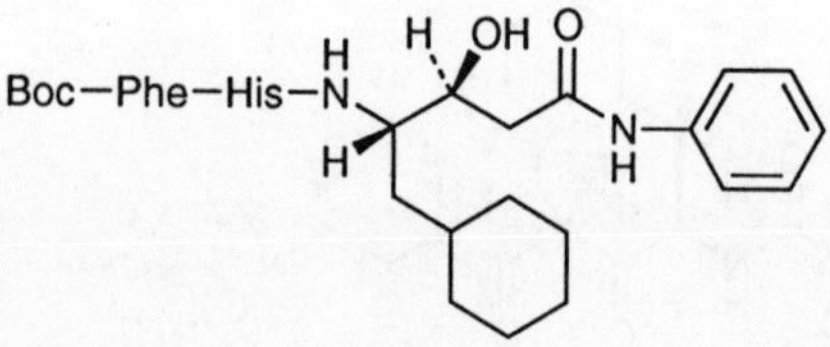

Boc—Phe—His—N ... (83)

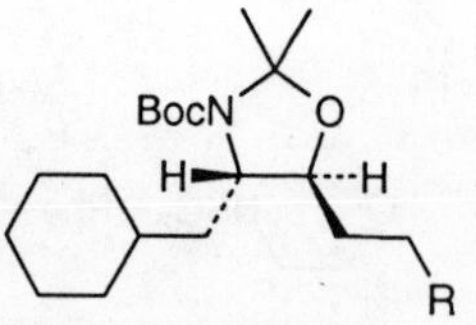

(84a) R = OMs

(84b) R = S–(2-pyridyl)

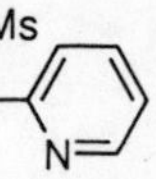

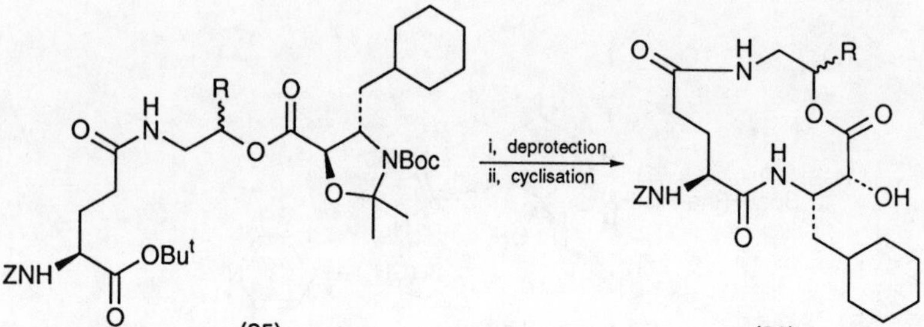

(85)

i, deprotection
ii, cyclisation

(86)

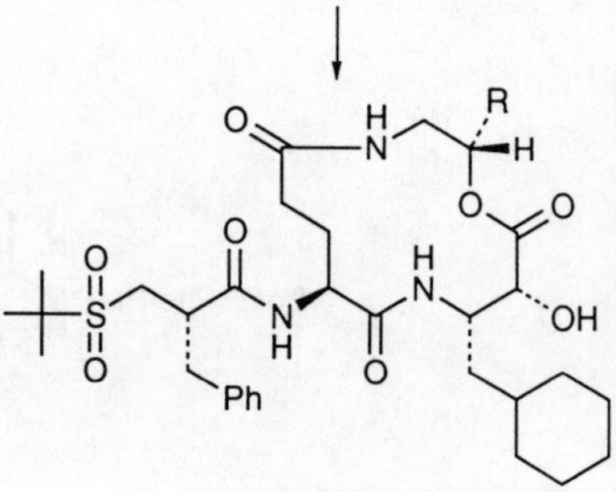

(87a) R = Bu

(87b) R = CH_2–N(morpholino)

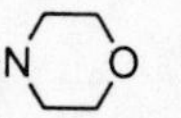

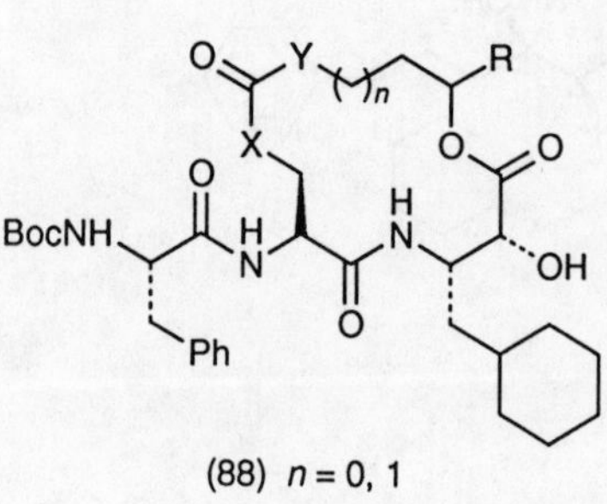

(88) $n = 0, 1$

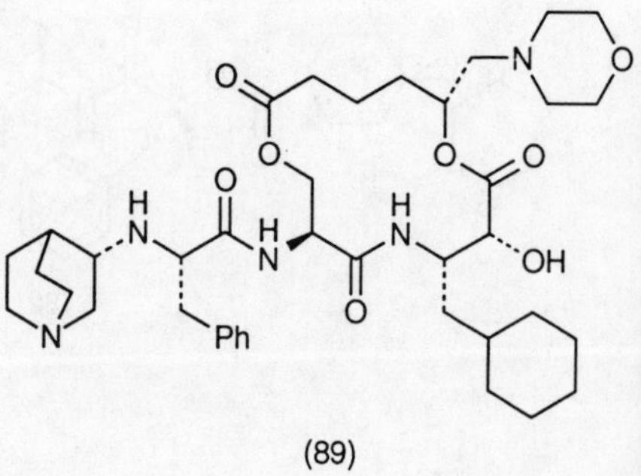

(89)

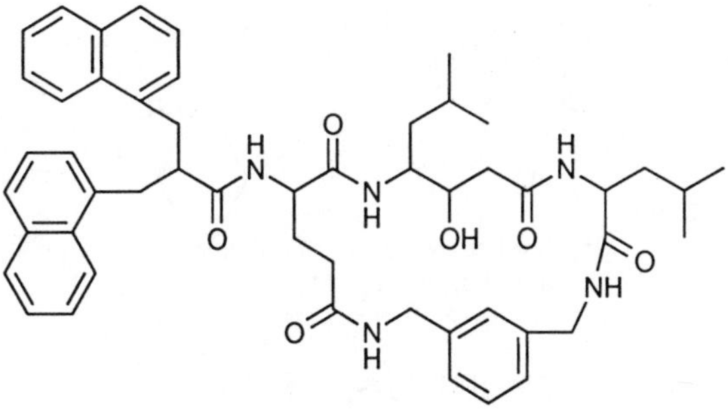

(90)

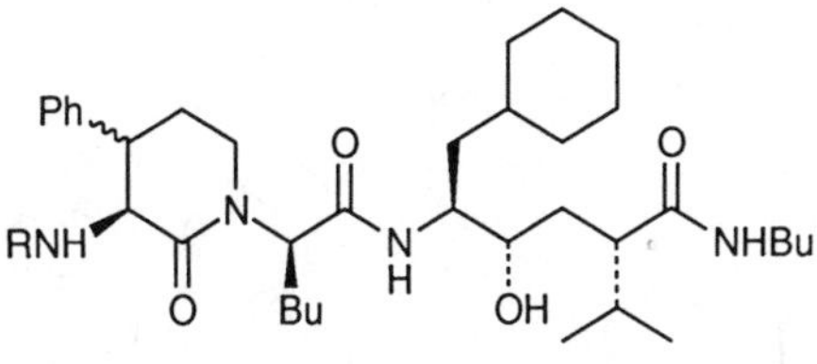

(91) R = Boc, Ac

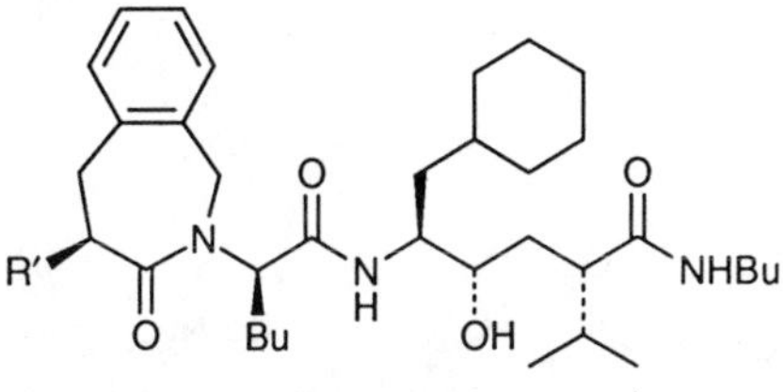

(92) R′ = PhthN, BocNH, AcNH

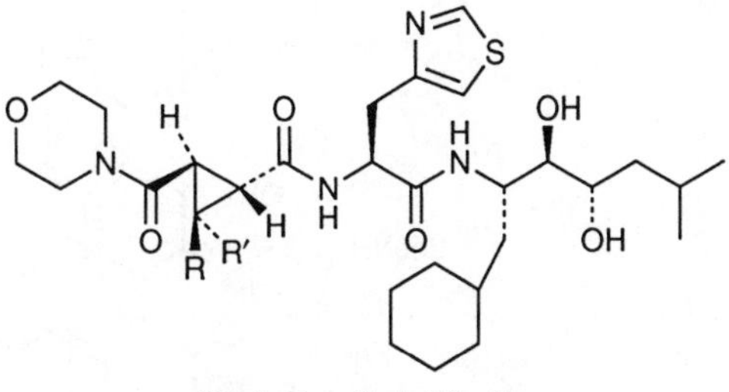

(93a) R = alkyl, R′ = H
(93b) R = H, R′ = alkyl

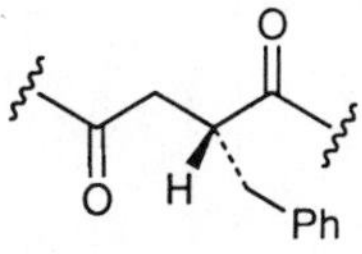

(94)

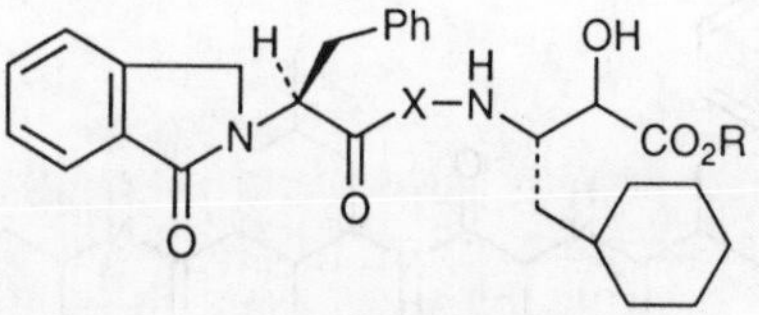

(95) X = Leu, Nva, Nle; R = $CHMe_2$
X = Leu, His; R = Me

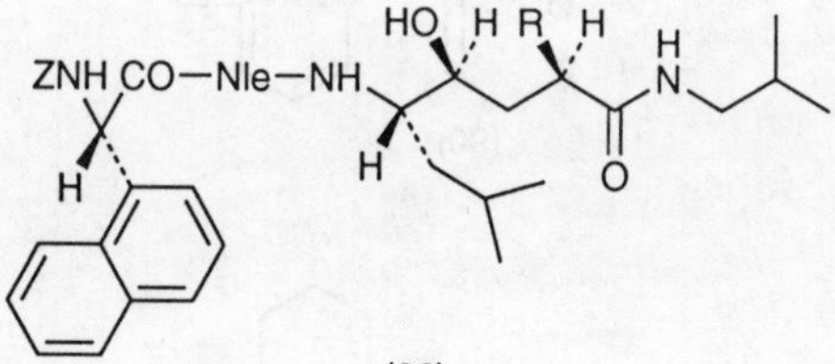

ZNH CO–Nle–NH

(96)

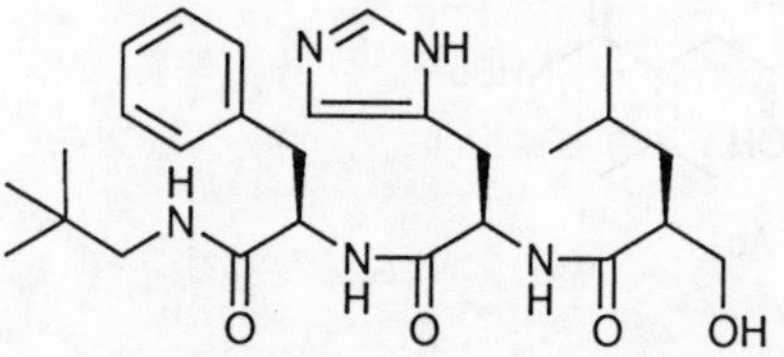

(97)

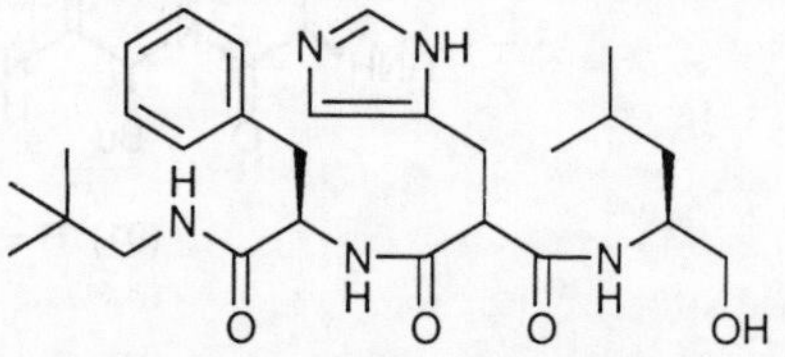

(98)

$\xrightarrow{Bu^nLi}$

camphor sulfonic acid

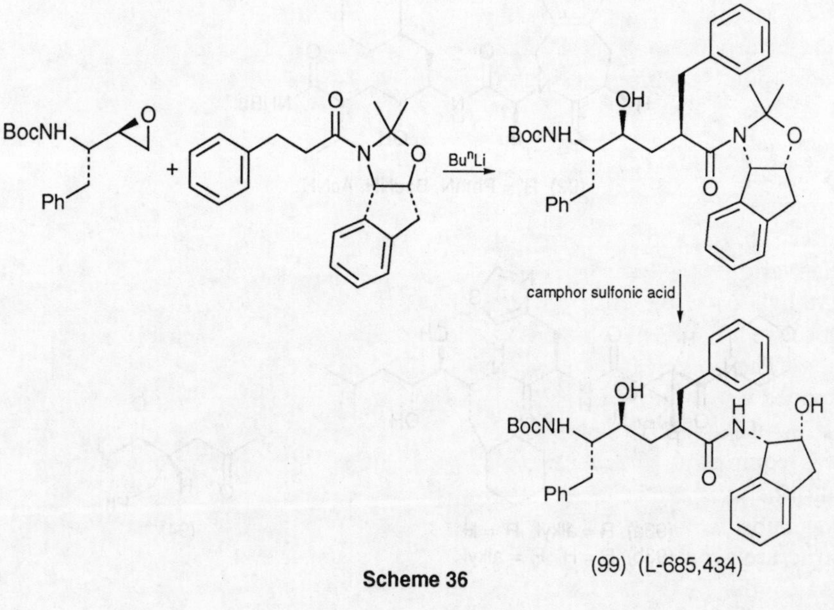

(99) (L-685,434)

Scheme 36

inhibitors [4*R*-(91), R=Ac] (IC_{50}=21nM) was 25-fold less potent than the control inhibitor with a phenylalanine residue at P_3. Further molecular modelling studies with compounds (91) and (92) suggested that the loss in inhibitor potency was due to conformational restrictions distorting the 3(*S*) centre from the geometry of the putative extended conformation present when the inhibitor is bound within the renin active site. The solution structure of the macrocycle, deduced from NMR and distance geometry calculations, included an unexpected hydrogen bond across the peptidomimetic ring. This bond caused a significant conformational change relative to that predicted by the initial modelling studies and the authors concluded that this apparently hindered the desired interactions in the renin active site.

A localised extended β-strand conformation could be enforced along a section of the peptide backbone by incorporation of a 1,2,3-trisubstituted cyclopropane moiety.[187] More potent inhibition was observed in (93) when the side chain substitutent of the cyclopropane was *syn* to the N-terminal functionality and the most active member of this group was (93a, R=Ph, R^1=H) (IC_5=0.7nM). The observation that the biological activities of this compound and the flexible analogue (*i.e.* 94) were virtually identical suggested that the spatial arrangement of the substituents on the rigid cyclopropane ring appeared to mimic closely the three-dimensional orientation of these groups in the biologically active conformation of (94).

Modified heterocyclic phenylalanine analogues were synthesised and incorporated as the P_3-P_4 moiety in the inhibitor series (95).[188] The most potent compound of the series was the X=Nva, R=$CHMe_2$ analogue with an IC_{50} of 8.9nM. Furthermore all of these compounds were very resistant to chymotrypsin degradation and, for example, (95, X=Leu, R=Me) remained >60% intact after a 24h exposure to the protease whereas the Boc-Phe analogue was nearly completely degraded after 1h. Structure-activity studies on the homostatine series (96, R=various alkyl, hydroxyalkyl) concluded that an ethyl, a 2-hydroxyethyl, or a 3-hydroxylpropyl group at the P_1' position was more effective for increasing potency than the isopropyl group.[189]

The retro-inverso peptide backbone modification has been applied to the tripeptide alcohol inhibitor Boc-Phe-His-Leu-ol (IC_{50}=16μM). The complete retro-inverso peptide (97) (IC_{50}=20μM) was equipotent to the parent compound with normal amide linkages but the diastereoisomeric mixture of the partial retro analogue (98) was substantially less active (IC_{50}=1200μM).[190] A systematic study of structure-activity relationships of P_6 to P_4' renin substrate analogues led to the design and synthesis of the canine inhibitor Ac-paF-Pro-

Phe-Val-statine-Leu-Phe-paF-NH_2 (IC_{50} = 1.7nM; paF = *para*-amino phenylalanine).[191]

5.3 *HIV-1 Protease Inhibitors.*

The amide bond of the primary cleavage site of HIV-protease continues to be targeted for isosteric replacement, and potent inhibitors containing the hydroxyethylene and hydroxyethylamine isosteres have been described.

The Merck group have developed an alternative protocol for the synthesis of their indane-based inhibitor (99) (Scheme 36) to the '*trans*'-lactone route reported last year.[192] Coupling of the enolate with the epoxide was highly diastereoselective (> 99:1 in favour of the 2*R* adduct) as the electrophile approached from the least hindered face of the Z-enolate. Furthermore, reaction of two equivalents of the lithium enolate with one equivalent of 3-iodo-2(iodomethyl)-1-propene led to the preparation of the pseudo-C_2 symmetrical inhibitor (100). Incorporating a hydroxyalkyl group at the *para*-position of the benzyl P_1' side chain of (99) resulted in an increase in enzyme inhibitory activity.[193,194] More significantly, several of these compounds, for example, the 4-morpholinyl-ethoxy (101) and the 4-(3-hydroxypropyl) (102) analogues were approximately 15 times more potent than the parent inhibitor (99) as antiviral agents in cell culture assays.

Stereoselective syntheses of the Pheψ[CH(OH)CH_2]Phe[195] and Tyrψ[CH(OH)CH]Pro[196] dipeptide mimetics are outlined in Schemes 37 and 38 respectively. The allylic iodide in Scheme 37 proved to be a versatile intermediate and by utilising other reagents such as Grignards, sodium mercaptides, or magnesium alkoxides a variety of substituents could be introduced at position 4 of the hydroxyethylene isostere. The *trans*-[2*R*,3*R*,4*S*,5*S*]Tyr-Pro hydroxyethylene diastereoisomer (103) proved to be a more effective inhibitor of the HIV-1 protease than the *cis*-isomer with 3*S* stereochemistry [IC_{50} of 0.50μM for (103) compared with IC_{50} > 30μM for the *cis* isomer]. Incorporation of the 2(*S*)-hydroxy-1(*R*)-amino indane as the P_2' ligand enhanced the inhibitory potency 5-fold but this compound (104) was still approximately 300 times less potent than the indane-based inhibitor (99) (IC_{50} = 0.23nM).

On the basis of results obtained during the development of several potent hydroxyethylene HIV-1 protease inhibitors the Merck group incorporated a Boc group at the N-terminus and the 1(*S*)-amino-2(*R*)-hydroxyindane functionality at the C-terminus of a series of four isomeric hydroxyethylamine compounds (Scheme 39).[197] Enzyme inhibition studies indicated a clear preference for the diastereoisomer with the (*R*)-configuration at the transition state alcohol carbon and the (*S*)-

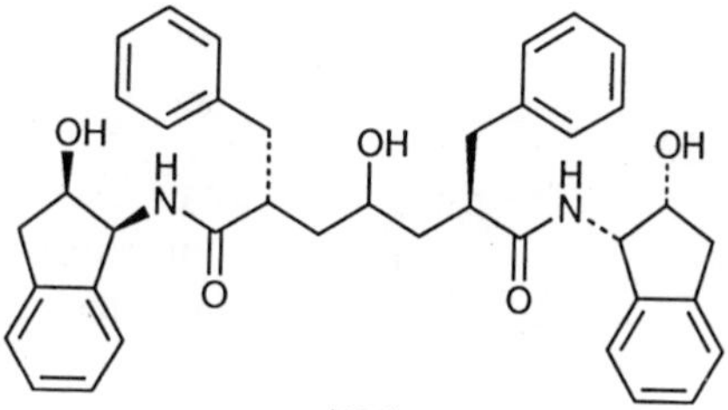

(100)

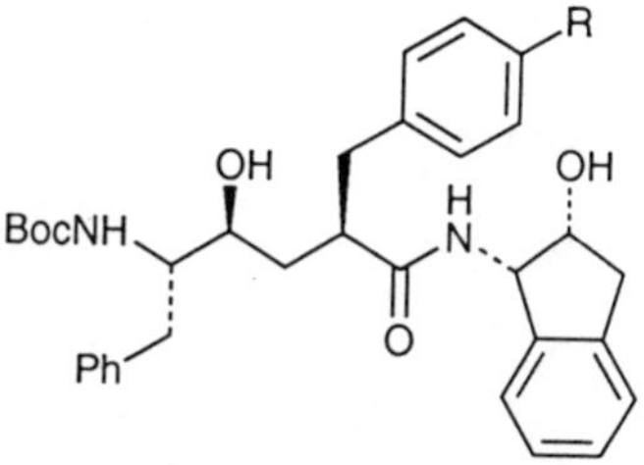

(101) R = OCH_2CH_2N(morpholino) (L-689,502)

(102) R = $CH_2CH_2CH_2OH$ (L-693,549)

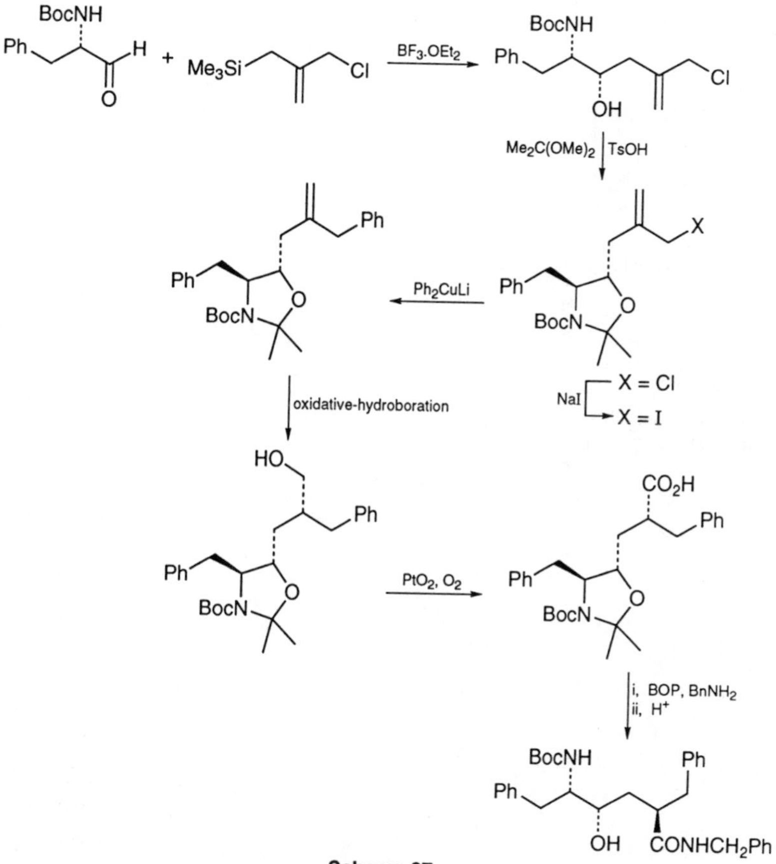

Scheme 37

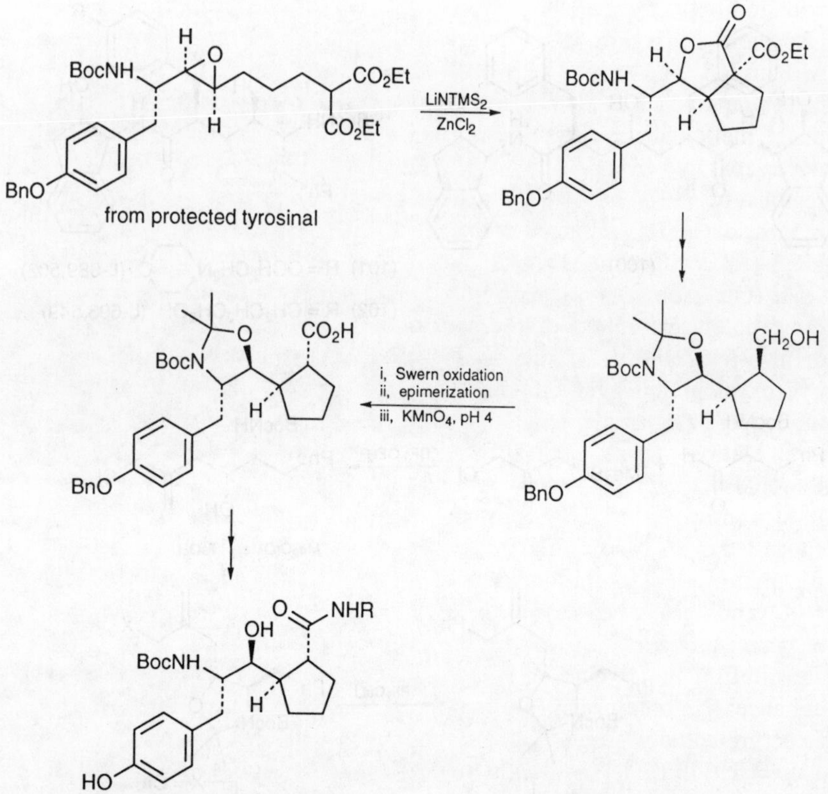

BocNH H O H CO_2Et CO_2Et BnO

from protected tyrosinal

$LiNTMS_2$
$ZnCl_2$

H O O CO_2Et BocNH H BnO

O CO_2H BocN H BnO

i, Swern oxidation
ii, epimerization
iii, $KMnO_4$, pH 4

O CH_2OH BocN H BnO

O NHR OH BocNH H HO

(103) $R = CH_2Ph$
(104) $R = 2(S)$-hydroxy-$1(R)$-indane

Scheme 38

configuration at the P_1' α-carbon atom. N-terminal extension of (105) into the P_2-P_3 region gave (106), the most potent inhibitor of the series with an IC_{50} of 5.4nM. The hydroxyethylamine isostere JG 365 (107) has been prepared as a 1:1 mixture of the epimeric *R/S* alcohols by a solid phase method and, after cleavage from the resin, the diastereoisomers were separated by HPLC.[198] Systematic modification of the P_3 and P_3' regions of (107) has led to smaller compounds that suppress viral replication in HIV-infected and SIV-infected cell cultures. Z-Asn-[Phe-HEA-(*S*)-Pro]-Ile-Phe-OMe and Qua-Asn-[Phe-HEA-(*S*)-Pro]-Ile-Phe-OMe (Qua = quinolin-2-ylcarbonyl, HEA = hydroxyethylamine) were the most potent inhibitors of the series with K_i values equal to 1.0 and 0.1nM respectively.[199] The design of both the hydroxyethylene and hydroxyethylamine inhibitors was to some extent aided by molecular modelling using crystal structures of enzyme-inhibitor complexes.

Data from the crystal structures of Ala-Ala-Pheψ[CH(OH)CH_2]Xaa-Val-Val-OMe (108) (Xaa = Gly, Ala, NorVal, Phe) complexed to HIV protease and from enzyme kinetic studies suggested that, for this class of compounds, the minimum requirement for tight binding inhibition was a network of 7 hydrogen bonds between (108) and the active site of the enzyme and occupation of the S_2-S_2' enzyme subsites.[200] Free energy calculations have predicted that substituting the glycine at P_1' position of (108) by norleucine should enhance the binding by approximately 1.7kcal/mol.[201] The result of structure-activity studies on fragment analogues of Val-Ser-Gln-Asn-Leuψ[CH(OH)CH_2]Val-Ile-Val indicated that the P_1-P_2' tripeptidyl sequence was the minimum chemical determinant for HIV-1 protease binding.[202] The sequences Suc-Ser-Leu-Asn-Tyr-Pro-Ile-NH^iBu and Suc-Val-Ser-Gln-Asn-Phe-Pro-Ile-NH^iBu have been prepared as specific substrates for HIV-1 protease.[203]

The most potent compound in a series of P_2-P_2' norstatine-based inhibitors was (109) (IC_{50} = 0.58nM, K_i = 0.4).[204] However substituting the pyrrolidine function at P_1' of Z-Asn-Pheψ[CH(OH)C(O)N]ProNHBu^t with a piperidine or saturated isoquinoline ring led to a reduction in inhibitory potency, a result which was in marked contrast to the enhancement in potency reported for the identical modification in the corresponding hydroxyethylamine series Z-Asn-Pheψ[CH(OH)CH_2N]ProNHBu^t. The authors speculated that the conformational constraints imposed on the complex by the hydroxamide carbonyl may cause less than optimum binding of the C-terminal functionality to the P_1' pocket.

Full experimental details of the synthesis of the C_2-symmetric diamino diol (110), the core unit of the Abbott inhibitor A-77003 (111)

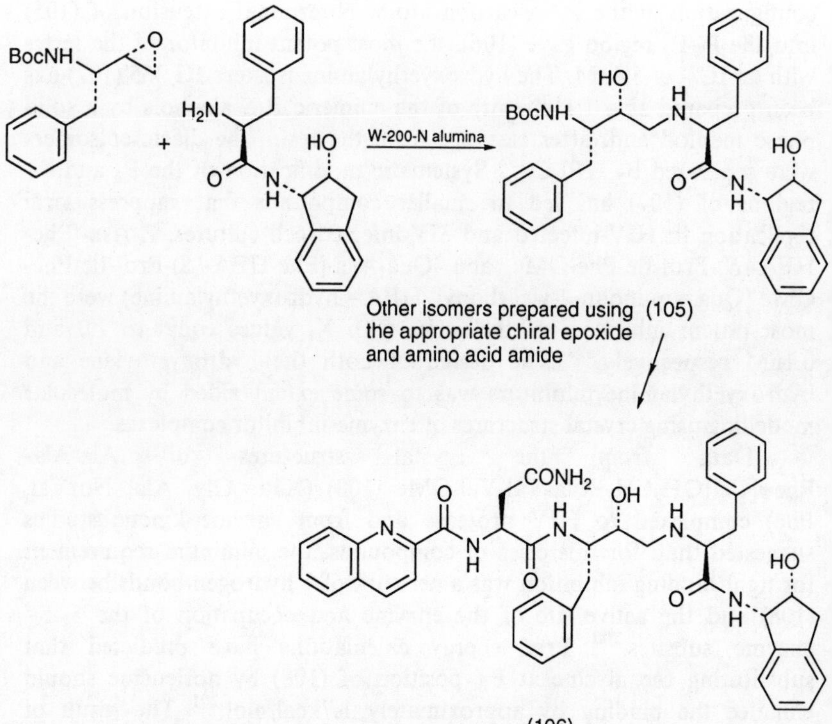

BocNH + H_2N HO W-200-N alumina HO BocNH HO

Other isomers prepared using (105) the appropriate chiral epoxide and amino acid amide

$CONH_2$ OH HO

(106)

Scheme 39

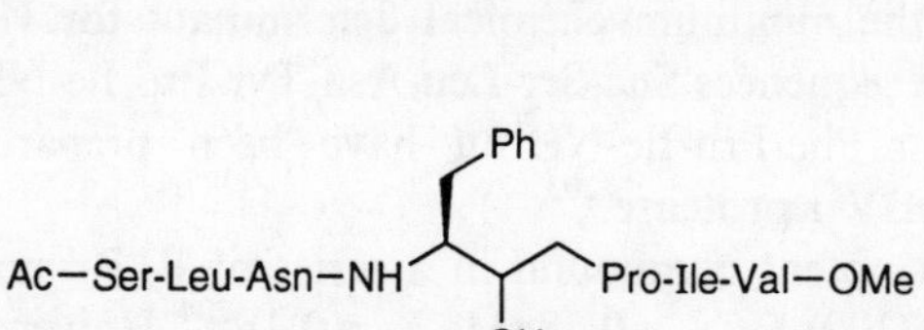

Ac—Ser-Leu-Asn—NH Ph Pro-Ile-Val—OMe OH

(107) (JG 365)

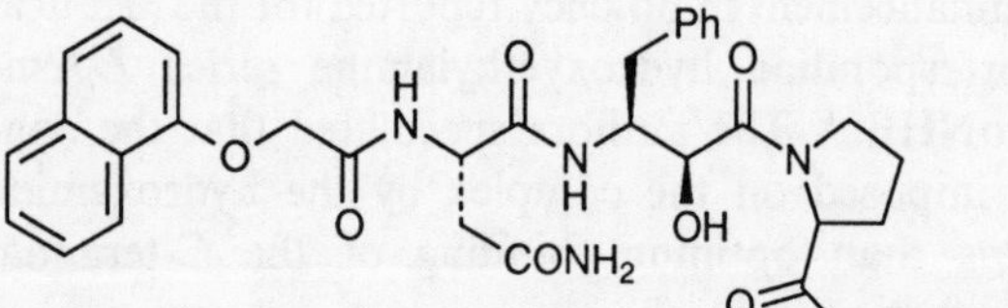

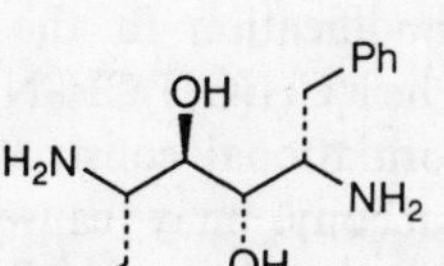

Ph O OH $CONH_2$ $NHBu^t$

(109)

Ph OH H_2N NH_2 OH Ph

(110)

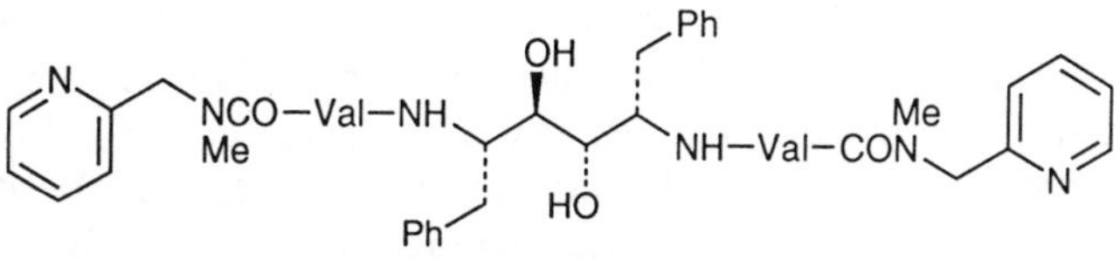

(111) (A-77003)

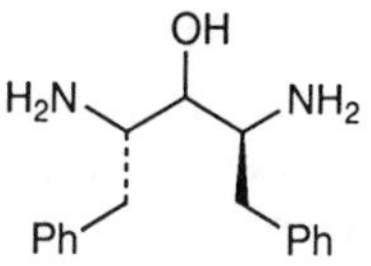

(112)

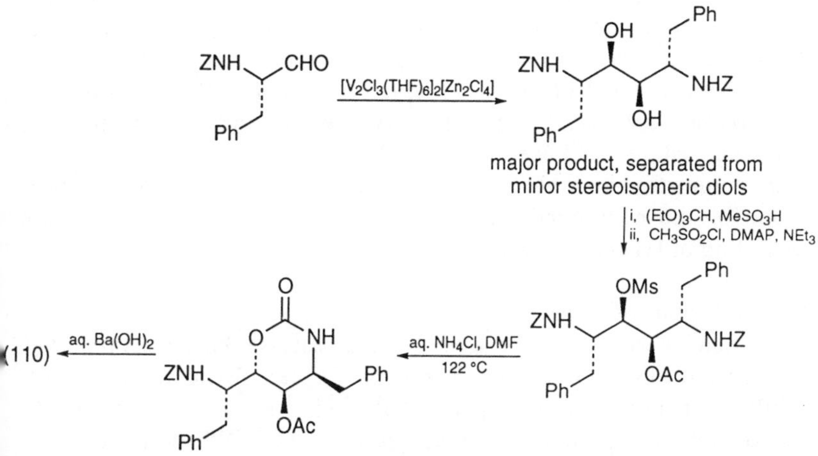

ZNH CHO Ph

$[V_2Cl_3(THF)_6]_2[Zn_2Cl_4]$

major product, separated from minor stereoisomeric diols

i, $(EtO)_3CH$, $MeSO_3H$
ii, CH_3SO_2Cl, DMAP, NEt_3

aq. NH_4Cl, DMF 122 °C

aq. $Ba(OH)_2$

(110)

Scheme 40

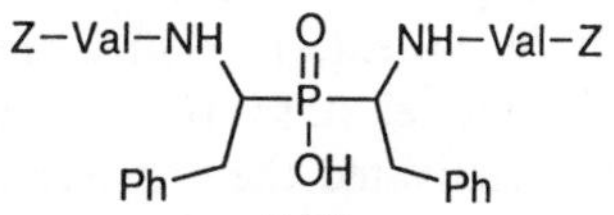

(113)

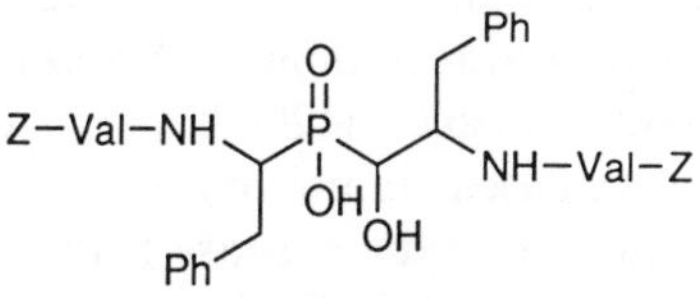

(114)

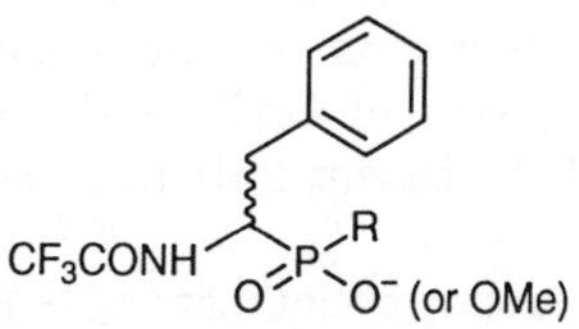

(115) R = Phe–NH_2 or Pro–NH_2

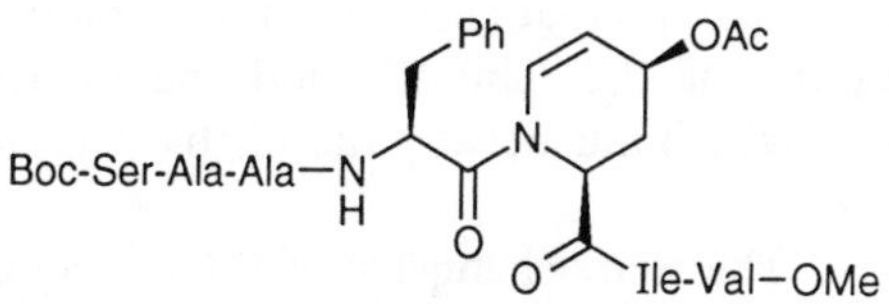

(116)

has been published.[205] Although several routes to (110) were described, the one outlined in Scheme 40 was the most practical for the preparation of multigram quantities. Methods were also developed for the synthesis of the three stereoisomeric diol analogues of (110) and the diamino alcohol (112). The phosphinic acid analogues of (110, 112) have also been prepared but as mixtures of stereoisomers. Coupling of Z-valine to the amino groups of the phosphinic acid core units gave (113, 114), which were potent inhibitors of HIV-1 protease (IC_{50}s in nanomolar range).[206,207] More modest inhibitory activity was observed for the phosphonamidate and phosphonamidate esters (115).[208]

The $\psi[PO_2^-CH_2N^+]$ moiety was designed as a 'late transition state/early product formation' isostere for amide bond hydrolysis in that it mimics both the tetrahedral hydrate and departing amine components. The pseudoheptapeptide Ac-Ser-Leu-Asn-(*R*)-Phe$\psi[PO_2^-CH_2N^+]$Pro-Ile-Val-OMe incorporating this novel isostere was a slow-binding inhibitor with a K_i* of 8nM and $t_{1/2}$(EI*) of 40mins.[209] Inactivation of the enzyme *via* a 'suicide mechanism' has been explored using substrates featuring either a 3-acetoxy-Δ-4,5-(L)-pipecolic acid (116)[210] or a sulphoniomethyl-aminobenzoic acid (117) residue[211] at the P_1' position. None of the compounds in the latter series showed any significant inhibition of HIV-1 protease.

5.4 *Inhibitors of Other Proteases.*

'Suicide' substrates for several serine proteases have been reported. The cyclopeptides (118) with X = Br or Cl were found to irreversibly inhibit α-chymotrypsin ($k_{inact}/K_i = 180M^{-1}min^{-1}$ for X = Cl) whereas the compounds with poorer leaving groups (X = OAc, OPh) were devoid of inactivating effect and only behaved as substrates of the enzyme.[212] Both trypsin and thrombin were rapidly inactivated by N^{α}-*N,N*-dimethylcarbamoyl)-α-azaornithine and α-azalysine esters (119).[213] The peptidyl-α-ketobenzoxazoles (120a,b) were found to be potent, reversible, mechanism-based inhibitors of human leukocyte elastase (HLE) and porcine pancreatic elastase (PPE) and they inactivated the enzymes by interacting with both the His and Ser of the catalytic triad.[214] X-ray analysis of the complex between PPE and (120b) clearly indicated the hemiketal linkage between the carbonyl carbon atom of the valine residue of the inhibitor and the hydroxyl group of Ser^{195} and the hydrogen bond betwen the nitrogen atom of the benzoxazole ring and His^{57}.

The stereochemistry of the thrombin inhibitor cyclotheonamide B was reassessed following total synthesis; the revised structure (121) contains *S* stereochemistry at both the arginine-like and vinylogous

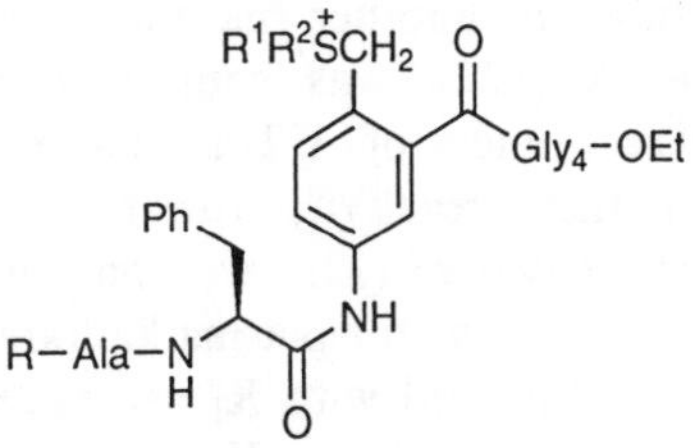

(117) R = CF_3CO, Ac–Gly-Ala; R^1 = Me; R^2 = Me, Ph

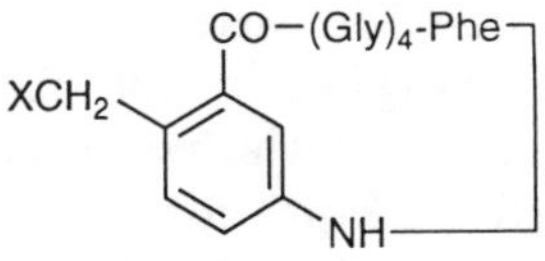

(118)

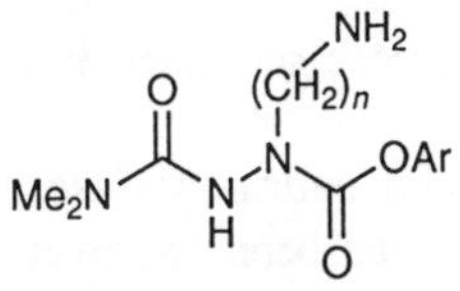

(119) Ar = C_6H_5, p-$NO_2C_6H_4$; n = 3, 4

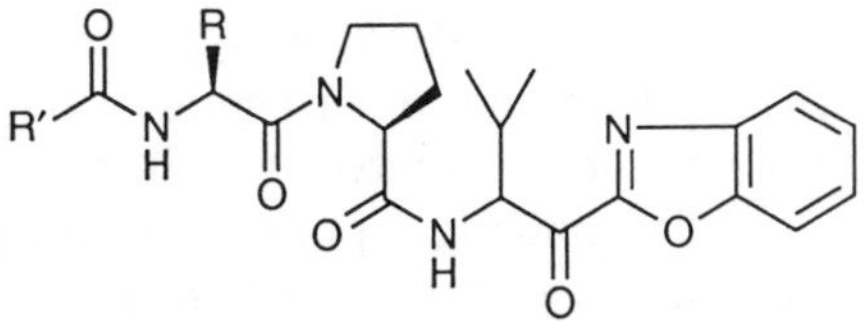

(120a) R = $CHMe_2$, R′ = OCH_2Ph
(120b) R = R′ = Me

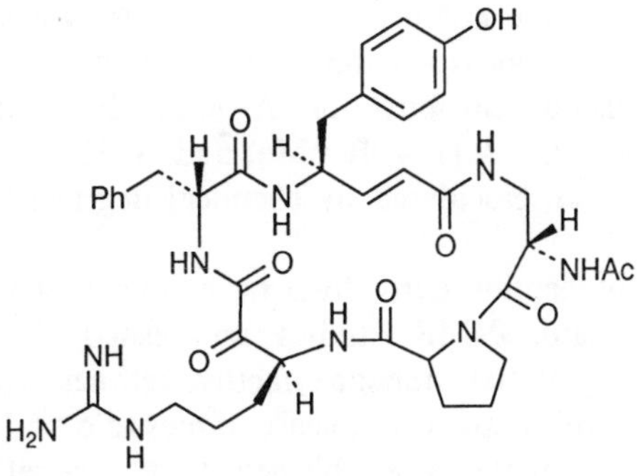

(121)

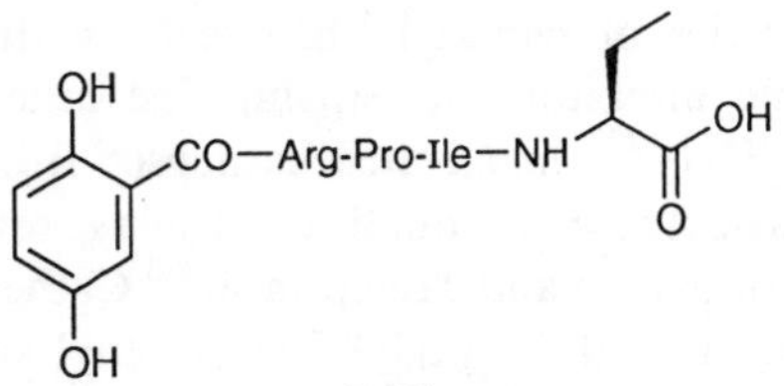

(122)

tyrosine residues.[215] The structure of another naturally occurring thrombin inhibitor, nazumamide A (122), was confirmed by total synthesis.[216] The DL and LL stereoisomers of *N*-Tos-Phe-(*m* or *p*-F)-Arg-OMe were synthesised and their reactivity towards thrombin investigated.[217] The phenylalanine derivative (123) was one of several inhibitors which were designed to be selective for plasma kallikrein[218,219] ($K_i = 0.81\mu M$ for plasma kallikrein compared with K_i values $> 200\mu M$ for glandular kallikrein, plasmin, urokinase, factor Xa, and thrombin). The 21 di- to heptapeptides which occur within the sequence Ser^{386} - Gln^{392} of bovine kininogen were synthesised and tested for inhibitory activity against tissue kallikrein.[220] The heptapeptide Ac-Ser-Pro-Phe-Arg-Ser-Val-Gln-NH_2 was the most effective inhibitor in the series ($K_i = 101\mu M$) but other peptides containing the core sequence Phe-Arg-Ser also showed inhibitory activity.

The finding, reported last year, that prolyl endopeptidase (PEP) was inhibited by fragments of human β-casein has been followed by a second paper describing the inhibitory activity of a series of analogues.[221] For example, extending the C-terminus of Ile-Tyr-Pro-Phe-Val-Glu-Pro-Ile ($IC_{50} = 8\mu M$) with a prolinal moiety led to Ac-Ile-Tyr(Bu^t)-Pro-Phe-Val-Glu(OBu^t)-Pro-Ile-Pro-H ($IC_{50} = 0.1\mu M$) which was a more potent PEP inhibitor than the octapeptide parent. The *N*-acyl derivatives of Gly-Pro-NH_2 were found to be more resistant to degradation by PEP than the *N*-benzyloxycarbonyl or *N*-phthalidyl analogues.[222] The structure of eurystatins A (124, R = Me) and B (124, R = Et), new PEP inhibitors, have been elucidated by chemical degradation and spectral studies.[223]

A variety of peptide derivatives have been prepared as potential inhibitors of papain. NMR studies established that Z-L-Phe-Gly-mesitoyloxy- and pentafluorophenoxymethyl ketones formed irreversible covalent adducts with papain *via* release of mesitoic acid or pentafluorophenol.[224] Kinetic methods established that the series of Michael acceptors Ac-Phe-$NHCH_2CH{=}CHE$ (E = electron withdrawing group) and $RNHCH_2CH{=}CHCO_2Me$ (R = *N*-Ac-amino acid, $PhCH_2CH_2CO$, PhCO)[225] and a series of peptidyl-2-haloacetyl hydrazines[226] were, in general, irreversible inhibitors of papain. The latter series, and also biotinyl-Phe-Ala-CHN_2,[227] inactivated cathepsin B. *N*-Peptidyl derivatives of D-glucosamine were found to display reversible inhibitory activity against both papain and cathepsin B.[228] Cathepsin L and calpain were both inactivated by the peptidyl fluoromethyl ketones Z-Leu-Leu-Tyr-CH_2F and Z-Leu-Tyr-CH_2F but only cathepsin L was inactivated by Z-Tyr-Ala-CH_2F.[229] An enzymic method of preparing the hydroxylamine cathepsin inhibitors R = NHOAc [R = Z-Ala-Asp(OCH_2Ph)-Phe,

H₂NCH₂ ... CO—L-Phe—NH— ... —CH₂CO₂H

(123)

(124)

(125)

(126)

(127a) $R^1 = NH_2$, $R^2 = H$
(127b) $R^1 = H$, $R^2 = NH_2$

(129)

H₂N ... CO—D-Pen-Gly-Phe-D-Pen—OH

(130)

H₂N ... CO—D-Pen-Gly-Phe-D-Pen—OH

(131)

Boc-Pro-Phe-Leu, Z-Ala-Phe, Boc-Pro-Phe, 4-MeOC_6H_4-CH_2O_2C-Phe] has been reported.[230] The chimeric enkephalin-cystatin peptide YGGFLQVVAGK-amide was found to be both a weak inhibitor of cathepsin L and papain and a ligand for opioid receptors.[231] The conformation of the cystatin mimicking peptide KVGGQVVCGAPWK (K_i for papain = 15μM) was studied using 2D NMR spectroscopy.[232]

The conformationally restricted analogues (125, X = spacer unit) of pepstatin were prepared by treating the linear bis-*S*-benzyl precursor peptide with Na/NH_3(l) followed by a dibromoalkane or alkene.[233] Molecular modelling studies indicated that the optimum number of carbon atoms needed to close the ring system without distorting the peptide backbone was between four and six. The *trans*-2-butene analogue (125, X = *trans*-CH_2-CH = CH-CH_2) was the most potent inhibitor with a K_i < 1nM against pepsin and 3.94nM against penicillopepsin.

The inhibitory potency of HS-CH_2-CH_2-CO-Pro-Yaa (Yaa = Ala, Leu, Nle) towards *Empedobacter collagenolyticum* collagenase was slightly increased by methylating the nitrogen of the Yaa residue.[234] NMR studies revealed that the *N*-methylation did not cause any important conformational effect and both the parent and modified peptides adopted a *trans*-conformation around the Pro-Yaa amide bond. Detailed kinetic and mechanistic studies for the inhibition of thermolysin by the peptide hydroxamic acid HONH-isobutylmalonyl-Ala-Gly-NH_2 found that the interaction was a complex process involving multiple intermediates.[235] Furthermore, the inhibitor was bound to thermolysin as the neutral un-ionized acid and not as an anion as suggested previously by other workers.

The α-keto amide analogue (126) of bestatin was found to be a slightly better inhibitor of several aminopeptidases than the hydroxy counterpart.[236] In contrast, replacing the hydroxyl group of bestatin or *epi*-bestatin with an amino function gave compounds (127a,b) which were less effective as aminopeptidase inhibitors.[237]

In vivo inhibition of the enkephalin-degrading enzymes, aminopeptidase N and neutral endopeptidase, was effected by i.v. coinjection of $(H_2NCHRCH_2S)_2$ [R = CH_2CH_2SMe, $CH_2CH_2S(O)Me$] with acetorphan [$CH_3COSCH_2CH(CH_2Ph)$-$CONHCH_2CO_2CH_2Ph$] or by combining these two compounds in one entity and administering the 'mixed inhibitor-prodrug' [H_2NCH (CH_2CH_2SMe) CH_2S-$SCH_2CH(CH_2Ph)$ $CONHCH(CH_2Ph)$ CO_2CH_2Ph] (128).[238,239] Ketomethylene pseudodipeptide analogues H-Lysψ[$COCH_2$]Xaa-OH (Xaa = Phe, Trp)[240] of arphamenine A and derivatives [R = Ac, $HS(CH_2)_nCO$, $MeO_2C(CH_2)_n$, n = 1,2][241] of the analgesic dipeptide R-Trp(Nps)-Lys-OMe {R = H,

Trp(Nps) = 2-[(2-nitrophenyl)sulphenyl]tryptophan} were synthesised and tested for inhibitory activity against enkephalin-degrading enzymes. Kelatorphan (129), a potent inhibitor of the enkephalin-degrading enzymes, was prepared in 7 steps and 35% overall yield from hydrocinnamyl chloride.[242] The structure of benarthin [2,3-$(HO)_2C_6H_3$CO-L-Arg-L-Thr-OH], an inhibitor of pyroglutamyl peptidase, was confirmed by total synthesis.[243]

6 Side Chain Interactions studied by Residue Substitution or Deletion and Similar Modifications

A comprehensive review of peptide drug design has been published.[244] The design of compounds such as kelatorphan (129) and the 'mixed inhibitor-prodrug' (128) of the enkephalin-degrading peptidases was discussed in a short review which also covered the design of selective ligands for neuropeptide receptors.[245]

6.1 *Peptides with 'Opioid Characteristics'.*

Various conformationally constrained analogues of tyrosine have been incorporated into δ-opioid receptor ligands to investigate the contribution of the tyrosine side chain to the backbone conformation. Replacing Tyr^1 of [D-Pen^2,D-Pen^5]enkephalin (DPDPE) with 2′,6′-dimethyltyrosine gave a ligand (130) with enhanced potency at the δ- and μ-receptors.[246,247] Although the binding affinity of [2′,6′-Me_2-Tyr^1]DPDPE (130) to the μ- and δ-receptors was increased relative to DPDPE (800 and 8-fold respectively) δ-selectivity was lost due to a marked increase in μ-potency. Further modification of (130) by replacing the Gly-Phe amide bond with a CH_2O, CH = CH, or CH_2CH_2 isostere gave analogues which showed activity in the *in vitro* assays but which lacked the *in vivo* activity of the parent compound. Another group of workers found that the selectivity for the δ-receptor was improved by replacing Tyr^1 by 2-Me′-Tyr^1 in DPDPE while in the series with β-methyl substituted/constrained Tyr^1 analogues both (131, 132) were potent and δ-receptor selective.[248] The *N,N*-disubstituted enkephalin analogues R_2Tyr-D-Met-Gly-Phe-εAhx-OMe [R_2 = *N,N*-diallyl, *N,N*-dibutyl, εAhx = $NH(CH_2)_5CO$] were both potent and highly selective for the μ-receptor whereas the corresponding compounds with an additional εAhx residue at the C-terminus were less active and not selective as agonists.[249] Introduction of a *N*-monoalkylated tyrosine at position 1 of [D-Pro^{10}]-dynorphin A-(1-11) gave analogues which retained the κ-receptor affinity of the parent compound but which bound poorly to the μ- and δ-receptors and were, therefore, very κ-selective (*N*-allyl analogue K_i ratio

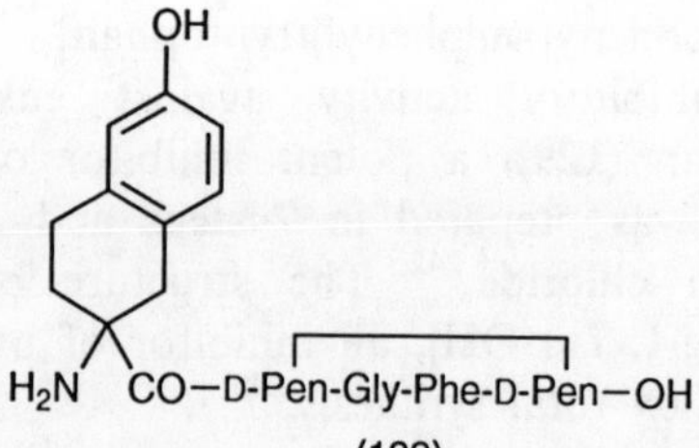
OH
H2N
CO—D-Pen-Gly-Phe-D-Pen—OH

(132)

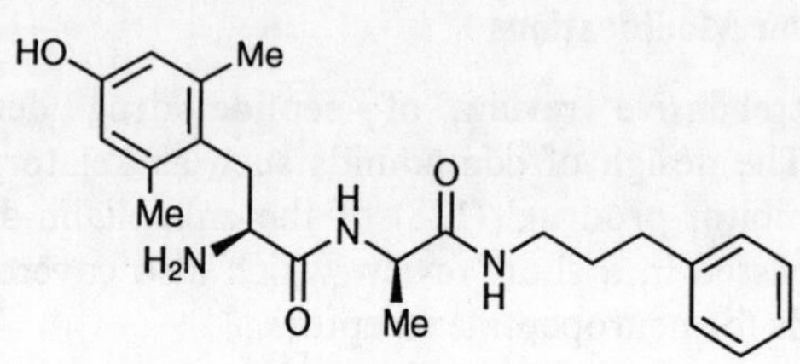
HO
Me
Me
H2N
H
N
O
N
H
O
Me

(133)

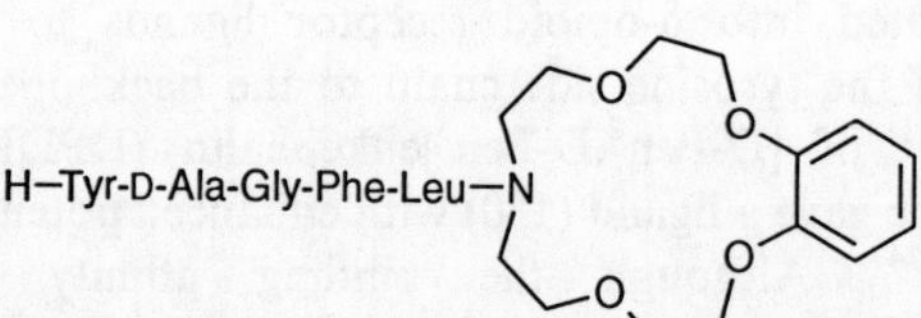
H—Tyr-D-Ala-Gly-Phe-Leu—N
O
O
O
O

(134)

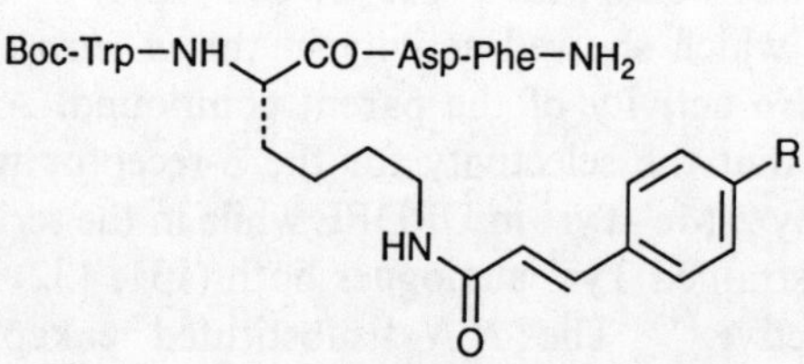
Boc-Trp—NH
CO—Asp-Phe—NH2
R
HN
O

(137)

κ/μ/δ = 1/122/9160; *N*-cyclopropylmethyl analogue K_i ratio 1/480/27900).[250]

Results of binding assays on a series of dynorphin A-(1-13) amide analogues in which the five basic residues were systematically substituted with N^{ε}-acetyllysine indicated that only Arg^9 could be replaced by a nonbasic residue without substantial loss of κ-opioid receptor selectivity.[251] Replacing Gly^2 of the tridecapeptide with L- or D- Asn, Lys, Met, and Ser led to a severe decrease in affinity and selectivity for the κ-receptor.[252] Although the L-amino acid substituent analogues generally retained some selectivity for the κ-opioid receptor the compounds containing a D-amino acid at position 2 were more potent and showed a preference for the μ-receptor. The solid-phase synthesis of the heptadecapeptide [Ala^8]-dynorphin A has been optimised.[253]

In the dermorphin and deltorphin families the side chains of Tyr^1 and Phe^3 are important for opioid receptor recognition and the effect of substitutions at these positions was central to several studies. A 3-fold improvement in δ-affinity and selectivity was observed by replacing Tyr^1 of the δ-selective dermorphin/deltorphin tetrapeptide analogue Tyr-D-Cys-Phe-D-Pen-OH (JOM-13) with L-*trans*-3-(4′-hydroxyphenyl) proline.[254] Incorporation of halogenated derivatives of Phe^3 into JOM-13 enhanced δ-receptor binding affinity whereas compounds with electron-donating substituents, such as alkyl or hydroxyl, or with a sterically bulky nitro group on the aromatic ring showed reduced receptor binding.[255] Analogues with π-excessive heteroaromatic side chains at position 3 of JOM-13 also showed diminished binding to both the μ- and δ-receptors.[256] Taken together these results indicated that the lipophilic, electronic, and steric properties of residue 3 all play a role in influencing binding interactions. Furthermore, results obtained with the bulky aromatic 1-Nal^3 and 2-Nal^3 [Nal = 3-(2-naphthyl)-L-alanine] substitutents in JOM-13 suggested that the shape of the receptor subsite with which the side chain of the aromatic residue interacts differs for δ- and μ-receptors. Both receptors could accomodate the steric bulk of the 1-Nal^3 side chain but only the δ-receptor could readily accept the more elongated 2-Nal^3 side chain. The analogue of JOM-13 with a Cha^3 residue [Cha = 3-(cyclohexyl)-L-alanine] was well tolerated in the binding assays which suggested that an aromatic side chain in this portion of the ligand was not required for δ-receptor binding.

The effect of incorporating a halogen, amino, or nitro group into the *para* position of Phe^3 of deltorphin C (H-Tyr-D-Ala-Phe-Asp-Val-Val-Gly-NH_2) was more ambiguous than in the JOM-13 study and the activity profile of the analogues was dependent upon the particular substitutent.[257] Substituting the conformationally restricted 2-aminoindan-2-carboxylic acid (Aic) or L- or D-2-aminotetralin-2-carboxylic acid (Atc) in place of Phe^3 in deltorphin C resulted in agonist compounds which retained the high δ-receptor selectivity of the parent peptide.[258] Furthermore, deltorphin-related analogues with a tetrahydroisoquinoline-3-carboxylic acid (Tic) residue in the 2-position of the peptide sequence showed even higher δ-selectivity and either antagonist or partial agonist properties at the δ-receptor. Binding data on a series of deltorphin A (H-Tyr-D-Met-Phe-His-Leu-Met-Asp-NH_2) analogues with D- for L-amino acid substitutions provided evidence that the side chains of Tyr^1, Phe^3, and Leu^5 were particularily important for high δ-affinity and selectivity.[259]

The [4′-^{3}H-Phe^3] and [4′-^{125}I-Phe^3] radiolabelled derivatives of the deltorphin H-Tyr-D-Ala-Phe-Glu-Val-Val-Gly-NH_2 were both highly selective ligands for the δ, relative to the μ, opioid receptors in rat brain.[260,261]

In the morphiceptin series H-Tyr-Pro-Phe-L-NHCH[$(CH_2)_n$Ph]$CONH_2$ (n = 1-4) the compound with the benzyl side chain at position 4 showed the highest affinity for the μ-receptor (3-fold more potent than morphiceptin).[262] The compounds with longer phenylalkyl side chains exhibited much weaker (12-26 fold) affinity than Phe^4-morphiceptin and the authors suggested that steric hindrance caused by the additional methylene groups prevented the aromatic ring in the higher homologues from adopting the optimum orientation for receptor binding. A HPLC study on β-casomorphin analogues[263] and a TLC study on dermorphin-related peptides[264] have been carried out.

Isosteric replacement of the D-Ala/phenylpropylamine amide bond of the analgesic dipeptide (133) produced a modest reduction of activity but modification of the dimethyltyrosine/D-Ala linkage gave compounds which were inactive in both *in vitro* and *in vivo* assays.[265] The heptapeptide H-Val-Val-Tyr-Pro-Trp-Thr-Gln-OH (valorphin) was isolated from bovine hypothalmic tissue and was found to exhibit some opiate activity in receptor binding assays and in the guinea pig ileum bioassay.[266] Among the products translated from mRNA extracted from the brains of a marine worm were several 60-70kDa polypeptides which were recognised by anti-dynorphin 1-17 and anti-α-neo-endorphin antibodies.[267]

The δ-receptor selectivity of Leu enkephalin was decreased by the addition of a sugar residue to the C-terminal carboxyl group.[268] In the radio-ligand receptor assay and biological assays, H-Tyr-D-Ala-Gly-Phe-Leu-Cys(Npys)-OH exhibited high affinity and selectivity for δ over μ receptor and, when incubated with mouse vas deferens tissue, this peptide covalently bound *via* a disulphide bond to the δ-receptors.[269] The crown ether (134) containing a Leu enkephalin fragment showed some analgesic activity.[270]

6.2 *Cholecystokinin Analogues.*

Structure-activity work with Ac-CCK-7 [Ac-Tyr(SO_3H)^{27}-Met^{28}-Gly^{29}-Trp^{30}-Met^{31}-Asp^{32}-Phe^{33}- NH_2] was reported in a series of papers from Roche laboratories. The studies focused on separately modifying Phe^{33}, Trp^{30}, Asp^{32} and the Met-Gly dipeptide moiety of Ac-CCK-7 and establishing their effects on the peripheral (CCK-A) receptor binding and on appetite suppression. The aromatic ring of the C-terminal phenylalanine could be replaced with a wide range of aryl or cycloalkyl groups without loss of receptor A activity.[271] Compounds with the cyclooctyl and 1-naphthyl modification were exceptionally potent in the binding assays with IC_{50s} of 0.00001nM in the pancreas (*cf* value of 0.60nM for Ac-CCK-7) and were approximately equipotent to CCK at stimulating amylase release. All of the compounds in the series derived from L-alanine were potent suppressors of food intake with several of them being an order of magnitude more potent than Ac-CCK-7 itself. The results of the Trp^{30} study were in marked contrast and only incorporation of closely related analogues of tryptophan, such as 2-naphthylalanine, resulted in compounds with an activity profile similar to that of the parent molecule.[272] The Asp^{32} residue could be replaced by a non-acidic (Asn, Pro, Aib) or a highly constrained amino acid (*R*-Dtc, Dtc=5,5-dimethyl-1,3-thiazolidine-4-carboxylic acid) with retention of potent binding to pancreatic receptors and suppression of food intake in rats.[273] The peptide incorporating γ-aminobutyric acid in place of the Met-Gly dipeptide unit was only 8 times less potent than Ac-CCK-7 in peripheral binding but was more potent than Ac-CCK-7 in meal feeding experiments.[274] Overall these studies suggested that the requirement for potent agonist activity at the peripheral receptor included an aryl or cycloalkylalanine in position 33, tryptophan or a closely related analogue in position 30 and an appropriate spacer to mimic the bend in the Met-Gly region of the peptide.

Systematic studies have been carried out to determine the effect of N-methylation of single peptide bonds on the biological activity and selectivity of analogues derived from Ac[Nle^{28,31}]CCK(26-33) (135)[275] and [desaminoTyr^{27},Nle^{28,31}]CCK(27-33) (136).[276] Although the pharmacological profiles of the two series were not directly comparable several trends were observed. N-Methylation at either the N- or C-terminus of (135) or at position 33 of (136) had only relatively minor effects on potency and selectivity whereas methylating the internal residues of both parent compounds had a more pronounced effect on receptor affinity.

Binding to both receptor subtypes was substantially decreased by N-methylation of Trp[30] and Nle[31] of (135) while in the heptapeptide series the drop in potency in CCK-A affinity caused by replacing Nle[31] with NMeLeu afforded a potent CCK-B ligand. The reverse situation was observed with [desamino-Tyr[27],Nle[28,31](*N*Me)-Asp[32]]CCK(27-33) which was found to be selective for the CCK-A receptor as a result of a drastic decrease in affinity for CCK-B receptors. N-Methylation of Asp[32] of the octapeptide conferred, to some extent, selectivity towards the CCK/gastrin receptor. Pancreatic binding of the highly selective CCK-A ligand Boc-Trp-Lys(Cbz)-Asp-Phe-NH_2 was significantly improved by incorporating an amide function at the ε-amino of lysine. For example, an approximately 30-fold increase in binding potency was observed with the 4-hydroxycinnamoyl derivative (137, R=OH).[277] Sulphation of the hydroxy group of (137, R=OH) to produce (137, R=OSO_3H) had little effect on the pancreatic activity.

Replacing the N-terminal Boc(Tyr)SO_3H residue in Boc[Nle^{28,31}]$CCK_{27\text{-}33}$ with Ac-Phe(*p*-CH_2CO_2H) led to slight decrease (~10 to 20-fold) in affinity for both the brain and pancreatic binding sites but the corresponding compound with the hydroxamic moiety Ac-Phe(*p*-$CH_2CONHOH$) at position 27 exhibited a much more pronounced drop in potency (~ 300-fold) at the CCK-A receptors.[278] The synthesis of tritiated derivatives {[^3H]propionyl-Tyr(SO_3Na)-gNle-mGly-Trp-(*N*Me)Nle-Asp-Phe-NH_2 and [cyclic]^3H[propionyl]-γ-D-Glu-Tyr(SO_3H)-Nle-D-Lys-Trp-Nle-Asp- Phe-NH_2} of two highly selective agonists for CCK-B receptors has been reported.[279]

The central amide bond and the C-terminal carboxylic acid of the Parke-Davis CCK-B 'dipeptoid' ligand (138a) (CCK-B IC_{50}=1.7nM) were both targeted for modification. Replacing the amide bond of the related compounds (139) (CCK-B IC_{50}=852nM) and (140a) (CCK-B IC_{50}=32nM) with a range of isosteres led to analogues which had lower affinity and selectivity for the CCK-B receptor then the parent compounds.[280] However the affinity was improved by appending a fumarate side chain to the phenethyl group. Thus the ester replacement (140b) had a 17-fold weaker CCK-B affinity than the amide (140a) but attaching the fumarate side chain gave compound (140c) (IC_{50}=38.8nM) which was approximately equiactive to (140a). In the carboxylic acid analogue series the compounds with a 3-hydroxyisoxazole (138b), a sulphoxide-linked azole (138c), and a sulphonic acid (138d) moiety all showed high binding at the CCK-B receptor with IC_{50} values of 2.6, 1.7, and 1.3nM respectively, which compared favourably with the parent carboxylic acid (138a) (IC_{50}=1.7nM).[281] Of all the analogues synthesised and tested in the latter series those with highest binding affinities for the

H N; 2-AdocN; H CH_3; CONH; H; NHCO; R

(138a) R = CH_2CO_2H

δ− OH; δ+; $N_{\delta-}$; O

(138b) R =

δ−; N; N; δ+; S; O; N; δ−

(138c) R = —S

(138d) R = $CH_2SO_3^-Na^+$

H N; HO; H; 2-AdocN; H H; CONH; Ph

(139)

H N; R; 2-AdocN; H Me; C; O; X; Ph

(140a) X = NH, R = H
(140b) X = O, R = H
(140c) X = O, R = *R*-NHCO—CH=CH—CO_2Me

R^1; O; H; Ph; N H; HN; NH; N H; O; O; NH; H; CH_2CO_2H; R^2; N; H; R^3; O

(141) R^1 = H or Me; $R^2 = R^3$ = H or $R^2 = R^3$ = cyclohexyl

CH_2—CO—D-Tyr(Et)-Phe-Val-Asn-Cys—X—NH—CH_2—CH_2—NHR
S———————————————————S

(142) X = Pro-Arg-Gly, Arg-Gly or Gly
R = H or retrolinked amino acid

CH_3 CH_2—CO—Phe(N_3)-Phe-Val-Asn-Cys-Pro-Lys-D-Tyr—NH_2
CH_3 S———————————————————S

(143)

CCK-B receptor had either an SO_3H group or an amide moiety which was planar and had a high charge distribution similar to that of the carboxylic acid (central relatively electropositive atom surrounded by two electronegative atoms). Macrocyclic analogues (141) of the 'dipeptoid' (138a) have also been prepared.[282] The binding data on this series was disappointing and the compounds all showed only low affinity for the CCK-B receptor (IC_{50s} in μM range). A Free-Wilson/Fujita-Ban analysis on (138a) and a series of analogues which differed at the N- and/or C-termini and/or the tryptophan moiety concluded that these three domains bound independently of each other and their effects on receptor affinity were therefore additive.[283]

6.3 *Angiotensin Analogues.*

Three analogues of angiotensin II have been prepared in which the tyrosine, isoleucine, and phenylalanine residues at positions 4, 5, and 8 respectively were replaced by the corresponding β-amino acid homologue [-NH-CH(R)-CH_2-CO-].[284] These three analogues were prepared using solution phase techniques whereas the solid phase method was adopted for the synthesis of a fourth compound, [2,4-methanoproline^7]angiotensin II.[285] This latter compound retained 26% of the binding affinity of angiotensin II and the authors discussed the implications of this result for the receptor-bound conformation of the neuropeptide. A similar type of proline substitution in bradykinin led to a drastic reduction in activity (see section 6.8).

6.4 *Oxytocin and Vasopressin Analogues.*

Several research groups have reported on the synthesis of vasopressin antagonists. Modifying the C-terminal moiety of arginine vasopressin antagonists was the subject of two comprehensive studies by Manning and co-workers. Solid phase techniques were adopted for the preparation of the series of V_2/V_{1a} (antidiuretic/vasopressor) antagonists in which all or part of the Pro-Arg-Gly-NH_2 side chain sequence was replaced by ethylenediamine.[286] Solution phase techniques were then utilised to acetylate these compounds with a variety of amino acids resulting in analogues with general structure (142). The synthesis of the second series involved replacing the C-terminal Gly-NH_2 of two V_1 antagonists with ethylenediamine, methylamine, or an amino acid amide.[287] All these modifications were also very well tolerated and the analogues were found to exhibit pharmacological properties similar to their parent molecules. Replacing Tyr^2 in [D-Har^8]vasopressin (D-Har = D-homoarginine) or in the corresponding deamino analogue, [Mpa^1,D-Har^8]vasopressin (Mpa = β-mercaptopropionic acid), with

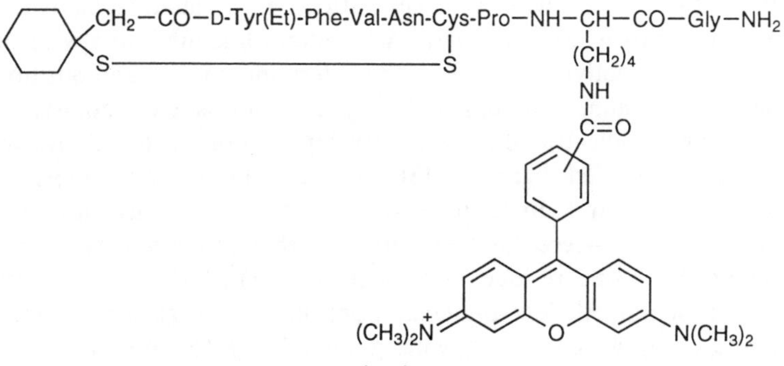

CH_2—CO—D-Tyr(Et)-Phe-Val-Asn-Cys-Pro—NH—CH—CO—Gly—NH_2

(144)

X—Cys-Y-Ile-Gln-Asn-Cys-Pro-Leu-Gly—NH_2

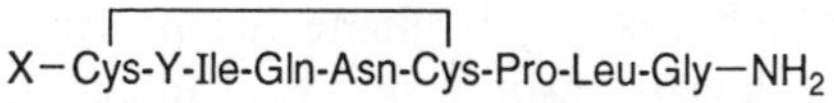

(145) X = H, Y = (*R* or *S*)-3-(*R* or *S*)-2′-dimethylphenylalanine
(146) X = (Gly)$_n$, n = 0–3, Y = D- or L-Phe(*p*-Et)

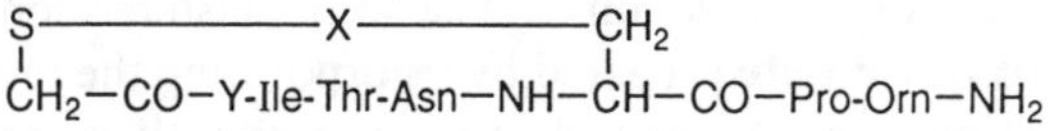

S——X——CH_2
CH_2—CO—Y-Ile-Thr-Asn—NH—CH—CO—Pro-Orn—NH_2

(147)

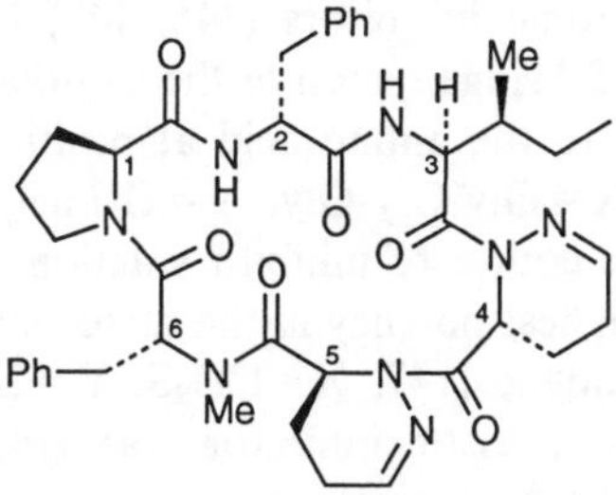

(148)

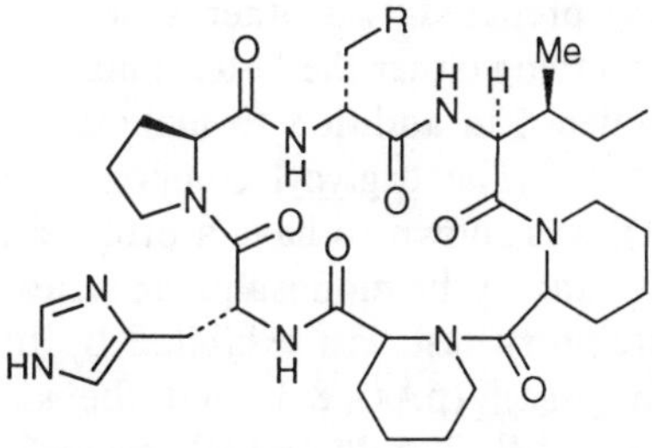

(149) R = 2-naphthyl or 3-indolyl

G W T L N S A G Y L L G P Q Q F F G L M amide
←galanin (1–13)→←substance P(5–11)→

(151)

modified tyrosine or with *ortho*-substituted phenylalanine derivatives gave compounds which, in general, were uterotonic inhibitors with very low antidiuretic activity.[288,289] Both the C-terminal moiety and position 2 of the cyclodesamino-β,β-dialkyl-Cys[1]-type vasopressin antagonists were targeted for modification during the development of labelled derivatives. The doubly labelled ligand (143) with a photoactivable moiety in position 2 and an iodination residue at the C-terminus was highly selective for the V_1-receptor.[290] Coupling a rhodamine moiety to the ε-amine of the lysine residue in [Ppa^1,D-Tyr(Et)^2,Val^4,Lys^8]vasopressin (Ppa = β,β-pentamethylene-β-mercaptopropionic acid) gave a fluorescent analogue (144) which bound with good affinity to both V_1 and V_2 receptors and which did not cause their activation.[291] Three heterofunctional V_{1a}-selective ligands were prepared by attaching biotin and azidosalicylate, either alone or in combination, to the ε-amino group of Lys^9 in [Ppa^1,Tyr(Me)^2,LysNH$_2{}^9$]arginine vasopressin.[292] In the case of derivatives of the linear vasopressin V_{1a} antagonist PhCH$_2$CO-D-Tyr(Et)-Phe-Gln-Asn-Lys-Pro-Arg-NH$_2$ the label was incorporated at the N-terminus by replacing the aromatic group with a phenolic moiety which was subsequently iodinated.[293]

Of the oxytocin analogues reported this year several have also involved modification at position 2. The four diastereoisomeric analogues (145) of oxytocin were prepared by incorporating the racemic amino acid into position 2 of the peptide sequence and then separating the stereoisomers on reverse phase HPLC.[294] All four analogues were inhibitors of the uterotonic activity of oxytocin with the most potent being the erthyro (2*R*,3*R*) isomer ($pA_2 = 8.3$). The same group of workers also prepared two other series of uterotonic inhibitors (146, 147) by modifying either the N-terminus or the S-S bridge region of the oxytocin sequence in addition to making changes to the amino acid at position 2.[295,296] The triglycyl compound [146, X = Gly-Gly-Gly, Y = D-Phe(*p*-Et)] was shown to have a prolonged time course of inhibitory action *in vivo* and in the monosulphide series the highest potency in the uterotonic inhibitory test was exhibited by the analogue [147, X = CH$_2$S, Y = D-Phe(*p*-Et)] ($pA_2 = 8.3$) and the strongest pressor inhibitor was [147, X = CH$_2$S, Y = Phe(*p*-Et)] ($pA_2 = 7.5$). The biological consequences of interchanging the leucine and proline residues in oxytocin analogues has been investigated.[297]

The cyclic oxytocin antagonist (148), a modified derivative of a natural product obtained from *Streptomyces silvensis*, has been chemically synthesised.[298] Structure-activity studies found that lipophilic amino acids at positions 2 and 3 of (148) and the unusual D-dehydropiperazic acid at position 4 were the most critical residues for obtaining good receptor affinity. However incorporating a larger aromatic amino acid at position 2 enhanced oxytocin receptor affinity whilst a basic amino acid at either of the less important 5- or 6- positions improved water solubility.[299] For example, the 2-Nal^{2},D-His^{6} or D-Trp^{2},D-His^{6} combinations (149) exhibited good oxytocin receptor affinity and selectivity as well as useful levels of aqueous solubility.

6.5 *Luteinising Hormone-releasing Hormone (LHRH) Analogues.*

Several LHRH antagonists with a hydrophobic Ac-D-Nal^{1}-D-Cpa^{2}-D-Pal^{3} (abbreviated to X) N-terminus and weakly basic N$^{\omega}$-triazolylornithine, -lysine, or -*p*-aminophenylalanine (Aph) residues at positions 5 and 6 were evaluated in a number of biological assays[300] [Nal = 3-(2-naphthyl)alanine, Cpa = 4-chlorophenylalanine, Pal = 3-(3-pyridyl)alanine]. Azaline B [X,Aph^{5}(atz),D-Aph^{6}(atz),ILys^{8},D-Ala^{10}]LHRH was found to inhibit LH secretion in the castrated male rat for three times as long as the analogue (Azaline A) with the Lys(atz)^{5,6} modification (72h compared with 24h). A similar long duration of action was observed for Antide [X, Lys^{5}(Nic),D-Lys^{6}(Nic),ILys^{8},D-Ala^{10}]LHRH [Lys(Nic) = lysine(*N*-ε-nicotinoyl]. An analogue obtained by translocating the basic residue at position 8 in [X-Ser-Lys(Pic)-D-Lys(Pic)-Leu-Arg-Pro-D-Ala-NH_2] [Lys(Pic) - lysine(*N*-ε-picolinoyl)] to position 7[301] (*i.e.* interchanging Leu and Arg) and also the three compounds mentioned above had low potency in the histamine release assay (low potency equates with a safer antagonist). The metabolic stability of the agonist [D-Leu^{6},Pro^{9}NHEt]LHRH (leuprolide) was improved by replacing Ser^{4} with the N-methylated derivative.[302] Analogues of LHRH with a N-terminal glutamine or glutamic acid residue have been prepared.[303]

6.6 *Tachykinin Analogues.*

The substance P (SP) antagonist Ac-Thr-D-Trp(CHO)-Phe-NHMeBzl (150) was developed in two stages from the known octapeptide SP antagonist H-D-Pro-Gln-Gln-D-Trp-Phe-D-Trp-D-Trp-Phe-NH_2. The first step involved the synthesis and biological evaluation of all the overlapping tripeptide fragments of the octapeptide and this identified the Gln-D-Trp-Phe sequence as the receptor binding element.[304] Boc-Gln-D-Trp(CHO)-Phe-OBzl was 7 times more potent than the octapeptide in inhibiting the binding of SP to guinea pig membranes. In a second step each component part of the protected tripeptide was optimised and these modifications led to (150).[305] Replacement of the D-Trp-Leu sequence in the nonselective neurokinin antagonist H-Arg-Gln-D-Trp-Phe-D-Trp-Leu-Nle-NH_2 by D-Pro-Pro combined with lipophilic N-terminal substitutions resulted in the potent and highly selective NK-2 antagonists X-Ala-D-Trp-Phe-D-Pro-Pro-Nle-NH_2 (X = PhCO-Ala, Boc-Arg).[306] Structure-activity studies on analogues of [Orn^6]SP_{6-11} obtained by replacing the SCH_3 of Met^{11} with secondary and tertiary amides concluded that a lipophilic side chain at this position was particularly important for activity at the NK-2 receptor.[307] For example, [Orn^6,Glu(NCH_3Ph)^{11}]SP_{6-11} was nearly 3 times as potent as the parent at the NK-2 receptor while the less lipophilic [Orn^6,Glu($NHCH_3$)^{11}]SP_{6-11} showed selectivity for the NK-1 receptor.

Scyliorhinin (H-Ala-Lys-Phe-Asp-Lys-Phe-Tyr-Gly-Leu-Met-NH_2), a deca-peptide belonging to the tachykinin family, and the Val^7 and Ile^7 analogues were found to be significantly less potent than SP in the guinea pig ileum assay.[308] Molecular dynamic simulations of the neurokinin A antagonist cyclo(Leuψ[CH_2NH_2]Leu-Gln-Trp-Phe-Gly) using nOe derived backbone interproton distances as constraints found that the molecule interconverted between three families of conformations each of which consisted of two fused β-turns.[309]

The findings of three studies, although not directly relevant to this section are, nevertheless, worth mentioning. The first of these involved the synthesis of a hybrid molecule (151) consisting of the N-terminal tridecapeptide of galanin and the C-terminal heptapeptide of SP.[310] This peptide (galantide) was found to recognize two classes of galanin receptors ($K_{D(1)}$ < 0.1nM and $K_{D(2)}$ ~ 6nM) and one type of SP receptor (K_D ~ 40nM) in the rat hypothalamus. The synthesis of hydroxamic acid derivatives of peptides related to fragments of SP were reported in the second study.[311] These compounds were found to inhibit the degradation of the radiolabelled substrate desamino-[3-^{125}I-Tyr^5]SP_{5-11} in rat hypothalmus preparations. The most potent of these inhibitors was desamino-Tyr-Phe-Phe-Gly-NHOH with an IC_{50} of 1.8μM. Finally, the hexapeptide H-Tyr-Pro-D-Phe-Phe-D-Trp-Met-NH_2 was reported to be a bifunctional pharmacophore in that it exhibited both casomorphin-like and SP antagonist-like activity.[312]

6.7 *Somatostatin Analogues.*

Two papers on somatostatin analogues both involved modifications to Merck's cyclic hexapeptide agonist cyclo[Pro^6-Phe^7-D-Trp^8-Lys^9-Thr^{10}-Phe^{11}] (152). Murray Goodman's group investigated the effect on conformation and receptor binding of replacing the Phe and D-Trp residues with the β-methylated derivatives[313] while Hirschmann, Nicolaou, and co-workers designed and synthesised the peptidomimetic (153).[314] Both studies confirmed the importance of the aromatic side

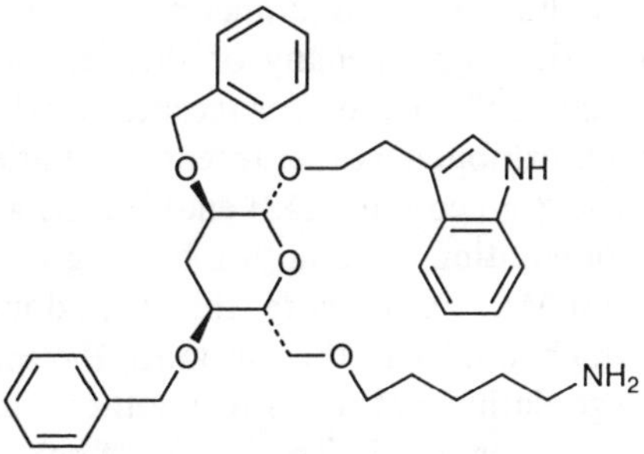

(153)

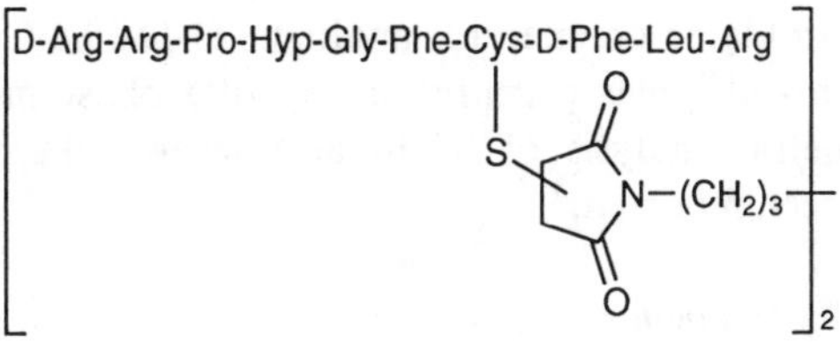

[D-Arg-Arg-Pro-Hyp-Gly-Phe-Cys-D-Phe-Leu-Arg, Cys–S–succinimide–N–$(CH_2)_3$–]$_2$

(154)

H–Arg-Cys-Cha-Gly-Gly-Arg-Ile-Asp-Arg-Ile-Phe-Arg-Cys–NH_2 (with a bridge drawn between the two Cys residues)

(155)

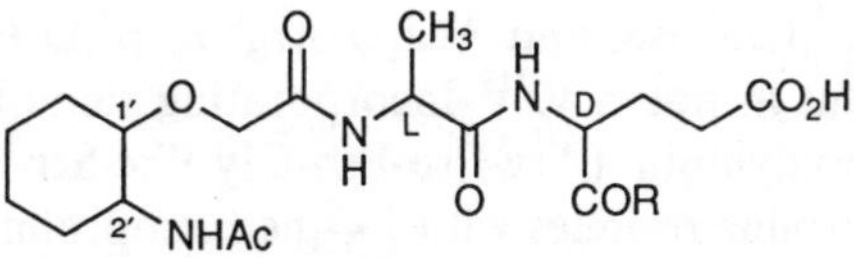

(156) *trans*-1′*R*,2′*R* or 1′*S*,2′*S*; R = NH_2, OH

chains in receptor binding. Steric hindrance caused by the β-methylation prevented the aromatic rings in many of the peptide analogues from adopting the optimum conformation for receptor binding with the result that these compounds exhibited weaker receptor affinities than the parent compound (152). The β-glucose in (153) enabled the aromatic functionality to adopt an orientation suitable for binding to the somatostatin receptor (IC_{50}=0.13μM). Futhermore this peptidomimetic was also found to display a high affinity to the substance P receptor with an IC_{50} of 0.18μM. The degradation pathways to another highly active Merck compound cyclo(*N*Me-Ala-Tyr-D-Trp-Lys-Val-Phe) have been elucidated.[315] The mechanisms of degradation under acidic or alkaline conditions appeared to be different but the first step under any pH was cleavage between methylalanine and phenylalanine to yield a linear fragment. The amidated octapeptide analogue H-D-Phe-Cys-Phe-D-Trp-Lys-Thr-Cys-Thr-NH_2 was prepared using solid phase methodology.[316] Mass spectrometric analysis of SS-14 and several alanine substituted analogues has been reported.[317]

6.8 *Bradykinin Analogues.*

Dimers of the bradykinin antagonist D-Arg^0-Arg^1-Pro^2-Hyp^3-Gly^4-Phe^5-Ser^6-D-Phe^7-Leu^8-Arg^9 have been prepared by substituting various amino acids with cysteine and then reacting the sulphydryl-containing momomers with a bismaleimidoalkane cross-linking reagent.[318] Several of the dimers showed greater potency and longer duration of action than the parent mononer antagonist and the optimum activity in *in vitro* assays was achieved with dimer (154). Attaching a bulky 1-adamantylacetyl (Aaa) or 1-adamantylcarbonyl (Aca) at the N-terminus of B_2 bradykinin antagonists to give, for example, Aaa[D-Arg^0,Hyp^3,D-Phe^7,Leu^8]BK and Aca[D-Arg^0,Hyp^3,D-Phe^7,Leu^8]BK led to an improvement in potency.[319] Incorporating steric bulk at positions 2, 3, or 7 of bradykinin (Arg-Pro-Pro-Gly-Phe-Ser-Pro-Phe-Arg) by substituting the proline residues with 2,4-methanoproline gave analogues which retained $<2\%$ of the binding affinity of bradykinin.[285] NMR nOe data indicated that analogues containing an α-MePro at position 3 or 7 adopted reverse-turn conformations at both Pro^2-Phe^5 and Ser^6-Arg^9 segments in aqueous solution whereas the natural hormone was largely disordered.[320,321] Enzymatic degradation studies on bradykinin and [D-Phe^7, Leu^8]BK have been carried out.[322]

6.9 *Miscellaneous Examples.*

By incorporating di-fatty-acyl-glyceryl moieties at the N-terminus of human gastrin(2–17), lipo-gastrin derivatives were obtained that behaved as synthetic lipids but which could bind, with high affinity, to membrane-associated gastrin/CCK receptors.[323] Lipophilic C-terminal amide analogues Boc-Trp-Leu-Asp-Phe-NHR (R = hexyldecyl, Bu, *iso*-Bu, *tert*-Bu) of tetragastrin have also been prepared.[324] Replacing the C-terminal phenylalanine amide of gastrin tetrapeptide (Boc-Trp-Leu-Asp-Phe-NH_2) with a phenethylester or phenethylamide moiety gave compounds which bound to gastrin receptors and exhibited antagonist activity on gastrin-induced acid secretion in rats.[325] Structure-activity studies on bombesin/GRP_{14-27} (GRP = gastrin releasing peptide) found that analogues with the desMet,Leuψ[$CH_2NHCOCH_3$] C-terminus were antagonists with no sign of agonist activity.[326] Several series of alkylated bombesin analogues have been prepared.[327,328] The best compromise between physicochemical parameters (solubility and stability in aqueous solutions) and biological properties (receptor binding and antagonist activity) was achieved with N-terminal 4-[bis(2-chloro-ethylamino) benzoyl] derivatives of bombesin (7-14) octapeptide carrying a reduced peptide bond between Leu^{13} and Met^{14}.

Systematically replacing each of the residues in NPY(18-36) (NPY = neuropeptide Y) with the D-isomer[329] and residues of oCRF (oCRF = ovine corticotropin releasing factor) with alanine[330] resulted in analogues with biological activity, in both series, ranging from $<1\%$ to $>200\%$ that of the parent peptides. The sensitivity of the N-terminus of both molecules to modification suggested the presence of a well-defined pharmacophore in NPY(18-36) while the side chains of residues 5-19 in oCRF were particularly important for receptor binding and activation.

Several structure-activity studies on GRF(1-29)NH2 (GRF = growth hormone-releasing factor) have been carried out. Results of the biological assays on an extended series of deletion analogues indicated that amino acids contained in the segment 13–21 were more important than those of 24–29 for high affinity receptor binding.[331] Replacing the alanine at position 19 with valine gave an analogue with increased GH-releasing activity and increased plasma stabitity *in vitro* but a similar increase in potency was not observed *in vivo*.[332] However several analogues with an agmatine [-$NH(CH_2)_4NHC(NH_2)=NH$] (Agm) residue at the C-terminus, eg, [desNH$_2$Tyr^1,Ala^{15},Nle^{27},Asp^{28}] GRF(1-28)Agm, showed increased potencies both *in vitro* and *in vivo* compared to the parent GRF(1-29)NH_2 compound.[333,334] Biotinylated derivatives such as [desNH$_2$Tyr^1,D-Ala^2,Ala^{15}]GRF(1-29)Gly-Gly-Cys-biotin retained the full bioactivity of the parent compound lacking the spacer/biotin moieties.[335] Each of these compounds were prepared by solid phase techniques whereas the potent analogue [desTyrNH$_2$^{1},D-

Ala^2,Ala^{15}]GRF(1-29)NH_2[336] and the native 1-44 amino acid peptide[337] were also prepared using enzymic semisynthetic methods.

Extensive structure-activity studies on atrial natriuretic peptide [ANP(1-28)] led to the preparation of a family of small analogues (eg 155) which exhibited the full range of ANP's actions.[338] The end result of structure-activity studies on endothelin was Ac-D-Dip-Leu-Asp-Ile-Ile-Trp-OH (D-Dip = D-diphenylalanine) which was an antagonist of endothelin-stimulated vasoconstriction.[339] Analogues D-Pro-D-Arg and D-Pro-L-Arg of the C-terminal dipeptide of tuftsin both showed analgesic activity in the tail flick immersion test whereas the tetrapeptide amides H-Thr-Lys-Pro-ArgNH$(CH_2)_nCH_3$ (n = 5,6) exhibited toxic effects.[340] The primary structure of seven peptides isolated from the glandular secretions of the Australian green tree frog *Litoria splendida* have been determined.[341]

Carbocyclic analogues (156) of muramyldipeptide (MDP) have been synthesised by two independent routes starting from either racemic *trans*-2-azidocyclohexanol or (1*R*,2*R*)-(1*S*,2*S*)-2-aminocyclohexanol.[342] Conjugates of hepatitis B surface antigen covalently bound to the D-isoglutamine residue of MDP have been prepared[343] as has a derivative of the dipeptide in which an acridine moiety was coupled to the sugar residue.[344] The conformation of a fourth analogue, *N*-acetylmuramyl-L-alanyl-3-carbomethoxymethyl-D-proline methyl ester, in DMSO was investigated using 2D NMR and molecular modelling techniques and the lowest energy conformer was found to contain two β-turns which gave the molecule an S-shape.[345]

7 Conformational Information Derived from Physical Methods

The numbers of papers cited in this section is not an accurate reflection of the interest in this area but is rather a consequence of the now common practice of combining conformational and synthetic studies in the same paper. Where appropriate therefore, the conformational properties of peptide analogues are discussed in other sections and this section concentrates on reporting the conformational properties of the native hormone/neuropeptides. The NMR of peptides and proteins is reviewed annually in another RSC SPR.[346]

Oxytocin (Cys-Tyr-Ile-Gln-Asn-Cys-Pro-Leu-Gly-NH_2)[347] and lysine vasopressin (Cys-Tyr-Phe-Gln-Asn-Cys-Pro-Lys-Gly-NH_2)[348] have both been the subject of detailed conformational analysis. A combination of 2D NMR techniques and energy calculations found that, in DMSO solution, the 20 membered ring portion of both hormones was well-defined with a β-turn structure between residues Tyr^2 and Asn^5 and that the tail regions were more flexible. A second β-turn involving the C-terminal sequence Cys^6-Pro^7-Leu^8-Gly^9 was observed in oxytocin but the corresponding region of lysine vasopressin was disordered. Another NMR study established that in aqueous solution approximately 10% of oxytocin and the vasopressins exist with the *cis* conformation across the Cys^6-Pro^7 amide bond.[349] The dynamic behaviour of lysine vasopressin derivatives in solution has been investigated by NMR relaxation studies.[350]

Cis/*trans* isomerism about Xaa-proline amide bonds was also detected in the 2D NMR spectra of neuropeptide Y,[351] N-Tyr-MIF-1 (Tyr-Pro-Leu-Gly-NH_2),[352] TRH (pGlu-His-Pro-NH_2),[353] and tuftsin (Thr-Lys-Pro-Arg).[354] The conformation of the major (all *trans*) isomer of neuropeptide Y in trifluoroethanol solution consisted of a well-defined α-helix extending from Arg^{19} to Gln^{34}. No regular structure was observed for the N-terminal half of the molecule. An inverse γ-turn centred on the Lys-Pro moiety was proposed for the *trans*-isomer of tuftsin in DMSO solution on the basis of nOes between the proline and adjacent lysine and arginine residues. The *cis* conformer, on the other hand, did not adopt a single preferred conformation in DMSO but existed as a mixture of extended structures. Conformational analysis of tuftsin using molecular dynamics calculations suggested the presence of a type IV β-turn at the Lys-Pro position.[355]

The results of NMR studies on angiotensin II (Asp-Arg-Val-Tyr-Ile-His-Pro-Phe) and its 1–7 and 1–6 fragments indicated that in H_2O or DMSO the peptides existed as a mixture of conformers with predominantly extended, β-sheet-like structures.[356] Tyrosine fluorescence of angiotensin II was measured in several alcoholic solvents and decay times of < 5ns were observed.[357] A conformational study of endothelins and sarafotoxins utilised CD spectroscopic techniques[358] while investigations into the structure of the N-terminal fragment (1–34) of human parathyroid hormone adopted 2D NMR techniques.[359]

Several peptide conformational studies have been carried out in simulated biological media. UV-visible[360] and NMR[361] data on substance P, bradykinin, and Met-enkephalin in the presence of SDS or lysophosphatidylcholine membranes indicated that aromatic residues of only the positively charged peptides (SP and BK) penetrated the hydrophobic core of the negatively charged SDS micelles. The opposite effects were observed in the presence of lysophosphatidylcholine membranes with only the Phe and Tyr moieties of the enkephalin inserting into the micellar structure. The authors discussed the implications of these results with respect to a possible binding mechanism for peptide-micelle interactions. The interaction of substance P with Ca^{2+} in membrane-mimetic solvents was investigated using CD spectroscopy.[362] A comprehensive conformational analysis of Met-enkephalin in both aqueous solution and in the presence of SDS has been carried put.[363] The viscosity of cytoplasm was reproduced for NMR experiments of deltorphin 1 by using cryoprotective mixtures. In 80 : 20 (v : v) DMSO/H_2O at 265K or 80 : 20 (v : v) ethylene glycol/H_2O at 298K deltorphin 1 was found to adopt folded conformations in preference to disordered ones.[364]

Structures of galanin, obtained from molecular dynamics simulations, were found to be similar to those determined experimentally using NMR techniques.[365] The conformational changes occurring in the water simulations were studied in detail and formation of a i-i+3 helical hydrogen bonding pattern followed by a series of turn-like conformations were observed during the unfolding. Molecular dynamics simulations of several cyclic dermorphin analogues have been carried out.[366] Low energy conformations of the enkephalins have been generated by molecular dynamics simulations,[367] a build-up procedure,[368] a random search and minimisation technique,[369] and a Monte-Carlo simulated annealing method.[370] The NMR structure of hirudin 56-65 bound to thrombin has been refined using a restrained electrostatically driven Monte-Carlo method.[371]

References

1. C. Unverzagt, A. Geyer, and H. Kessler, *Angew.Chem.Int.Ed.*, 1992, **31**, 1229.
2. J. Jurayj and M. Cushman, *Tetrahedron*, 1992, **48**, 8601.
3. L. Lankiewicz, C.Y. Bowers, G.A. Reynolds, V. Labroo, L.A. Cohen, S. Vonhof, A.-L. Siren, and A.F. Spatola, *Biochem.Biophys.Res.Commun.*, 1992, **184**, 359.
4. H. Kessler, A. Geyer, H. Matter, and M. Kock, *Int.J.Pept.Protein Res.*, 1992, **40**, 25.
5. H. Durr, M. Goodman, and G. Jung, *Angew.Chem.Int.Ed.*, 1992, **31**, 785.
6. O.E. Said-Nejad, E.R. Felder, D.F. Mierke, T. Yamazaki, P.W. Schiller, and M. Goodman, *Int.J.Pept.Protein Res.*, 1992, **39**, 145.
7. T. Yamazaki, D.F. Mierke, O.E. Said-Nejad, E.R. Felder, and M. Goodman, *Int.J.Pept.Protein Res.*, 1992, **39**, 161.
8. T. Yamazaki and M. Goodman, *Chirality*, 1991, **3**, 268.
9. F. De Angelis, *NATO ASI Ser.,Ser.C*, 1992, **353**, 357.
10. S. Doulut, M. Rodriguez, D. Lugrin, F. Vecchini, P. Kitabgi, A. Aumelas, and J. Martinez, *Pept.Res.*, 1992, **5**, 30.
11. S.L. Harbeson, S.A. Shatzer, T.-B. Le, and S.H. Buck, *J.Med.Chem.*, 1992, **35**, 3949.
12. N.G.J. Delaet, P.M.F. Verheyden, D. Tourwe, G. Van Binst, P. Davis, and T.F. Burks, *Biopolymers*, 1992, **32**, 957.
13. S. Salvadori, R. Guerrini, P.A. Borea, and R. Tomatis, *Int.J.Pept.Protein Res.*, 1992, **40**, 437.
14. W.M. Kazmierski, R.D. Ferguson, R.J. Knapp, G.K. Lui, H.I. Yamamura, and V.J. Hruby, *Int.J.Pept.Protein Res.*, 1992, **39**, 401.

15. T. Ibuka, *Organomet.News*, 1991, 96.
16. T. Ikuba, H. Yoshizawa, H. Habashita, N. Fujii, Y. Chounan, M. Tanaka, and Y. Yamamoto, *Tetrahedron Lett.*, 1992, **33**, 3783.
17. K.M. Bol and R.M.J. Liskamp, *Tetrahedron*, 1992, **48**, 6425.
18. M. Rodriguez, A. Heitz, and J. Martinez, *Int.J.Pept.Protein Res.*, 1992, **39**, 273.
19. D. Tourwe, J. Couder, M. Ceusters, D. Meert, T.F. Burks, T.H. Kramer, P. Davis, R. Knapp, H.I. Yamamura, J.E. Leysen, and G. Van Binst, *Int.J.Pept.Protein Res.*, 1992, **39**, 131.
20. H. Jaspers, D. Tourwe, G. Van Binst, H. Pepermans, P. Borea, L. Ucelli, and S. Salvadori, *Int.J.Pept.Protein Res.*, 1992, **39**, 315.
21. Y.-L. Li, K. Luthman, and U. Hacksell, *Tetrahedron Lett.*, 1992, **33**, 4487.
22. M.J. Dominguez, R. Gonzalez-Muniz, and M.T. Garcia-Lopez, *Tetrahedron*, 1992, **48**, 2761.
23. I. Gomez-Monterrey, R. Gonzalez-Muniz, C. Perez-Martin, M. Lopez de Ceballos, J. Del Rio, and M.T. Garcia-Lopez, *Arch.Pharm.(Weinheim, Ger.)*, 1992, **325**, 261.
24. R.V. Hoffman and H.-O. Kim, *Tetrahedron Lett.*, 1992, **33**, 3579.
25. L. Cheng C.A. Goodwin, M.F. Schully, V.V. Kakkar, and G. Claeson, *J.Med.Chem.*, 1992, **35**, 3364.
26. J. DiMaio, B. Gibbs, J. Lefebvre, Y. Konishi, D. Munn, S.Y. Yue, and W. Hornberger, *J.Med.Chem.*, 1992, **35**, 3331.
27. C. Yuan and G. Wang, *Phosphorus, Sulfur Silicon Relat.Elem.*, 1992, **71**, 207.
28. C. Yuan and S. Chen, *Synthesis*, 1992, 1124.
29. P. Dumy, R. Escale, J.P. Girard, J. Parello, and J.P. Vidal, *Synthesis*, 1992, 1226.
30. P. Coutrot, C. Grison, and C. Charbonnier-Gerardin, *Tetrahedron*, 1992, **48**, 9841.
31. R. Neidlein and P. Greulich, *Helv.Chim.Acta*, 1992, **75**, 2545.
32. C.-L.J. Wang, T.L. Taylor, A.J. Mical, S. Spitz, and T.M. Reilly, *Tetrahedron Lett.*, 1992, **33**, 7667.
33. S. Elgendy, J. Deadman, G. Patel, D. Green, N. Chino, C.A. Goodwin, M.F. Scully, V.V. Kakkar, and G. Claeson, *Tetrahedron Lett.*, 1992, **33**, 4209.
34. J.F. Schwieger and B. Unterhalt, *Arch.Pharm.(Weinheim, Ger.)*, 1992, **325**, 709.
35. M. Cushman and E.-S. Lee, *Tetrahedron Lett.*, 1992, **33**, 1193.
36. R. Chen, S. Deng, and B. Cai, *Gaodeng Xuexiao Huaxue Xuebao*, 1991, **12**, 1335.
37. R. Chen, Y. Zhang, and M. Cheng, *Gaodeng Xuexiao Huaxue Xuebao*, 1992, **13**, 611.
38. W.J. Moree, G.A. van der Marel, and R.M.J. Liskamp, *Tetrahedron Lett.*, 1992, **33**, 6389.
39. A. Calcagni, E. Gavuzzo, F. Mazza, F. Pinnen, G. Pochetti, and D. Rossi, *Gazz.Chim.Ital.*, 1992, **122**, 17.
40. M.R. Angelastro, J.P. Burkhart, P. Bey, and N.P. Peet, *Tetrahedron Lett.*, 1992, **33**, 3265.
41. P. Talaga and W. Koenig, *Tetrahedron Lett.*, 1992, **33**, 609.
42. W. Hong, L. Dong, Z. Cai, and R. Titmas, *Tetrahedron Lett.*, 1992, **33**, 741.
43. R.P. Robinson and K.M. Donahue, *J.Org.Chem.*, 1992, **57**, 7309.
44. J.W. Skiles, C. Miao, R. Sorcek, S. Jacober, P.W. Mui, G. Chow, S.M. Weldon, G. Possanza, M. Skoog, J. Keirns, G. Letts, and A.S. Rosenthal, *J.Med.Chem.*, 1992, **35**, 4795.
45. J.W. Skiles, V. Fuchs, C. Miao, R. Sorcek, K.G. Grozinger, S.C. Mauldin, J. Vitous, P.W. Mui, S. Jacober, G. Chow, M. Matteo, M. Skoog, S.M. Weldon, G. Possanza, J. Keirns, G. Letts, and A.S. Rosenthal, *J.Med.Chem.*, 1992, **35**, 641.
46. N. Galeotti, C. Montagne, J. Poncet, and P. Jouin, *Tetrahedron Lett.*, 1992, **33**, 2807.
47. P. Wipf and C.P. Miller, *Tetrahedron Lett.*, 1992, **33**, 6267.

48. B.H. Kim, Y.J. Chung, G. Keum, J. Kim, and K. Kim, *Tetrahedron Lett.*, 1992, **33**, 6811.
49. J. Zabrocki, J.B. Dunbar, Jr., K.W. Marshall, M.V. Toth, and G.R. Marshall, *J.Org.Chem.*, 1992, **57**, 202.
50. A.D. Abell, D.A. Hoult, and E.J. Jamieson, *Tetrahedron Lett.*, 1992, **33**, 5831.
51. J. Hlavacek and V. Kral, *Collect.Czech.Chem.Commun.*, 1992, **57**, 525.
52. M. Hagihara, N.J. Anthony, T.J. Stout, J. Clardy, and S.L. Schreiber, *J.Am.Chem.Soc.*, 1992, **114**, 6568.
53. A.B. Smith, III, T.P. Keenan, R.C. Holcomb, P.A. Sprengeler, M.C. Guzman, J.L. Wood, P.J. Carroll, and R. Hirschmann, *J.Am.Chem.Soc.*, 1992, **114**, 10672.
54. M.T. Reetz, J. Kanand, N. Griebenow, and K. Harms, *Angew.Chem.Int.Ed.*, 1992, **31**, 1626.
55. V. Moretto, G. Valle, M. Crisma, G.M. Bonora, and C. Toniolo, *Int.J.Biol.Macromol.*, 1992, **14**, 178.
56. M. Crisma, M. Anzolin, G.M. Bonora, C. Toniolo, E. Benedetti, B. Di Blasio, V. Pavone, M. Saviano, A. Lombardi, F. Nastri, and C. Pedone, *Gazz.Chim.Ital.*, 1992, **122**, 239.
57. B. Di Blasio, V. Pavone, M. Saviano, A. Lombardi, F. Nastri, C. Pedone, E. Benedetti, M. Crisma, M. Anzolin, and C. Toniolo, *J.Am.Chem.Soc.*, 1992, **114**, 6273.
58. A.M. Piazzesi, R. Bardi, M. Pantano, F. Formaggio, M. Crisma, and C. Toniolo, *Z.Kristallogr.*, 1992, **200**, 83.
59. M. Vlassi, H. Brueckner, and M. Kokkinidis, *Z.Kristallogr.*, 1992, **202**, 89.
60. V. Pavone, A. Lombardi, F. Nastri, M.Saviano, B. Di Blasio, F. Fraternali, C. Pedone, and T. Yamada, *J.Chem.Soc., Perkin Trans.2*, 1992, 971.
61. V. Pavone, B. Di Blasio, A. Lombardi, C. Isernia, C. Pedone, E. Benedetti, G.Valle, M. Crisma, C. Toniolo, and R. Kishore, *J.Chem.Soc., Perkin Trans.2*, 1992, 1233.
62. G. Basu and A. Kuki, *Biopolymers*, 1992, **32**, 61.
63. C. Aleman, J.A. Subirana, and J.J. Perez, *Biopolymers*, 1992, **32**, 621.
64. R.M. Brunne and D. Leibfritz, *Int.J.Pept.Protein Res.*, 1992, **40**, 401.
65. I.L. Karle, J.L. Flippen-Anderson, M. Sukumar, and P. Balaram, *J.Med.Chem.*, 1992, **35**, 3885.
66. P. Balaram, *Pure Appl.Chem.*, 1992, **64**, 1061.
67. F.S. Nandel, B. Singh, and A. Saran, *Int.J.Quantum Chem.*, 1992, **42**, 1669.
68. S. Kuwata, A. Nakanishi, T. Yamada, and T. Miyazawa, *Tetrahedron Lett.*, 1992, **33**, 6995.
69. B. Di Blasio, F. Rossi, E. Benedetti, V. Pavone, M. Saviano, C. Pedone, G. Zanotti, and T. Tancredi, *J.Am.Chem.Soc.*, 1992, **114**, 8277.
70. M. Saviano, F. Rossi, V. Pavone, B. Di Blasio, and C. Pedone, *J.Biomol.Struct.Dyn.*, 1992, **9**, 1045.
71. B. Di Blasio, V. Pavone, C. Isernia, C. Pedone, E. Benedetti, C. Toniolo, P.M. Hardy, and I.N. Lingham, *J.Chem.Soc., Perkin Trans.2*, 1992, 523.
72. B. Di Blasio, A. Lombardi, F. Nastri, M. Saviano, C. Pedone, T. Yamada, M. Nakao, S. Kuwata, and V. Pavone, *Biopolymers*, 1992, **32**, 1155.
73. Z. Galdecki, B. Luciak, K. Kaczmarek, M.T. Leplawy, and A.S. Redlinski, *Monatsh.Chem.*, 1992, **123**, 993.
74. R. Bardi, A.M. Piazzesi, M. Crisma, C. Toniolo, Sudhanand, R. Balaji Rao, and P. Balaram, *Z.Kristallogr.*, 1992, **202**, 302.
75. K.-H. Altmann, E. Altmann, and M. Mutter, *Helv.Chim.Acta*, 1992, **75**, 1198.
76. C. Toniolo, F. Formaggio, M. Crisma, G.M. Bonora, S. Pegoraro, S. Polinelli,

W.H.J. Boesten, H.E. Schoemaker, Q.B. Broxterman, and J. Kamphuis, *Pept.Res.*, 1992, **5**, 56.
77. B. Kaptein, W.H.J. Boesten, Q.B. Broxterman, H.E. Schoemaker, and J. Kamphuis, *Tetrahedron Lett.*, 1992, **33**, 6007.
78. Z.P. Tian, P. Edwards, and R.W. Roeske, *Int.J.Pept.Protein Res.*, 1992, **40**, 119.
79. D.D. Smith, J. Slaninova, and V.J. Hruby, *J.Med.Chem.*, 1992, **35**, 1558.
80. H.-L. Ruthrich, G. Grecksch, R. Schmidt, and K. Neubert, *Peptides*, 1992, **13**, 483.
81. J. Rizo, S.C. Koerber, R.J. Bienstock, J. Rivier, A.T. Hagler, and L.M. Gierasch, *J.Am.Chem.Soc.*, 1992, **114**, 2852.
82. J. Rizo, S.C. Koerber, R.J. Bienstock, J. Rivier, L.M. Gierasch, and A.T. Hagler, *J.Am.Chem.Soc.*, 1992, **114**, 2860.
83. G. Osapay and J.W. Taylor, *J.Am.Chem.Soc.*, 1992, **114**, 6966.
84. M. Bouvier and J.W. Taylor, *J.Med.Chem.*, 1992, **35**, 1145.
85. M.T. Reymond, L. Delmas, S.C. Koerber, M.R. Brown, and J.E. Rivier, *J.Med.Chem.*, 1992, **35**, 3653.
86. D.C. Fry, V.S. Madison, D.N. Greeley, A.M. Felix, E.P. Heimer, L. Frohman, R.M. Campbell, T.F. Mowles, V. Toome, and B.B. Wegrzynski, *Biopolymers*, 1992, **32**, 649.
87. M. Marastoni, S. Salvadori, G. Balboni, V. Scaranari, S. Spisani, E. Reali, A.L. Giuliani, and R. Tomatis, *Eur.J.Med.Chem.*, 1992, **27**, 383.
88. H. Kessler and B. Haase, *Int.J.Pept.Protein Res.*, 1992, **39**, 36.
89. K. Ishikawa, T. Fukami, T. Nagase, K. Fujita, T. Hayama, K. Niiyama, T. Mase, M. Ihara, and M. Yano, *J.Med.Chem.*, 1992, **35**, 2139.
90. M. North, *Tetrahedron*, 1992, **48**, 5509.
91. C. Linget, D.G. Stylianou, A. Dell, R.E. Wolff, Y. Piemont, and M.A. Abdallah, *Tetrahedron Lett.*, 1992, **33**, 3851.
92. A. Polinsky, M.G. Cooney, A. Toy-Palmer, G. Osapay, and M. Goodman, *J.Med.Chem.*, 1992, **35**, 4185.
93. T. Kataoka, D.D. Beusen, J.D. Clark, M. Yodo, and G.R. Marshall, *Biopolymers*, 1992, **32**, 1519.
94. I. Fric, M. Lebl, and V.J. Hruby, *Collect.Czech.Chem.Commun.*, 1992, **57**, 614.
95. G. Byk and C. Gilon, *J.Org.Chem.*, 1992, **57**, 5687.
96. J. Saulitis, D.F. Mierke, G. Byk, C. Gilon, and H. Kessler, *J.Am.Chem.Soc.*, 1992, **114**, 4818.
97. M.J. Deal, R.M. Hagan, S.J. Ireland, C.C. Jordan, A.B. McElroy, B. Porter, B.C. Ross, M. Stephens-Smith, and P. Ward, *J.Med.Chem.*, 1992, **35**, 4195.
98. T. Inoue, S. Kishida, M. Ohsaki, and H. Kimura, *Bull.Chem.Soc.Jpn.*, 1992, **65**, 1728.
99. P.D. Bailey and G.A. Crofts, *Tetrahedron Lett.*, 1992, **33**, 3207.
100. P.D. Bailey, S.R. Carter, D.G.W. Clarke, G.A. Crofts, J.H.M. Tyszka, P.W. Smith, and P. Ward, *Tetrahedron Lett.*, 1992, **33**, 3211.
101. P.D. Bailey, S.R. Carter, D.G.W. Clarke, G.A. Crofts, P.W. Smith, and P. Ward, *Tetrahedron Lett.*, 1992, **33**, 3215.
102. K. Maruyama, M. Hashimoto, and H. Tamiaki, *J.Org.Chem.*, 1992, **57**, 6143.
103. C.M. Falcomer, Y.C. Meinwald, I. Choudhary, S. Talluri, P.J. Milburn, J. Clardy, and H.A. Scheraga, *J.Am.Chem.Soc.*, 1992, **114**, 4036.
104. J.W. Bean, K.D. Kopple, and C.E. Peishoff, *J.Am.Chem.Soc.*, 1992, **114**, 5328.
105. A.N. Stroup, A.L. Rockwell, and L.M. Gierasch, *Biopolymers*, 1992, **32**, 1713.
106. Z.-P. Liu and L.M. Gierasch, *Biopolymers*, 1992, **32**, 1727.

107. H.A. Nagarajaram, P.K.C. Paul, K. Ramanarayanan, K.V. Soman, and C. Ramakrishnan, *Int.J.Pept.Protein Res.*, 1992, **40**, 383.
108. S.H. Gellman, D.R. Powell, and J.M. Desper, *Tetrahedron Lett.*, 1992, **33**, 1963.
109. A. Lecoq, G. Boussard, M. Marraud, and A. Aubry, *Tetrahedron Lett.*, 1992, **33**, 5209.
110. G.-B. Liang, C.J. Rito, and S.H. Gellman, *J. Am.Chem.Soc.*, 1992, **114**, 4440
111. V. Pavone, A. Lombardi, G. D'Auria, M. Saviano, F. Nastri, L. Paolillo, B. Di Blasio, and C. Pedone, *Biopolymers*, 1992, **32**, 173.
112. H.R. Wyssbrod and M. Diem, *Biopolymers*, 1992, **32**, 1237.
113. V.M. Naik, *Spectrosc.Lett.*, 1992, **25**, 1231.
114. M.J. Genin and R.L. Johnson, *J.Am.Chem.Soc.*, 1992, **114**, 8778.
115. S. Chen, R.A. Chrusciel, H. Nakanishi, A. Raktabutr, M.E. Johnson, A. Sato, D. Weiner, J. Hoxie, H.U. Saragovi, M.I. Greene, and M. Kahn, *Proc.Natl.Acad.Sci.USA*, 1992, **89**, 5872.
116. R. Gonzalez-Muniz, M.J. Dominguez, and M.T. Garcia-Lopez, *Tetrahedron*, 1992, **48**, 5191.
117. J.L. Krstenansky, M. del Rosario-Chow, and B.L. Currie, *J.Heterocycl.Chem.*, 1992, **29**, 707.
118. D.J. Kyle, L.M. Green, P.R. Blake, D. Smithwick, and M.F. Summers, *Pept.Res.*, 1992, **5**, 206.
119. M. Sato, J.Y.H. Lee, H. Nakanishi, M.E. Johnson, R.A. Chrusciel, and M. Kahn, *Biochem.Biophys.Res.Commun.*, 1992, **187**, 999.
120. J.E. Baldwin, V. Lee, and C.J. Schofield, *Heterocycles*, 1992, **34**, 903.
121. K.D. Moeller and C.E. Hanau, *Tetrahedron Lett.*, 1992, **33**, 6041.
122. K.D. Moeller and S.L. Rothfus, *Tetrahedron Lett.*, 1992, **33**, 2913.
123. G.A. Flynn, T.P. Burkholder, E.W. Huber, and P. Bey, *Bioorg.Med.Chem.Lett.*, 1991, **1**, 309.
124. W.P. Nolan, G.S. Ratcliffe, and D.C. Rees, *Tetrahedron Lett.*, 1992, **33**, 6879.
125. B. Rzeszotarska, *Wiad.Chem.*, 1991, **45**, 689.
126. V. Busetti, M. Crisma, C. Toniolo, S. Salvadori, and G. Balboni, *Int.J.Biol.Macromol.*, 1992, **14**, 23.
127. V.S. Chauhan and K.K. Bhandary, *Int.J.Pept.Protein Res.*, 1992, **39**, 223.
128. M.R. Ciajolo, A. Tuzi, C.R. Pratesi, A. Fissi, and O. Pieroni, *Biopolymers*, 1992, **32**, 717.
129. K.R. Rajashankar, S. Ramakumar, and V.S. Chauhan, *J.Am.Chem.Soc.*, 1992, **114**, 9225.
130. A. Bharadwaj, A. Jaswal, and V.S. Chauhan, *Tetrahedron*, 1992, **48**, 2691.
131. A.A. Karapetyan, V.O. Topuzyan, and Y.T. Struchkov, *Zh.Struck.Khim.*, 1992, **33**, 151.
132. M. Souhassou, C. Lecomte, N.-E. Ghermani, M.M. Rohmer, R. Wiest, M. Bernard, and R.H. Blessing, *J.Am.Chem.Soc.*, 1992, **114**, 2371.
133. G. Pietrzynski, B. Rzeszotarska, and Z. Kubica, *Int.J.Pept.Protein Res.*, 1992, **40**, 524.
134. E. Ciszak, G. Pietrzynski, and B. Rzeszotarska, *Int.J.Pept.Protein Res.*, 1992, **39**, 218.
135. L. El-Masdouri, A. Aubry, G. Boussard, and M. Marraud, *Int J.Pept.Protein Res.*, 1992, **40**, 482.
136. B. Padmanabhan, S. Dey, B. Khandelwal, G.S. Rao, and T.P. Singh, *Biopolymers*, 1992, **32**, 1271.
137. D.E. Palmer, C. Pattaroni, K. Nunami, R.K. Chadha, M. Goodman, T. Wakamiya,

K. Fukase, S. Horimoto, M. Kitazawa, H. Fujita, A. Kubo, and T. Shiba, *J.Am.Chem.Soc.*, 1992, **114**, 5634.
138. D. Ranganathan, K. Shah, and N. Vaish, *J.Chem.Soc., Chem.Commun.*, 1992, 1145.
139. F. Berti, C. Ebert, and L. Gardossi, *Tetrahedron Lett.*, 1992, **33**, 8145.
140. A. Hammadi, J.M. Nuzillard, J.C. Poulin, and H.B. Kagan, *Tetrahedron Asymmetry*, 1992, **3**, 1247.
141. M. Laghmari, D. Sinou, A. Masdeu, and C. Claver, *J.Organomet.Chem.*, 1992, **438**, 213.
142. P. Kaur, G.K. Patnaik, R. Raghubir, and V.S. Chauhan, *Bull.Chem.Soc.Jpn.*, 1992, **65**, 3412.
143. U. Schmidt, H. Griesser, V. Leitenberger, A. Lieberknecht, R. Mangold, R. Meyer, and B. Riedl, *Synthesis*, 487.
144. Z. Li and G. Simchen, *Youji Huaxue*, 1992, **12**, 294.
145. Y. Nakamura, C.-G. Shin, K. Umemura, and J. Yoshimura, *Chem.Lett.*, 1992, 1005.
146. R.W. Armstrong, J.E. Tellew, and E.J. Moran, *J.Org.Chem.*, 1992, **57**, 2208.
147. N.L. Subasinghe and R.L. Johnson, *Tetrahedron Lett.*, 1992, **33**, 2649.
148. B. Ornik, B. Stanovnik, and M. Tisler, *J.Heterocycl.Chem.*, 1992, **29**, 1241.
149. T.J. Blacklock, R.F. Shuman, J.W. Butcher, P. Sohar, T.R. Lamanec, W.E. Shearin, and E.J.J. Grabowski, *Chem.Ind.(Dekker)*, 1992, **47**, 45.
150. W.R. Baker and S.L. Condon, *Tetrahedron Lett.*, 1992, **33**, 1577.
151. M. Vincent, C. Pascard, M. Cesario, G. Remond, J.-P. Bouchet, Y. Charton, and M. Laubie, *Tetrahedron Lett.*, 1992, **33**, 7369.
152. J.P. Bouchet, J.P. Volland, M. Laubie, M. Vincent, B. Marchand, and N. Platzer, *Magn.Reson.Chem.*,1992, **30**, 1186.
153. S.V.S. Mariappan and D.L. Rabenstein, *J.Org.Chem.*, 1992, **57**, 6675.
154. S.S. Ghosh, O. Said-Nejad, J. Roestamadji, and S. Mobashery, *J.Med.Chem.*, 1992, **35**, 4175.
155. A. Arcadi, S. Cacchi, F. Marinelli, V. Adovasio, and M. Nardelli, *Gazz.Chim.Ital.*, 1992, **122**, 127.
156. G. Yang, S. Lu, X. Hu, C. Guo, S. Shen, and Y. Chen, *Lanzhou Daxue Xuebao, Ziran Kexueban*, 1991, **27**, 137.
157. M. Bessodes, M. Saiah, and K. Antonakis, *J.Org.Chem.*, 1992, **57**, 4441.
158. K.L. Rinehart, R. Sakai, V. Kishore, D.W. Sullins, and K.-M. Li, *J.Org.Chem.*, 1992, **57**, 3007.
159. I. Ojima, Y.H. Park, C.M. Sun, T. Brigaud, and M. Zhao, *Tetrahedron Lett.*, 1992, **33**, 5737.
160. Y. Kobayashi, Y. Takemoto, T. Kamijo, H. Harada, Y. Ito, and S. Terashima, *Tetrahedron*, 1992, **48**, 1853.
161. R.W. Dugger, J.L. Ralbovsky, D. Bryant, J. Commander, S.S. Massett, N.A. Sage, and J.R. Selvidio, *Tetrahedron Lett.*, 1992, **33**, 6763.
162. B.H. Norman and M.L. Morris, *Tetrahedron Lett.*, 1992, **33**, 6803.
163. F. Matsuda, T. Matsumoto, M. Ohsaki, Y. Ito, and S. Terashima, *Bull.Chem.Soc.Jpn.*, 1992, **65**, 360.
164. M.A. Poss and J.A. Reid, *Tetrahedron Lett.*, 1992, **33**, 1411.
165. P.G.M. Wuts, A.R. Ritter, and L.E. Pruitt, *J.Org.Chem.*, 1992, **57**, 6696.
166. A. Dondoni and D. Perrone, *Tetrahedron Lett.*, 1992, **33**, 7259.
167. M. Sakurai, F. Saito, Y. Ohata, Y. Yabe, and T. Nishi, *J.Chem.Soc., Chem.Commun.*, 1992, 1562.
168. F. D'Aniello, S. Gehanne, and M. Taddei, *Tetrahedron Lett.*, 1992, **33**, 5621.
169. F. Rehders and D. Hoppe, *Synthesis*, 1992, 859.

170. F. Rehders and D. Hoppe, *Synthesis*, 1992, 865.
171. W.R. Baker and S.L. Condon, *Tetrahedron Lett.*, 1992, **33**, 1581.
172. M.F. Chan and C.-N. Hsiao, *Tetrahedron Lett.*, 1992, **33**, 3567.
173. W.R. Baker, A.K.L. Fung, H.D. Kleinert, H.H. Stein, J.J. Plattner, Y.-L. Armiger, S.L. Condon, J. Cohen, D.A. Egan, J.L. Barlow, K.M. Verburg, D.L. Martin, G.A. Young, J.S. Polakowski, S.A. Boyd, and T.J. Perun, *J.Med.Chem.*, 1992, **35**, 1722.
174. S.A. Boyd, A.K.L. Fung, W.R. Baker, R.A. Mantei, Y.-L. Armiger, H.H. Stein, J. Cohen, D.A. Egan, J.L. Barlow, V. Klinghofer, K.M. Verburg, D.L. Martin, G.A. Young, J.S. Polakowski, D.J. Hoffman, K.W. Garren, T.J. Perun, and H.D. Kleinert, *J.Med.Chem.*, 1992, **35**, 1735.
175. W.C. Patt, H.W. Hamilton, M.D. Taylor, M.J. Ryan, D.G. Taylor, Jr., C.J.C. Connolly, A.M. Doherty, S.R. Klutchko, I. Sircar, B.A. Steinbaugh, B.L. Batley, C.A. Painchaud, S.T. Rapundalo, B.M. Michniewicz, and S.C. Olson, *J.Med.Chem.*, 1992, **35**, 2562.
176. H.D. Kleinert, S.H. Rosenberg, W.R. Baker, H.H. Stein, V. Klinghofer, J. Barlow, K. Spina, J. Polakowski, P. Kovar, J. Cohen, and J. Denissen, *Science*, 1992, **257**, 1940.
177. J.T. Repine, J.S. Kaltenbronn, A.M. Doherty, J.M. Hamby, R.J. Himmelsbach, B.E. Kornberg, M.D. Taylor, E.A. Lunney, C. Humblet, S.T. Rapundalo, B.L. Batley, M.J. Ryan, and C.A. Painchaud, *J.Med.Chem.*, 1992, **35**, 1032.
178. H.-W. Kleemann, H. Heitsch, R. Henning, W. Kramer, W. Kocher, U. Lerch, W. Linz, W.- U. Nickel, D. Ruppert, H. Urbach, R. Utz, A. Wagner, R. Weck, and F. Wiegand, *J.Med.Chem.*, 1992, **35**, 559.
179. A.M. Doherty, I. Sircar, B.E. Kornberg, J. Quinn, III, R.T. Winters, J.S. Kaltenbronn, M.D. Taylor, B.L. Batley, S.R. Rapundalo, M.J. Ryan, and C.A. Painchaud, *J.Med.Chem.*, 1992, **35**, 2.
180. W.T. Ashton, C.L. Cantone, L.C. Meurer, R.L. Tolman, W.J. Greenlee, A.A. Pachett, R.J. Lynch, T.W. Schorn, J.F. Strouse, and P.K.S. Siegl, *J.Med.Chem.*, 1992, **35**, 2103.
181. W.T. Ashton, C.L. Cantone, R.L. Tolman, W.J. Greenlee, R.J. Lynch, T.W. Schorn, J.F. Strouse, and P.K.S. Siegl, *J.Med.Chem.*, 1992, **35**, 2772.
182. P. Raddatz, A. Jonczyk, K.-O. Minck, F. Rippenmann, C. Schittenhelm, and C.J. Schmitges, *J.Med.Chem.*, 1992, **35**, 3525.
183. D.S. Dhanoa, W.H. Parsons, W.J. Greenlee, and A.A. Patchett, *Tetrahedron Lett*, 1992, **33**, 1725.
184. A.E. Weber, M.G. Steiner, P.A. Krieter, A.E. Colletti, J.R. Tata, T.A. Halgren, R.G. Ball. J.J. Doyle, T.W. Schorn, R.A. Stearns, R.R. Miller, P.K.S. Siegl, W.J. Greenlee, and A.A. Patchett, *J.Med.Chem.*, 1992, **35**, 3755.
185. M.D. Reily, V. Thanabal, E.A. Lunney, J.T. Repine, C.C. Humblet, and G. Wagner, *FEBS Lett*, 1992, **302**, 97.
186. S.E. de Laszlo, B.L. Bush, J.J. Doyle, W.J. Greenlee, D.G. Hangauer, T.A. Halgren, R.J. Lynch, T.W. Schorn, and P.K.S. Siegl, *J.Med.Chem.*,1992, **35**, 833.
187. S.F. Martin, R.E. Austin, C.J. Oalmann, W.R. Baker, S.L. Condon, E. deLara, S.H. Rosenberg, K.P. Spina, H.H. Stein, J. Cohen, and H.D. Kleinert, *J.Med.Chem.*, 1992, **35**, 1710.
188. T.D. Ocain, D.D. Deininger, R. Russo, N.A. Senko, A. Katz, J.M. Kitzen, R. Mitchell. G. Oshiro, A. Russo, R. Stupienski, and R.J. McCaully, *J.Med.Chem.*, 1992, **35**, 823.
189. S. Atsuumi, M. Nakano, Y. Koike, S. Tanaka, K. Matsuyama, M. Nakano, and H. Morishima, *Chem.Pharm.Bull.*, 1992, **40**, 364.

190. D.V. Patel and D.E. Ryono, *Bioorg.Med.Chem.Lett.*, 1992, **2**, 1089.
191. K.Y. Hui, H.M. Siragy, and E. Haber, *Int.J.Pept.Protein Res.*, 1992, **40**, 152.
192. D. Askin, M.A. Wallace, J.P. Vacca, R.A. Reamer, R.P. Volante, and I. Shinkai, *J.Org.Chem.*, 1992, **57**, 2771.
193. W.J. Thompson, P.M.D. Fitzgerald, M.K. Holloway, E.A. Emini, P.L. Darke, B.M. McKeever, W.A. Schleif, J.C. Quintero, J.A. Zugay, T.J. Tucker, J.E. Schwering, C.F. Homnick, J. Nunberg, J.P. Springer, and J.R. Huff, *J.Med.Chem.*, 1992, **35**, 1685.
194. S.D. Young, L.S. Payne, W.J. Thompson, N. Gaffin, T.A. Lyle, S.F. Britcher, S.L. Graham, T.H. Schultz, A.A. Deana, P.L. Darke, J. Zugay, W.A. Schleif, J.C. Quintero, E.A. Emini, P.S. Anderson, and J.R. Huff, *J.Med.Chem.*, 1992, **35**, 1702.
195. F. D'Aniello and M. Taddei, *J.Org.Chem.*, 1992, **57**, 5247.
196. W.J. Thompson, R.G. Ball, P.L. Darke, J.A. Zugay, and J.E. Thies, *Tetrahedron Lett.*, 1992, **33**, 2957.
197. T.J. Tucker, W.C. Lumma, Jr., L.S. Payne, J.M. Wai, S.J. de Solms, E.A. Giuliani, P.L. Darke, J.C. Heimbach, J.A. Zugay, W.A. Schleif, J.C. Quintero, E.A. Emini, J.R. Huff, and P.S. Anderson, *J.Med.Chem.*, 1992, **35**, 2525.
198. P.F. Alewood, R.L. Brinkworth, R.J. Dancer, B. Garnham, A. Jones, and S.B.H. Kent, *Tetrahedron Lett.*, 1992, **33**, 977.
199. D.H. Rich, J.V.N.V. Prasad, Vara, C.-Q. Sun, J. Green, R. Mueller, K. Houseman, D. MacKenzie, and M. Malkovsky, *J.Med.Chem.*, 1992, **35**, 3803.
200. G.B. Dreyer, D.M. Lambert, T.D. Meek, T.J. Carr, T.A. Tomaszek, Jr., A.V. Fernandez, H. Bartus, E. Cacciavillani, A.M. Hassell, M. Minnich, S.R. Petteway, Jr., B.W. Metcalf, and M. Lewis, *Biochemistry*, 1992, **31**, 6646.
201. B.G. Rao, R.F. Tilton, and U.C. Singh, *J.Am.Chem.Soc.*, 1992, **114**, 4447.
202. T.K. Sawyer, D.J. Staples, L. Liu, A.G. Tomasselli, J.O. Hui, K. O'Connell, H. Schostarez, J.B. Hester, J. Moon, W.J. Howe, C.W. Smith, D.L. Decamp, C.S. Craik, B.M. Dunn, W.T. Lowther, J. Harris, R.A. Poorman, A. Wlodawer, M. Jaskolski, and R.L. Heinrikson, *Int.J.Pept.Protein Res.*, 1992, **40**, 274.
203. B.K. Handa and C. Kay, *Int.J.Pept.Protein Res.*, 1992, **40**, 363.
204. T.F. Tam, J. Carriere, I.D. MacDonald, A.L. Castelhano, D.H. Pliura, N.J. Dewdney, E.M. Thomas, C. Bach, J. Barnett, H. Chan, and A. Krantz, *J.Med.Chem.*, 1992, **35**, 1318.
205. D.J. Kempf, T.J. Sowin, E.M. Doherty, S.M. Hannick, L. Codavoci, R.F. Henry, B.E. Green, S.G. Spanton, and D.W. Norbeck, *J.Org.Chem.*, 1992, **57**, 5692.
206. A. Peyman, K.-H. Budt, J. Spanig, B. Stowasser, and D. Ruppert, *Tetrahedron Lett.*, 1992, **33**, 4549.
207. B. Stowasser, K.-H. Budt, L. Jian-Qi, A. Peyman, and D. Ruppert, *Tetrahedron Lett.*, 1992, **33**, 6625.
208. D.A. McLeod, R.I. Brinkworth, J.A. Ashley, K.D. Janda, and P. Wirsching, *Bioorg.Med.Chem.Lett.*, 1991, **1**, 653.
209. S. Ikeda, J.A. Ashley, P. Wirsching, and K.D. Janda, *J.Am.Chem.Soc.*, 1992, **114**, 7604.
210. J.-P. Mazaleyrat, I. Rage, J. Xie, J. Savrda, and M. Wakselman, *Tetrahedron Lett.*, 1992, **33**, 4453.
211. J. Xie, J.P. Mazaleyrat, J. Savrda, and M. Wakselman, *Bull.Soc.Chim.Fr.*, 1992, **129**, 642.
212. M. Wakselman, J.P. Mazaleyrat, J. Xie, J.J. Montagne, A.C. Vilain, and M. Reboud-Ravaux, *Eur.J.Med.Chem.*, 1991, **26**, 699.
213. R. Ferraccioli, P.D. Croce, C. Gallina, V. Consalvi, and R. Scandurra, *Farmaco*, 1991, **46**, 1517.

214. P.D. Edwards, E.F. Meyer, Jr., J. Vijayalakshmi, P.A. Tuthill, D.A. Andisik, B. Gomes, and A. Strimpler, *J.Am.Chem.Soc.*, 1992, **114**, 1854.
215. M. Hagihara and S.L. Schreiber, *J.Am.Chem.Soc.*, 1992, **114**, 6570.
216. K. Hayashi, Y. Hamada, and T. Shioiri, *Tetrahedron Lett.*, 1992, **33**, 5075.
217. S.A. Poyarkova, V.P. Kukhar, M.T. Kolicheva, S.N. Khrapunov, and A.I. Dragan, *Biopolim.Kletka*, 1992, **8**, 20, 76.
218. N. Teno, K. Wanaka, Y. Okada, Y. Tsuda, U. Okamoto, A. Hijikata-Okunomiya, T. Naito, and S. Okamoto, *Chem.Pharm.Bull.*, 1991, **39**, 2930.
219. K. Wanaka, Y. Okada, Y. Tsuda, U. Okamoto, A. Hijikata-Okunomiya, and S. Okamoto, *Chem.Pharm.Bull.*, 1992, **40**, 1814.
220. M.S. Deshpande and J. Burton, *J.Med.Chem.*, 1992, **35**, 3094.
221. M. Asano, N. Nio, and Y. Ariyoshi, *Biosci.Biotech.Biochem.*, 1992, **56**, 976.
222. J. Moess and H. Bundgaard, *Int.J.Pharm.*, 1992, **82**, 91.
223. S. Toda, C. Kotake, T. Tsuno, Y. Narita, T. Yamasaki, and M. Konishi, *J.Antibiot.*, 1992, **45**, 1580.
224. V.J. Robinson, H.W. Pauls, P.J. Coles, R.A. Smith, and A. Krantz, *Bioorg.Chem.*, 1992, **20**, 42.
225. S. Liu and R.P. Hanzlik, *J.Med.Chem.*, 1992, **35**, 1067.
226. C. Giordano, R. Calabretta, C. Gallina, V. Consalvi, and R. Scandurra, *Farmaco*, 1991, **46**, 1497.
227. B. Walker, B.M. Cullen, G. Kay, I.M. Halliday, A. McGinty, and J. Nelson, *Biochem.J.*, 1992, **283**, 449.
228. C. Giordano, C. Gallina, V. Consalvi, and R. Scandurra, *Eur.J.Med.Chem.*, 1991, **26**, 753.
229. H. Angliker, J. Anagli, and E. Shaw, *J.Med.Chem.*, 1992, **35**, 216.
230. S.T. Chen, S.L. Lin, S.C. Hsiao, and K.T. Wang, *Bioorg.Med.Chem.Lett.*, 1992, **2**, 1685.
231. N. Marks, M.J. Berg, R.C. Makofske, J. Swistok, E.J. Simon, D. Ofri, K. Del Compare, and W. Danho, *Pept.Res.*, 1992, **5**, 194.
232. C. Samsoen, E. Lebrun, R. van Rapenbusch, D. Davoust, and G. Lalmanach, *Magn.Reson.Chem.*, 1992, **30**, 992.
233. Z. Szewczuk, K.L. Rebholz, and D.H. Rich, *Int.J.Pept.Protein Res.*, 1992, **40**, 233.
234. V. Dive, A. Yiotakis, C. Roumestand, B. Gilquin, J. Labadie, and F. Toma, *Int.J.Pept.Protein Res.*, 1992, **39**, 506.
235. M. Izquierdo-Martin and R.L. Stein, *J.Am.Chem.Soc.*, 1992,**114**,325.
236. T.D. Ocain and D.H. Rich, *J.Med.Chem.*, 1992, **35**, 451.
237. R. Herranz, S. Vinuesa, J. Castro-Pichel, C. Perez, M.T. Garcia-Lopez, *J.Chem.Soc., Perkin Trans.1*, 1992, 1825.
238. M.-C. Fournie-Zaluski, P. Coric, S. Turcaud, L. Bruetschy, E. Lucas, F. Noble, and B.P. Roques, *J.Med.Chem.*, 1992, **35**, 1259.
239. M.-C. Fournie-Zaluski, P. Coric, S. Turcaud, E. Lucas, F. Noble, R. Maldonada, and B.P. Roques, *J.Med.Chem.*, 1992, **35**, 2473.
240. M.T. Garcia-Lopez, R. Gonzalez-Muniz, J.R. Hartoa, I. Gomez-Monterrey, C. Perez, M.L. De Ceballos, E. Lopez, and J. Del Rio, *Arch.Pharm.(Weinheim, Ger.)*, 1992, **325**, 3.
241. R. Gonzalez-Muniz, J.R. Harto, M.L. De Ceballos, J. Del Rio, and M.T. Garcia-Lopez, *Arch.Pharm.(Weinheim, Ger.)*, 1992, **325**, 743.
242. S.W. Djuric, *Synth.Commun.*, 1992, **22**, 871.
243. M. Hatsu, M. Tuda, Y. Muraoka, T. Aoyagi, and T. Takeuchi, *J.Antibiot.*, 1992, **45**, 1088.

244. J.L. Fauchere and C. Thurieau, *Adv.Drug Res.*, 1992, **23**, 127.
245. B.P. Roques, *Biopolymers*, 1992, **32**, 407.
246. D.W. Hansen, Jr., A. Stapelfeld, M.A. Savage, M. Reichman, D.L. Hammond, R.C. Haaseth, and H.I. Mosberg, *J.Med.Chem.*, 1992, **35**, 684.
247. N.S. Chandrakumar, A. Stapelfeld, P.M. Beardsley, O.T. Lopez, B. Drury, E. Anthony, M.A. Savage, L.N. Williamson, and M. Reichman, *J.Med.Chem.*, 1992, **35**, 2928.
248. G. Toth, K.C. Russell, G. Landis, T.H. Kramer, L. Fang, R. Knapp, P. Davis, T.F. Burks, H.I. Yamamura, and V.J. Hruby, *J.Med.Chem.*, 1992, **35**, 2384.
249. R. Paruszewski, R. Matusiak, G. Rostafinska-Suchar, S.W.G. Gumulka, K. Misterek, and A. Dorociak, *Pol.J.Pharmacol.Pharm.*, 1991, **43**, 381.
250. H. Choi, T.F. Murray, G.E. DeLander, V. Caldwell, and J.V. Aldrich, *J.Med.Chem.*, 1992, **35**, 4638.
251. K.R. Snyder, S.C. Story, M.E. Heidt, T.F. Murray, G.E. DeLander, and J.V. Aldrich, *J.Med.Chem.*, 1992, **35**, 4330.
252. S.C. Story, T.F. Murray, G.E. DeLander, and J.V. Aldrich, *Int.J.Pept.Protein Res.*, 1992, **40**, 89.
253. N.A. Sole and G. Barany, *J.Org.Chem.*, 1992, **57**, 5399.
254. H.I. Mosberg and H.B. Kroona, *J.Med.Chem.*, 1992, **35**, 4498.
255. D.L. Heyl and H.I. Mosberg, *J.Med.Chem.*, 1992, **35**, 1535.
256. D.L. Heyl and H.I. Mosberg, *Int.J.Pept.Protein Res.*, 1992, **39**, 450.
257. S. Salvadori, C. Bianchi, L.H. Lazarus, V. Scaranari, M. Attila, and R. Tomatis, *J.Med.Chem.*, 1992, **35**, 4651.
258. P.W. Schiller, G. Weltrowska, T.M.-D. Nguyen, B.C. Wilkes, N.N. Chung, and C. Lemieux, *J.Med.Chem.*, 1992, **35**, 3956.
259. L.H. Lazarus, S. Salvadori, G. Balboni, R. Tomatis, and W.E. Wilson, *J.Med.Chem.*, 1992, **35**, 1222.
260. B. Buzas, G. Toth, S. Cavagnero, V.J. Hruby, and A. Borsodi, *Life Sci.*, 1992, **50**, PL75.
261. L.Fang, R.J. Knapp, T. Matsunaga, S.J. Weber, T. Davis, V.J. Hruby, and H.I. Yamamura, *Life Sci.*, 1992, **51**, PL189.
262. K. Sakaguchi, T. Costa, H. Sakamoto, and Y. Shimohigashi, *Bull.Chem.Soc.Jpn.*, 1992, **65**, 1052.
263. F. Kalman, T. Cserhati, K. Valko, and K. Neubert, *Anal.Chim.Acta*, 1992, **268**, 247.
264. V.Y. Davydov and Z. Zhao, *Beijing Daxue Xuebao, Ziran Kexueban*, 1992, **28**, 178.
265. N.S. Chandrakumar, P.K. Yonan, A. Stapelfeld, M. Savage, E. Rorbacher, P.C. Contreras, and D. Hammond, *J.Med.Chem.*, 1992, **35**, 223.
266. J. Erchegyi, A.J. Kastin, J.E. Zadina, and X.-D. Qiu, *Int.J.Pept.Protein Res.*, 1992, **39**, 477.
267. T. Dugimont, S. Guissi-Kadri, and J.-J. Curgy, *Int.J.Pept.Protein Res.*, 1992, **39**, 300.
268. L. Varga-Defterdarovic, S. Horvat, N.N. Chung, and P.W. Schiller, *Int.J.Pept.Protein Res.*, 1992, **39**, 12.
269. R. Matsueda, T. Yasunaga, H. Kodama, M. Kondo, T. Costa, and Y. Shimohigashi, *Chem.Lett.*, 1992, 1259.
270. N.G. Luk'yanenko, S.S. Basok, N.V. Kulikov, T.L. Karaseva, and Z.N. Tsapenko, *Khim.- Farm.Zh.*, 1992, **26**, 63.
271. J.W. Tilley, W. Danho, S.-J. Shiuey, I. Kulesha, R. Sarabu, J. Swistok, R. Makofske, G.L. Olson, E. Chiang, V.K. Rusiecki, R. Wagner, J. Michalewsky,

J. Triscari, D. Nelson, F.Y. Chiruzzo, and S. Weatherford, *Int.J.Pept.Protein Res.*, 1992, **39**, 322.

272. W. Danho, J.W. Tilley, S.-J. Shiuey, I. Kulesha, J. Swistok, R. Makofske, J. Michalewsky, R. Wagner, J. Triscari, D. Nelson, F.Y. Chiruzzo, and S. Weatherford, *Int.J.Pept.Protein Res.*, 1992, **39**, 337.
273. J.W. Tilley, W. Danho, V. Madison, D. Fry, J. Swistok, R. Makofske, J. Michalewsky, A. Schwartz, S. Weatherford, J. Triscari, and D. Nelson, *J.Med.Chem.*, 1992, **35**, 4249.
274. J.W. Tilley, W. Danho, S.-J. Shiuey, I. Kulesha, J. Swistok, R. Makofske, J. Michalewsky, J. Triscari, D. Nelson, S. Weatherford, V. Madison, D. Fry, and C. Cook, *J.Med.Chem.*, 1992, **35**, 3774.
275. D. Ron, C. Gilon, M. Hanani, A. Vromen, Z. Selinger, and M. Chorev, *J.Med.Chem.*, 1992, **35**, 2806.
276. M.W. Holladay, M.J. Bennett, M.D. Tufano, C.W. Lin, K.E. Asin, D.G. Witte, T.R. Miller, B.R. Bianchi, A.L. Nikkel, L. Bednarz, and A.M. Nadzan, *J.Med.Chem.*, 1992, **35**, 2919.
277. K. Shiosaki, C.W. Lin, H. Kopecka, R.A. Craig, B.R. Bianchi, T.R. Miller, D.G. Witte, M. Stashko, and A.M. Nadzan, *J.Med.Chem.*, 1992, **35**, 2007.
278. I. Mccort-Tranchepain, D. Ficheux, C. Durieux, and B.P. Roques, *Int.J.Pept.Protein Res.*, 1992, **39**, 48.
279. P.J. Corringer, C. Durieux, M. Ruiz-Gayo, and B.P. Roques, *J.Labelled Compd.Radiopharm.*, 1992, **31**, 459.
280. C.I. Fincham, M. Higginbottom, D.R. Hill, D.C. Horwell, J.C. O'Toole, G.S. Ratcliffe, D.C. Rees, and E. Roberts, *J.Med.Chem.*, 1992, **35**, 1472.
281. M.J. Drysdale, M.C. Pritchard, and D.C. Horwell, *J.Med.Chem.*, 1992, **35**, 2573.
282. E. Didier, D.C. Horwell, and M.C. Pritchard, *Tetrahedron*, 1992, **48**, 8471.
283. M. Higginbottom, C. Kneen, and G.S. Ratcliffe, *J.Med.Chem.*, 1992, **35**, 1572.
284. K. Plucinska, W. Gumulka, and E.I. Wisniewska, *Pol.J.Chem.*, 1991, **65**, 1251.
285. P. Juvvadi, D.J. Dooley, C.C. Humblet, G.H. Lu, E.A. Lunney, R.L. Panek, R. Skeean, and G.R. Marshall, *Int.J.Pept.Protein Res.*, 1992, **40**, 163.
286. M. Manning, J. Przybylski, Z. Grzonka, E. Nawrocka, B. Lammek, A. Misicka, L.L. Cheng, W.Y. Chan, N.C. Wo, and W.H. Sawyer, *J.Med.Chem.*, 1992, **35**, 3895.
287. M. Manning, S. Stoev, K. Bankowski, A. Misicka, B. Lammek, N.C. Wo, and W.H. Sawyer, *J.Med.Chem.*, 1992, **35**, 382.
288. M. Zertova, Z. Prochazka, J. Slaninova, J. Skopkova, T. Barth, and M. Lebl, *Collect.Czech.Chem.Commun.*, 1992, **57**, 604.
289. M. Zertova, Z. Prochazka, J. Slaninova, T. Barth, P. Majer, and M. Lebl, *Collect.Czech.Chem.Commun.*, 1992, **57**, 1103.
290. D. Barbeau, S. Guay, W. Neugebauer, and E. Escher, *J.Med.Chem.*, 1992, **35**, 151.
291. W. Lutz, J.M. Londowski, M. Sanders, J. Salisbury, and R. Kumar, *J.Biol.Chem.*, 1992, **267**, 1109.
292. J. Howl, D.C. New, and M. Wheatley, *J.Mol.Endocrinol.*, 1992, **9**, 123.
293. M. Manning, K. Bankowski, C. Barberis, S. Jard, J. Elands, and W.Y. Chan, *Int.J.Pept.Protein Res.*, 1992, **40**, 261.
294. M. Lebl, G. Toth, J. Slaninova, and V.J. Hruby, *Int.J.Pept.Protein Res.*, 1992, **40**, 148.
295. B.C. Pal, J. Slaninova, T. Barth, J. Trojnar, and M. Lebl, *Collect.Czech.Chem.Commun.*, 1992, **57**, 1345.

296. Z. Prochazka, J. Slaninova, T. Barth, A. Stierandova, J. Trojnar, P. Melin, and M. Lebl. *Collect.Czech.Chem.Commun.*, 1992, **57**, 1335.
297. R. Jezek, J. Franc, J. Slaninova, and M. Lebl, *Collect.Czech.Chem.Commun.*, 1992, **57**, 621.
298. D.S. Perlow, J.M. Erb, N.P. Gould, R.D. Tung, R.M. Freidinger, P.D. Williams, and D.F. Veber, *J.Org.Chem.*, 1992, **57**, 4394.
299. P.D. Williams, M.G. Bock, R.D. Tung, V.M. Garsky, D.S. Perlow, J.M. Erb, G.F. Lundell, N.P. Gould, W.L. Whitter, J.B. Hoffman, M.J. Kaufman, B.V. Clineschmidt, D.J. Pettibone, R.M. Freidinger, and D.F. Veber, *J.Med.Chem.*, 1992, **35**, 3905.
300. J. Rivier, J. Porter, C. Hoeger, P. Theobald, A.G. Craig, J. Dykert, A. Corrigan, M. Perrin, W.A. Hook, R.P. Siraganian, W. Vale, and C. Rivier, *J.Med.Chem.*, 1992, **35**, 4270.
301. G. Flouret, K. Mahan, and T. Majewski, *J.Med.Chem.*, 1992, **35**, 636.
302. F. Haviv, T.D. Fitzpatrick, C.J. Nichols, R.E. Swenson, E.N. Bush, G. Diaz, Cybulski, and J. Greer, *J.Med.Chem.*, 1992, **35**, 3890.
303. E. Masiukiewicz, B. Rzeszotarska G. Fortuna, and K. Kochman, *J.Prakt.Chem.*, 1991, **333**, 573.
304. D. Hagiwara, H. Miyake, H. Morimoto, M. Murai, T. Fujii, and M. Matsuo, *J.Med.Chem.*, 1992, **35**, 2015.
305. D. Hagiwara, H. Miyake, H. Morimoto, M. Murai, T. Fujii, and M. Matsuo, *J.Med.Chem.*, 1992, **35**, 3184.
306. A.B. McElroy, S.P. Clegg, M.J. Deal, G.B. Ewan, R.M. Hagan, S.J. Ireland, C.C. Jordan, B. Porter, B.C. Ross, P. Ward, and A.R. Whittington, *J.Med.Chem.*, 1992, **35**, 2582.
307. M. Antoniou, C. Poulos, and T. Tsegenidis, *Int.J.Pept.Protein Res.*, 1992, **40**, 395.
308. K. Rolka, G. Kupryszewski, P. Janas, J. Myszor, and Z.S. Herman, *Pol.J.Pharmacol.Pharm.*, 1992, **44**, 79.
309. J.A. Malikayil and S.L. Harbeson, *Int.J.Pept.Protein Res.*, 1992, **39**, 497.
310. U. Langel, T. Land, and T. Bartfai, *Int.J.Pept.Protein Res.*, 1992, **39**, 516.
311. A. Ewenson, R. Laufer, J. Frey, M. Chorev, Z. Selinger, and C. Gilon, *Eur.J.Med.Chem.*, 1992, **27**, 179.
312. A.W. Lipkowski and K. Misterek, *Pol.J.Pharmacol.Pharm.*, 1992, **44**, 25.
313. Z. Huang, Y.-B. He, K. Raynor, M. Tallent, T. Reisine, and M. Goodman, *J.Am.Chem.Soc.*, 1992, **114**, 9390.
314. R. Hirschmann, K.C. Nicolaou, S. Pietranico, J. Salvino, E.M. Leahy, P.A. Sprengeler, G. Furst, A.B. Smith, III, C.D. Strader, M.A. Cascieri, M.R. Candelore, C. Donaldson, W. Vale, and L. Maechler, *J.Am.Chem.Soc.*, 1992, **114**, 9217.
315. R. Krishnamoorthy and A.K. Mitra, *Pharm.Res.*, 1992, **9**, 1314.
316. W. Voelter, G. Breipohl, C. Tzougraki, and E. Jungfleisch-Turgut, *Collect.Czech.Chem.Commun.*, 1992, **57**, 1707.
317. A.G. Craig and J.E. Rivier, *Org.Mass.Spectrom.*, 1992, **27**, 549.
318. J.C. Cheronis, E.T. Whalley, K.T. Nguyen, S.R. Eubanks, L.G. Allen, M.J. Duggan, S.D. Loy, K.A. Bonham, and J.K. Blodgett, *J.Med.Chem.*, 1992, **35**, 1563.
319. B. Lammek, Y. Ito, I. Gavras, and H. Gavras, *Collect.Czech.Chem.Commun.*, 1992, **57**, 1960.
320. O. Zerbe, J.H. Welsh, J.A. Robinson, and W. von Philipsborn, *Magn.Reson.Chem.*, 1992, **30**, 683.
321. J.H. Welsh, O. Zerbe, W. von Philipsborn, and J.A. Robinson, *FEBS Lett.*, 1992, **297**, 216.

322. C. Choi, *Han'guk Nonghwa Hakhoechi*, 1991, **34**, 334.
323. R. Romano, H.-J. Musiol, E. Weyher, M. Dufresne, and L. Moroder, *Biopolymers*, 1992, **32**, 1545.
324. Y.M. Taskaeva, G.A. Korshunova, and Y.P. Shvachkin, *Zh.Obshch.Khim.*, 1992, **62**, 717.
325. J.C. Galleyrand, P. Fulcrand, J.P. Bali, M. Rodriguez, R. Magous, J. Laur, and J. Martinez, *Peptides*, 1992, **13**, 519.
326. M. Mokotoff, K. Ren, L.K. Wong, A.V. LeFever, and P.C. Lee, *J.Med.Chem.*, 1992, **35**, 4696.
327. P. Lucietto, R. De Castiglione, and L. Gozzini, *Farmaco*, 1991, **46**, 1111.
328. R. De Castiglione, L. Gozzini, M. Galantino, F. Corradi, M. Ciomei, F. Roletto, and F. Bertolero, *Farmaco*, 1992, **47**, 855.
329. R.D. Feinstein, J.H. Boublik, D. Kirby, M.A. Spicer, A.G. Craig, K. Malewicz, N.A. Scott, M.R. Brown, and J.E. Rivier, *J.Med.Chem.*, 1992, **35**, 2836.
330. W.D. Kornreich, R. Galyean, J.-F. Hernandez, A.G. Craig, C.J. Donaldson, G. Yamamoto, C. Rivier, W. Vale, and J. Rivier, *J.Med.Chem.*, 1992, **35**, 1870.
331. P. Gaudreau, L. Boulanger, and T. Abribat, *J.Med.Chem.*, 1992, **35**, 1864.
332. A.R. Friedman, A.K. Ichhpurani, W.M. Moseley, G.R. Alaniz, W.H. Claflin, D.L. Cleary, M.D. Prairie, W.C. Krueger, L.A. Frohman, T.R. Downs, and R.M. Epand, *J.Med.Chem.*, 1992, **35**, 3928.
333. M. Zarandi, P. Serfozo, J. Zsigo, L. Bokser, T. Janaky, D.B. Olsen, S. Bajusz, and A.V. Schally, *Int.J.Pept.Protein Res.*, 1992, **39**, 211.
334. M. Zarandi, P. Serfozo, J. Zsigo, A.H. Deutch, T. Janaky, D.B. Olsen, S. Bajusz, and A.V. Schally, *Pept.Res.*, 1992, **5**, 190.
335. R.M. Campbell, Y. Lee, T.F. Mowles, K.W. McIntyre, M. Ahmad, A.M. Felix, and E.P. Heimer, *Peptides*, 1992, **13**, 787.
336. J. Bongers, T. Lambros, W. Liu, M. Ahmad, R.M. Campbell, A.M. Felix, and E.P. Heimer, *J.Med.Chem.*, 1992, **35**, 3934.
337. J. Bongers, A.M. Felix, R.M. Campbell, Y. Lee, D.J. Merkler, and E.P. Heimer, *Pept.Res.*, 1992, **5**, 183.
338. T.W. von Geldern, T.W. Rockway, S.K. Davidsen, G.P. Budzik, E.N. Bush, M.Y. Chu- Moyer, E.M. Devine, Jr., W.H. Holleman, M.C. Johnson, S.D. Lucas, D.M. Pollock, J.M. Smital, A.M. Thomas, and T.J. Opgenorth, *J.Med.Chem.*, 1992, **35**, 808.
339. W.L. Cody, A.M. Doherty, J.X. He, P.L. DePue, S.T. Rapundalo, G.A. Hingorani, T.C. Major, R.L. Panek, D.T. Dudley, S.J. Haleen, D. LaDouceur, K.E. Hill, M.A. Flynn, and E.E. Reynolds, *J.Med.Chem.*, 1992, **35**, 3301.
340. E. Nawrocka-Bolewska, A. Kubik, Z. Szewczuk, I.Z. Siemion, E. Obuchowicz, K. Golba, and Z.S. Herman, *Pol.J.Pharmacol.Pharm.*, 1991, **43**, 281.
341. D.J.M. Stone, R.J. Waugh, J.H. Bowie, J.C. Wallace, and M.J. Tyler, *J.Chem.Soc., Perkin Trans.1*, 1992, 3173.
342. D. Kikelj, J. Kidric, P. Pristovsek, S. Pecar, U. Urleb, A. Krbavcic, and H. Honig, *Tetrahedron*, 1992, **48**, 5915.
343. Z. Mackiewicz, H. Swiderska, M. Kalmanowa, Z. Smiatacz, A. Nowoslawski, and G. Kupryszewski, *Collect.Czech.Chem.Commun.*, 1992, **57**, 204.
344. K. Dzierzbicka, A. Kolodziejczyk, and B. Wysocka-Skrzela, *Pol.J.Chem.*, 1991, **65**, 2293.
345. Y. Boulanger, Y. Tu, V. Ratovelomanana, E. Purisima, and S. Hanessian, *Tetrahedron*, 1992, **48**, 8855.
346. H.G. Parkes, *Nucl.Magn.Reson.*, 1992, **21**, 276.

347. R. Bhaskaran, L.-C. Chuang, and C. Yu, *Biopolymers*, 1992, **32**, 1599.
348. C. Yu, T.-H. Yang, C.-J. Yeh, and L.-C. Chuang, *Can.J.Chem.*, 1992, **70**, 1950.
349. C.K. Larive, L. Guerra, and D.L. Rabenstein, *J.Am.Chem.Soc.*, 1992, **114**, 7331.
350. G. Zieger and H. Sterk, *Magn.Reson.Chem.*, 1992, **30**, 387.
351. D.F. Mierke, H. Durr, H. Kessler, and G. Jung, *Eur.J.Biochem.*, 1992, **206**, 39.
352. M. Petersheim, R.L. Moldow, H.N. Halladay, A.J. Kastin, and A.J. Fischman, *Int.J.Pept.Protein Res.*, 1992, **40**, 41.
353. V.Y. Gorbatyuk, Y.E. Shapiro, A.A. Mazurov, V.G. Zhuravlev, S.A. Andronati, T.I. Korotenko, and P.Y. Romanovskii, *Bioorg.Khim.*, 1992, **18**, 235.
354. A. D'Ursi, M. Pegna, P. Amodeo, H. Molinari, A. Verdini, L. Zetta, and P.A. Temussi, *Biochemistry*, 1992, **31**, 9581.
355. S.D. O'Connor, P.E. Smith, F. Al-Obeidi, and B.M. Pettitt, *J.Med.Chem.*, 1992, **35**, 2870.
356. J.A. Cushman, P.K. Mishra, A.A. Bothner-By, and M.S. Khosla, *Biopolymers*, 1992, **32**, 1163.
357. K.J. Willis and A.G. Szabo, *J.Fluoresc.*, 1992, **2**, 1.
358. H. Tamaoki, Y. Kyogoku, K. Nakajima, S. Sakakibara, M. Hayashi, and Y. Kobayashi, *Biopolymers*, 1992, **32**, 353.
359. V. Wray and K. Nokihara, *Shimadzu Hyoron*, 1991, **48**, 259.
360. J.K. Young, W.H. Graham, D.J. Beard, and R.P. Hicks, *Biopolymers*, 1992, **32**, 1061.
361. R.P. Hicks, D.J. Beard, and J.K. Young, *Biopolymers*, 1992, **32**, 85.
362. V.S. Ananthanarayanan and S. Orlicky, *Biopolymers*, 1992, **32**, 1765.
363. W.H. Graham, E.S. Carter, II, and R.P. Hicks, *Biopolymers*, 1992, **32**, 1755.
364. P.A. Temussi, D. Picone, G. Saviano, P. Amodeo, A. Motta, T. Tancredi, S. Salvadori, and R. Tomatis, *Biopolymers*, 1992, **32**, 367.
365. H. De Loof, L. Nilsson, and R. Rigler, *J.Am.Chem.Soc.*, 1992, **114**, 4028.
366. B.C. Wilkes and P.W. Schiller, *Int.J.Pept.Protein Res.*, 1992, **40**, 249.
367. J.J. Perez, G.H. Loew, and H.O. Villar, *Int.J.Quantum Chem.*, 1992, **44**, 263.
368. G.M. Ciuffo and E.A. Jauregui, *An.Asoc.Quim.Argent.*, 1991, **79**, 225.
369. A.G. Michel, C. Ameziane-Hassani, and N. Bredin, *Can.J.Chem.*, 1992, **70**, 596.
370. Y. Okamoto, T. Kikuchi, and H. Kawai, *Chem.Lett.*, 1992, 1275.
371. D.R. Ripoll and F. Ni, *Biopolymers*, 1992, **32**, 359.

4
Cyclic, Modified and Conjugated Peptides

By J.S. DAVIES

1 Introduction

As for the last two years of these Reports, the coverage in the cyclic peptide sections of this Chapter has been restricted mostly to naturally-occurring cyclic and modified peptides, but does include cyclic structures investigated for structural/conformational purposes. This rationale eliminates as far as possible the overlap with sections in Chapter 3 which concentrate on cyclic analogues of biologically significant domains, and peptide surrogates. Phosphorylated and glycosylated peptides, an expanding area, remain topics within this Chapter.

The core of the review has depended on papers accessed from CA Selects[1] on Amino Acids, Peptides and Proteins, produced by Computer Search. (Abstracts up to Issue 12 1993 were reviewed.) However, no attempt at covering the patent literature has been made. To attempt as comprehensive a coverage as possible, mainstream Journals in the subject were also scanned manually. During the preparation of this Chapter, proceedings of the 22nd European Peptide Symposium at Interlaken[2] come to hand, but while this volume contains very relevant subject matter for this Report, the usual practice of concentrating on refereed papers has been adhered to. Two very useful specialist review books[3,4] have also appeared while this Chapter was in embryo.

A significant increase (25%) in the number of papers reviewed can be recorded for the 1992 coverage mainly due to the increased activity in glycopeptides. Structural elucidation of naturally-occurring molecules remains very active with 2D-nmr techniques, Mass Spectrometry and occasionally X-ray-diffraction appearing as 'routine'

2 Cyclic Peptides

2.1 *General Considerations.*

Nmr and computational methods have made powerful inroads into the determination of peptide conformations. The current 'state of the art' has been reviewed both in general[5] and in the context of cyclic peptides involved in the regulation of the immune system.[6]

Key to the efficient synthesis of a cyclic peptide is a high yielding cyclisation step. Both the traditional high dilution conditions using a range of activation methods, and the more recent methodology of cyclisation on the polymer support have been widely used. Representative of the former approach are the attempts[7] to optimise the macrolactamisation step from a C-terminal aspartyl precursor in the preparation of cyclic analogues of a CCK-B receptor antagonist such as (1). Of the five reagents tried, diphenylphosphoryl azide (DPPA) 54% yield, and the BOP reagent 49% yield proved the best. 2-Chloro-1-methylpyridinium iodide, and DCC/HOBt only achieved modest yields while the phosgene precursor bis(trichloromethyl) carbonate proved the least satisfactory. It is not always easy to predict the most efficient position to choose for the final link. For (Asu[1,7])-calcitonin fragment derivative (2) past efforts had used position I as the cyclisation point. However, the cyclisation yields[8] as shown in Table 1 can vary over the different positions used and also the temperature.

Table 1

Cyclisation Position	Cyclisation Yields (%) Room Temperature	Cyclisation Yields (%) Low Temperature
I	21	20
II	37	46
III	59	86
IV	72	81
V	57	80
VI	65	85

It is possible to directly compare this result with the synthesis[9] of almost an identical analogue of the cyclic portion of calcitonin using the solid phase equivalent utilising the Kaiser oxime resin as summarised in Scheme 1. The cyclisation yield quoted was 65%, with a 56% overall yield achieved for a [Asu[1,6]]-oxytocin analgue. The same 'oxime resin' approach proved successful in preparation of the neuropeptide Y analogues (3)[10] and the highly α-helical bridged uncosapeptide (4), obtained in 54% yield.[11]

In this Chapter last year two examples were given of head to tail cyclisation on solid supports, where the growing linear peptide is first attached to the resin *via* the side-chains of aspartic or glutamic acid. In a third example[12], *cyclo*-(Lys-Arg-Ser-Gly-X) (with X as Asp or Asn) have been synthesised starting with attachment of Fmoc-Asp-O-allyl ester to the appropriate resin. After 'on-resin' cyclistion with TBTU the allyl group can be removed utilising $Pd°[P(C_6H_5)_3]_4$.

In the synthesis of calcitonin (5) and other similar molecules it was

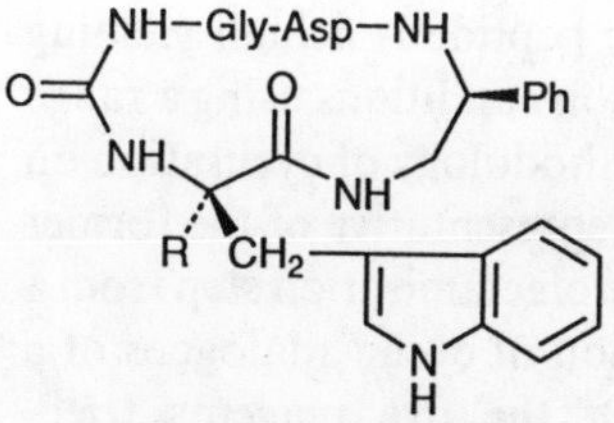

NH–Gly-Asp–NH

O=C O Ph

NH NH

R CH_2

N H

(1) R = H, or Me

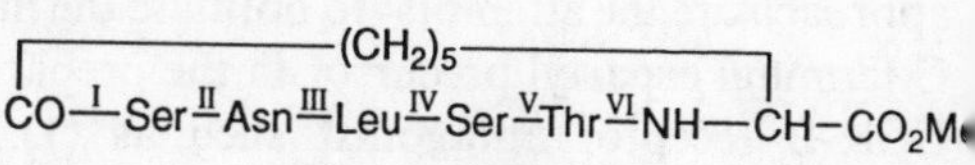

$(CH_2)_5$

CO–I–Ser–II–Asn–III–Leu–IV–Ser–V–Thr–VI–NH–CH–CO_2Me

(2)

NO_2 P

N

O

OCO

$(CH_2)_5$

BocNH CONHNHZ

NO_2 P

N

O

OCO

$(CH_2)_5$

H–Ser(Bzl)-Asn-Leu-Ser(Bzl)-Thr(Bzl)NHCH–CONHNHZ

$Et_3N/HOAc$

$(CH_2)_5$

CO–Ser(Bzl)-Asn-Leu-Ser(Bzl)-Thr(Bzl)–NH CONHNHZ

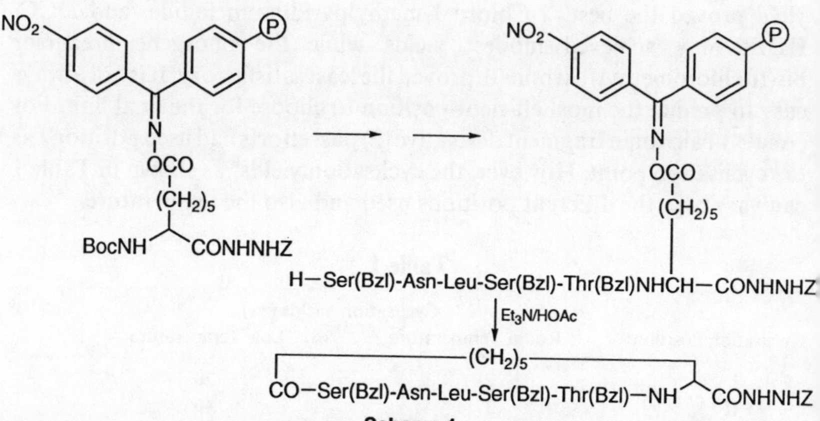

Scheme 1

R–Lys-Arg-Tyr-Tyr-Asp–R′

(3)

H–(Lys-Leu-Lys-Glu-Leu-Asp)$_3$–OH

(4)

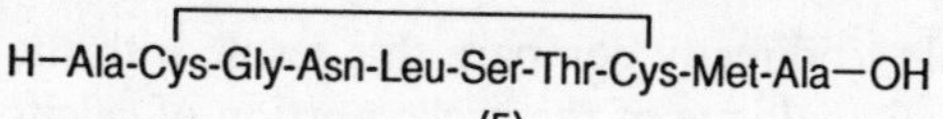

H–Ala-Cys-Gly-Asn-Leu-Ser-Thr-Cys-Met-Ala–OH

(5)

CH_2

O_2C–C–CH_2OH

Me H

H CH_2 O

$HOCH_2$ O O N $CHMe_2$

N O

Me_2CH CH_2 O CH_2OH

O H

H Me

$HOCH_2CCO_2$

CH_2

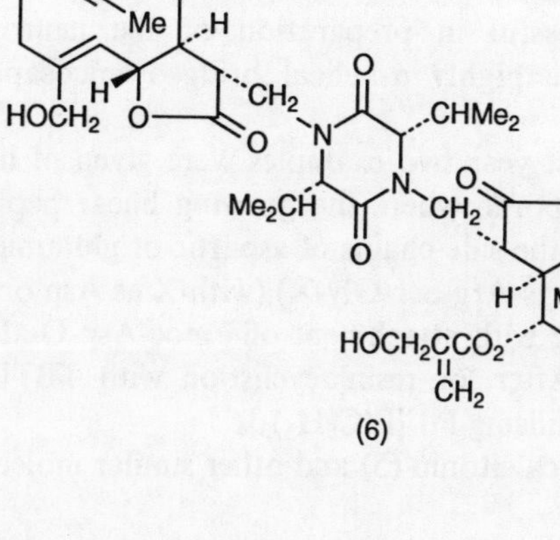

(6)

shown[13] that cyanogen iodide, ICN, gave a more efficient synthesis of the disulfide bond than iodine. The improved solubility of ICN in MeOH/ H_2O mixtures was given as an explanation for improved product yields, in one case raised to 69%, when I_2 only manged 6%.

2.2 *Naturally occurring Dioxopiperzines (Cyclic Dipeptides).*

Nature has produced a rather interesting di-N-alkylated cyclo(-Val-Val) in the form of the germacranolide-adduct dimer (6) from *Centaurea aspera.*[14] Some further information on the mode of action of the antibiotic bicyclomycin (7) has been revealed[15] by the reactivities of the exocyclic double bond towards sulfur nucleophiles in the analogues (8) and (9). Analogue (8) displayed enhanced reactivity with sodium ethanethiolate when compared with (7) and (9), implying that the C(1′)-OH gp has a bearing on the attack on the double bond by protein nucleophiles. The dioxopiperazine (10) precursor (10) of bicyclomycin has been constructed[16] by quite a novel route from (11) *via* the intramolecular addition of an amide to an α,β diketo amide. This could have more general use for making unsymmetrical dioxopiperazines.

2.3 *Other Dioxopiperazines.*

The active conformation of *cyclo(-(S)* Phe-*(S)*His), referred to as Inoue's catalyst for the asymmetric addition of HCN to aldehydes, has been studied by nmr and molecular modelling and FT-IR techniques[17a,b]. The major conformer of two existing in d_6DMSO or CD_3OD, is designated the form shown in (12) with the imidazole ring being kept essentially rigid by an intramolecular H-bond between it and the His-CO group, with the phenyl ring above the dioxopiperazine ring. Addition of HCN made little change to the spectra, so it is suggested that the interaction between HCN and the catalyst is H-bonding rather than protonation. In a separate study[18] the nature of the interaction was calculated using a model system when the nmr and ir evidence was not conclusive. When the tyrosyl analogue *cyclo* (Tyr-His) was attached[9] to chloromethylated polystyrene and to polysiloxane *via* spacer groups catalysis was maintained but the enantioselectivity of the conversion of aromatic aldehydes to cyanohydrins was low.

Kinetic studies[20] on the hydrolysis of *p*-nitrophenyl acetate catalysed by *cyclo* (L-His-L-His) and *cyclo* (D-His-L-His) showed the latter to be better than the former. Yet when a mono-histidyl analogue is attached[21] to a crown ether as in (13), kinetic data on hydrolysis of *p*-nitrophenyl esters of protonated amino acids indicated that the more crowded L-L form (13) is better than the D-L form. For a series of histidyl dioxopiperazines *cyclo* (X-His) where X was Gly, Ala, Val, Leu,

(7)

(8) X = Y = OH
(9) X,Y = $OCMe_2O$

(10) R = MeO–C_6H_4–CH_2–

(11) R = MeO–C_6H_4–CH_2–

(12)

(13)

(14) R = H
(15) R = Me

(16)

(17)

(18)

(19)

Pro, Phe or His, thermodynamic proton parameters and nmr data confirmed[22] that protonation of the imidazole ring causes the ring to move away from the region directly above the dioxopiperazine ring.

^{1}H and ^{13}C chemical shifts and ^{13}C-relaxation times have been compared[23] for *cyclo* (L-Pro-L-MePhe) and its non-methylated analogue. The phenyl group in both cases folds over the dioxopiperazine ring which is a flattened half-chain with the Pro pyrrolidine ring puckered, a conformation also seen in an X-ray study of the compound. When the partner amino acid to proline is tyrosine as in *cyclo* (Tyr-Pro), X-ray crystallography[24] again shows the phenyl ring above the flattened chain conformation of the dioxopiperazine ring, with the Pro ring in a pseudo C_2 symmetrical twist. However the nmr spectra of a solution of *cyclo* (Tyr-Pro) showed a conformation with an extended aromatic side-chain and a dioxopiperazine ring adopting the boat conformation.

When the four diastereoisomers of *cyclo* (Asp-Val) were synthesised[25] none of them fitted the physical data and properties of cairomycin A. Out of the ten possible stereoisomers of *cyclo* (L-Ile-L-Ile), when the amino acid was condensed[26] in ethane-1,2-diol three isomers predominated in the reaction mixture, namely *cis-(3S,6S)* and cis- and *trans-(3R,6R)*, i.e., *cyclo* (L-Ile-L-Ile), *cyclo* (L-Ile-D-aIle) and *cyclo* (D-aIle-D-aIle). The ability of many amino acids to sublime without decomposition, which include glycine, alanine, valine, leucine and proline, is a necessary requirement when *cyclo*-dipeptides are produced[27] in the presence of macroporous silica at reduced pressure and a temperature of 160-220°C. When Ac-Asp-Glu-OH was subjected[28] to methylation with diazomethane CH_2N_2, *cyclo*-[Asp(OMe)-Glu(OMe)] was readily formed which could be hydrolysed to the hitherto unreported *cyclo* (Asp-Glu).

The dioxopiperazine ring finds itself an ubiquitous structure in many situations. Chirally substituted dioxopiperazine (14) when enolised with $LiN(SiMe_3)_2$ followed by alkylation with methyl iodide afforded[29] the (*3S,6S*) dimethyl derivative (15) in 98% diastereomeric excess. With 57% hydrogen iodide this breaks down to *(S)*-alanine, and the opposite enantiomer resulted from starting with the 3*(R)* form of (14). The asymmetric synthesis[30] of (+)-*(1R,2S)*-allocoronamic acid (17) can be mediated *via* the reaction of CH_2N_2 with dioxopiperazine (16) followed by photolysis in benzene and subsequent hydrolysis. A number of substituted aromatic amino acids have been made[31] *via* the 3,6-diarylidene-2,5 piperazindione (18) formed from diacetyl-2,5-piperazindione under solid-liquid phase transfer catalytic conditions. Zinc powder and conc. hydrochloric acid converted the compounds to substituted amino acids. Enzyme catalysed conversion of Ac-Leu-Lys-Gly-Acp-OEt (Acp = 1-aminocyclopentane carboxylic acid) into the spiro compound (19) has been proven[32] *in vitro* with purified peptidase enzymes and after administration of urokinase *in vivo*. Cyclodipeptide formation is an unwanted side-reaction which causes cleavage of dipeptides off the resin when the Kaiser oxime resin is used. In a study[33] of the rates of cyclisation-cleavage of dipeptides on this resin it was shown that *cyclo* (Ala-Pro) with a k_{app} of 1.5 min^{-1} was much more rapidly formed than *cyclo* (Ala-Ala) (k_{app} 1.8 $\times$ $10^{-1}min^{-1}$). Cyclodipeptides from H-Ala-Val-resin and H-Val-Ala-resin had k_{app} of 2.2 $\times$ 10^{-2} and 5.5 $\times$ $10^{-2}min^{-1}$, respectively. In the synthesis of tripeptides on this resin using symmetrical anhydride or DCC/HOBt coupling agents, significant cleavage at the dipeptide stage was observed, which was successfully prevented by BOP reagent, except in the case of H-Ala-Pro-resin.

2.4 *Cyclotripeptides and Cyclotetrapeptides.*

Although in strict nomenclature, a tripeptide conformationally constrained *via* an amide bridge is not truly a cyclotripeptide it is sufficiently interesting to note the type of side-chain required[34] to place a tripeptide in an accessible conformation for receptor recognition, and in their ability to accommodate a β-turn. Cyclisation through SCH_2, S-S and CONH have been evaluated, and it was found that cyclic tripeptides having *cis*- or *trans*-4-mercaptoproline at residue 3 could not accommodate any type of β-turn. Thioether and amide-bridge peptides yielded results similar to the disulfide-bridged peptides.

For the smaller cyclopeptides choosing the correct linear precursor is very important, since dimerisation often competes with cyclisation. So prediction of sequence best able to cyclise, and the best position of cyclisation is therefore crucial. It is therefore interesting to note the results of modelling calculations[35] using the GenMol programme, in predicting the course of action for the synthesis of the 4-Ala-chlamodycin analogue *cyclo* (-Aib-Phe-D-Pro-Ala). The transition-state energy rather than the dimerisation reaction proved the more determining factor in ring closure. The results clearly indicated that the linear precursor to choose would be H-Ala-Aib-Phe-D-Pro-OPh since it was calculated to have a transition state energy of 61.7 kcal/mol compared with 106, 103 and 92.7 for the other options. Of even more significance is the actual results obtained for a series of sequences tested[36] for the synthesis of *cyclo* (Arg-Gly-Asp-Phg) on a Kaiser oxime resin using Scheme 2. A summary of the results are given in Table 2.

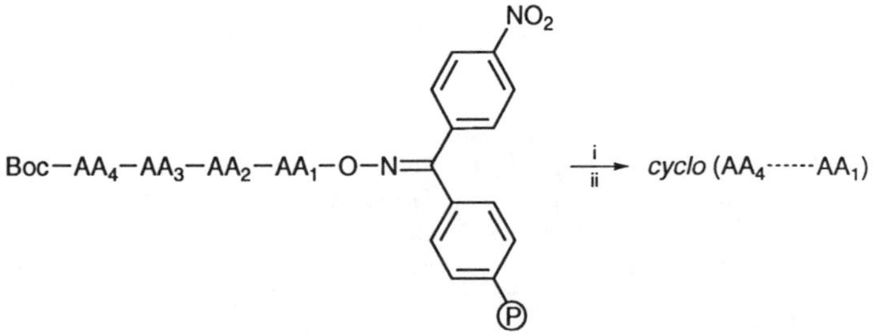

$Boc{-}AA_4{-}AA_3{-}AA_2{-}AA_1{-}O{-}N{=}$ $\xrightarrow[ii]{i}$ *cyclo* (AA_4······AA_1)

NO_2

Ⓟ

Reagents: i, TFA; ii, $Et_3N/AcOH$

Scheme 2

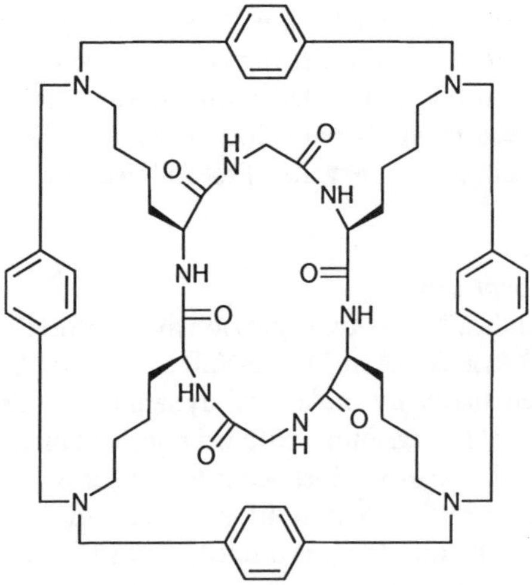

(20)

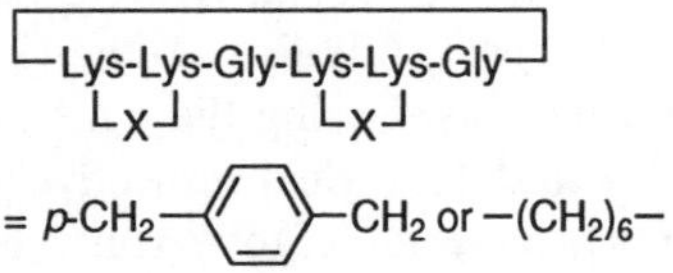

Lys-Lys-Gly-Lys-Lys-Gly

(21) X = *p*-CH_2–C_6H_4–CH_2 or $-(CH_2)_6-$

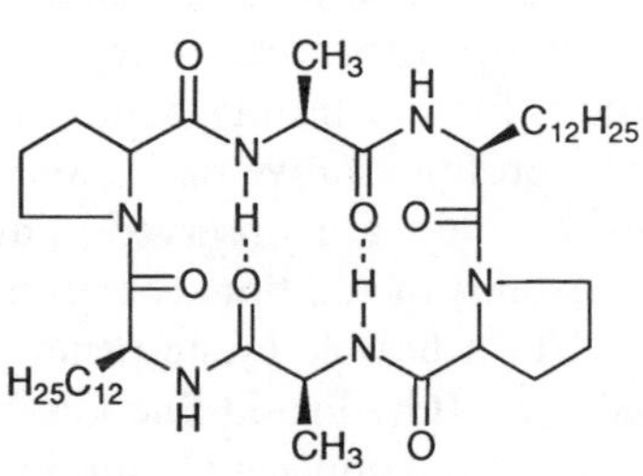

(22)

Table 2

AA_4	AA_3	AA_2	AA_1	Yield %	Monomer %	Dimer %
1. Arg(Tos)	Gly	Asp(cHex)	Phg*	65	75	20
2. Phg	Arg(Tos)	Gly	Asp(cHex)	65	50	25
3. Asp(cHex)	Phg	Arg(Tos)	Gly	70	40	40
4. Gly	Asp(cHex)	Phg	Arg(Tos)	53	10	Not detected

* Phg = phenylglycine

Sequence 1 obviously gave the best result possibly because of its tendency towards a β-turn even at the linear stage but the poor yield given by sequence 4 can be explained by the Boc-Arg(Tos)-oxime resin readily forming a lactam as fast as the cleavage of cyclic peptides from the resin. Improved solubilisation of peptides in apolar solvents can be achieved[37] by N-perbenzylation using P_4-phosphazene as base. For *cyclo* (Leu-Sar-Sar-Gly) the Gly nitrogen benzylated first followed by the NH of the Leu residue.

2.5 *Cyclopentapeptides.*

Further studies[38] on two previously synthesised cyclic RGD analogues, *cyclo* (Arg-Gly-Asp-D-Phe-Val) and *cyclo* (Arg-Gly-Asp-Phe-D-Val) have been made by molecular dynamics simulations using the GROMOS force-field programme with nmr derived interproton distances as constraints. Both *cyclo*-peptides seem to have a βII′-γ turn motif, but the position of the Arg-Gly-Asp motif is different in the two compounds. In L-Val analogue the motif forms a tight γ-turn with Gly central, but in the D-Val analogue the Arg and Asp side-chains are more separated pointing in opposite directions. It is suggested that the former conformer is more in line with the laminin receptor, while the latter would be a better fit for the vitronectin receptor. *Cyclo*-(Pro-Pro-Ala-Ala-Ala) has been used[39] for illustrating some of the advantages of applying a penalty function (quadrative or skewed harmonic) for each pair of protons for which nOe's are observed. This will become more important as the availability of additional heteronuclear coupling constants increases.

Two flexible cyclic pentapeptides, *cyclo* (Gly-Pro-D-Phe-Gly-Ala) and *cyclo* (Gly-Pro-D-Phe-Gly-Val), synthesised using the *p*-nitrophenyl ester method under high dilution (53% and 43% yield respectively using the Pro-D-Phe as the cyclisation link), have undergone study by nmr techniques[40]. Due to the presence of the Gly residue both have the DLDDL sequence, and in $CDCl_3$ solution (5% d_6DMSO added) and in neat d_6DMSO, both show a distorted β-turn around the Pro-D-Phe section which includes a γ-turn around D-Phe. These have been confirmed[41] by combined nmr and molecular dynamics simulations. In addition to the presence of different H-bonded all *trans* conformers there is *cis-trans* interconversion around the Gly-Pro bond with both theory and practical offering an activation energy of conversion in the 18.8-21 kcal/mol range for the rotational barrier. The effect of interfacial environments on the conformation of *cyclo* (D-Phe-Pro-Gly-D-Ala-Pro), known to be present as a single rigid conformation (in $CDCl_3$ or d_6DMSO) and the more flexible *cyclo* (Gly-Pro-D-Phe-Gly-Val) has been explored[42] in sodium dodecyl sulfate (SDS) micelles. The more rigid conformer does not seem to change in the presence of the micelles but in the latter of the two cyclopentapeptides there is an increase in the proportion of the *cis* conformer around Gly-Pro bond, and the *trans* form seems to be averaging between two conformers around the D-Phe-Gly bond. An ir (vibrational) cd spectral investigation[43] of *cyclo* (Gly-Pro-Gly-D-Ala-Pro) in non-aqueous media failed to generate agreement between collected spectra and ones reproduced computationally using the extended coupled oscillator (ECO). The formation of β-turns has been put forward as an explanation.

Modifications[44] to the naturally occurring inhibitors from *Streptomyces misakiensis*, *cyclo* (D-Trp-D-Glu-Ala-D-Val-Leu) and *cyclo* (D-Trp-D-Glu-Ala-D-aIle-Leu) have yielded several potent and highly endothelin A receptor antagonists. As an example *cyclo* (D-Trp-D-Asp-Pro-D-Val-Leu) inhibits [^{125}I]-endothelin-1 binding to aortic smooth muscle membranes with an IC_{50} of 22nM, but only weakly inhibits binding to porcine cerebellum membranes with an IC_{50} of 18 μM. The synthesis utilised the azide methodology between a Leu-$NHNH_2$ residue and the D-Trp *N*-terminal residue giving a 62% yield of the cyclopentapeptide.

2.6 *Cyclohexapeptides.*

A cyclic hexapeptide base ring has been used[45] in an interesting way to create doubly cross-linked polycyclic constrained structures such as (20), ultimately with the intention of designing peptide enzyme mimetics. Cyclodimerisation of H-Lys(Z)-Lys(Z)-Gly-OPfp to *cyclo* (Lys(Z)-Lys(Z)-Gly)$_2$ using Cs_2CO_3 proved[46] the most efficient (38% overall yield) of constructing the cyclohexapeptide ring. In an interim study[47] of only linking two pairs of the Lys side chains to give (21), the key discovery involved side chain alkylation of tosyl or trifluoroacetyl derivatives of the lysine residues using α,ω-dibromoalkenes in presence of Cs_2CO_3 at the pre-cyclodimerisation stage. Compound (20) was prepared[48], after a great deal of painstaking chromatography in 10% yield, by double alkylation of *cyclo* (Lys-Lys-Gly)$_2$ with p-$BrCH_2C_6H_4CH_2Br$ in trifluoroethanol/DMF. Analogues with a mixture of bridging links were also synthesised, with nmr and molecular modelling confirming the presence of two unusual β-turns in the peptide structure. Another 'custom-built' cyclohexapeptide reported[49] was the amino-myristic acid containing lipophilic molecule (22). The *N*-hydroxy-succinimide ester of the linear precursor under high dilution was used in the synthesis (36% yield) and the conformation proven to be (22) by cd and 400 MHz nmr techniques. The compound gave almost an ideal Nernstian slope when made into a pvc membrane electrode with a preference towards Ca^{2+} ion selectivity. *Cyclo* (L-Ala-L-Pro-D-Amy)$_2$ where Amy = α-amino myristic acid in the same work showed no response to any cations, a situation explained in terms of its type II β-turn/γ-turn conformation.

Two names have now been allocated to the structure (23), a highly modified cyclohexapeptide. It was first isolated[50] from *Lissoclinum bistratum*, a marine ascidian and given the name cycloxazoline, but another marine organism *Westiellopsis prolifica* has yielded[51] the same compound named westiellamide, which has also been synthesised in 20% yield from (24) as in Scheme 3. Structure-activity profiles[52] of the analogues (such as 26 and 27) of the oxytocin antagonist L-365-209 (25) derived from *Streptomyces silvensis* have been investigated. Lipophilic amino acids at positions 2 and 3 and the unusual amino acid D-ΔPiz at position 4 were the most critical for obtaining good oxytocin receptor affinity. In the syntheses a Fmoc strategy was adopted because of the acid sensitivity of the residues. Efficient cyclisation was carried out at the 1-2 position using DPPA, and Fmoc-amino acid chlorides proved useful[53]. In a touch coupling of Fmoc-L-Ile with N^{δ}-Z-piperazic acid, the acid chloride only gave 30% yield but addition of AgCN in toluene raised this to 76% presumably due to the acid chloride's conversion to oxazolone (28)

All cyclic hexapeptides seem to contain at least one Gly, Pro or D-residues, presumably in order to stabilise the turn structures. However, in the synthesis of these it does not seem to matter in what position the D-residue is situated in the linear precursor. For the synthesis[54] of *cyclo* (Leu-Tyr-Leu-Glu-Ser-Leu) using the Leu^6-Leu^1 bond for the final link and the azide method, varying each residue for a D-residue in turn, only produced a spread of less than 10% in the average 55% yield. So the yield did not seem to be influenced by location of the D-residue, but the latter was still needed, as the all-L precursor only gave a 1% yield. Phase-transfer catalysis using tetrabutylammonium hydrogen sulfate in THF with saturated K_2CO_3 has been used[55] for preparing partially protected linear sequences for conversion to a series of *cyclo* (Xxx-Pro-Gly-Yyy-Pro-Gly)

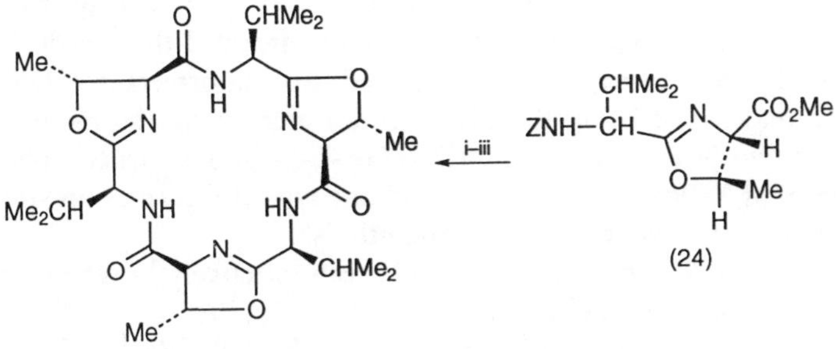

(23) ← i–iii — (24)

Reagents: i, H_2/Pd; ii, NaOH/MeOH; iii, 3 mol. DPPA/DMF

Scheme 3

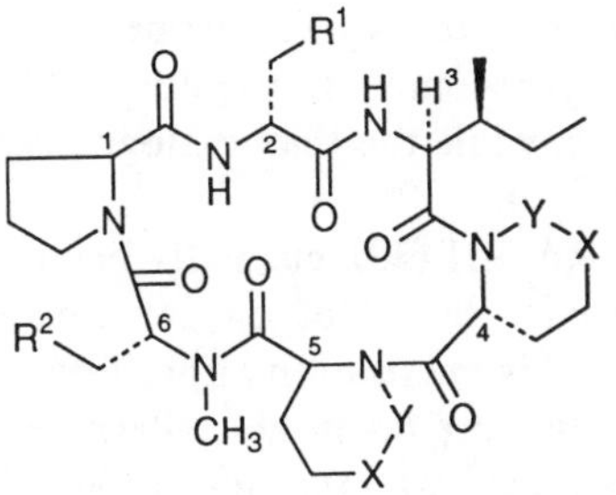

(25) L-365-209, $R^1 = R^2$ = Ph, X–Y = CH=N
Analogue $R^1 = R^2$ = Ph, X–Y = CH_2CH_2

(26) L-366-682, R^1 = Ph, R^2 = 3-indolyl

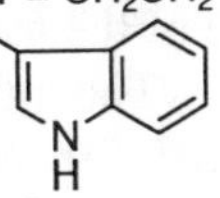

(27) L-366-948, R^1 = naphthyl, R^2 = imidazolyl

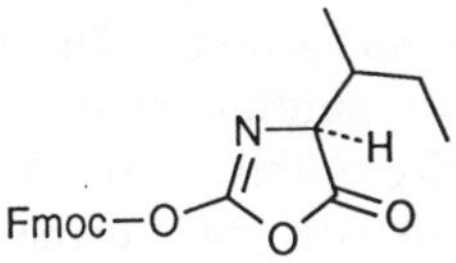

(28) 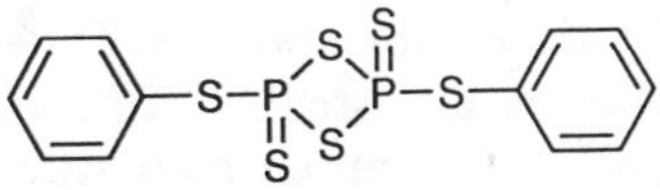 (29)

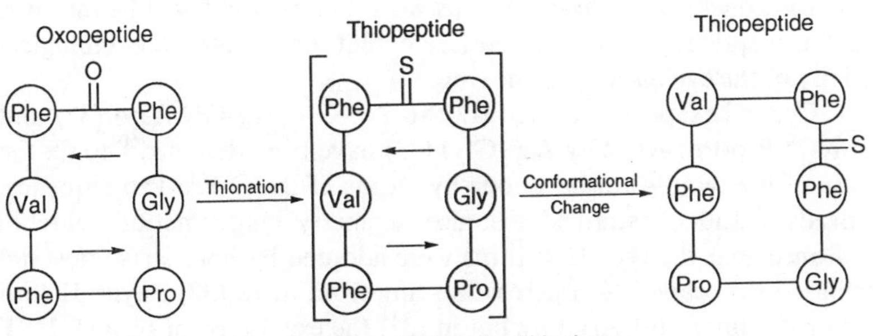

⟶ represent H-bonds within rings

Scheme 4

with Xxx = Yyy = Glu; Xxx = Arg, D-Arg, Yyy = Tyr; or Xxx = Lys, Yyy = Glu. High field nmr techniques suggested that *cyclo* (L-L-Gly-L-L-Gly) sequences generally displayed two conformers in d_6DMSO with the major one being the bis-*cis* conformer, while the minor one contained two β-turns. For *cyclo* (D-L-Gly-L-L-Gly) sequences, one *cis* and one *trans* X-Pro bond was seen, and one type II β-turn as previously predicted from the work of Kopple and others.

Yokoyama's thionation reagent (29) can selectively[56] exchange one CO group at the Phe^5-Phe^6 position of *cyclo* ((Gly^1-Pro^2-Phe^3-Val^4-Phe^5-Phe^6) to give *cyclo* (Gly^1-Pro-Phe^3-Val^4-$Phe^5\Psi$[CSNH]Phe^6). Nmr and restrained molecular dynamics reveal conformation changes in the analogue and changes in the H-bonding pattern as summarised in Scheme 4. Both the oxo- and thio-compounds exist as two slowly exchanging isomers with two separate sets of nmr signals (due to *cis-trans* rotational isomers about Gly-Pro), but the cyclic oxopeptide shows[57] only a weak inhibition of trypanosomal triosephosphate isomerase, a key enzyme in sleeping sickness, while the thio-analogue is one of the most potent inhibitors (IC_{50}[μm] = 6).

The antitumour bicyclic hexapeptide RA-VII (30), currently being tested for clinical use has been isomerised[58] by base to give four derivatives. Two of the derivatives represented isomerisation at positions 3 and 5 which corresponds to the energetically more favourable alternate D and L sequence. In the other two forms Tyr^5 was the only residue isomerised. Only derivatives shown to have cis N-methylated amide between Ala^2-Tyr^3 showed *in vivo* antitumour activity. The original significance of a *cis* bond between Tyr^5-Tyr^6 for antitumour activity does not seem to be an essential requirement from this work. Yet in a comparison[59] of the nmr characteristics of (30) with the N-demethylated analogue (31) both forms seem to show predominantly the *cis*-conformation with the analogue (31) showing twice the potent cytotoxic activity of (30). The conclusion is drawn in this work that it could well be that it is the tetrapeptide part of the molecule that potentiates the biological activity of the cycloisodityrosine ring.

Cyclic hexapeptides *cyclo* (D-Pro-Pro-Gly-Arg-Gly-Asp) (32) and *cyclo* (D-Pro-Pro-Arg-Gly-Asp-Gly) (33) have been designed[60] to fix the phase of the two β-turn backbone by means of the D-Pro-Pro sequence. Nmr data and constrained distance geometry conformation searches confirmed that the type II′ β-turns were adopted by both structures and stabilised the second turn across the ring from it. In (32) a type II turn was predominant for Arg-Gly but in (33) the exact type of turn (I, II II′ and III) around Gly-Arg could not be distinguished. The effect of glycosylation on the conformation of cyclohexapeptide (35) has been

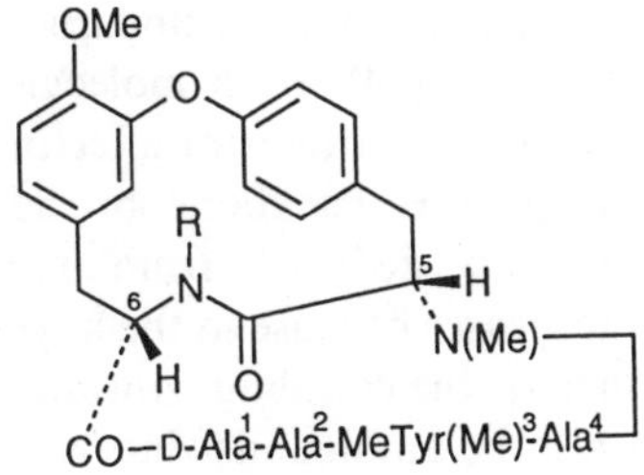

CO—D-Ala¹-Ala²-MeTyr(Me)³-Ala⁴

(30) R = Me
(31) R = H

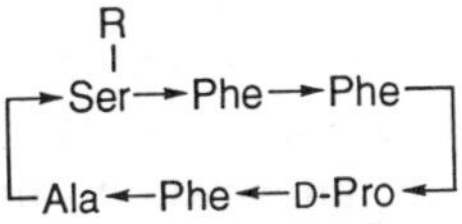

cyclo(Ser(R)→Phe→Phe→D-Pro→Phe→Ala)

(34) R = [disaccharide]
(35) R = H

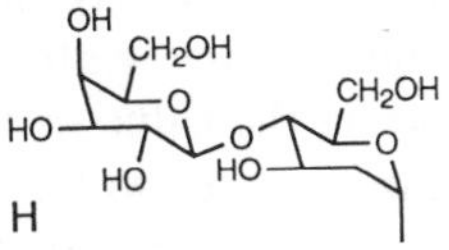

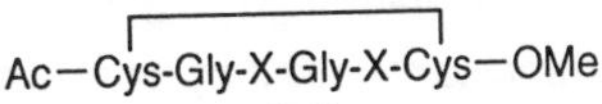

Ac—Cys-Gly-X-Gly-X-Cys—OMe

(36)

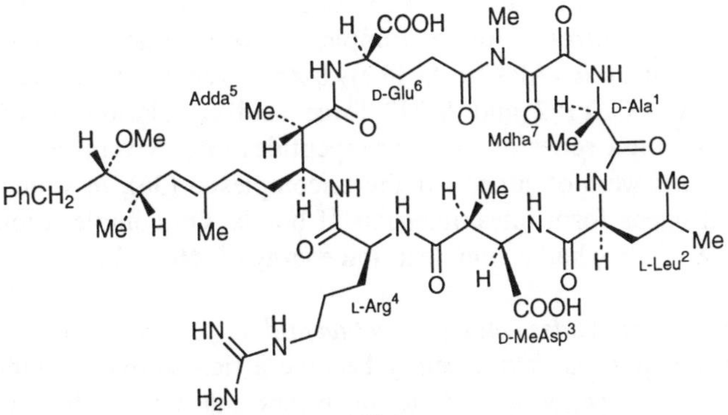

Adda⁵, D-Glu⁶, Mdha⁷, D-Ala¹, L-Leu², D-MeAsp³, L-Arg⁴

(37)

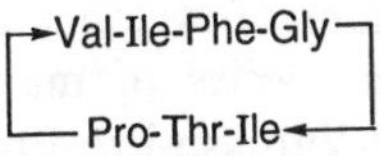

cyclo(Val-Ile-Phe-Gly-Ile-Thr-Pro)

(38)

cyclo(Leu-Pro-Pro-Pro-Ile-Phe-Gly-Gly)

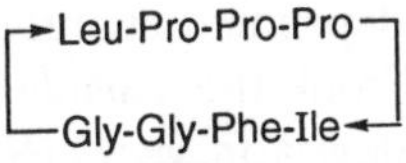

(39)

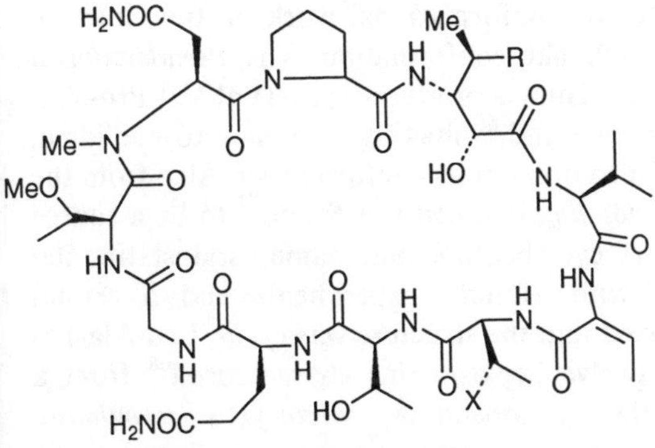

(40) R = $(CH_2)_9CO(CH_2)_3CH_3$, X = OH
(41) R = $(CH_2)_9CO(CH_2)_3CH_3$, X = H
(42) R = $(CH_2)_{11}CH_3$, X = OH

studied[61] by X-ray and nmr techniques of the parent (35) in comparison to its 2-deoxy-D-lactopyranosyl-α-(1-3) derivative (34). Both molecules seem to exist as two conformers, with the O-glycosylation not affecting the conformation or overall shape of the peptide backbone or side chains. The information about conformation deduced from nmr coupling constants has only been used in a few cases because in the larger molecules the line width is often in the range of the coupling constants. Converting the value to a bond angle can also be ambiguous since each coupling constant will have four dihedral angles corresponding to it *via* the Karplus type equation. However, using *cyclo*-[D-Pro-Phe-Phe-Lys(Z)-Trp-Phe] it has been demonstrated[62] that combined use of homonuclear and heteronuclear coupling constants two of the four coupling constants that define the ϕ angle on the peptide backbone can be determined quantitatively. The Local States (LS) method for estimating the entropy for a sample of conformations has been developed[63] for *cyclo* $(Ala\text{-}Pro\text{-}D\text{-}Phe)_2$ both in vacuum and in a crystal environment. ECEPP/2 and MM2 force field calculations have been carried out[64] on a series of cyclic hexapeptides (36). Several low energy conformations were obtained for the uncomplexed (36), in which most amide hydrogens form intramolecular H-bonds. On complexation with Na^+ (36, X = Pro) had a complexation energy of -60 kcal/mol.

2.7 *Cycloheptapeptides and Cyclo-octapeptides.*

Marine sponges have recently become a rich source of important classes of cyclic peptides and depsipeptides but up to this year no cycloheptapeptides had been found. Yet within a few months of one another, three independent characterisations of cycloheptapeptides have been reported. From the *Pseudoxinyssa* species of marine sponge in Borneo came Malaysiatin, *cyclo* (Val-Val-Val-Asn-Pro-Pro-Phe) with an all L (or all D) configuration.[65] 2D Nmr techniques and mass spectrometry contributed greatly to the conformational work on the molecule which suggests a folded, saddle-like conformation. The *Pseudoaxinella massa* sponge from Papua New Guinea produces *cyclo* (Val-Val-Pro-Val-Asn-Pro-Phe) as an all L-compound,[66] making it similar to evolidine, and also possesses a *cis*-Val-Pro bond in its conformation. Also from the South Pacific Ocean the *Stylotella aurantum* was found[67] to be a source of new cell growth inhibitory cycloheptapeptide named stylostatin. 500 MHz Nmr measurements with 2D-nmr experiments and a crystal structure determination proved that the structure was *cyclo* (Leu-Ala-Ile-Pro-Phe-Asn-Ser). Of the twelve hepatotoxins characterised[68] from a Homer Lake bloom of the cyanobacteria, *Microsystis aeroginosa*, *M.viridis* and *M.wesenbergii*, nine of the structures were found to be novel, and found to be analogues of microcystin LR (37). Amongst the new structures wer [DMAdda^5]-microcystin LR, the first to contain a 9-O-demethyl-Adda unit, [Dha^7]microcystin LR without the *N*-methyl on the dehydroalanine residue, microcystin FR, microcystin AR, microcystin M(O)R and microcystin WR with Phe, Ala[Met→O] and Trp residues respectively instead of Leu at position 2. The LD_{50} ranged from 90-800 μg/Kg for these new microcystins. The roots of *Pseudostelleria heterophylla* have yielded[69] both the cycloheptapeptide (38) and the cyclo-octapeptide (39).

Two cyclo-octapeptides, *cyclo* [Lys(Z)-Lys(Z)-Gly-Phe-Phe-X-Sar-X] with X = Gly or Sar and *cyclo* [Glu(Bzl)-Glu-Bzl)-Gly-Phe-Phe-Sar-Sar-Sar] have been prepared[70] from linear precursor using solution phase operations and a water soluble carbodiimide/HOBt in the presence of alkali metal cations for cyclisation. Transport experiments through an organic liquid membrane showed that both cyclic peptides transported H-D-PheOMe-HCl more efficiently than other guest salts. The ion-binding characteristics and conformation of *cyclo* (Ala-Leu-Pro-Gly)$_2$ in a lipophilic solvent (CH_3CN), studied[71] by cd and nmr measurements showed that there was preferential binding of divalent cations Ca^{2+}, Mg^{2+} and Ba^{2+}, with the conformation of the free and complexed cyclopeptide showing well-defined β- and γ-turns.

2.8 *Cyclononapeptides.*

The only report[72] under this category this year is on the [Aib^{5,6}, D-Ala^8] cyclolinopeptide A, *cyclo* (Pro-Pro-Phe-Phe-Aib-Aib-Ile-D-Ala-Val). This was designed as an analogue containing changes at residues away from positions 1-4 which are believed to be involved in cyclolinopeptide A's ability to inhibit uptake of cholate by hepatocytes. Insertion of the Aib and D-Ala constraints effectively quenched the high flexibility of the peptide and lowered its biological activity from IC_{50} of 0.84 μM to 30 μM.

2.9 *Cyclodecapeptides.*

Structural information previously reported for puwainaphycins C and D, which are cardioactive cyclic peptides from *Anabaena* sp BQ-16-1, has now been augmented[73] by data on other congeners in the group. The structures (40-42) have been determined for puiwainaphycins A, B and E respectively. A strong antifungal cyclodecapeptide, calophycin (43) has been extracted and characterised[74] from blue-green alga *Calothridx fusca*. Calophycin is obviously structurally related to the puwainophycins, but its anti-fungal activity is not shown by the latter.

In order to study the interaction between peptides and membranes

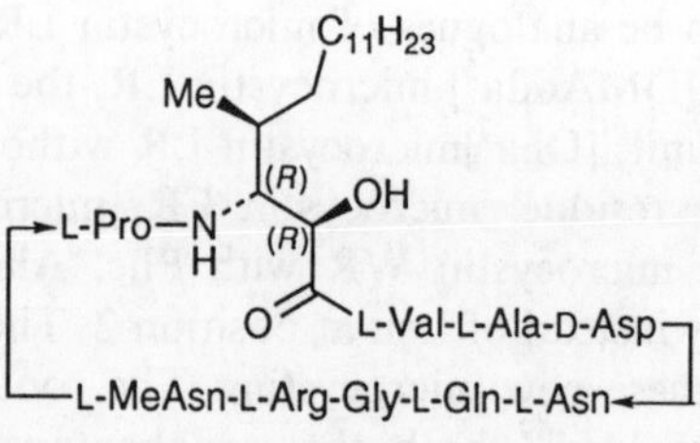

L-Pro–N(H)–(R)-CH[(R)-CH(Me)CH$_2$C$_{11}$H$_{23}$]–(R)-CH(OH)–CO–L-Val-L-Ala-D-Asp–L-MeAsn-L-Arg-Gly-L-Gln-L-Asn (cyclic)

(43)

Ala[7]-D-Ala[8]-MeLeu[9]-MeLeu[10]-MeVal[11]–N(Me)–CH(1)–CO–L-Abu[2]-Sar[3]-MeLeu[4]-Val[5]-MeLeu[6] (cyclic); MeBmt[1] side chain with XCH$_2$, Y, Me

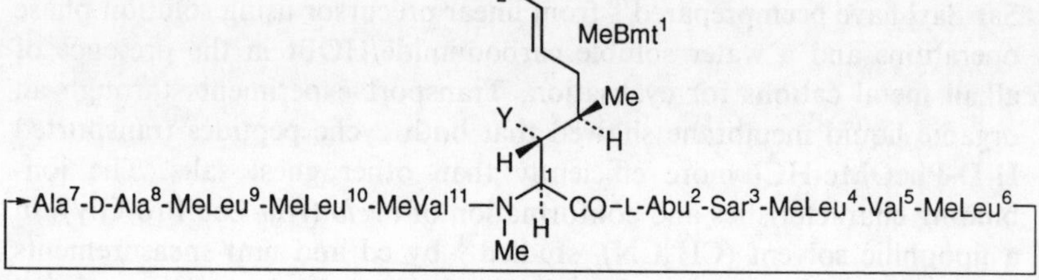

(44) Y = OH, X = H
(45) Y = AcO, X = H
(46) Y = OH, X = OH

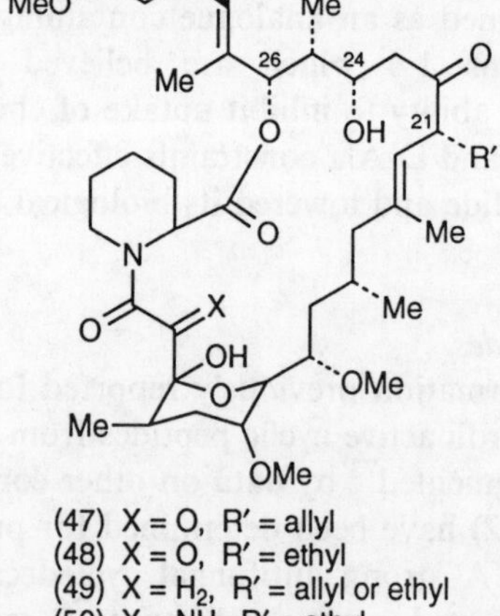

(47) X = O, R′ = allyl
(48) X = O, R′ = ethyl
(49) X = H$_2$, R′ = allyl or ethyl
(50) X = NH, R′ = ethyl

the L-α-aminomyristic acid (Amy) residue with its lipophilic $C_{12}H_{25}$ side-chain has been incorporated[75] into the 3 and 3′ positions of gramicidin S. [Amy$^{3,3'}$]-Gramicidin S showed enhanced affinity for membranes without change in the conformation of the gramicidin S peptide backbone as confirmed by cd and ^{1}H nmr. The *N*-hydroxysuccinimide ester under high dilution was used in the solution phase synthesis. The Kaiser oxime resin has again found utility in the synthesis[76] of [D-pyrenylalanine$^{4,4'}$]gramicidin S with an overall yield of 45% quoted. The pyrenylalanyl residue was incorporated as a fluorescent probe, and it was shown that it did not change the natural conformation of the gramicidin S backbone in MeOH.

2.10 *Higher Cyclic Peptides.*

Activity in this section was again dominated by studies on the immunosuppressants such as cyclosporin A (CsA) (44) and FK506. Structurally, it is difficult to justify discussion of the latter in the present context, but both tend to be associated closely with one another as probes of cellular functions[77] and the identification of a new family of protein receptors - the immunophilins. A model of the cyclophillin/cyclosporin A complex has been generated[78] from the nmr-determined bond conformation of cyclosporin A (published 1991) and the X-ray structure of free cyclophilin (also published 1991). Consistent with previous nmr results and structure/activity relationships the model shows that the MeLeu^9-MeLeu^{10}-MeVal^{11} and MeBmt^1 portion of CsA binds to the cyclophilin.

The main metabolite of CsA (44) in animals and humans is the allylic alcohol (46). This has now been synthesised[79] by treatment of acetyl derivative (46) with *N*-bromosuccinimide followed by hydrolysis. Other metabolites formed by oxidation at positions 4 and 9, namely [γ-(OH)MeLeu^9]CsA and [Leu^4, γ(OH)MeLeu^9]CsA have also been synthesised[80], using aspartic acid as the starting point for the γ(OH)MeLeu residues. The D-Ala^8 position has also been modified[81] to a [D-Ala^8]CsA by treatment of [2-deutero-3-fluoro-D-Ala^8]CsA with $LiN(CHMe_2)_2$ in THF at low temperature. Analogues with D-Ala side-chain replaced by $CH_3(SOMe)$ and $CH_3(SCH_2CH_2OH)$ had immunosuppressant activity comparable to that having the F(CHO substituents instead of D-Ala^8. Novel conjugates of cyclosporin A have been prepared[82], *via* a modified MeBmt^1 side chain. Acetylated derivative (45) was cleaved at the double bond using periodate/permanganate to form a carboxyl group at one end of the double bond, which was subsequently reacted with 5-aminoacetamide-fluorescein and poly(L-lysine) in the presence of 1-ethyl-3-[3-(dimethylamino)propyl]carbodiimide to give the novel conjugates.

Two reports confirm that the crucial conformation of the N-methylated bond between $MeLeu^9$ and $MeLeu^{10}$ changes to *trans* in the presence of LiCl solution. Isotope-edited nmr techniques revealed[83] this conformation when CsA was dissolved in d_8-THR containing 0.47M LiCl. In a separate study[84] on lithium-complexed CsA, using nOe build-up rates at 600 MHz, and restrained molecular dynamics calculations, it was shown that the complex had all its transannular H-bonds disrupted as well as the flip-over of the $MeLeu^9$-$MeLeu^{10}$ bond to *trans*. No transannular H-bond was observed in the complex, although a new βII'-turn-like structure was observed between residues 7 and 10 without a real H-bond between residues i and i + 3. Features in the conformation of the complex were very similar to that found when CsA is bound to the cyclophilin receptor, which could also explain why addition of LiCl leads to a significant increase in peptidyl *cis-trans* isomerase (PPIase) inhibition. Evidence has also been obtained[85] from nmr measurements that Ca^{2+} and Mg^{2+} form 1:1 complexes with CsA in acetonitrile solution, with conformations very different from the uncomplexed form. As plasma has readily available Ca^{2+} and Mg^{2+} ions, these complexes could be relevant to the active forms of CsA. Most reports on the conformation of CsA in non-polar solvents reflect the same features as the one observed in $CDCl_3$. It has now been shown[86] that the same major conformation exists in 50% aq.MeOH.

Modifications to the backbone structures of FK-506 (47) and its 21-ethyl analogue, ascomycin (48) have involved: (a) removing the carbonyl group from the binding domain as in (49) to yield compounds about 4 times weaker than the parent compound[87], (b) forming the imine instead of the 9-CO group, as in (50) or replacement of C-10 hydroxyl by an amino group, both conversions not affecting activity[88], and (c) the isolation[89] of an isomer of FK-506 from *Streptomyces tsukubaenis* where the lactone ring has moved over to the 24-OH position, a migration which can be mimicked *in vitro* in the presence of imidazole.

The blue-green alga *Anabaena laxa* FK-1-2 have yielded[90] a mixture of antifungal and cytotoxic cyclopeptides given the family name laxaphycins. Nmr and mass spectral studies showed that the two major constituents A and B had structures (51) and (53) while the minor constituents O and E were allocated structures (54) and (52) respectively. Laxaphycins A and B displayed an interesting biological synergism, in that when A was tested on its own it was inactive and B had greatly diminished activity. Yet as a mixture they showed a broad spectrum of antifungal activity. The tropical cyanobacterium *Hormothamnion enter-*

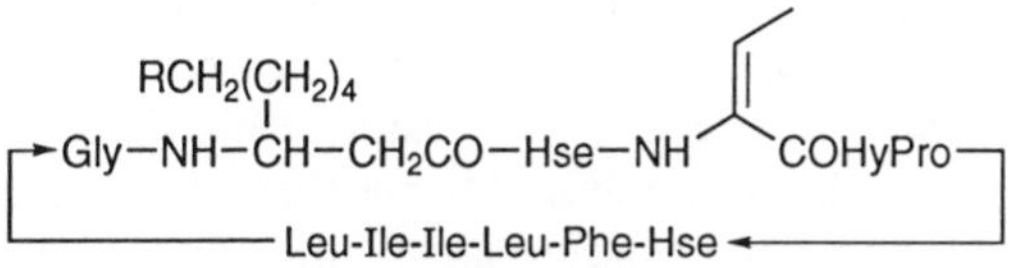

RCH$_2$(CH$_2$)$_4$
→Gly—NH—CH—CH$_2$CO—Hse—NH COHyPro
Leu-Ile-Ile-Leu-Phe-Hse←

(51) R = H
(52) R = Et

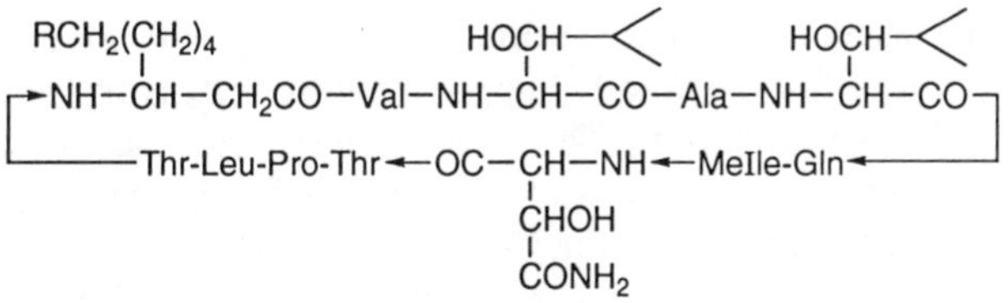

RCH$_2$(CH$_2$)$_4$ HOCH HOCH
→NH—CH—CH$_2$CO—Val—NH—CH—CO—Ala—NH—CH—CO
Thr-Leu-Pro-Thr←OC—CH—NH←MeIle-Gln←
CHOH
CONH$_2$

(53) R = Et
(54) R = H

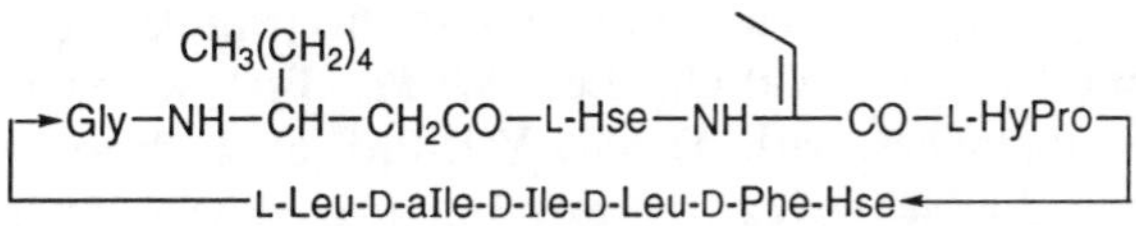

CH$_3$(CH$_2$)$_4$
→Gly—NH—CH—CH$_2$CO—L-Hse—NH—CO—L-HyPro
L-Leu-D-aIle-D-Ile-D-Leu-D-Phe-Hse←

(55)

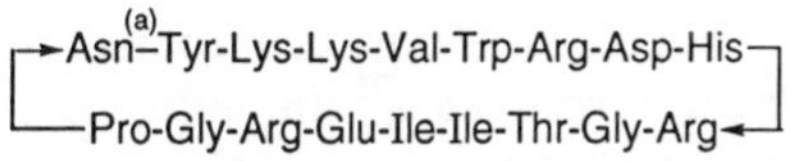

→Asn$^{(a)}$-Tyr-Lys-Lys-Val-Trp-Arg-Asp-His
Pro-Gly-Arg-Glu-Ile-Ile-Thr-Gly-Arg←

(56) (a) represents cyclization position

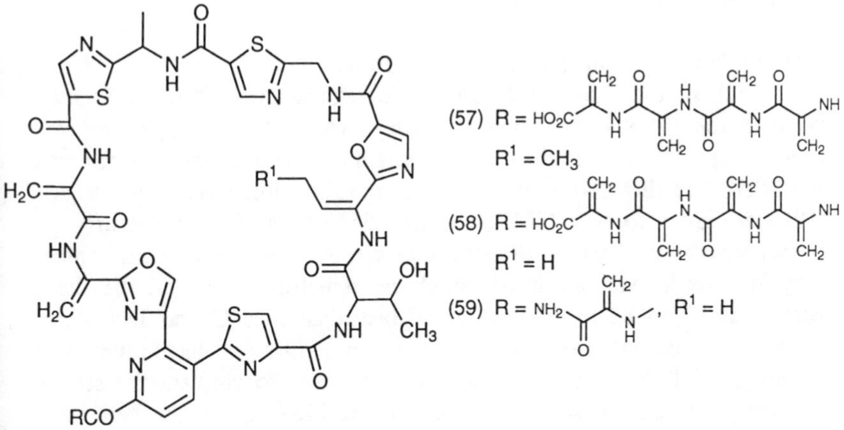

O N S N H O S N N H O O N O NH R^1 N H_2C O HN NH O OH O H_2C N S N NH CH_3 N N O RCO

(57) R = HO_2C CH$_2$ O N H CH$_2$ H N O CH$_2$ O N H CH$_2$ NH ; R^1 = CH$_3$

(58) R = HO_2C CH$_2$ O N H CH$_2$ H N O CH$_2$ O N H CH$_2$ NH ; R^1 = H

(59) R = NH_2 O CH$_2$ N H , R^1 = H

omorphoides produces a mixture of cytotoxic and antimicrobiol substances, and the most lipophilic component hormothamnin A has been deduced[91] to have structure (55) as a result of high field nmr, FAB-MS and chemical degradative studies. Hormothamnin A can be seen to possess the same backbone residues as laxaphycin A (51) yet individual residues differ in their stereochemistry.

An 18-residue cyclic peptide (56) corresponding to a loop involved in the curaremimetic action of a snake toxic protein has been synthesised[92] using solid phase techniques, starting with attaching Boc-Asp-OFm (Fm = fluorenylmethyl gp) to a methylbenzhydryl amine resin *via* its β-carboxylic function. The BOP reagent was used to close the cyclic ring between the Tyr and the α-COOH of the Asp-residue. Treatment with HF converted the C-terminal Asp to Asn, while cleaving off the peptide from the resin. The complete structure of the 62-residue peptide, eclosion hormone (EH), isolated from *Manduca sexta*, has been shown[93] to have three disulfide bonds at positions Cys^{14}-Cys^{38}, Cys^{18}-Cys^{34} and Cys^{21}-Cys^{49}. When synthetic cyclic peptides related to the antibiotic gratisin, *cyclo* $(Val-Orn-Leu-Phe-Pro-Tyr)_2$ were analysed[94] on hplc, some analogues with D-X-D-Y-L-Pro or L-Pro-D-X-D-Y sequences gave double peaks, representing two slowly interconvertible conformers in equilibrium with each other at low temperature. A ring structure consisting of twelve residues was necessary for this phenomenon to be noticed. The constitution and configuration of R_o09-0198 (cinnamycin) has been investigated[95] in different environments. The conformation was drastically changed in going from water solutions to SDS micelles. The molecule oriented itself into the lipid bilayers with its hydrophobic part inside the bilayer.

2.11 *Peptides containing Thiazole Type Rings.*

Each of the thiopeptide antibiotics characterised[96] from *Streptomyces gardneri* has been shown to have a large monocyclic peptide core to which is attached a multiple Δ-Ala chain of variable length. Antibiotic A,10255B was shown to have the structure (57), while structures (58) and (59) were given to A,10255G and A,10255J respectively. Structure (57) shows several similarities to berninamycin A, sulfomycin I and thioxamycin. The heavily modified backbone structures (60) and (61) have been found[97] to be representative of bistralsamides C and D isolated from the ascidian *Lissoclinum* bistratum. 2D-Nmr techniques and ozonolysis followed by derivatisation were used to confirm the stereochemistry of each residue. The oxazole residues have not been seen previously in this genus of ascidians. Molecular dynamics simulations predict a planar structure for (60) as a result of the constraints imposed

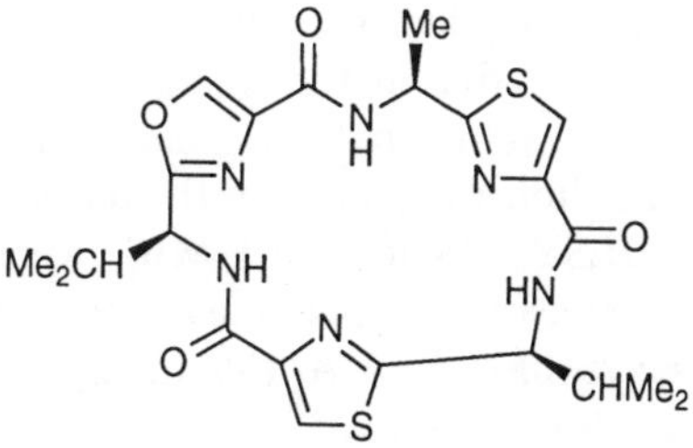

(60)

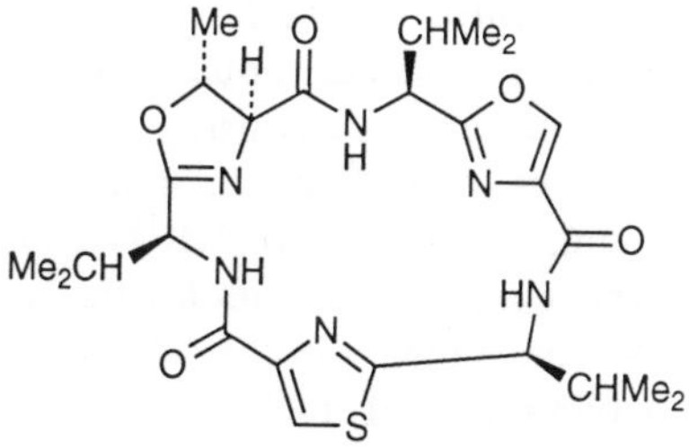

(61)

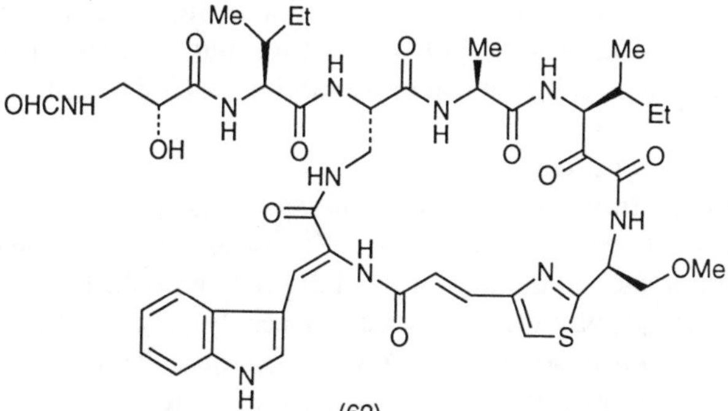

(62)

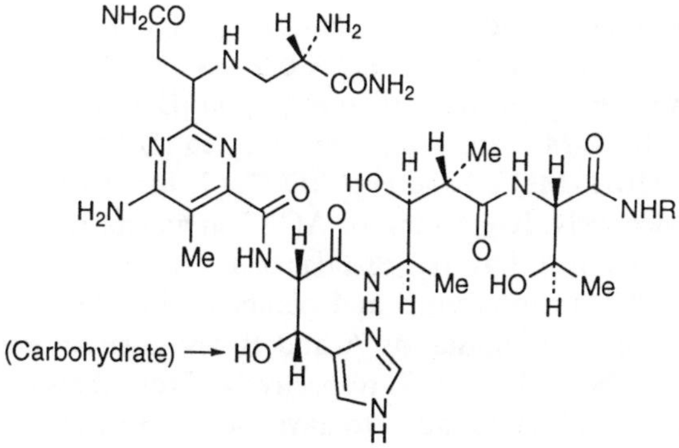

(Carbohydrate) → HO

(63) R =

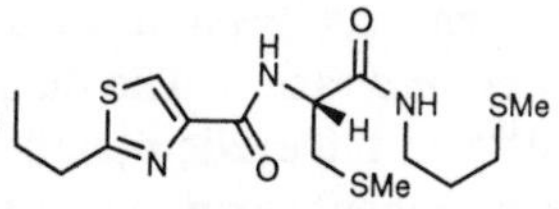

(64) R =

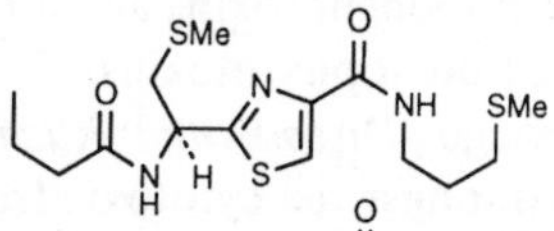

(65) R =

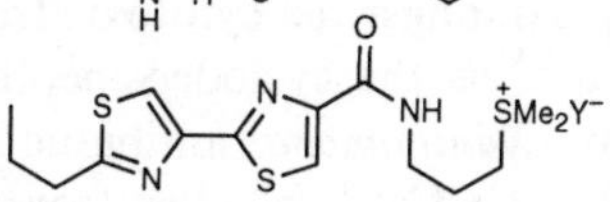

by the three 'aromatic' residues. These cyclopeptides showed depressant effects in the i.c. mouse assay but were inactive in the human colon tumour cytotoxicity assay. *(R)*-Isoseine and (*O*-methylseryl)thiazole are features found[98] for the first time in keramimide F (62) from the marine sponge *Theonella* sp. ^{1}H-^{1}H COSY, NOESY, HSQC and HMBC were used as nmr techniques for the structural elucidation. The role of the two thiazole moieties in the side-chain of deglycobleomycin A_2 (65) in oxygen activation and DNS degradation has been explored[99] through the synthesis of two mono-thiazole containing analogues (63) and (64) with S-methyl-L-cysteine *in lieu* of each thiazole in turn. Both (63) and (64) proved to be good catalysts for oxygenation but were less effective than bleomycin or deglycobleomycin in promoting DNA degradation.

2.12 *Cyclodepsipeptides.*

A richer trawl than usual of novel depsipeptide structures have been reported this year. Space does not allow discussion of the structural elucidation of each in detail, but most approaches reflect the routine use of sophisticated physical methods, especially nmr techniques, mass spectrometry and in some cases, X-ray techniques. This précis therefore cannot really reflect the effort involved in unravelling these diverse structures.

Enniatins D, E and F from *Fusarium* sp. FO-1305 detected and identified[100] as inhibitors of acyl-CoA: cholesterol acyltransferase (ACAT) have been given the structures, *cyclo* [D-Hiv-L-MeLeu-D-Hiv-L-MeVal-D-Hiv-L-MeVal], a mixture of *cyclo* [D-Hiv-L-MeIle-D-Hiv-L-MeLeu-D-Hiv-L-MeVal] and *cyclo* [D-Hiv-L-MeLeu-D-MeIle-D-Hiv-L-MeIle] respectively. IC_{50} vaues for ACAT in an enzyme assay were 87, 57 and 40 μM, respectively. *Pseudomonas syringae pv syringae*, an organism which damages crops and plants worldwide has yielded two main components, syringostatins A and B whose structures have been elucidated[101] to be (66) and (67) respectively. From the same organism, syringomycin 1 has been found[102] to have the structure (68). The cause of leaf spot disease in corn has been associated with a BZR-cotoxin produced by *Bipolaris zeicola* race 3. The main component of this toxin has been characterised[103] as the cyclic nonadepsipeptide (69), known as BZR-cotoxin II. The component toxins again showed little activity but in combination they were potent phytotoxins.

The marine sponge *Discodermia kiiensis* has afforded several metabolites already. Amongst the cytotoxic fractions, discokiolides A-D have been identified[104] as the cyclodepsipeptides (70)-(73) respectively. Xanthostatin, an anti-*Xanthomonas* antibiotic has been shown[105] to have structure (74), while the Ca^{2+} blocker from *Hapsidospora irregularis*,

OH
Me(CH2)nCHCHCH2CONHCH—NH—CH—CONH—CH—CONH—CH—CO
R1
R (CH2)2NH2 (CH2)2Y
CH2
O
CO
NH
CH—(CH2)3NHB
CO
R2CH2CHOH—CH—NHCO—CH—NH—CO NH—(A)
HO2C—CHOH
Me

(66) $n = 9$, $R = CH_2CH_2NH_2$, $R^1 = H$, $R^2 = Cl$, $Y = OH$, $B = H$, $A = Thr$
(67) $n = 9$, $R = CH_2CH_2NH_2$, $R^1 = OH$, $R^2 = Cl$, $Y = OH$, $B = H$, $A = Thr$
(68) $n = 7$, $R = CH_2OH$, $R^1 = H$, $R^2 = Cl$, $Y = NH_2$, $B = C{=}NH(NH_2)$, $A = Phe$

Me—CHCH2OH
CH2CHMe2 CH2
CH2CHMe2
O—CH—CONH—COO CONMe CO
CO
NMe
Me—CH
CHCHMe2
Me—N—CO—N—CO OCO—CH—NH—CO
CHMe2
Me OH

(69)

Me
Ph
O
H
RCH2
N
CH
N
Me
CH
O
Δ8
Δ7
Me
CO2H O
O
N
Me
O
MeIle-Asn-Pro—NH—CH—CO
CH—R1
Ph

(70) $R = H$, $R^1 = MeO$
(71) $R = Me$, $R^1 = MeO$, Δ^8
(72) $R = Me$, $R^1 = H$
(73) $R = Me$, $R^1 = MeO$, Δ^7

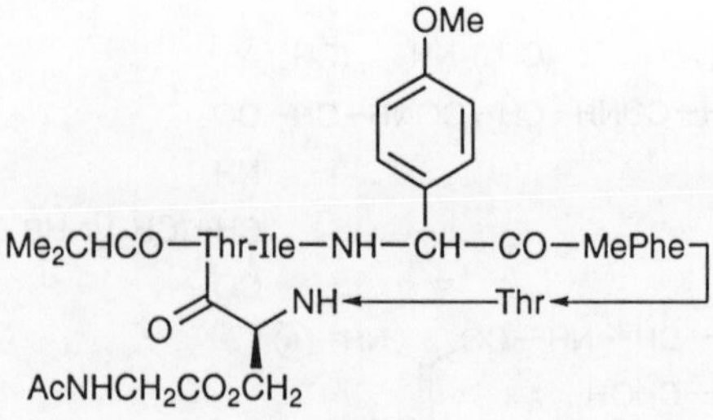

OMe

Me_2CHCO—Thr-Ile—NH—CH—CO—MePhe
NH←Thr
O
$AcNHCH_2CO_2CH_2$

(74)

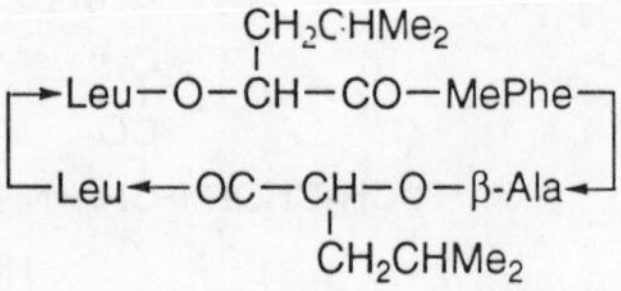

CH_2CHMe_2
→Leu—O—CH—CO—MePhe
Leu←OC—CH—O—β-Ala←
CH_2CHMe_2

(75)

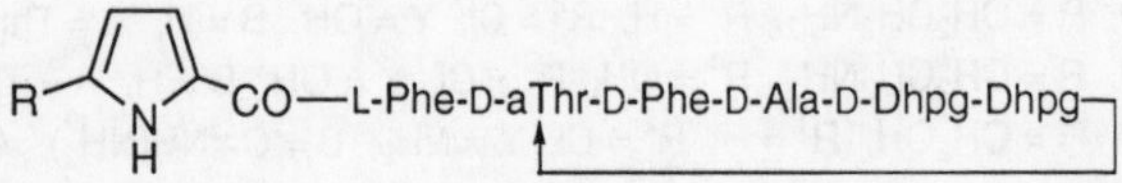

R—(pyrrole, NH)—CO—L-Phe-D-aThr-D-Phe-D-Ala-D-Dhpg-Dhpg

(76)

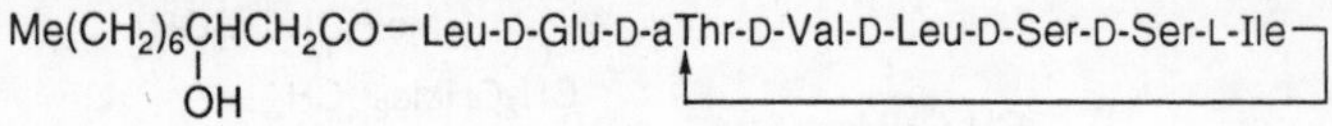

$Me(CH_2)_6CH(OH)CH_2CO$—Leu-D-Glu-D-aThr-D-Val-D-Leu-D-Ser-D-Ser-L-Ile

(77)

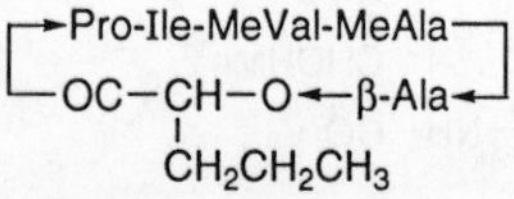

→Pro-Ile-MeVal-MeAla
OC—CH—O←β-Ala←
$CH_2CH_2CH_3$

(78)

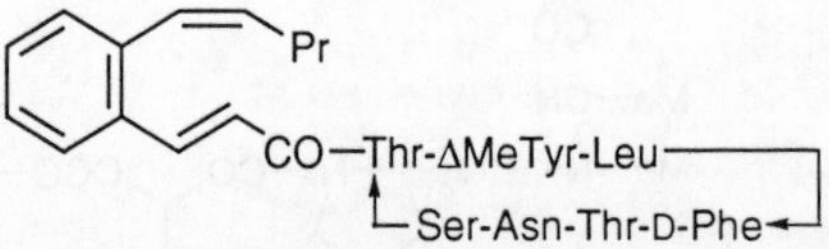

Pr
CO—Thr-ΔMeTyr-Leu
Ser-Asn-Thr-D-Phe←

(79)

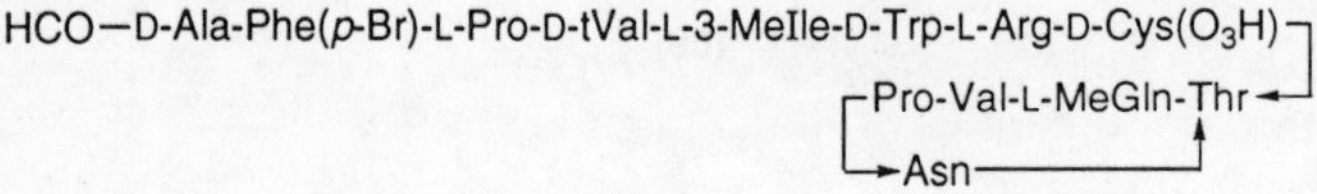

HCO—D-Ala-Phe(*p*-Br)-L-Pro-D-tVal-L-3-MeIle-D-Trp-L-Arg-D-Cys(O_3H)
Pro-Val-L-MeGln-Thr←
→Asn

(80)

H—Ile-Dhb-D-Ala-Ile-Dha-Leu-Ala[7] (S bridge) — D-Abu-Pro-Gly-Ala-Lys[12] (S bridge)—
Segment I Segment II

—D-Abu[13]-Gly-Ala-Leu-Met-Gly-Ala-Asn-Met[21] (S bridge) — Lys-D-Abu-Ala-D-Abu-Ala-His-Ala[28] (S bridges)
Segment III Segment IV

—Ser[29]-Ile-His-Val-Dha-Lys[34]—OH
Segment V

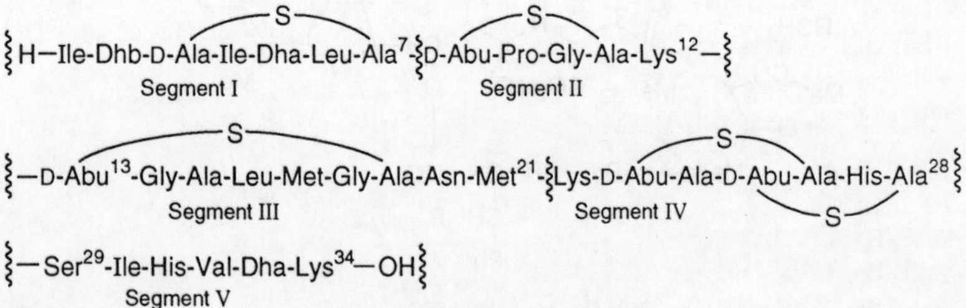

(81)

Ala-S-Ala = *meso*-lanthionine
Abu-S-Ala = *threo*-methyllanthionine

named leualacin, has been proved[106] to be a cyclic-didepsipeptide (75). Endothelin antagonists from *Microbispora* sp. ATCC55140, the cochinmicins, have all been identified[107] as having the same basic sequence but the individual congeners differ in the stereochemistry of 3,5-dihydroxyphenyl glycine (Dhpg) and whether or not the pyrrole ring is substituted by Cl. The general formula (76) summarises the main structural backbone. The white-line-inducing principle from *Pseudomonas reactans* has been shown[108] by crystallography to have the structure (77). The fungus *Metarhizium anisopliae*, grown on media containing α-amino-4-pentenoic acid or norvaline, produced[109] dihydrodestruxin A (78), while *Streptomyces violaceus-niger* produced[110] WS9326A, a novel tachykinin antagonist with the structure (79). Polydiscamide A, from the marine sponge *Discodermia*, has been shown[111] to have quite novel features, e.g. 3-methylisoleucine, in its structure (80).

Valinomycin *cyclo* [(L-Val-D-Hyv-D-Val-L-Lac)$_3$] still commands a great deal of interest. The crystal structure of the valinomycin monohydrate crystallised from aqueous dioxan, contained[112] two independent molecules, both of the octahedral cage conformation previously described, with the cages containing a trapped water molecule. The valinomycin analogue *cyclo* [(D-Val-Hyi-Val-D-Hyi)$_3$], crystallised from aqueous dioxan showed[113] a conformation analogous to one established for free meso-valinomycin from other organic solvents. It is a bracelet conformation stabilised by six intramolecular 4→1 type H-bonds. The smaller synthetic analogue *cyclo* [(D-Val-L-Hyi-L-Val-D-Hyi)$_2$], a cyclo-octadepsipeptide showed[114] by X-ray analysis, a different conformation depending on whether it had been crystallised from dioxan or chloroform - the more polar the solvent of crystallisation the fewer intramolecular H-bonds formed. Table 3 summarises the relationships deduced between the solvent of crystallisation and the number of H-bonds in dodeca- and octadepsipeptides.

Table 3

Compound	Crystallisation Solvent	Intramolecular Bond Type 4→1	5→1	3→4
(D-Val-L-Hyi-L-Val-D-Hyi)$_2$	dioxan			2
(D-Val-L-Hyi-L-Val-D-Hyi)$_2$	$CHCl_3$	4		
(D-Ile-L-Lac-L-Ile-D-Hyi)$_2$	DMF			2
(L-Val-D-Hyi-D-Val-L-Lac)$_3$	DMSO	2		
(L-Val-D-Hyi-D-Val-D-Lac)$_3$	octane/acetone or EtOH/H_2O	4		2
(D-Val-L-Hyi-L-Val-D-Hyi)$_3$	DMF/PetEther	6		
(D-Ile-L-Lac-L-Ile-D-Hyi)$_3$	heptane	5	1	
(D-Ile-L-Lac-L-Ile-D-Hyi)$_4$	EtOH/H_2O	8		(ref. 115)

The hexadecaisoleucinomycin analogue *cyclo*-(D-Ile-L-Lac-L-Ile-D-Hyi)$_4$ crystallised from ethanol/water, when investigated[115] by X-ray techniques revealed an elongated bracelet form with a 2-fold axis of pseudo-symmetry, stabilised by eight 4→1 H-bonds. This larger valinomycin analogue has the capacity to bind larger ions such as Cs^+, tetramethylammonium and acetylcholine.

Valinomycin has undergone detailed analysis by a number of mass spectrometric techniques[116]. Four sector tandem ms of the molecular and quasi molecular ions were compared. FAB tandem mass spectra of $[M + Na]^+$ and $[M + K]^+$ ions showed greater degrees of fragmentation than the $[M + H]^+$ ions. A separate field-desorption mass spectral study[117] of valinomycin has also been reported.

In order to add proof to a previously published empirical Yule (Kato et al.) stating 'that ester bonds in cyclic tetradepsipeptides always take the *trans* form', the cyclodepsipeptide *cyclo* (L-Ala-L-2-hydroxy-3-methylbutanoyl)$_2$ was synthesised[118] and subjected to both nmr studies and theoretical calculations. Both ester and peptide bonds in the molecule take distorted *trans* conformations, with the esters showing larger distortions to compensate for the strain of the 12-membered cyclic backbone. Due to this probable strain in the all-L configuration only a 4.7% yield of the compound was obtained when the final depside link was made from its linear precursor using the water soluble carbodiimide in the presence of DMAP.

2.13 *Cyclic Peptides containing Other Non-Protein Ring Components.*

Just a few lines in a Report of this kind does not do justice to the effort and tenacity required to undertake a total synthesis of a demanding natural product structure. Such is the case in the total synthesis[119] of the food preservative nisin (81) by the Shiba group. The strategy involved segment condensation in the solution phase at the points indicated in the structure (81).

The naturally occurring resin and ACE inhibitors, lyciumins A (82) and B (83) have a unique ring structure in that both N atoms of tryptophan are incorporated in the ring. Their synthesis[120] *via* a linear precursor involved cyclisation using pentafluorophenyl ester activation at C-terminal glycine to link with *N*-terminal serine. The side-chains, Pyr-Pro-Tyr and Pyr-Pro-Trp- were then added as the last stage using TBTU as coupling reagent. The thrombin inhibitor cyclotheonamide B (84) has been synthesised[121] and as a result the stereochemistry has been re-assigned as given in (84). Final cyclisation to make the ring was carried out between the carboxyl group of proline and the argininyl amino group using pentafluorophenyl ester for activation. The details have also been

(82) R = Pyr-Pro-Tyr-
(83) R = Pyr-Pro-Trp-

(84)

(85)

(86)

(87)

reported[122] on the synthesis of the bottom segment of (84), from the diaminopropanoic acid residue through to arginine. In this Chapter last year, preliminary details were included on the synthesis of biphenomycin B which contains a biphenyl unit. Full details of the synthesis have now appeared[123], together with the total synthesis of biphenomycin A which also relied on a linear pentafluorophenyl ester for the cyclisation stage. This mode of activation also found favour[124] in activation of the hydroxy-proline carboxyl to form the rigid 14-membered ring in the peptide alkaloid nummularine F (85). Cyclisation was carried out on a silylated saturated precursor of (85), which could be eliminated at the final stage to generate the enamide double bond. The highly toxic amatoxins, from amanita mushrooms, bind strongly to RNA polymerase II, thus preventing transcription of DNA to hn-RNA. The side chains at positions 6, 2 and 3 have been implicated in the binding process, and to clarify the role of the L-Ile side chain at position 3, analogues of the *S*-deoxoamaninamide form of (86) have been investigated[125]. The L-*allo* Ile^3, *(2S,3R)*- or *(2S,3S)*-2-amino-4-hydroxy-3-methylbutyryl^3, or D-Ile^3 anlogues all gave weaker affinities for RNA polymerase II, thus pointing to the importance of the *(R)*-configuration at the β-carbon atom in position 3.

Attempts at improving the poor oral absorption and short duration of activity of renin inhibitors have justified an assessment of cyclic analogues such as (87)[126] and (88)[127]. Although (87) was custom-built from molecular modelling, it displayed poor binding affinity, but as the solution structures derived from nmr data differed in H-bonding characteristics from the model, it has been used as a possible explanation for its ineffectiveness. In the synthesis of (88), macrocyclisation was achieved either by treatment of a protected linear precursor with water soluble carbodiimide/HOBt in THF (30-70% yield) or by using the same diimide with DMAP using a syringe pump technique, and refluxing methanol-free chloroform for 20 hrs (55-65% yield). An analogue of (88) with a 2*(R)*-t-butylsulfonylmethyl-3-phenylpropionyl side chain instead of the Phe has provided an entry into a potent class (IC_{50} = 5nM) of macrocyclic renin inhibitors. A variation on the theme of naturally-occurring vinylogous polypeptides to yield an alternative peptide back-bone has been explored through the synthesis[128] of (89).

The Papua New Guinea sponge *Theonella swinhoe* Gray has produced motuporin (90), which inhibits protein phosphatase-1 at < 1 n molar activity level. High field nmr techniques have shown[129] that its structure corresponds to (90) and is believed to be the first cyclic pentapeptide of this type from a marine source. X-Ray crystallo-graphy[130] has been instrumental in elucidating the structure of ustiloxin

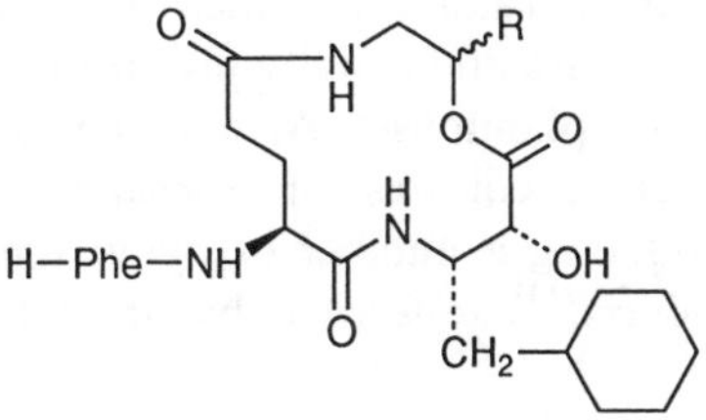

(88)

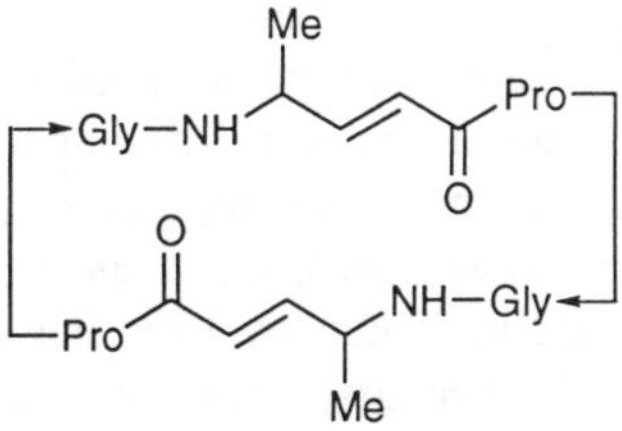

(89)

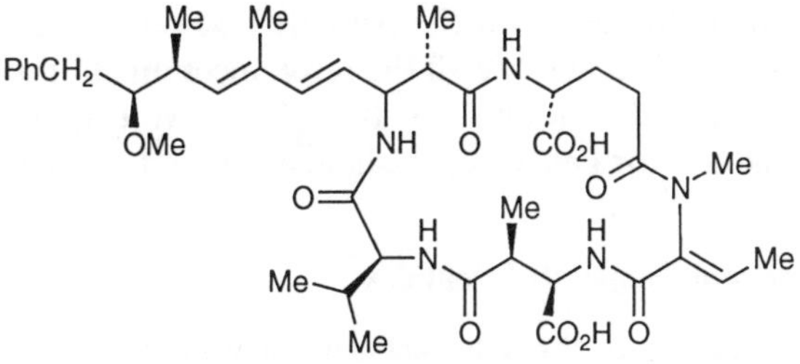

(90)

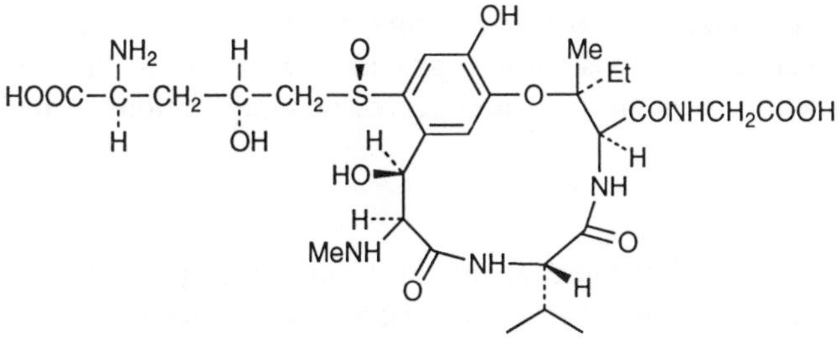

(91)

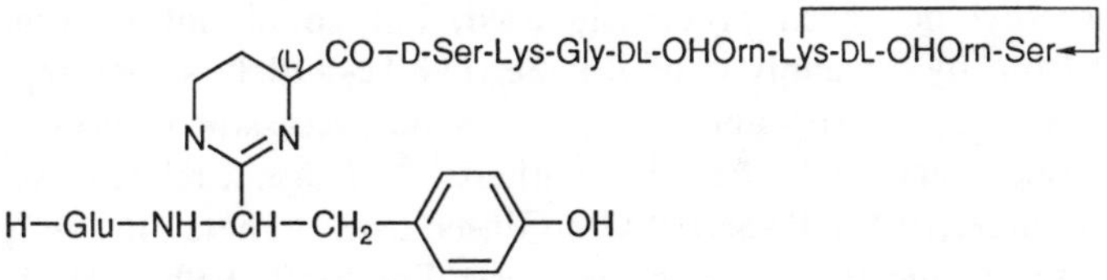

(92)

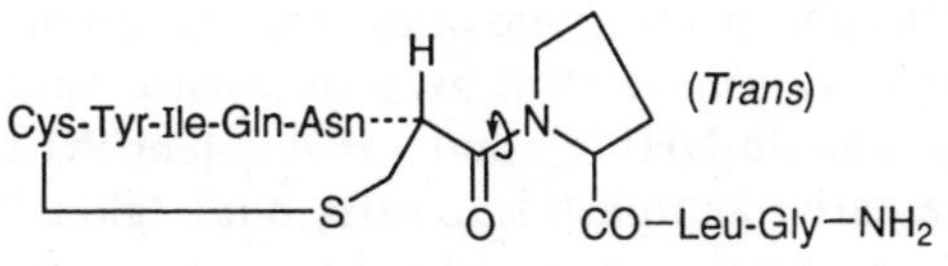

(93)

(91) from a pathogen, *Ustilaginoidea virens* Takahishi, which produces false smut balls on rice panicles. The structure and activity profile of the compound are very similar to those of phomopsin A, a mycotoxin known to cause lupinosis. Iron-deficient cultures of *Pseudomonas fluorescens* ATCC13525 as well as producing a number of pyoverdins, azobactins and azoverdin, has also produced[131] desferriferribactin (92), a possible precursor of the pyoverdins.

Previous investigations of the conformations of oxytocin, Arg-vasopressin or Leu-vasopressin (93) have only detected the presence of a *trans* Cys^6-Pro conformer, with the possible cis-isomer obscured. It has now been possible[132] to identify all the resonances due to the *cis*-form in the nmr spectra and amounts to 10% for oxytocin, 9% for Arg-vasopressin and 8% for Leu-vasopressin in aqueous solutions.

3 Modified and Conjugated Peptides

Regular readers of this Chapter will have perceived that the usual sub-heading and sub-division at this juncture have been changed or more accurately, combined. Since most of the peptide backbone modifications and surrogate insertions are now being exclusively covered in Chapter 3, a more rational sub-division is to provide a platform to bring together peptides whose structural features represent modifications in their side-chains.

3.1 *Phosphopeptides.*

The potential application in the food industry of oligophosphoserine sequences as found in bovine casein, has justified detailed studies towards the goal of suitable phosphoserine derivatives. For use in peptide synthetic strategies. Boc-Ser(PO_3Tc_2)OH, where Tc = 2,2,2-trichloroethyl has been previously used, but could not be prepared readily from hydrogenolysis of Boc-Ser(PO_3Tc_2)OBzl. A new report[133] reveals the last hydrogenolysis stage can be successful in acetic acid/trifluoroacetic acid (1:1). For the synthesis[134] of casein related, Ac-Glu-Ser(P)-Leu-Ser(P)-Ser(P)-Ser(P)-Glu-Glu-NHMe (where P = phosphate), the serine derivative chosen was Boc-Ser(PO_3Ph_2)OH. It was incorporated using a mixed anhydride coupling schedule, with final deprotection of the phosphate group using hydrogenation in the presence of platinum oxide. A similar strategy proved successful[135] in making H[Ser(PO_3H_2)]$_3$-Glu-Glu-NHMe and H-Asp-[Ser(PO_3H_2)]$_2$-Glu-Glu-NHMe. The alternative approach is to carry out a 'global' phosphorylation of the residues after the sequence has been constructed. Thus H-Ala-Thr(PO_3H_2)-Tyr(PO_3H_2)-Ser-Ala-OH was synthesised[136] using Fmoc-strategy on a solid phase resin but incorporating Thr and Tyr residues as Fmoc derivatives without side-chain hydroxyl protection. Global bis-phosphorylation of the peptide resin was carried out by treatment with di-*tert*-Bu^t,*N,N*-diethylphosphoramidite/^{1}H-tetrazole followed by *m*-chloroperoxy benzoic acid. Release of modified peptide from the resin using 5% anisole/TFA gave the di-phosphorylated peptide in high yield and purity. When fewer equivalents of the phosphorylating agent was used it was found that the Tyr residue phosphorylated before the threonyl. Fmoc-Tyr[$PO_3(CH_2Ph)_2$]-OH was the derivative of choice in the synthesis[137] of the triphosphorylated peptide, H-Gly-Asp-Phe-Gly-Met-Thr-Asp-Ile-Tyr (PO_3H_2)-Glu-Thr-Asp-Tyr (PO_3H_2)-Tyr (PO_3H_2)-Arg-Lys-Gly-Gly-Lys-Gly-OH which corresponds to the autophosphorylation site of the insulin receptor. Phosphorylation appeared to give a rigidification of the peptide backbone when the tyrosyl side chains bore negative charges. In a comparison[138] between multi-phosphorylation using the 'global' and 'building block' approaches for the insulin receptor (1142-1153) dodecapeptide fragment, H-Thr-Arg-Asp-Ile-Tyr-Glu-Thr-Asp-Tyr-Tyr-Arg-Lys-OH, the building block approach yielded the better results. This was in contrast to a report last year where monophosphorylation had been better using the 'global' approach. A very useful O-phosphotyrosine analogue is the enzymatically and chemically stable *p*-($CH_2PO_3H_2$)Phe residue. For studies[139] on the 344-357 sequence of the β_2-adrenergic receptor the Phe analogue was introduced as Boc-*p* ($CH_2PO_3Et_2$)Phe-OH. Deprotection of the diethylphosphate was carried out by trimethylsilylbromide, which gave both the fully and partially deprotected peptides, the latter being useful for increased permeability of cellular membranes. The mass spectra of Boc-Ser(PO_3R_2)-OH derivatives where R = Me,Et,Bu^t,Ph and CH_2PH have been studied[140] and an efficient phosphite derivatisation[141] of Boc-Ser-OMe with $Cl_3CMe_2COPCl_2$ and a glycerol derivative followed by oxidation, gave H-Ser[P(O)OH-O-CH_2CHOH-$CH_2O(CH_2)_5CO_2H$]-OH, a phosphoserine ether glycero-lipid.

There is as yet no short and easy way to synthesise derivatives of H-L-Abu(PO_3H_2)-OH, i.e. NH_2-CH($CH_2CH_2PO_3H_2$)COOH. An improved synthesis[142] of Boc-L-Abu(PO_3Me_2)OH from Boc-Asp-OBu^t in 42% yield has been reported, but still demands seven synthetic steps-initiated by a reduction ($NaBH_4$)/oxidation (NaOCl) to form the aldehyde from the β-COOH of Asp. Phosphonylation of the aldehyde with $(MeO)_2POSiMe_3$, reaction with thiocarbonyldiimidazole followed by radical deoxygenation with Bu_3SnH, was necessary to reach a suitably protected derivative Boc-L-Abu(PO_3Me_2)-OH was successfully introduced into the strategy for making H-Glu-Abu(PO_3H_2)-Leu-OH using the solution and solid phase[143] strategies. Syntheses[144] of the same tripeptide using Z-Abu(PO_3Et_2)OH, was not completely successful as the deprotection of the -PO_3Et_2 group could not be optimised, although success had been achieved[145] in the synthesis of AcAbu(PO_3H_2)NH-Me using the same derivative. Fmoc-Abu(PO_3Me_2)OH has been prepared in high yield[146] *via* the same 7-step pathway from Boc-Asp-OBu^t as previously discussed. After removal of the Boc and Bu^t groups the Fmoc group was introduced using Fmoc-succinimide.

A brief overview of the chemical synthesis of phospho-peptides has appeared[147].

3.2 *Glycopeptide Antibiotics.*

The synthetic problems associated with the non-peptidic segments of the vancomycin family of antibiotics have been under active consideration. For example, the first synthesis[148] of the C-terminal biphenyl moiety (94) of vancomycin has been reported. This segment of the molecule plays a crucial role in binding with peptides, but is difficult to synthesise, e.g. (94) required 14 synthetic steps. When the approach[149] outlined in Scheme 5 was utilised, the biphenyl derivative (95) failed to cyclise to the quivalent ring system in vancomycin, but another analogue did cyclise to the cyclic biphenyl segment (96) of the biphenomycin antibiotics. Model studies[150] have shown that it is possible to synthesise the macrocyclic lactone equivalent (97) of the biaryl ether segment in vancomycin, the lactonisation link being secured by the use of water soluble carbodiimide. The macrocyclic lactam equivalent has not been prepared yet by these methods. An alternative synthesis[151] of the biaryl ether moiety has involved coupling a ruthenium complex (formed from L-*p*-Cl-phenylalanine derivatives and [cyclopentadienyl-Ru($MeCN)_3$]PF_6) with 3-hydroxyphenylglycine derivatives to yield a ruthenium biaryl ether complex which could be decomposed by photolysis. However the macrolactamisation step again failed.

K-13 (98) a potent non-competitive inhibitor of angiotensin converting enzyme has been synthesised several times while awaiting the more demanding breakthrough for vancomycin. To optimise the making of the biaryl ether moiety *via* the Ullmann method (NaH/CuBr.Me_2S) it has been shown[152] that a nitro group, *ortho* to the aryl halide facilitates the synthesis. A first synthesis[153] of the *N*-terminal 14-membered ring (99) of teicoplanin has been reported. The biarylether was formed from the coupling of 3-methoxy-5-(methoxycarbonyl)phenoxide and 5,5′-diformyl-2,2′-dimethoxydiphenyl iodonium bromide. After building up the side chains on both aryl rings, the final lactamisation was successfully

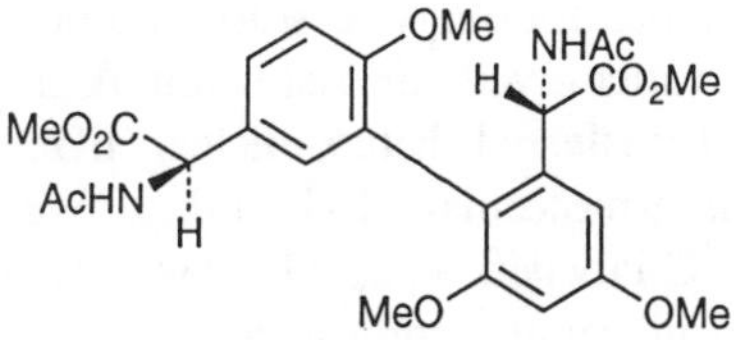

(94)

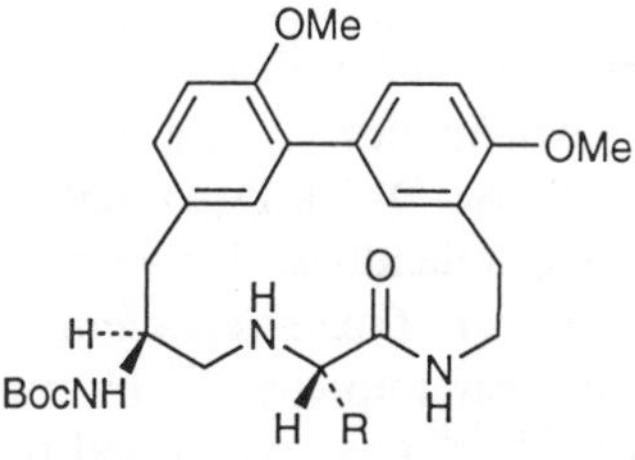

(96) R = Me, $(CH_2)_3NHBoc$

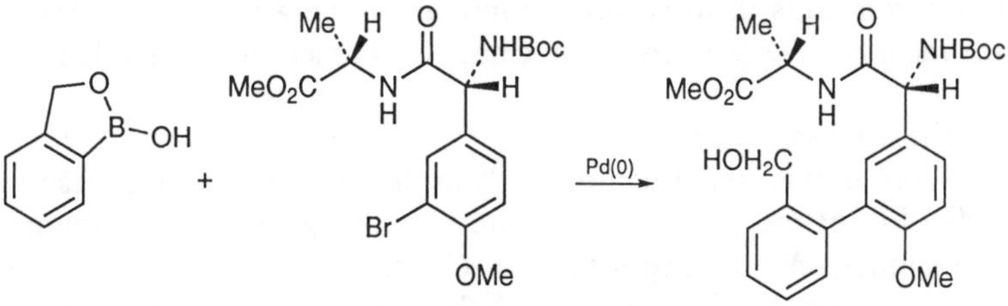

Pd(0)

(95)

Scheme 5

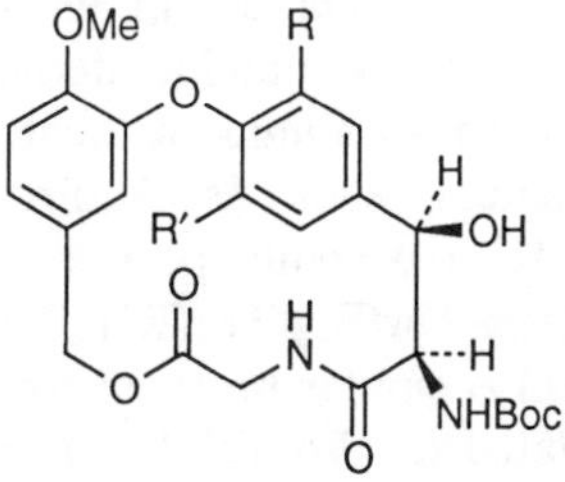

(97) R = Cl, R′ = H; R = H, R′ = Cl

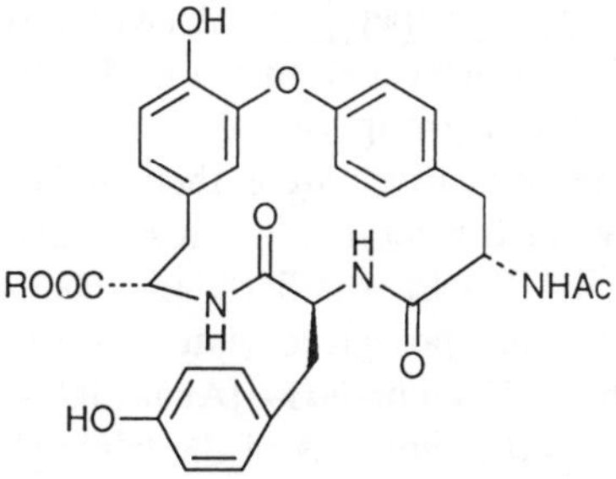

(98) R = H or Et

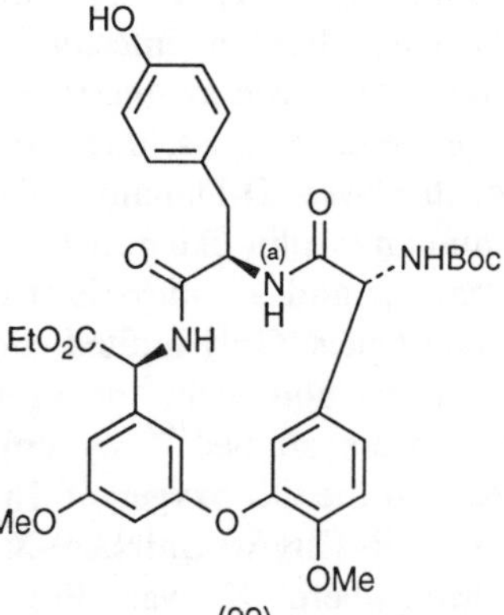

(99)

carried out at position (a) in structure (99) using pentafluorophenyl ester activation (50% yield).

Application of the ^{13}C-^{1}H nOe methodology to study the binding between Ac-D-Ala-D-Ala-OH and vancomycin or ristocetin A is not without difficulties. However, the transferred heteronuclear nOe approach for the assignment of intermolecular H-bonding can be significantly improved[154] by using ^{13}C-labelled dipeptide. Two separate reports[155,156] have confirmed that capillary electrophoresis is a sensitive technique for studying molecular recognition in low M.W. receptors. Binding constants deduced for the interaction between Ac-Ala-Ala-OH and vancomycin were in good agreement with other literature values.

3.3 *Other Glycopeptides.*

Work in this area is continuously expanding and now justifies some sub-division of the reviewed literature to those covering interests in *O*-glycopeptides, *N*-glycopeptides and *C*-glycopeptides. A recent overview[157] and a specialist review[158] of the synthesis of glycopeptides are useful compilations as background to the current literature.

O-Glycopeptides - A swift way has been found[159] to link the sugar unit to a protected amino acid active ester as summarised in Scheme 6. The azido group can then be converted to the acetamido derivative. These derivatives were then incorporated *via* an Fmoc-protocol on a polydimethylacrylamide resin into sequences, such as Ac-Pro-Thr-Ser(R^1)-Thr-Pro-Ile-Ser-Thr-NH_2 where R^1 represents part structure (100), or polyglycosylated Ac-Ile-Thr(R^1)-Thr(R^1)-Thr(R^1)-Thr(R^1)-Thr(R^1)-Val-Thr-NH_2. [Again R^1 = (100).] A similar protocol could be used in the synthesis of the triple glycosylated Ac-Pro-Thr(R^2)-Thr(R^2)-Thr(R^2)-Pro-Ile-Ser-Thr-NH_2 where R^2 could be the disaccharide (101) or a combination of (100) and (101). Forty different *O*-glycopeptides representing sequences from human intestinal mucin and porcine submaxillary gland mucin have been synthesised[160] using simultaneous multiple-column solid phase techniques, and using the Fmoc-strategy described in the approach above. O-Dimannosylated heptadecapeptide analogues (102) of the human insulin-like growth-factor 1 have also been synthesised[161] in this manner and a comparison made using nmr with their non-glycosylated analogues. Only limited effects due to glycosylation could be detected. When models for the antifreeze glycoprotein, a mucin-type glycoprotein, were studied[162] by nmr, intramolecular H-bonds between Gal-NAc and the CO oxygen of Thr could be detected in compounds such as R-Thr(Ac_3GalNAc)-X-Ala-OMe or R-Thr(GalNAc)-X-Ala-OMe where X was Pro or Ala. Fmoc-Ser-OCH_2COPh, glycosylated with a derivative of the disaccharide

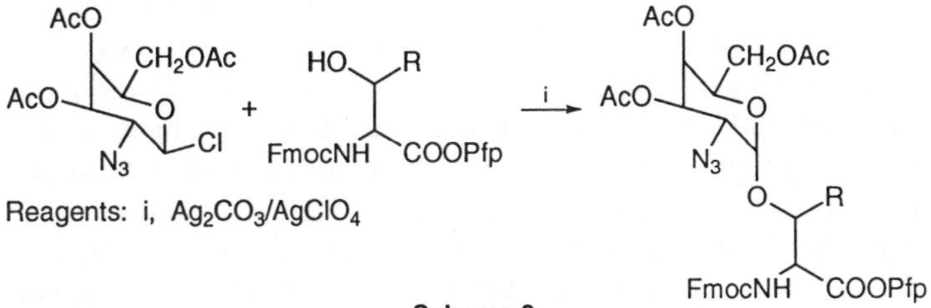

AcO, CH_2OAc, AcO, O, N_3, Cl + HO, R, FmocNH, COOPfp $\xrightarrow{i}$ AcO, CH_2OAc, AcO, O, N_3, O, R, FmocNH, COOPfp

Reagents: i, $Ag_2CO_3/AgClO_4$

Scheme 6

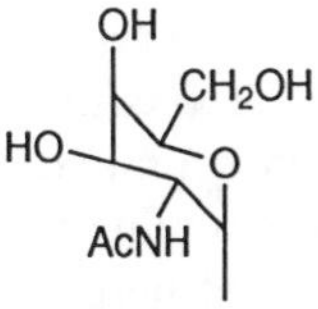

(100)

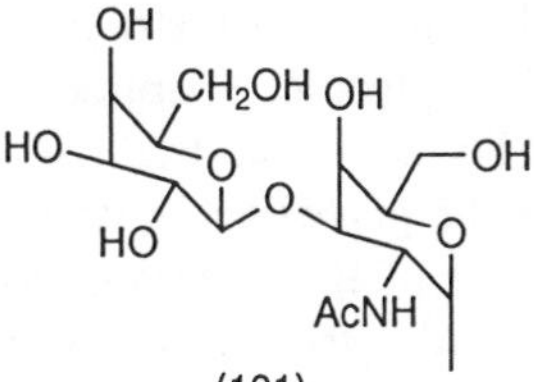

(101)

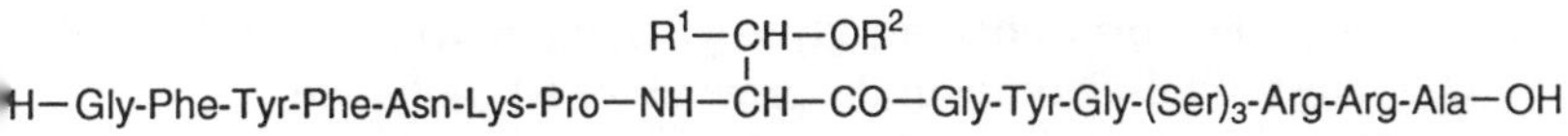

H—Gly-Phe-Tyr-Phe-Asn-Lys-Pro—NH—CH(R^1—CH—OR^2)—CO—Gly-Tyr-Gly-(Ser)$_3$-Arg-Arg-Ala—OH

(102) R^1 = H or Me; R^2 = α-D-Man*p*(1→2)α-D-Man*p*(1→)

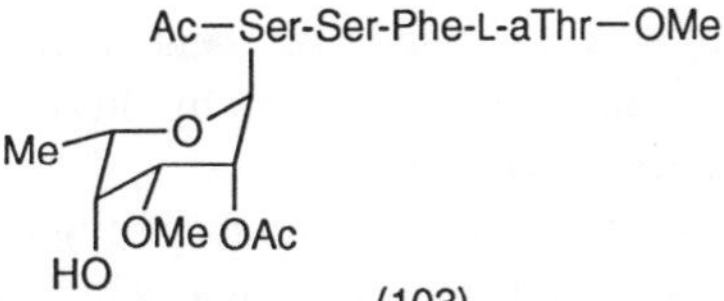

Ac—Ser-Ser-Phe-L-aThr—OMe

(103)

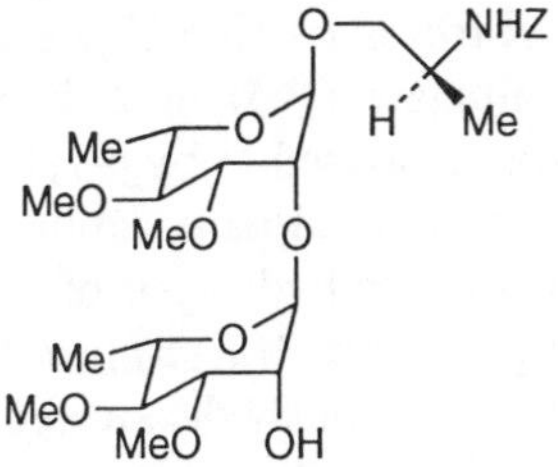

(104)

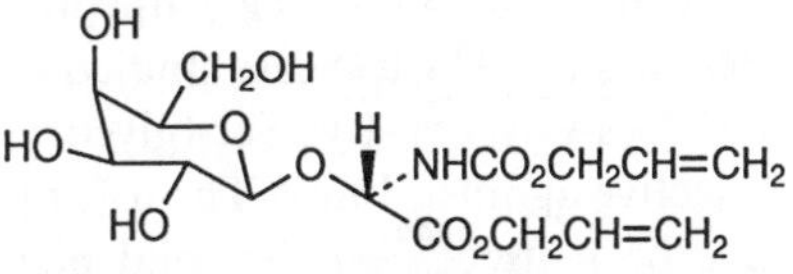

$NHCO_2CH_2CH{=}CH_2$, $CO_2CH_2CH{=}CH_2$

(105)

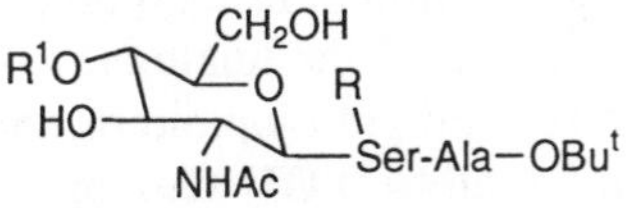

R^1O, CH_2OH, HO, NHAc, R, Ser-Ala—OBu^t

(106) R^1 = H

(107) R^1 = HO (sugar), CH_2OH

CH_2OR, H,OH, NHAc, MeCHCONH—D-Ala-D-Glu—NH_2

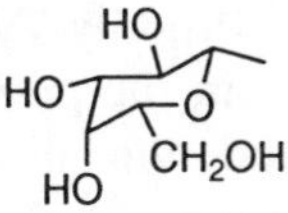

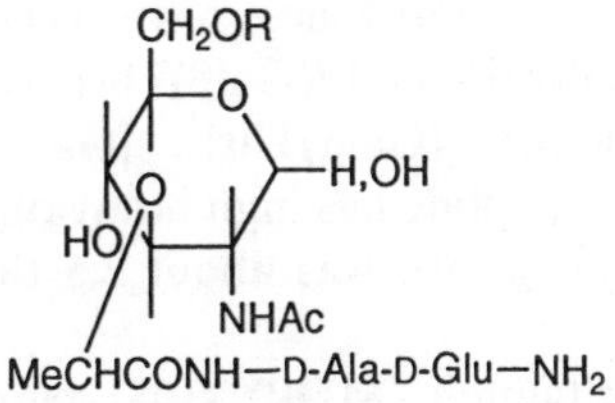

(108) R = H

α-D-Xyl*p*-(1→3)-β-D-Glcp, and then treated with Zn to remove the phenacyl group, has been used[163] successfully in the synthesis of blood clotting factor IX glycopeptide fragments. The same disaccharide, as its trichloroacetimidate, together with a trisaccharide analogue α-D-Xyl*p* (1→3)-α-D-Xyl*p* (1→3)-D-Glcp, have been linked[164] to Z-Ser-OBzl to provide derivatives for elucidation of their biological function in the context of their relevance to human blood clotting factors VII, IX and protein Z. The tetrasaccharide, β-D-GlcA-(1→3)(SO_3Na→4)-β-D-Gal-(1→3)-β-D-Gal-(1→4)-β-D-Xyl-(1→3), as its trichloroacetimidate has also been linked[165] to Z-Ser-OBzl to study the linkage region of cell-surface glycans. The glycotetrapeptide (103), from *Mycobacterium xenopi* has an unusual Ser-Ser link, but has been successfully synthesised[166] through reacting the glycoside as its trichloroacetimidate, with Ac-Ser-OBzl in the presence of BF_3 etherate, prior to coupling with DCC/HOBt on to the pre-made tripeptide. The part structure (104) of the glycopeptidolipid from *Mycobacterium fortuitum* has also been synthesised[167] using the corresponding glycosyl trichloroacetimidate but this time coupling with Z-alaninol in the presence of BF_3-etherate.

In the difficult task of chemical synthesis of glycopeptides, any possible assistance from enzymes would be useful. Galactosidation of allyloxycarbonyl-serine allyl ester using *E.* coli β-galactosidase has given glycoconjugate (105) in 28% yield[168] and *O*-glycosylation of (106) to (107) was catalysed[169] by β-1,4-galactosyltransferase. Enzymatic cleavage of esters of acid/base labile conjugates can also be beneficial and examples of the hydrolysis of t-butyl[170] and heptyl[171] esters by thermitase and a lipase from *Mucor javanis* have been recorded.

The previously discovered increase in the nucleophilicity of Ser and Thr side-chain hydroxyls when the amino-acids are protected as schiff bases has been used[172] in the first stages of the synthesis of *O*-linked glycopeptides either in solution or using solid phase techniques. The test case studied was the synthesis of the analogue H-Tyr-D-Cys-Gly-Phe-Cys-Ser-(glycosyl)-Gly-NH_2 a potent δ-opioid receptor selective agonist. Fmoc-Thr(α-(or β-) Gal-β-3-GalNAc)-OH derivatives have been used[173] in the solid phase synthesis of the fibronectin fragments H-Val-Thr(α-(or β-)GalNAc-β-3-Gal)-His-Pro-Gly-Tyr-OH, and similar approaches have been successful[174] in the preparation of $[(Gal\alpha)Thr]^3$- and $[(Gal\beta)Thr]^3$-vespulakinin and the di-glycosylated analogue $[(Gal\alpha)Thr]^3$, $[Gal\alpha)Thr]^4$-vespulakinin. The parent glycoheptadecapeptide has high bradykinin-like activity and the diglycosylated analogue above was about 2.5 times as active as bradykinin.

N-Acetylmuramyl-L-Ala-D-isoglutamine (MDP) (108) represents the smallest structure to express the activity of Freunds complete adjuvant and its conformation has been the subject of many studies. However, it has now been compared[175] with a more constrained analogue MAP, having a D-Pro-OMe in the *C*-terminal position. High field nmr studies and an extensive computer search (SYBYL) of the optimal structures of the two glycopeptides have confirmed a β-turn in both glycopeptides formed by H-bonding between the *N*-acetyl CO group and the NH of L-Ala. A second stable β-turn in MDP between D-Lac CO and the D-isoglutamine was not consistent with the data but there was evidence of this in MAP which gave it a S-shape which might explain why it is biologically inactive. The R group in structures such as (108) have been modified[176] with acridinyl carboxylic acids for testing as potential antitumour agents. A diastereomerically pure muramyl dipeptide analogue (109) containing a carbocyclic ring has been prepared[177] and found to be immunologically active. Studies on the synthesis[178] and immunogenicity of *N*-acetylmuramyl-L-Ala-Glu-NH_2 and analogues covalently attached to the hepatitis B virus antigen (14-32)Pre-S2 region have been reported.

N-Glycopeptides - Developments to enable *N*-glycopeptide synthesis to be carried out in the solid phase are well in hand. Fmoc-Asn(sugar)-OH derivatives appear to be excellent candidates for insertion into sequences. Synthesis of seven different derivatives based on (110) have been reported[179], the key step in the syntheses being the reaction of Fmoc-Asp-OBu^t, activated in the side-chain as the pentafluorophenyl ester, with a β-amino sugar. The anomeric configuraton of the *N*-glycosyl bond in each case was shown to β. Another stereoselective route to Asn-linked glycopeptides involved[180] the rather intriguing sulfonamido shift as depicted in Scheme 7. The total synthesis of nephritogenoside (111) first isolated in 1988, has been achieved[181]. The allyloxycarbonyl group (Aloc) with its mild removal conditions (Pd complex), proved a successful protecting group as derivative (112). The original plan was to couple (112) to the remaining 2-21 eicosapeptide, but the formation of cyclic succinimide was a major side-reaction. So the Pro residue had to be added to give (113) using water soluble carbodiimide prior to coupling to the rest of the peptide. Three *N*-glycoconjugates of [Leu^5]-enkephalin, H-Tyr-Gly-Gly-Phe-Leu-NHR, where RNH_2 = 6-(or 2-)amino-6-(or 2-) deoxy-D-glucopyranose, or β-D-glucopyranosamine β-D-glucopyranosylamine have been synthesised[182] using active ester and mixed anhydride methodology. All three *N*-glycoconjugates showed higher potency in the guinea pig ileum assay, but lower potency in the mouse vas deferens assay indicating a decrease in δ-opioid receptor selectivity. The synthesis[183] of the Lewis^a antigen side-chain bearing an amino group (114) provided a key step for condensation with Teoc-Asp-OAllyl in the

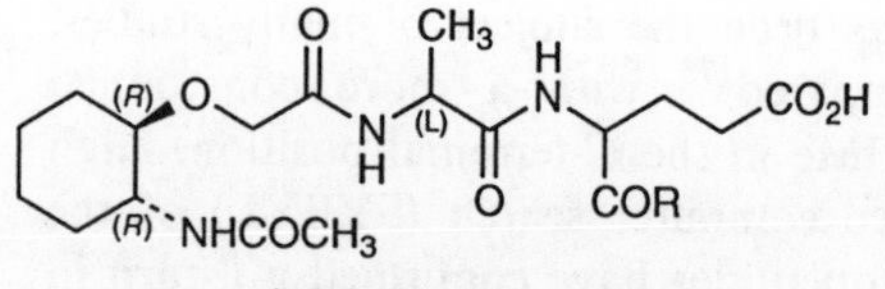

(109) R = NH_2 or OH

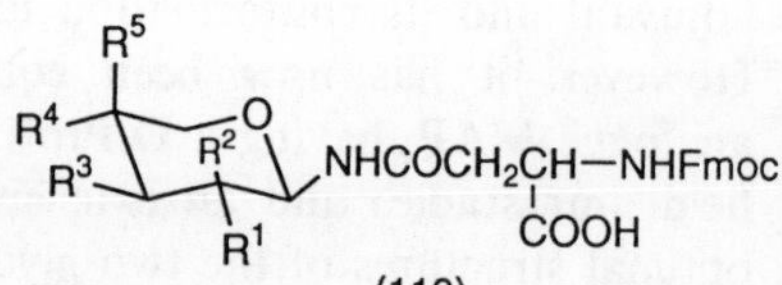

(110)

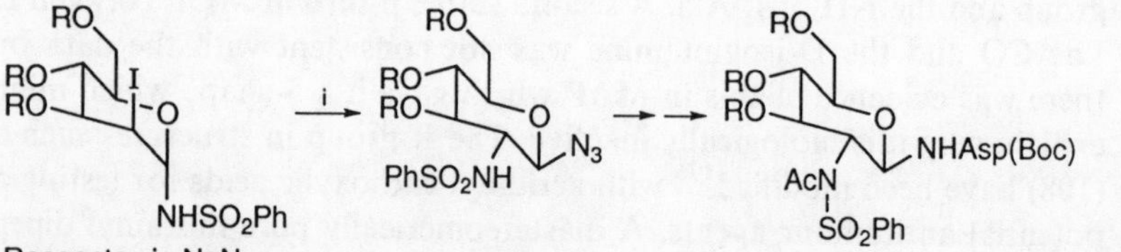

Reagents: i, NaN_3

Scheme 7

CO−Pro-Leu-Phe-Gly-Ile-Ala-Gly-Glu-Asp-Gly-Pro-Thr-Gly-Pro-Ser-Gly-Ile-Val-Gly-Gln−OH

(111)

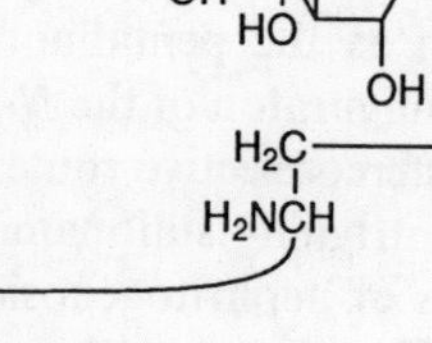

(112) X = OSu
(113) X = Pro-OSu

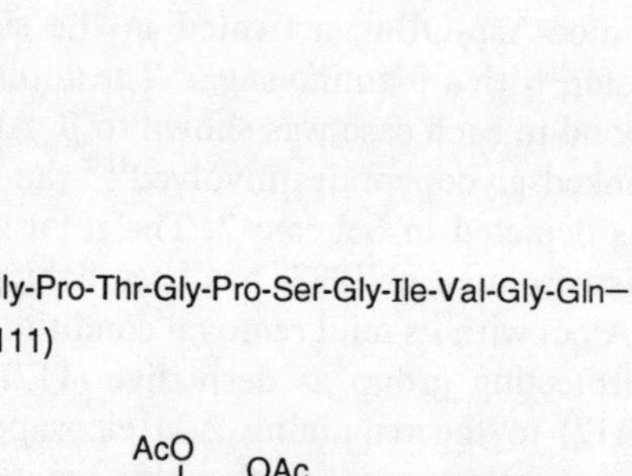

(114) R = H
(115) R = $COCH_2CH(NHCOCH_2CCl_3)CO_2CH_2CH{=}CH_2$

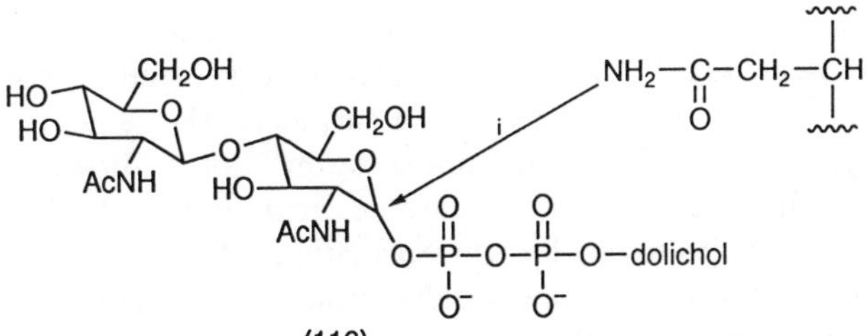

(116)

Reagent: i, enzyme catalysed – OST (oligosaccharide transferase)

Scheme 8

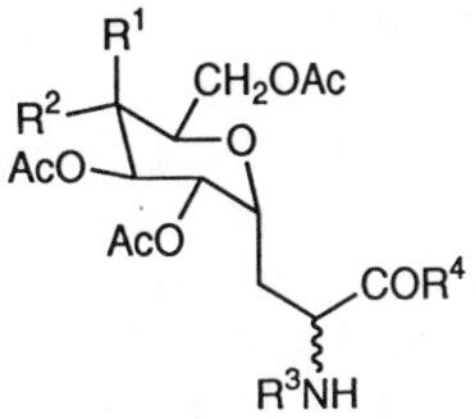

(117)

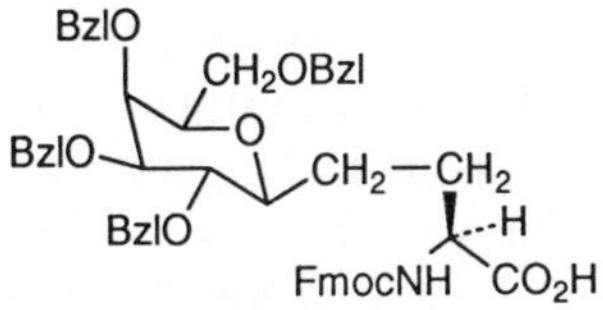

(118)

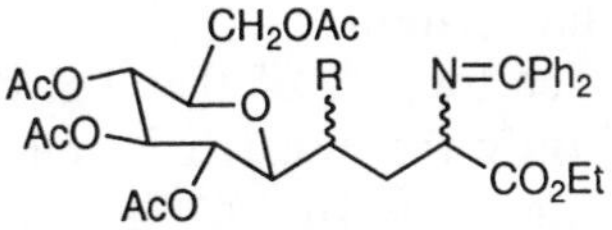

(119) R = NO_2
(120) R = H

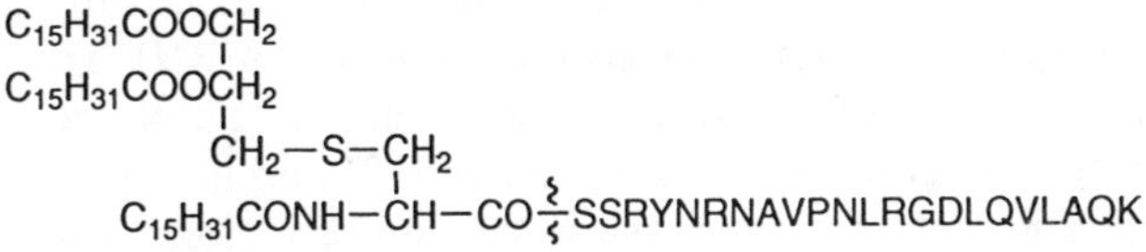

$C_{15}H_{31}COOCH_2$
$C_{15}H_{31}COOCH_2$
CH_2-S-CH_2
$C_{15}H_{31}CONH-CH-CO$-SSRYNRNAVPNLRGDLQVLAQK

(121)

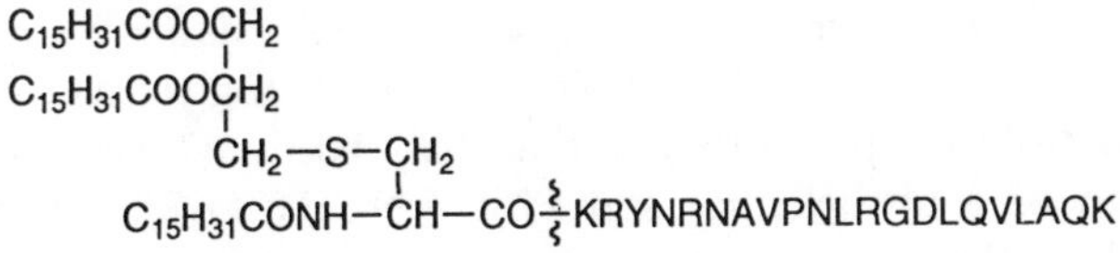

$C_{15}H_{31}COOCH_2$
$C_{15}H_{31}COOCH_2$
CH_2-S-CH_2
$C_{15}H_{31}CONH-CH-CO$-KRYNRNAVPNLRGDLQVLAQK

(122)

$Me(CH_2)_{12}COO$ … O … CO—Lys-Arg-Thr-Leu-Arg—OH
$Me(CH_2)_{12}COO$

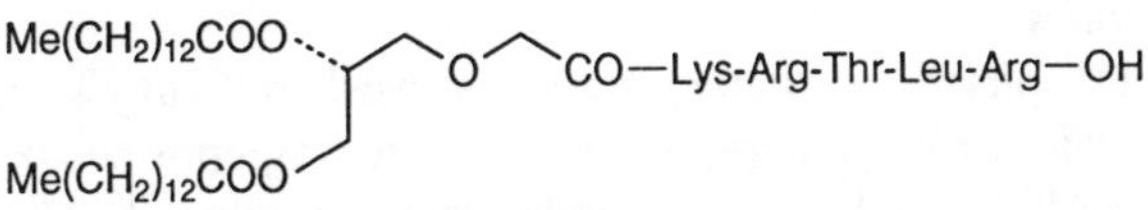

(123)

presence of isobutyl-2-isobutyloxy-1,2-dihydroquinoline-1-carboxylate (IIDQ) to yield (113), which on selective deprotection could be incorporated[184] into the HIV peptide T sequence H-Ala-Ser-Thr-Thr-Thr-Asn(glycoside)-Tyr-Thr-OH. Condensation[185] of N^{ε}-Boc-L-Lys-Bu^t with 2-azidoethylglycosides of glucuronic acid and β-D-Glc*p* NAc-(1→3)-β-D-Glc*p* A, followed by transformation into 2-acrylamidoethyl glycosides and deprotection, have yielded monomers for conversion into high M.W. co-polymer neo-glycoconjugates with acrylamide.

The synthesis[186] of the lipid disaccharide (116) provided a substrate for an enzyme-catalysed link to be made with the side chain amide of Asn in Bz-Asn-Leu-Thr-NH_2 as summarised in Scheme 8. The product was compared with a sample prepared chemically from the amino sugar and the protected aspartyl peptide and found to be identical. In order to prove whether the conformation in the environs of the asparagine residue had an effect on the OST catalysis, *cyclo* (Pro-D-Ser-Asn-Gly-Thr-Val) and *cyclo* (Pro-D-Phe-Asn-Asn-Ala-Thr) were used[187] to bias the Asn-Xaa-Thr configuration into type I β-turn. However, the enzyme kinetic analysis showed that this was not a favoured 'recognition' conformation.

C-Glycopeptides - The fragility of the *O*-glycosyl and *N*-glycosyl bonds in glycopeptides justify the research on mimetics which are not subject to ready deglycosylation *in vivo*. *C*-Glycosidine compounds can fulfil this role and have been synthesised in Europe and the U.S. The example (115) was synthesised[188] by radical addition of the protected glycosyl bromides to suitably protected dehydroalanine derivatives in the presence of Bu_3SnH and α,α'-azobisisobutyronitrile (AlBN). The stereochemistry of the adducts were shown to be exclusively α at the sugar link. The *C*-glycosyl unit (116) has been synthesised[189] and incorporated into a 17 residue α-helical peptide to assess the conformational effect of *C*-glycosylation on an order secondary structure. Substitution of one *C*-glycosyl unit showed a marked decrease in α-helix stability similar to effects seen by *O*- and *N*-linked glycopeptides. A *C*-glycosyl analogue (119) of *O*-(β-D-xylopyranosyl) serine has been synthesised[190] by condensation of a nitroethanol and $Ph_2C{=}NCH_2CO_2Et$. Denitration with Bu_3SnH/AlBNB gave (120) a precursor to useful *C*-glycosyl serine.

3.4 *Lipopeptides.*

The solid phase synthetic strategies used by Jung's group at Tubingen to synthesise lipopeptide vaccines eliciting epitope-specific B-, T-helper and T-killer cell response have been compiled[191]. The advantages of ion-spray mass spectrometry in characterising[192] Pam_3Cys-Ser-Ser-[VP1(135-154)] (121) and Pam_3Cys-Lys[VP1(135-154)]-Ala (122) has been demonstrated. VP1 is the O_1K foot and mouth disease virus. Molecular weights of up to 6000 units can be considered routine for the ion spray method method D-Lac-Ala-γ-D-Glu-(L)-*meso*-2,6-diaminopimelyl-(L)-Gly-OH (FK156) and heptanoyl-γ-D-Glu-(L)-*meso*-2,6-diaminopimelyl-(L)-D-Ala-OH (FK565) enhance the host defence ability aginst microbial infections and exhibit strong antiviral activity. An improved synthesis[193] of the diamino pimelic acid derivatives, using selective enzyme hydrolysis to elicit differentiation between the two chiral centres in the molecule, has provided a convenient route to FK-156 and 565. A potential inhibitor (123) of protein kinase C was synthesised[194] by acylation of the *N*-terminal Lys peptide on the resin, using the pentafluorophenyl ester derivative of 1,2-dimyristoyl-*sn* glycerol.

The echinocandins, well known lipopeptides with potent antifungal activity against *Candida* species are still being actively 'modified' for potential clinical use. An analogue L-688,786 (124), which is an antifungal and antipneumocystis lipopeptide, has been modified[195] at the homotyrosyl phenolic group to enhance its water solubility, and to review potential as pro-drugs. It was deduced that the phenolic group may be important in antifungal activity, and it was only the phosphate ester derivative (124) $R^2 = PO_3H_2$ that was equivalent to the parent drug in the *in vivo* models used. Echinocandin B (125) when treated[196] with $NaCNBH_3$ and $NaB(OAc)_3$ in TFA can be converted to its congener echinocandin C (126). Earlier structure/activity studies on the fatty acid side chain of the echinocandins has led to the development of cilofungin (125) which is undergoing clinical assessment. A series of simplified analogues of (125) have been synthesised[197] by solid phase methodology (and cyclisation in which by diphenylphosphoryl azide) many of the unusual amino acids have been replaced by natural analogues, e.g. L-Pro instead of *(2S,3S,4S)*-3-hydroxy-4-MePro, L-Tyr instead of 3,4-dihydroxyhomotyrosine and L-Orn for 3,4-dihydroxyornithine. As expected the new analogue was inactive. By systematically re-introducing the structural complexity the activity studies have shown that the 3-HO-4MePro enhances activity, but the L-homo Tyr residue is crucial for both antifungal activity and for the inhibition of β-1,3-glucan biosynthesis.

Lipidic amino acid conjugates of hydrophilic compounds have been synthesised[198] to improve the transmembrane absorption of poorly-absorbed compounds. Examples included (128-130) and nmr studies revealed that these compounds formed aggregates or micelles at high concentration but existed as monomers in dilute solution. Ceramide analogues $Me(CH_2)_nCH(CH_2OH)NHR$, where n could be 9, 11, 13 or 15 and R was Ac, Boc, $COCH(NHBoc)(CH_2)_{13}Me$, have been synthesised[199] by chemoselective reduction ($NaBH_4$) of the corresponding mixed

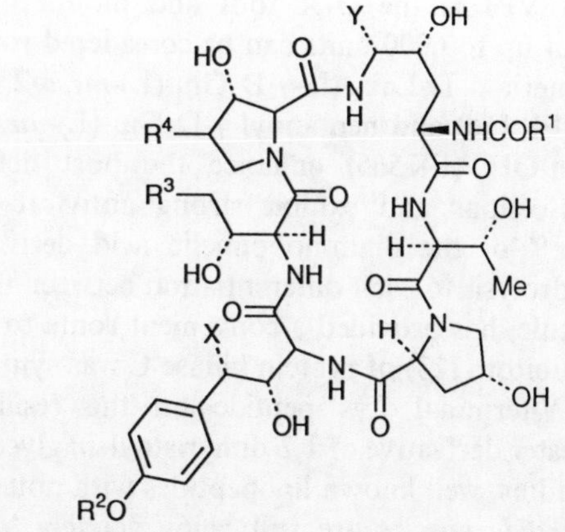

(124) L688,786, $R^1 = (CH_2)_8(CHMeCH_2)_2Me$, $R^2 = R^4 = H$, $R^3 = CONH_2$, X = Y = OH

(125) $R^1 = (CH_2)_7CH{=}CH{-}CH_2{-}CH{=}CH(CH_2)_4CH_3$, $R^2 = R^3 = H$, $R^4 = Me$, X = Y = OH

(126) $R^1 = (CH_2)_7CH{=}CH{-}CH_2{-}CH{=}CH(CH_2)_4CH_3$, $R^2 = R^3 = H$, $R^4 = Me$, X = H, Y = OH

(127) $R^1 = C_6H_4{-}O(CH_2)_7CH_3$, $R^2 = R^3 = H$, X = Y = OH

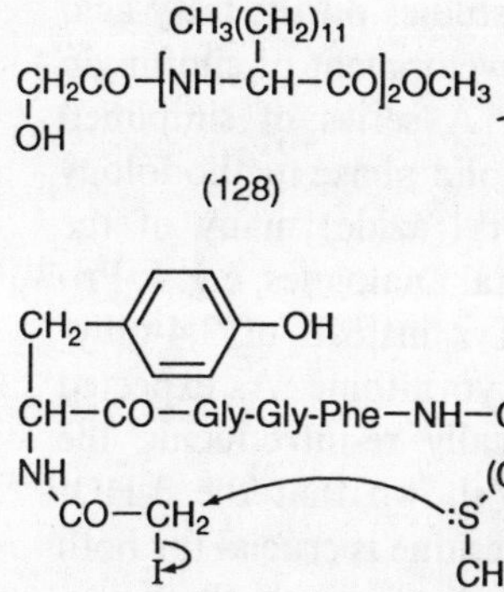

$HOCH_2CO{-}[NH{-}CH(C_{12}H_{25}){-}CO]_2OCH_3$

(128) 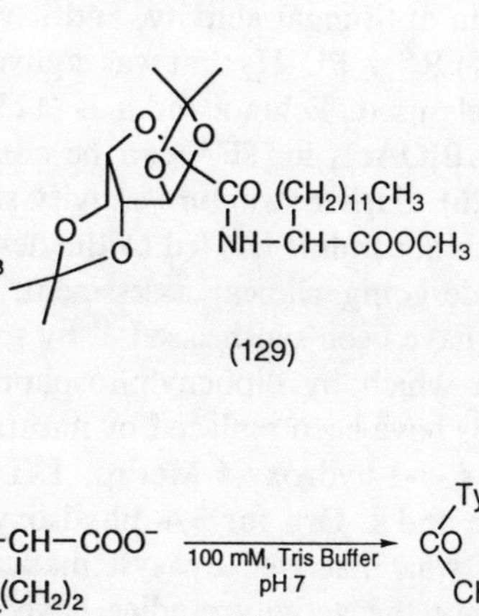 (129) 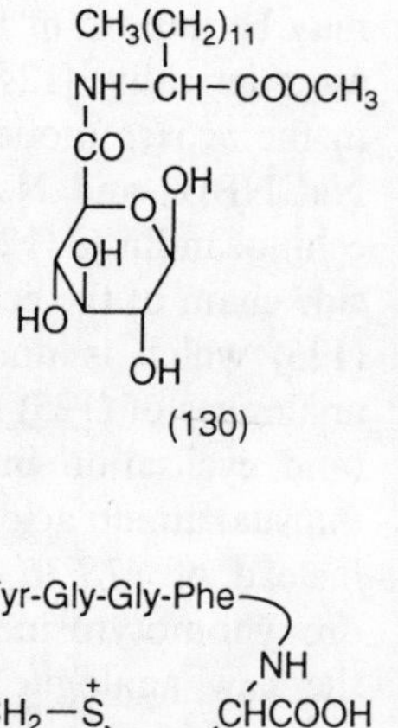 (130)

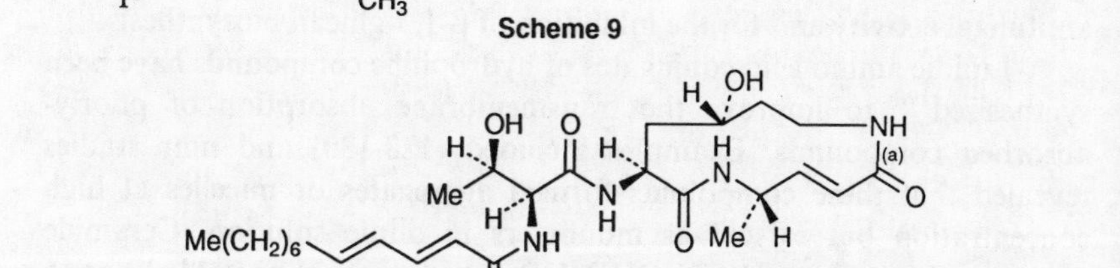

Scheme 9

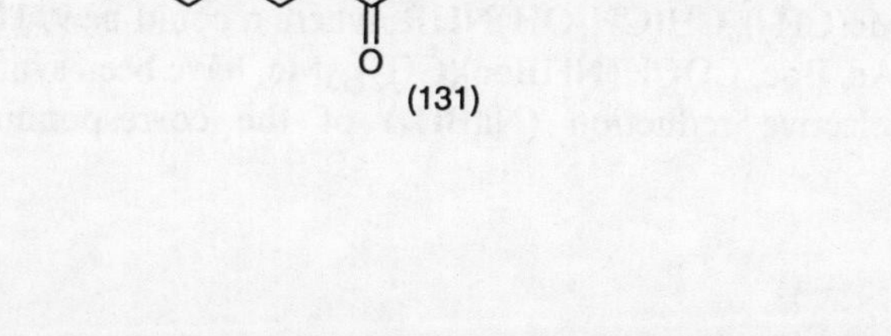

(131)

anhydrides of the precursor amino acid derivatives. The sphingosine analogues, with n = 9, 11 and 13 and R = H were obtained by deprotection of the Boc derivatives.

4 Miscellaneous Examples

Covalent cyclisation for introducing conformational constraints into peptides have been the main theme of this Chapter, and demands intensive synthetic expertise. Yet some useful preliminary information could be derived if constraints were to be introduced into native biologically active peptides, without recourse to extensive synthesis. Novel cyclisation chemistry with this theme in mind has been developed[200] for native peptides and is summarised in Scheme 9. For example, the *N*-terminal residue was modified to an iodoacetyl derivative which could then react with a nucleophilic side-chain, e.g. methionine as in Scheme 9, or His and Lys would also work. To be useful the products would have to be sufficiently stable for biological evaluation. The strategy for the synthesis[201] of the antibiotic glidobactin A (131) involved activation of the carboxylic precursor (position (a) in the structure) as a pentafluorophenyl ester, and high dilution conditions in the presence of 4-pyrrolidinyl pyridine. A cyclisation yield of 20% was achieved.

References

1. CA Selects on Amino Acids, Peptides and Proteins, published by the American Chemical Society and Chemical Abstracts Service, Columbus, Ohio.
2. "Peptides 1992" Proc. of the 22nd European Peptide Symposium, eds. C.H. Schneider and A.N. Eberle, ESCOM, Leiden, 1993.
3. "Biomedical Appl. Biotechnology: Vol. 1, Biol Active Peptides; Design Synthesis and Utilisation", eds. W.V. Williams and D.B. Weiner, Technomic Publ. Co., Pennsylvania, 1993.
4. "Medical Chemistry for the 21st Century", eds. C.G. Wermuth, with N. Koga, H. Kōnig and B.W. Metcalfe, IUPAC/Blackwell Scientific Publications, 1992.
5. M.P. Williamson and J.P. Waltho, *Chem.Soc.Rev.*, 1992, **21**, 227.
6. H. Kessler, *J.Prakt.Chem/Chem.-Ztg.*, 1992, **334**, 549.
7. E. Didier, D.C. Horwell and M.C. Pritchard, *Tetrahedron*, 1992, **48**, 8471.
8. T. Inoue, S. Kishida, M. Ohsaki and H. Kimura, *Bull.Chem.Soc.Jpn.*, 1992, **65**, 1728.
9. N. Nishino, M. Xu, H. Mihara, T. Fujimoto, M. Ohba, Y. Ueno and H. Kumagai, *J.Chem.Soc.Chem.Commun.*, 1992, 180.
10. M. Bovier and J.W. Taylor, *J.Med.Chem.*, 1992, **35**, 1145.
11. G. Osapay and J.W. Taylor, *J.Amer.Chem.Soc.*, 1992, **114**, 6966.
12. A. Trzeciak and W. Bannwarth, *Tetrahedron Lett.*, 1992, **33**, 4557.
13. P. Bishop and J. Chmielewski, *Tetrahedron Lett.*, 1992, **33**, 6263.
14. J.A. Marco, J.F. Sanz, A. Yuste and J. Jakupovic, *Tetrahedron Lett.*, 1991, **32**, 5193.
15. M.A. Vela and H. Kohn, *J.Org.Chem.*, 1992, **57**, 5223.

16. H.H. Wasserman, V.M. Rotello and G.B. Krause, *Tetrahedron Lett.*, 1992, **33**, 5419.
17. M. North, *Tetrahedron*, 1992, **48**, 5509; (b) W.R. Jackson, H.A. Jacobs and H.J.Kim, *Aust.J.Chem.*, 1992, **45**, 2073.
18. D. Callant, B. Coussens, T. Van der Maten, J.G. De Vries and N.K. De Vries, *Tetrahedron:Asymmetry*, 1992, **3**, 401.
19. H.J. Kim and W.R. Jackson, *Tetrahedron:Asymmetry*, 1992, **3**, 1421.
20. L.L. Constanzo, S. Guiffrida, S. Sortino and G. Pappalardo, *J.Chem.Res.Synop.*, 1992, 118.
21. S. Janus and E. Sonveaux, *Tetrahedron Lett.*, 1992, **33**, 3757.
22. G. Arena, G. Impellizzeri, G. Maccarrone, G. Pappalardo, D. Sciotto and E. Rizzarelli, *J.Chem.Soc.Perkin Trans.2*, 1992, 371.
23. M. Budesinsky, J. Symersky, J. Jecny, J. Van Hecke, N. Hosten, M. Anteurio and F. Borremans, *Int.J.Pept.Protein Res.*, 1992, **39**, 123.
24. P.J. Milne, D.W. Oliver and H.M. Roos, *J.Crystallogr.Spectrosc.Res.*, 1992, **22**, 643.
25. T. Ueda, K. Kiyohara, S. Lee, H. Aoyagi and N. Izumiya, *J.Antibiot.*, 1992, **45**, 235.
26. B. Cook, R.R. Hill and G.E. Jeffs, *J.Chem.Soc. Perkin Trans. 1*, 1992, 1199.
27. V.A. Basyuk, T. Yu. Gromovoi, A.A. Chuiko, V.A. Soloshonok and V.P. Kukhar, *Synthesis*, 1992, 449.
28. A.V. Reddy and B. Ravindranath, *Int.J.Pept.Protein Res.*, 1992, **40**, 472.
29. M. Orena, G. Porzi and S. Sandri, *J.Org.Chem.*, 1992, **57**, 6532.
30. C. Alcaraz, A. Herrero, J.L. Marco, E. Fernandez-Alvarez and M. Bernabe, *Tetrahedron Lett.*, 1992, **33**, 5605.
31. Z. Du, X. Zhou, Y. Shi and H. Hu, *Chin.J.Chem.*, 1992, **10**, 82.
32. E. Kasafirek, M. Rybak, I. Krejci, A. Sturc, E. Krepela and A. Sedo, *Life Sci.*, 1992, **50**, 187.
33. N. Nishimo, M. Xu, H. Mihara and T. Fujimoto, *Bull.Chem.Soc.Jpn.*, 1992, **65**, 991.
34. T. Kataoka, D.D. Beusen, J.D. Clark, M. Yodo and G.R. Marshall, *Biopolymers*, 1992, **32**, 1519.
35. F. Cavelier-Frontin, G. Pèpe, J. Verducci, D. Siri and R. Jacquier, *J.Amer.Chem.Soc.*, 1991, **114**, 8885.
36. N. Nishoni, M. Xu, H. Mihara, T. Fujimoto, Y. Ueno and H. Kumagai, *Tetrahedron Lett.*, 1992, **33**, 1479.
37. T. Pietzonka and D. Seebach, *Angew.Chem.Int.Ed.*, 1992, **31**, 1481.
38. G. Mueller, M. Gurrath, H. Kessler and R. Timpl' *Angew.Chem.Int.Ed.* 1992, **31**, 358.
39. D.F. Mierke and H. Kessler, Biopolymers, 1992, **32**, 1277.
40. A.N. Stroup, A.L. Rockwell and L.M. Gierasch, *Biopolymers*, 1992, **32**, 1713.
41. Z.-P. Liu and L.M. Gierasch, Biopolymers, 1992, 32, 1727.
42. M.D. Bruch, J. Rizo and L.M. Gierasch, *Biopolymers*, 1992, **32**, 1741.
43. H.R. Wyssbrod and M. Diem, *Biopolymers*, 1992, **32**, 1237.
44. K. Ishikawa, T. Fukami, T. Nagase, K. Fujita, T. Hayama, K. Niyama, T. Mase, M. Ihara and M. Yano, *J.Med.Chem.*, 1992, **35**, 2139.
45. P.D. Bailey, D.G.W. Clarke and G.A. Crofts, *J.Chem.Soc.Chem.Commun.*, 1992, 658.
46. P.D. Bailey and G.A. Crofts, *Tetrahedron Lett.*, 1992, **33**, 3207.
47. P.D. Bailey, S.R. Carter, D.G.W. Clarke, G.A. Crofts, J.H.M. Tyszka, P.W. Smith and P. Ward, *Tetrahedron Lett.*, 1992, **33**, 3211.
48. P.D. Bailey, S.R. Carter, D.G.W. Clarke, G.A. Crofts, P.W. Smith and P. Ward, *Tetrahedron Lett.*, 1992, **33**, 3215.
49. N. Nishino, T. Shiroishi, T. Muraoka and T. Fujimoto, *Chem.Lett.*, 1992, 665.

50. T.W. Hanbley, C.J. Hawkins, M.F. Lavin, A. Van der Brenk and D.J. Watters, *Tetrahedron*, 1992, **48**, 341.
51. P. Wipf and C.P. Miller, *J.Amer.Chem.Soc.*, 1992, **114**, 10975.
52. P.D. Williams, M.G. Bock, R.D. Tung, V.M. Gorsky, D.S. Perlow, J.M. Erb, G.F. Lundell, N.P. Gould, W.L. Whitter, J.B. Hoffman, M.J. Kaufman, B.V. Glinesch-midt, D.J. Pettibone, R.M. Friedinger and D.F. Veber, *J.Med.Chem.*, 1992, **35**, 3905.
53. D.S. Perlow, J.M. Erb, N.P. Gold, R.D. Tung, R.M. Friedinger, P.D. Williams and D.F. Veber, *J.Org.Chem.*, 1992, **507**, 4394.
54. H. Kessler and B. Haase, *Int.J.Pept.Protein Res.*, 1992, **39**, 36.
55. A.F. Spatola, M.K. Anwer and M.N. Rao, *Int.J.Pept.Protein Res.*, 1992, **40**, 322.
56. H. Kessler, A. Geyer, H. Matter and M. Köck, *Int.J.Pept.Protein Res.*, 1992, **40**, 25.
57. H. Kessler, H. Maltes, A. Geyer, H-J. Diehl, M. Köck, G. Kurz, F.R. Opperdoes, M. Callens and R.K. Wierenga, *Angew.Chem.Int.Ed.*, 1992, **31**, 328.
58. H. Itokawa, H. Morita, K. Kondo, Y. Hitotsuyanagi, K. Takeya and Y. Itaka, *J.Chem.Soc.Perkin Trans.2*, 1992, **16**, 35.
59. D.L. Boger, D. Yohannes and J.B. Myers, Jr., *J.Org.Chem.*, 1992, **57**, 1319.
60. J.W. Bean, K.D. Kopple and C.E. Peishoff, *J.Amer.Chem.Soc.*, 1992, **114**, 5328.
61. H. Kessler, H. Matter, G. Gemmecker, M. Kottenhahn and J.W. Bats, *J.Amer.Chem.Soc.*, 1992, **114**, 4805.
62. P. Schmieder and H. Kessler, *Biopolymers*, 1992, **32**, 435.
63. H. Meirovitch, D.H. Kitson and A.T. Hagler, *J.Amer.Chem.Soc.*, 1992, **114**, 5386.
64. K. Yang, I.S. Koo, I. Lee and C.K. Sohn, *J.Korean Chem.Soc.*, 1992, **36**, 523.
65. R. Fernandez, S. Omar, M. Feliz, E. Quinoa and R. Riguera, *Tetrahedron Lett.*, 1992, **33**, 6017.
66. F. Kong, D.L. Burgoyne, R.J. Andersen and T.M. Allen, *Tetrahedron Lett.*, 1992, **33**, 3269.
67. G.R. Pettit, J.K. Srirangam, D.L. Herald, K.L. Erickson, D.L. Doubek, J.M. Schmidt, L.P. Tackett and G.J. Bakus, *J.Org.Chem.*, 1992, **57**, 7217.
68. M. Namikoshi, K.L. Rinehart, R. Sakai, R.R. Stotts, A.M. Dahlem, V.R. Beasley, W.W. Carmichael and W.R. Evans, *J.Org.Chem.*, 1992, **57**, 866.
69. N. Tan, J. Zhou, S. Zhao, H. Zhang, D. Wang, C. Chen and X. Liu, *Chin.Chem.Lett.*, 1992, **3**, 629.
70. H. Kataoka, T. Hanawa and T. Katagi, *Chem.Pharm.Bull.*, 1992, **40**, 570.
71. D.S. Seetharama Jois, K.R.K. Easwaran, M. Bednarek and E.R. Blout, *Biopolymers*, 1992, **32**, 993.
72. B.D. Blasio, F. Rossi, E. Benedetti, V. Pavone, M. Saviano, C. Pedone, G. Zanotti and T. Tancredi, *J.Amer.Chem.Soc.*, 1992, **114**, 8277.
73. J.M. Gregson, J-L. Chen, G.M.L. Patterson and R.E. Moore, *Tetrahedron*, 1992, **48**, 3727.
74. S.S. Moon, J.L. Chen, R.E. Moore and G.M.L. Patterson, *J.Org.Chem.*, 1992, **57**, 1097.
75. H. Mihara, N. Nishino, H.I. Ogawa, N. Izumiya and T. Fujimoto, *Bull.Chem.Soc.Jpn.*, 1992, **65**, 228.
76. M. Xu, N. Nishino, H. Mihara, T. Fujimoto and N. Izumiya, *Chem.Lett.*, 1992, 191.
77. M.K. Rosen and S.L. Schreiber, *Angew.Chem.Int.Ed.*, 1992, **31**, 384.
78. S.W. Fesik, P. Neri, R. Meadows, E.T. Olejnicjak and G. Gemmecker, *J.Amer.Chem.Soc.*, 1992, **114**, 3165.
79. M.K. Eberle and F. Nuninger, *J.Org.Chem.*, 1992, **57**, 2689.
80. R.M. Wenger, K. Martin, C. Timbers and A. Tromelin, *Chimia*, 1992, **46**, 314.

81. A.A. Patchett, D. Taub, O.D. Hensens, R.T. Goegelman, L. Yang, F. Dumont, L. Peterson and N.H. Sigal, *J.Antibiot.*, 1992, **45**, 94.
82. P.A. Paprica, A. Margaritis and N.O. Petersen, *Bioconjugate Chem.*, 1992, **3**, 32.
83. J.L. Kafron, P. Kuzmie, V. Kishore, G. Gemmecker, S.W. Fesik and D.H. Rich, *J.Amer.Chem.Soc.*, 1992, **114**, 2670.
84. M. Koeck, H. Kessler, D. Seebach and A. Thaler, *J.Amer.Chem.Soc.*, 1992, **114**, 2676.
85. J.A. Carver, N.H. Rees, D.L. turner, S.J. Senior and B.Z. Chowdhry, *J.Chem.Soc.Chem.Commun.*, 1992, 1682.
86. S.Y. Ko and C. Dalvit, *Int.J.Pept.Protein Res.*, 1992, **40**, 380.
87. G. Emmer and S. Weber-Roth, Tetrahedron, 1992, **48**, 5861.
88. P. Nussbaumer, M. Grassberger and G. Schulz, *Tetrahedron Lett.*, 1992, **33**, 3845.
89. M.A. Grassberger, Th. Fehr, A. Horvath and G. Schulz, *Tetrahedron*, 1992, **48**, 413.
90. W.P. Frankmolle, G. Knubel, R.E. Moore and G.M.L. Patterson, *J.Antibiot.*, 1992, **45**, 1458; W.P. Frankmolle, L.K. Larsen, F.R. Caplan, G.M.L. Patterson, G. Knubel, I.A. Levine and R.E. Moore, *ibid.*, 1992, **45**, 1451.
91. W.H. Genwick, Zh.D. Jiang, S.K. Agarwal and B.T. Farmer, *Tetrahedron, 1992*, **48**, 2313.
92. A. Tromelin, M-H Fulachier, G. Mowier and A. Menez, *Tetrahedron Lett.*, 1992, **33**, 5197.
93. H. Kataoka, J.P. Li, A.S.T. Lui, S.J. Kramer and D.A. Schooley, *Int.J.Pept.Protein Res.*, 1992, **39**, 29.
94. M. Tamaki, S. Akabori and I. Muramatsu, *J.Chromatogr.*, 1992, **574**, 65.
95. H. Kessler, D.F. Mierke, J. Saulitis, S. Seip, S. Steuernagel, T. Wein and M. Will, *Biopolymers*, 1992, **32**, 427.
96. M. Debono, R.M. Malloy, J.L. Occolowitz, J.W. Paschal, A.H. Hunt, K-H. Michel and J.W. Mrtin, *J.Org.Chem.*, 1992, **57**, 5200.
97. M.P. Foster, G.P. Concepcion, G.B. Caraan and C.M. Ireland, *J.Org.Chem.*, 1992, **57**, 6671.
98. F. Itagaki, H. Shigemori, M. Ishibashi, T. Nakamura, T. Sasaki and J. Kobayashi, *J.Org.Chem.*, 1992, **57**, 5540.
99. N. Hamamichi, A. Natrajan and S.M. Hecht, *J.Amer.Chem.Soc.*, 1992, **114**, 6278.
100. H. Tomoda, H. Nishida, X.H. Huang, R. Masuma, Y.K. Kim and S. Omura, *J.Antibiot.*, 1992, **45**, 1207.
101. N. Fukuchi, A. Isogai, J. Nakayama, S. Takayama, S. Yamashita, K. Suyama and A. Suzuki, *J.Chem.Soc. Perkin Trans. 1*, 1992, 875.
102. N. Fukuchi, A. Isogai, J. Nakayama, S. Takayama, S. Yomashita, K. Suyama, J.Y. Takemoto and A. Suzuki, *Int.J.Pept.Protein Res.*, 1992, 1149.
103. K. Ueda, J.Z. Xiao, N. Doke and S. Nakatsuka, *Tetrahedron Lett.*, 1992, **33**, 5377.
104. H. Tada, T. Tozyo, Y. Terui and F. Hayashi, *Chem.Lett.*, 1992, 431.
105. S. Kim, M. Ubukata, K. Kobayashi and K. Isono, *Tetrahedron Lett.*, 1992, **33**, 2561.
106. K. Hamano, M. Kinoshita, K. Tanzawa, K. Yoda, Y. Ohki, T. Nakamura and T. Kinoshita, *J.Antibiot.*, 1992, **45**, 906.
107. D. Zink, O.D. Hensens, Y.K.T. Lam, R. Reamer and J.M. Liesch, *J.Antibiot.*, 1992, **45**, 1717.
108. F. Han, R.J. Mortishire-Smith, P.B. Bainey and D.H. Williams, *Acta Crystallog., Sect.C:Cryst.Struct.Commun.*, 1992, **C48**, 1965.
109. A. Jegorov, V. Matha, P. Sedmera and D.W. Roberts, *Phytochemistry*, 1992, **31**, 2669.
110. K. Hayashi, M. Hashimoto, N. Shigematsu, M. Nishikawa, M. Ezaki, M.

Yamashita, S. Kiyoto, M. Okuhara, M. Kohsaka and H. Imanaka, *J.Antibiot.*, 1992, **45**, 1055.
111. N.K. Gulavita, S.P. Gunasekera, S.A. Pomponi and E.V. Robinson, *J.Org.Chem.*, 1992, **57**, 1767.
112. D.A. Langs, R.H. Blessing and W.L. Duax, *Int.J.Pept.Protein Res.*, 1992, **39**, 291.
113. V.Z. Pletnev, I.N. Tsygannik, Yu.A. Fonarev, V.T. Ivanov, D.A. Langs, P. Grochulski and W.L. Duax, *Biorg.Khim*, 1992, **18**, 794.
114. P. Grochulski, G.D. Smith, D.A. Langs, W.L. Duax, V.Z. Pletnev and V.T. Ivanov, *Biopolymers*, 1992, **32**, 757.
115. V.Z. Pletnev, V.T. Ivanov, D.A. Langs and P. Strong, *Biopolymers*, 1992, **32**, 819.
116. J.M. Curtis, C.D. Bradley, P.J. Derrick and M.M. Sheil, *Org.Mass.Spectrom.*, 1992, **27**, 502.
117. M.M. Shiel and P.J. Derrick, *Org.Mass.Spectrom.*, 1992, **27**, 1000.
118. T. Kato, H. Mizuno, S. Lee, H. Aoyagi, H. Kodama, N. Go and T. Kato, *Int.J.Pept.Protein Res.*, 1992, **39**, 485.
119. K. Fukase, M. Kitazawa, A. Sane, K. Shimbo, S. Horimoto, H. Fujita, A. Kubo, T. Wakamiya and T. Shiba, *Bull.Chem.Soc.Jpn.*, 1992, **65**, 2227.
120. U. Schmidt and F. Stäbler, *J.Chem.Soc.Chem.Commun.*, 1992, 1353.
121. M. Hagihara and S.L. Schreiber, *J.Amer.Chem.Soc.*, 1992, **114**, 6570.
122. P. Wipf, H.J. Kim and Y. Hong, *Tetrahedron Lett.*, 1992. **33**, 4275.
123. U. Schmidt, R. Meyer, V. Leitenberger, H. Griesser and A. Lieberknecht, *Synthesis*, 1992, 1025; U. Schmidt, V. Leitenberger, H. Griesser, J. Schmidt and R. Meyer, *ibid.*, 1992, 1248.
124. R.J. Heffner, J. Jiang and M.M. Joullié, *J.Amer.Chem.Soc.*, 1992, **114**, 10181.
125. G. Zanotti, G. Petersen and Th. Wieland, *Int.J.Pept.Protein Res.*, 1992, **40**, 551.
126. M.D. Reily, V. Thanabal, E.A. Lunney, J.T. Repine, C.C. Humblet and G. Wagner, *FEBS Letters*, 1992, **302**, 97.
127. D.S. Dhanoa, W.H. Parsons, W.J. Greenlee and A.A. Patchett, *Tetrahedron Lett.*, 1992, **33**, 1725.
128. M. Hagihara, N.J. Anthony, T.J. Stour, J. Clardy and S.L. Schreiber, *J.Amer.Chem.Soc.*, 1992, **114**, 6568.
129. E.D. de Silva, D.E. Williams, R.J. Anderson, H. Klix, C.F.B. Holmes and T.M. Allen, *Tetrahedron Lett.*, 1992, **33**, 1561.
130. Y. Koiso, M. Natori, S. Iwasaki, S. Sato, R. Sonada, Y. Fiyita, H. Yaegashi and Zenji Sato, *Tetrahedron Lett.*, 1992, **33**, 4157.
131. C. Linget, D.G. Stylianou, A. Dell, R.E. Wolff, Y. Piemont and M.A. Abdallah, *Tetrahedron Lett.*, 1992, **33**, 3851.
132. C.K. Larive, L. Guerra and D.L. Rabenstein, *J.Amer.Chem.Soc.*, 1992, **114**, 7331.
133. A. Paquet, *Int.J.Pept.Protein Res.*, 1992, **39**, 82.
134. J.W. Perich, R.B. Johns and E.C. Reynolds, *Aust.J.Chem.*, 1992, **45**, 385.
135. J.W. Perich and D.P. Kelly, *Int.J.Pept.Protein Res.*, 1992, **40**, 81.
136. J.W. Perich, *Int.J.Pept.Protein Res.*, 1992, **40**, 134.
137. A. Chavanieu, H. Naharisoa, F. Heitz, B. Calas and F. Grigorescu, *Biorg.Med.Chem.Lett.*, 1991, **1**, 299.
138. W. Bannwarth and E.A. Kitas, *Helv.Chim.Acta*, 1992, **75**, 707.
139. C. Garbay-Jaureguiberry, D. Fieheux and B.P. Roques, *Int.J.Pept.Protein Res.*, 1992, **39**, 523.
140. J.W. Perich and R.B. Johns, *Aust.J.Chem.*, 1992, **45**, 1759.
141. A.B. Kazi and J. Hajdu, *Tetrahedron Lett.*, 1992, **33**, 2291.
142. G. Tong, J.W. Perich and R.B. Johns, *Aust.J.Chem.*, 1992, **45**, 1225.

143. J.W. Perich and E.C. Reynolds, *Aust.J.Chem.*, 1992, **45**, 1765.
144. J.W. Perich, R.M. Valerio and R.B. Johns, *Aust.J.Chem.*, 1992, **45**, 919.
145. R.M. Valerio, J.W. Perich, P.F. Alewood, G. Tong and R.B. Johns, *Aust.J.Chem.*, 1992, **45**, 777.
146. J.W. Perich, *Synlett*, 1992, 595.
147. L. Otvos and M. Hollosi in "Biologically Active Peptides: Design, Synthesis and Utilization", eds. W.V. Williams and D.B. Weiner, Technomic Publ. Co., Pennsylvania, USA, 1993, Chapter 7, p. 172.
148. A.V.R. Rao, T.K. Chakraborty and S.P. Joshi, *Tetrahedron Lett.*, 1992, **33**, 4045.
149. A.G. Brown, M.J. Crimmin and Peter D Edwards, *J.Chem.Soc. Perkin Trans. 1*, 1992, 123.
150. R.B. Lamont, D.G. Allen, I.R. Clemens, C.E. Newall, M.V.J. Ramsay, M. Rose, S. Fortt and T. Gallagher, *J.Chem.Soc.Chem.Commun.*, 1992, 1693.
151. A.J. Pearsond and J.G. Park, *J.Org.Chem.*, 1992, **57**, 1744.
152. A.V.R. Rao, T.K. Chakraborty, L.K. Reddy and A.S. Rao, *Tetrahedron Lett.*, 1992, **33**, 4799.
153. T.K. Chakraborty and G.V. Reddy, *J.Org.Chem.*, 1992, **57**, 5462.
154. G. Batta, K.E. Kövər, Z. Székely and F. Sztaricskai, *J.Amer.Chem.Soc.*, 1992, **114**, 2757.
155. Y-H. Chu and G.M. Whitesides, *J.Org.Chem.*, 1992, **57**, 3524.
156. J.L. Carpenter, P. Camilleri, D. Dhanak and D. Goodall, *J.Chem.Soc.Chem. Commun.*, 1992, 804.
157. Ref. 47, p. 158.
158. H. Kunz and W.K.D. Brill, *Trends Glycosci, Glycotechnol*, 1992, **4**, 71.
159. T. Bielfeldt, S. Peters, M. Meldal, K. Block and H. Paulsen, *Angew.Chem.Int.Ed.*, 1992, **31**, 857; *Tetrahedron Lett.*, 1992, **33**, 6445.
160. S. Peters, T. Bielfeldt, M. Meldal, K. Bock and H. Paulsen, *J.Chem.Soc. Perkin Trans. 1*, 1992, 1163.
161. A.M. Jansson, M. Meldal and K. Bock, *J.Chem.Soc. Perkin Trans. 1*, 1992, 1699.
162. Y. Mimura, Y. Yamamoto, Y. Inoue and R. Chujo, *Int.J.Biol.Macromol.*, 1992, **14**, 242.
163. B. Luening, T. Norberg and J. Tejbrant, *J.Carbohydrate Chem.*, 1992, **11**, 933.
164. K. Fukase, S. Hase, T. Ikenaka and S. Kusumoto, *Bull.Chem.Soc.Jpn.*, 1992, **65**, 436.
165. F. Goto and T. Ogawa, *Tetrahedron Lett.*, 1992, **33**, 5099.
166. M.K Gurjar and U.K. Saha, *Tetrahedron Lett.*, 1992, **33**, 4979.
167. M.K. Gurjar and U.K. Saha, *Tetrahedron*, 1992, **48**, 4039.
168. E.W. Holla, M. Schudok, A. Weber and M. Zulauf, *J.Carbohydrate Chem.*, 1992, **11**, 659.
169. M. Schultz and H. Kunz, *Tetrahedron Lett.*, 1993, 5319.
170. M. Schultz, P. Hermann and H. Kunz, *Synlett*, 1992, 37.
171. P. Braun, H. Waldmann and H. Kunz, *Synlett*, 1992, 39.
172. R. Polt, L. Szabo, J. Treiberg, Y. Li and V.J. Hruby, *J.Amer.Chem.Soc.*, 1992, **114**, 10249.
173. B. Leuning, T. Norberg, C. Rivera-Baeza and J. Tejbrant, *Glycoconjugate J.*, 1991, **8**, 450.
174. M. Gobbo, L. Biondi, F. Filira, B. Scolaro, R. Rocchi and T. Pick, *Int.J.Pept.Protein Res.*, 1992, **40**, 54.

175. Y. Boulanger, Y. tu, V. Ratovelomanana, E. Purisima and S. Hanessian, *Tetrahedron*, 1992, **48**, 8855.
176. K. Dzierzbicka, A. Kolodziejczyk and B. Wysocka-Skrzela, *Pol.J.Chem.*, 1991, **65**, 2293.
177. D. Kikelj, J. Kidrič, P. Pristovšek, S. Pečar, U. Urleb, A. Krbaučič and H. Hönig, *Tetrahedron*, 1992, **48**, 5915.
178. Z. Mackiewicz, H. Swiderska, M. Kalmanowa, Z. Smiatacz, A. Nowoslawski and G. Kupryszewski, *Collect.Czech.Chem.Commun.*, 1992, **57**, 204.
179. L. Urge, L. Otvos, Jr., E. Lang, K. Wroblewski, I. Laczko and M. Hollosi, *Carbohydrate Res.*, 1992, **235**, 83.
180. F.E. McDonald and S.J. Danishefsky, *J.Org.Chem.*, 1992, **57**, 7007.
181. T. Teshima, K. Nakajima, M. Takahashi and T. Shiba, *Tetrahedron Lett.*, 1992, **33**, 363.
182. L. Varga-Defterdarovic, S. Horvat, N.N. Chung and P.W. Schiller, *Int.J.Pept.Protein Res.*, 1992, **39**, 12.
183. J. Maerz and H. Kunz, *Synlett*, 1992, 589.
184. H. Kunz and J. März, *Synlett*, 1992, **591**.
185. Ya. A. Chernyak, L.O. Kononov, P.R. Krishna, N. Kochetkov and A.V.R. Rao, *Carbohydrate Res.* 1992, **225**, 279.
186. J. Lee and J.K. Coward, *J.Org.Chem.*, 1992, **57**, 4126.
187. B. Imperiali, K.L. Shannon and K.W. Rickert, *J.Amer.Chem.Soc.*, 1992, **114**, 7942.
188. H. Kessler, V. Wittmann, M. Köck and M. Kottenhahn, *Angew.Chem.Int.Ed.*, 1992, **31**, 902.
189. C.R. Bertozzi, P.D. Hoeprich, Jr. and M.D. Bednarski, *J.Org.Chem.*, 1992, **57**, 6092.
190. L. Petrus, J.N. BeMiller, *Carbohydrate Res.*, 1992, **230**, 197.
191. K-H. Wiesmüller, W.G. Bessler and G. Jung, *Int.J.Pept.Protein Res.*, 1992, **40**, 255.
192. J.W. Metzger, W. Beck and G. Jung, *Angew.Chem.Int.Ed.*, 1992, **31**, 226.
193. A.M. Kolodziejczyk, A.S. Kolodziejczyk and S. Stoev, *Int.J.Pept.Protein Res.*, 1992, **39**, 382.
194. H.B.A. De Bont, J.H. Van Boom and R.M.J. Liskamp, *Recl.Trav.Chim.Pay-Bas*, 1992, **111**, 222.
195. J.M. Balkovec, R.M. Black, M.L. Hammond, J.V. Heck, R.A. Zambias, G. Abruzzo, K. Bartizal, H. Kropp, C. Trainor, R.E. Schwartz, D.C. McFadden, K.H. Nollstadt, L.A. Pittarelli, M.A. Powles and D.M. Schmatz, *J.Med.Chem.*, 1992, **35**, 198.
196. J.M. Balkovec and R.M. Black, *Tetrahedron Lett.*, 1992, **33**, 4529.
197. R.A. Zambias, M.L. Hammond, J.V. Heck, K. Bartizal, C. Trainor, G. Abruzzo, D.M. Schmatz and K.M. Nollstadt, *J.Med.Chem.*, 1992, **35**, 2843.
198. I. Toth, G.J. Anderson, R. Hussain, I.P. Wood, E.O. Fernandez, P. Ward and W.A. Gibbons, *Tetrahedron*, 1992, **48, 923.**
199. G. Kokotos, V. Constantinou-Kokotou, E.D.O. Fernandes, I. Toth and W.A. Gibbons, *Liebigs Ann.Chem.*, 1992, 961.
200. S.J. Wood and R. Wetzel, Int.J.Pept.Protein Res., 1992, **39**, 533.
201. U. Schmidt, A. Kleefeldt and R. Mangold, *J.Chem.Soc.Chem.Commun.*, 1992, 1687.

5
Current Trends in Protein Research

by J. A. LITTLECHILD

1 Introduction

The subject of this chapter is to cover some trends in protein research over the year 1992. To cover all of the literature appearing during this year and to put it into context with that appearing in earlier years when this specialist periodical report did not cover protein research is a job that is beyond the space allocated to this chapter. I have however tried to address the special areas of interest in protein research starting with protein sequence and structure prediction methods, factors affecting protein stability and folding, and ending with discussion on special groups of important proteins, and some catalytic mechanisms.

I must apologize to groups working in those areas of protein research that I have not been able to cover in this report.

2 Protein Sequencing

In many cases protein sequencing is being carried out indirectly by DNA sequencing techniques. Important developments of gas phase sequencing from liquid samples and protein bands electro-blotted on to polyvinyl difluoride (PDVF), from sodium dodecyl sulphate acrylamide gels now allows routinely 25-50 residues of N-terminal sequence to be obtained from as little as 10-100 pmol of protein. The ability to take bands blotted from pre-stained acrylamide gels enables proteins to be sequenced from complex mixtures without prior purification.

Tryptic or other proteolytic peptide fragments from purified proteins can also be readily sequenced where the information is required to construct oligonucleotide probes for gene isolation. Mass spectrometry is not only being used to study protein modification (see below) but also for protein sequencing. New technology for protein sequencing is discussed by Granlundmoyer[1] and Wittmann-Liebold[2]. The ninth international conference on Methods of Protein Sequence Analysis was held in Japan in September 1992. The short communications presented at this meeting are reported in a special issue of the Journal of Protein Chemistry[3]. Capillary zone electrophoresis offers an interesting analytical approach for peptides[4]. Bergman has shown that it is feasible to obtain sufficient amounts for sequence analysis with this method[5]. Moritz and Simpson have used capillary reversed-phase high performance liquid-chromatography to obtain high sensitivity protein sequence analysis[6]. Sequencing of proteins is carried out by Edman degradation of N-terminal residues. A paper has described a new sensitization method involving the aminolysis of the reaction intermediate, 2-anilino-5-thiazolinone to PTC-amino acid amide with 4-aminofluorescin instead of the rearrangement to PTH-amino acid[7]. The same group have developed a method for C-terminal sequencing using prefluorinated carboxylic acids in monohydrate form and carboxylic acid anhydrides. These reagents produce the mixtures of C-terminal successive degradation molecules which are able to be analysed by FAB or ESI-mass spectrometry[8]. A C-terminal sequencer has been designed by Bradaczek and Wittmann-Liebold[9] and Nokihara *et al.*[10], and in the future we should expect to see C-terminal sequencers commercially available. Bailey *et al.*[11], have reported the automated carboxy-terminal sequence analysis of proteins and peptides using isothiocyanate reagents to produce amino acid thiohydantoin derivatives. The peptides are covalently coupled to a novel polyethylene solid support[12].

A large number of proteins are N-terminally blocked. Of these blocking groups, the acetyl group is the most common, but formyl and proglutamyl groups are often found in many proteins and prevent them from being sequenced by Edman degradation. Tsunasawa and Hirano have developed a method of microsequencing such N-blocked proteins[13] by deblocking after blotting the protein onto a PVDF membrane. The membrane is exposed to trifluoroacetic acid vapour for acetyl serine/threonine, incubation in 0.6 M HCl at 25°C for 24 hours to remove formyl groups or treatment with 0.5% polyvinylpyrrolidone-40, 0.1 M acetic acid at 37°C for 20 minutes followed by pyroglutamyl peptidase to remove pyroglutamic acid. Alternative methods are still required for proteins blocked with myristoyl and alkyl groups.

2.1 *Mass Measurements of Proteins.*

Most new protein sequences are now obtained from DNA sequence of the respective gene. This means that protein amino acid modification such as ε-N-acetylation of lysines[14] and specific oligosaccharide modifications[15] are not detected. For this reason the improvements recently made in mass measurement of proteins has been of vital importance.

Two recent developments are electrospray (ESMS)[16] and matrix-assisted laser desorption-time of flight (LD-TOF)[17-19] which allow

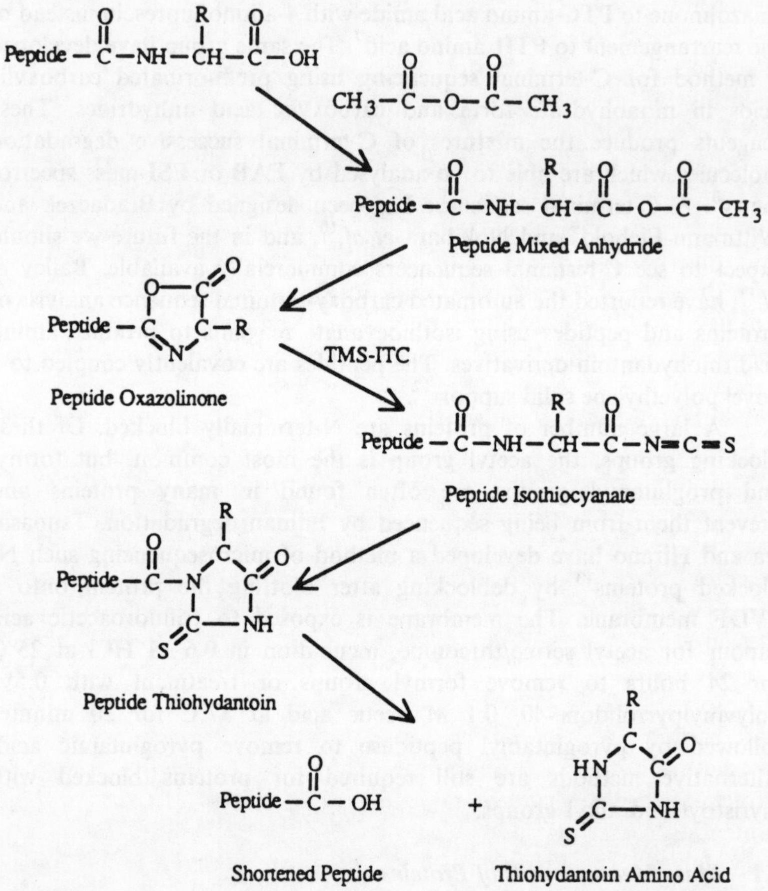

Peptide — C(=O) — NH — CH(R) — C(=O) — OH

CH_3—C(=O)—O—C(=O)—CH_3

Peptide — C(=O) — NH — CH(R) — C(=O) — O — C(=O) — CH_3

Peptide Mixed Anhydride

Peptide Oxazolinone

TMS-ITC

Peptide — C(=O) — NH — CH(R) — C(=O) — N=C=S

Peptide Isothiocyanate

Peptide Thiohydantoin

Peptide — C(=O) — OH + Thiohydantoin Amino Acid

Shortened Peptide

Figure 1. **Reaction scheme for the sequential C-terminal degradation of peptides using the thiocyanate chemistry (reproduced with permission from reference 11, Cambridge University Press).**

accurate measurement of a wide molecular-mass range that encompasses most proteins (40→300 K Daltons). The increased accuracy allows mass measurements to be made within 0.005–0.02% of those calculated from amino acid sequences. This is accurate enough to detect the loss of a single amino acid (truncation), reduction of a few disulphide bridges, significant oxidation of a single amino acid or single point mutations. An important role for this technique is also in the characterisation and quality control of chemical conjugates of therapeutic proteins such as described in a recent report for the modification of proteins with polyethylene glycol to increase their lifetimes in plasma[20]. Mass spectrometry is useful to study N-terminally blocked peptides or proteins.

The rate of a reaction such as proteolysis can be followed by mass measurement. This has been used to follow the limited proteolysis of non-denatured proteins[21,22]. In the future we may look to using mass spectrometry to sequence proteins directly[23,24] rather than using conventional gas phase sequencing methods. A paper by Johnson and Walsh describes the sequence analysis of peptide mixtures by automated integration of Edman and mass-spectroscopic data[25]. New sequencing reagents have been made which yield amino acids with enhanced detectability by mass-spectrometry[26].

The technique of electrospray MS is not only limited to covalent modifications of proteins but has recently been shown to be sufficiently mild that non-covalent complexes between proteins and specific ligands can be observed. This approach has been applied to the study of receptor[27] drug interactions and to protein-protein interactions[28].

Two reviews on the potential and future developments in mass measurement of peptides and proteins have been written in 1992[29,30].

Several papers have appeared in 1992 which discuss organisation of the protein-sequence data base[31,32]. The SWISSPROT. protein sequence data-bank has been documented by Bairoch and Boeckmann[33] and PIR-international protein-sequence database by Barker *et al.*[34]. A database of Protein-Sequence Alignments is presented by Barker *et al.*[35]. The major protein databases are accessible through SEQNET facility at the Daresbury laboratory, Daresbury, U.K.[36]. These include GENEBANK, EMBL, NBRF, SWISSPROT, and others.

3 Protein Sequence and Structure

The number of protein primary sequences now available is increasing dramatically. With the ability to sequence proteins at the DNA level this is likely to increase at an alarming rate. There are now over 60,000 known protein amino acid sequences. How can we translate this information into protein structure? It is now clear that there can be many different protein sequences which can be folded into the same overall shape. This means that very different proteins can have different amino acid sequences but can fold up in a similar way.

Protein structures are stored at the Protein Data Bank in Brookhaven[37]. As of January 1993 this contained 1,100 structures determined from X-ray crystallographic and solution NMR studies. Of this data one can identify over 120 distinct folding patterns. It appears however that the number of possible protein folds should be limited to about 1,000. This is discussed in an article by Chothia[38]. There are still great difficulties in predicting protein structure from sequence alone. The best method of prediction remains the use of known protein folding motifs for modelling unknown protein structures. Modelling unknown structures into a known protein structure which is known to be structurally related is still the best approach to structure prediction. Several software packages are available for this purpose, for example Homology Biosym[39], and Quanta, Polygen[40]. Recently this kind of approach has been improved by the group of Blundell[41]. In this method sequences are aligned to sequences, but the scoring table used is different for each position and depends on the environment of the known structure.

A different approach to thinking about this problem has recently been extended by the group of Eisenberg[42,43]. This method is called inverted structure prediction. This measures the compatibility of a sequence with a structure. The 3D profile method classifies amino acid residues into 19 'environmental classes' on the basis of solvation preference and also considering the polarity of the environment and the secondary structure. The pioneering work of Ponder and Richards[44] made the reasonable assumption that residues in the interior of the protein (apolar) determine the protein fold. In general the interior of a protein is made of apolar amino acids and the exterior by polar amino acids although polar residues can sometimes be found in fully buried sites so long as their hydrogen bonding groups are satisfied by interaction with nearby polar atoms. Sippl and co-workers have used the inverted protein folding approach using defined semi-empirical energy functions for interactions of protein side chains[45]. Using this approach, globin sequences with very low similarity to globins of known structure were correctly matched to one of the globins in the structural data base. The group of Thornton[46] have recently extended the above procedures involved in inverse folding by using a 'threading' procedure to thread the amino acid sequence into a known structure and selecting the structures with the lowest energy. This approach looks very promising and reports the correct recognition of distantly related folds for several proteins. The method was able to select the correct fold for proteins as distantly related as phycocyanin and globins (7% sequence identity) stellacyanin and azurin (6% sequence identity). Another variation of the inverted folding algorithm has been devised by Godzik and Skolnick[47] using the topology fingerprint approach extended to protein fragments. Finally a more theoretical approach to this problem has been reported by Goldstein *et al.*[48,49], using concepts from spin-glass theory to approach the inverted folding problem.

3.1 *Structural Motifs.*

As discussed in the section above over 120 distinct protein structural motifs are reported in the protein data bank at Brookhaven. This review article cannot begin to discuss all of them.

3.1.1 a/β Barrel. – From both personal interest and evolutionary consideration the eight stranded αβ barrel fold will be discussed since twenty one proteins are now known to have this fold. Several of these structures were reported in the last year or so. These include aldose reductase[50], the first αβ barrel protein to bind the cofactor NADPH; cyclodextrin glycosyltransferase[51], *Aspergillus niger* Acid α amylase[52]; Soyabean β-amylase[53]; adenosine deaminase[54], Tobacco ribulose 1,5 bisphosphate carboxylase/oxygenase[55], *Escherichia coli* phosphoribosyl-anthranilate Isomerase: Indoleglycerolphosphate Synthase[56], Human muscle aldolase[57], *Drosphilia melanogaster* fructose bisphosphate aldolase[58] and Mandelate racemase[59]. These enzymes clearly have completely different activities and the question is still not clear as to how they have evolved: whether it be by convergent or divergent evolution. Up until recently all α/β barrel proteins were enzymes but the structure of narbonin, a seed storage protein which has no activity has been reported[60]. In this paper it is suggested this molecule is proof of convergent evolution to a stable fold. However other α/β barrel proteins such as enolase has evolved into τ-crystallin in duck lens[61]. There is an apparent lack of sequence homology between proteins of this group. However, sequence homology at the phosphate binding sites of nine of the α/β barrels has been found[62]. Similarities and differences between human cyclophilin A and other β-barrel structures has been discussed in the light of recent structural refinement at 1.63Å[63]. It appears to have differences to other proteins of this group, one of which is that it is a 'closed' barrel so that neither the immunosuppressive drug cyclosporin A, nor the proline containing substrate (in its proline isomerase activity) can bind to the hydrophobic core of the barrel. In this case cyclophilin A is thought to be neither functionally or evolutionary related to other β barrel proteins. The active sites of the αβ barrel enzymes are not all at the same position in the barrel. The human muscle aldolase has an additional N-terminal helix which closes off the amino end of the α/β barrel. The active site comprising the side chain of the Schiff base-forming lysine residue is in a partially buried environment with the ε-amino group of the lysine pointing up from the floor of the cavity formed across the carboxyl end of the α/β barrel. In contrast the active sites of both triose phosphate isomerase and pyruvate kinase, other β-barrel proteins, occur much more towards the carboxyl ends of the β strands. The NADPH cofactor in aldose reductase is bound in the same position in the barrel as the flavin cofactors of flavocytochrome b[64] and glycolate oxidase[65]. Adenosine deaminase54 is the first α/β barrel known to bind zinc, and does not appear to be closely related to the other proteins. The above facts provide evidence for divergent evolution.

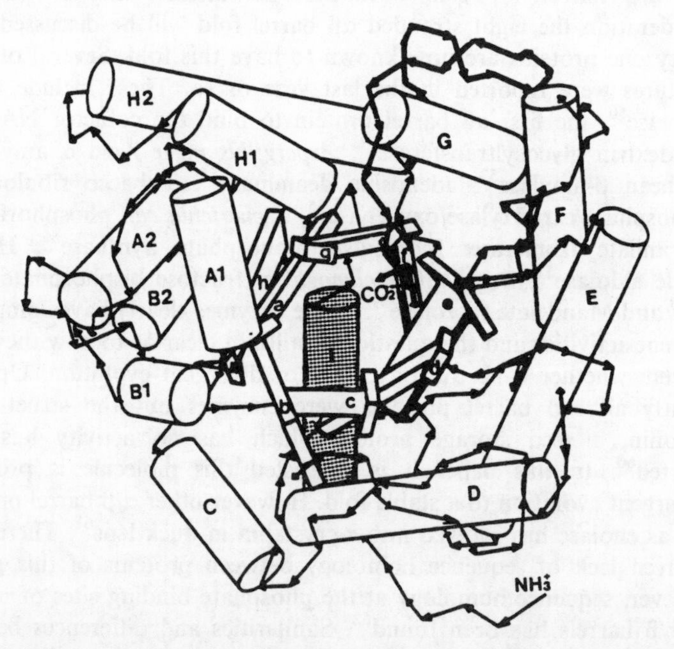

Figure 2. A diagram of the α/β barrel structure of one of the four identical human muscle aldolase subunits produced using the programme devised by Lesk and Hardman showing the helices as barrels and the β sheets as arrows. The C-terminal tail which extends from the end of the helix H2 to the centre of the α/β barrel is drawn in bold. A tyrosine residue at the C-terminus of the protein is important for activity of the enzyme. The additional N-terminal helix is shaded. The Schiff base lysine residue is located on strand f.[57]

The prediction of α/β barrel proteins has been extended and reviewed recently by the group of Sternberg[66]. They concluded that it was very difficult to predict this motif using the inverse folding problem. Further work on the folding pathways of α/β barrel proteins has been reported in 1992[67,68] including the reversible unfolding and refolding of a monomeric aldolase from *Staphlococcus aureus*. The group of Kirscher has continued their work with the folding of the protein N-(5′phosphoribosyl) anthranilate isomerase. Two protein fragments of this protein have been made, one with a six stranded domain and the other with a two stranded fragment and the reconstituted protein has nearly identical activity to the wild-type enzyme[69]. In addition, the same group has engineered a disulphide bond between α helices 1 and 8 of the same protein[70] and found this to increase its stability by 1 kcal mol^{-1}. Scheerlinck *et al.*,[71] find that the loops in α/β barrel proteins cannot be modified, although they are away from the active site. Kirschner and Urfer with N-(5-phosphoribosyl) anthranilate isomerase find that loops at the amino terminal of the barrel are more important for stability of the protein than loops at the carboxy-terminal end of the barrel[72].

3.1.2 αβ Hydrolase Fold. – Another new protein fold that has been reported in 1992 is the α/β hydrolase fold[73]. A group of enzymes have been compared with each other that have different sequences, substrates and physical properties. These are acetyl cholinesterase from *Torpedo california*[74], carboxypeptidase II from wheat[75,76], dienelactone hydrolase from *Pseudomonas* sp. B13,[77] haloalkane dehalogenase from *Xanthobacter autotrophicus*[78] and lipase from *Geotrichum candidum*[79]. Some of the proteins are monomers, some are dimers; some are glycosylated others not and they vary in molecular weight from 25-60 KDaltons. All of the five enzymes have a central catalytic domain of unique topology and a three dimensional structure which has been called the 'α/β hydrolase fold'. They all have a 'catalytic triad' (nucleophile usually serine, an acidic residue, glutamic or aspartic acid, and a histidine residue) which reside at the same topological location. The nucleophile is located at the crossover point of the parallel β-sheet. A distinct 'nucleophile elbow' is found which is made up of the nucleophile located as the central residue in a sharp turn between a specific β strand and α helix. The sharpness of the turn results in the nucleophile backbone phi and psi angles being in an unfavourable region of the Ramachandran plot. The 'elbow bend' is quite different from the peptide surrounding the nucleophile in papain. Also the sequence around the nucleophile must be Sm-X-Nu-X-Sm-Sm (Sm - small residue) this is different from G-X-Nu-G-G seen in the trypsin like proteases. The catalytic triad residues in the α/β hydrolase fold always appear in the order, nucleophile, acid, histidine, and this is different from other proteins forming the 'catalytic triad'. In the α/β hydrolase fold enzymes the twist of the sheet imposes a 'handedness' which is approximately a 'mirror image' of that seen in the serine proteases. In all of the α/β hydrolase fold enzymes there is an oxyanion hole that appears to be well designed to stabilize the tetrahedral intermediate. The results presented in reference 73 suggest that the five enzymes have evolved from a common ancestor so as to preserve the positions of key catalytic components, which can cause a hydrolysis reaction.

3.2 *Loop Regions in Proteins.*

Much study has gone into the structure of loop regions in proteins. These are the segments connecting the α helices and β strands of the protein. Short loop regions can be classified into groups[80]. This helps with model building of unknown protein structures since a data base of the loop regions of all proteins in the data base can be used as a search library[81]. A different approach to loop classification has been described very recently[82] which groups the loop structures into three main classes, linear, non-linear and flat and globular. Longer loops have more conformational possibilities and adopt unique conformations on tertiary templates. These are more difficult to model however rules can be applied in specific cases i.e. antibody loops.

A novel three dimensional structure has been reported by Efimov[83] called a 3β-corner. This can be represented by an antiparallel triple stranded β-sheet folded on itself so that the two β-β hairpins are packed approximately orthogonally in different layers and the central strand bends at 90° in the right-handed direction when passing from one layer to the other. In all triple-stranded corners observed in proteins, the first β-β hairpins are right-handed and the second ones are left-handed when viewed from the concave sides of the corners.

3.3 *Protein Modules.*

The idea that many proteins are made up of structurally similar modules is an area of increasing awareness. These proteins were difficult to crystallise so until solution NMR was used little was known of their structure. Proteins of this group include those associated with blood coagulation, fibrinolysis, neural development and cell adhesion. The DNA coding for many eukaryotic proteins has introns, which is additional DNA found between the DNA coding for the protein which is called an exon. The idea that each exon represents a module is often the case[84]. The total protein is constructed of many either identical or

Enzyme	*Substrate*
Dienelactone hydrolase	Dienelactone 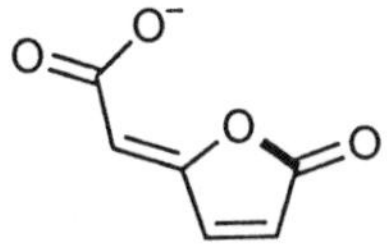
Haloalkane dehalogenase	1,2-Dichloroethane 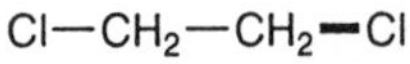
Carboxypeptidase II	Peptide 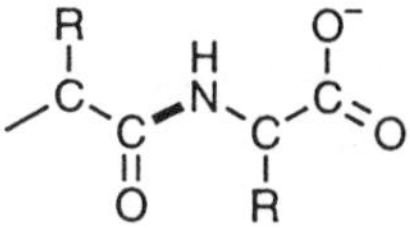
Acetylcholinesterase	Acetylcholine 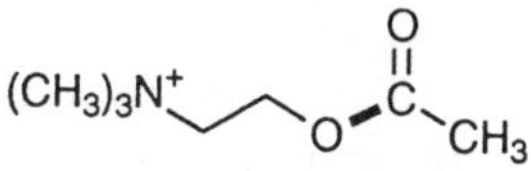
Lipase	Triacylglycerol 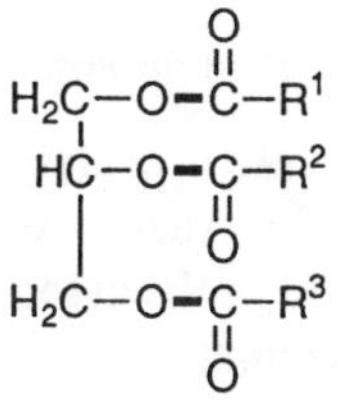

Figure 3. Substrates of the α/β fold enzymes. The bonds drawn with thick lines are cleaved by the enzyme (figure modified from reference 73 by permission of Oxford University Press).

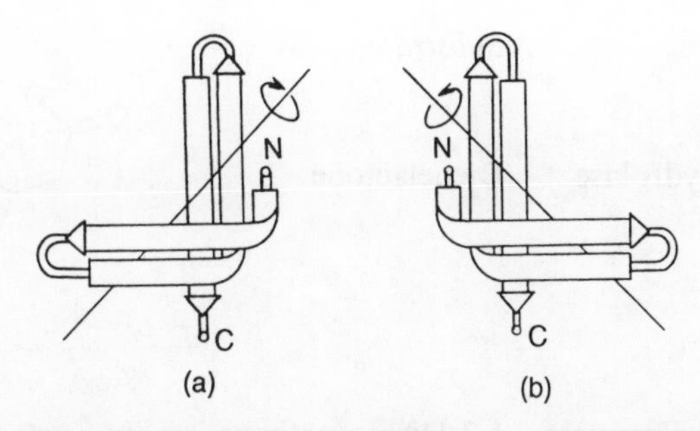

Figure 4. A schematic representation of right-handed a), left handed b) triple strand corners (figure reproduced with permission from reference 83).

different domains or modules. In 1992 two papers reported the epidermal growth factor (EGF)- like module, one of the earliest examples of this type found together with an immunoglobulin like domain in one protein[85,86]. This year also saw the structure of mouse EGF[87], human EGF[88], and three EGF's from factor X and IX89-91 reported from NMR studies.

Papers in this area are increasing so rapidly that a complete survey of the literature in 1992 is beyond the scope of this review.

4 Protein Folding

How proteins fold has been a major topic of investigation in the last few years. As proteins are made on the ribosome they are either folded in the cytoplasm or are exported from the cell via the endoplasmic reticulum.

It has become clear that in many cases additional proteins aid the folding process making sure that misfolding of proteins does not occur. The first of these proteins to be discussed is protein disulphide isomerase which catalyses the formation of disulphide bridge formation in proteins and ensures that the correct disulphide bridges occur. This enzyme is discussed in detail in a recent article[93]. The enzyme is a homodimer of 114-140 KDaltons and is highly homologous to the protein thioredoxin. The enzyme is found in bacteria where it is compartmentalised in the periplasmic space and in eukaryotes where it is found in the endoplasmic reticulum. This enzyme has recently been found to have other functions including proline hydroxylation, binding of thyroid hormone and triacylglycerol and transfer. These functions and the subcellular distribution of the enzyme are discussed in a recent review by Noiva and Lennarz[94]. It is suggested that the various catalytic functions of the enzyme are connected with separate active sites. Site-directed mutagen-

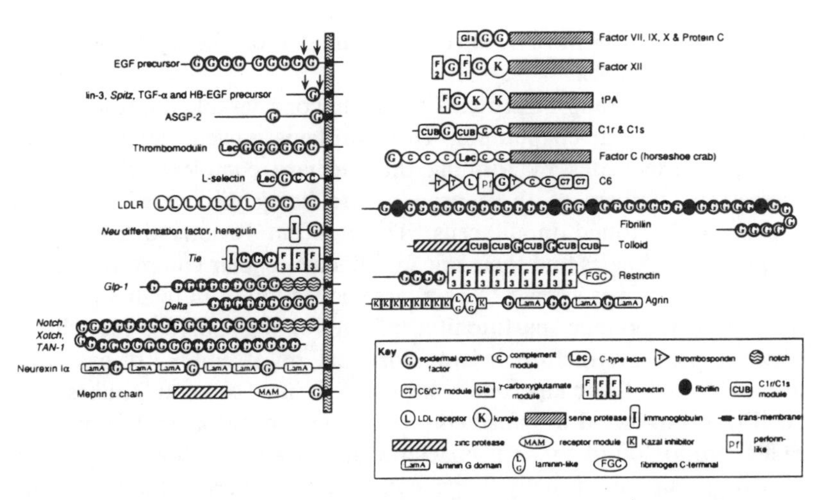

Figure 5. An illustration of some of the diverse proteins which contain EFG modules. EFG, epidermal growth factor, (reproduced with permission from reference 92, Current Science).

esis studies with human protein disulphide isomerase[95] has shown that the two active sequences in the polypeptide chain operate independently to one another.

It has been suggested for many years that the *cis-trans* isomerization of the amino acid, proline, is a rate limiting step in protein folding. There are now two families of structurally unrelated enzymes that are known to catalyse this reaction. These are the cyclophilins[96] and the FK506 binding proteins[97]. They are named after the immunosuppressants cyclosporin A and FK506 that inhibit the peptidyl proline isomerase activity.

Site-directed mutagenesis of cyclophilin has allowed the proline isomerase activity to be separated from cyclosporin A binding and calcineurin inhibition[98]. The FK506 binding proteins appear to be associated with the glucocorticoid receptor[99-101]. The location of proline isomerase activity has been found to be on the endoplasmic reticulum[102].

Models for mechanism of proline isomerisation have been made using the folding of RNase A and RNase TI as examples[103,104]. In the case of disulphide containing proteins the protein disulphide isomerase and the prolyl isomerase are found to have synergistic effects[105]. Kinetic studies have revealed information regarding the X-Pro bond which is twisted out of planarity by the prolyl isomerase enzyme, resulting in a non-polar transition state in which the single bond character of the peptide bond is increased[106,107]. NMR and X-ray studies have been carried out which have resulted in a model of the cyclosporin A/ Cyclophilin Complex[108-110].

The third group of proteins receiving considerable interest are the so called molecular chaperones. These proteins prevent aggregation of folding intermediates by forming intermediary complexes with them. Increasing numbers of these proteins are being discovered and they appear to be found in all cells. Three recent reviews discuss these chaperone molecules and their role in different cellular compartments[111-113]. The chaperones appear to be a large collection of unrelated proteins that mediate kinetic partitioning of (re)folding polypeptide chains between proper structure information and misassembly. It has been proposed that the chaperonins be divided into two subfamilies: the GroEL group found in eubacteria, plasmids and mitochondria and the TCP1 group found in archaebacteria and eukaryotic cytosol[114]. The GroEL group are usually large complex molecules with a 'double-doughnut' structure observed from electron microscopy studies[115]. These usually appear as cylindrical decatetramers of two stacked rings although mammalian mitochondrial chaperonin 60 shows it to have a heptameric, single ring structure[116]. Studies to elucidate how complexes are produced by GroEL (cpn 60) and GroES (cpn 10) in relation to the protein substrate have been made[117,118]. The GroES binds to one end of the GroEL cylinder which accommodates the protein substrate within its central cavity.

This is consistent with the idea that one decatetramer binds one ligand. A report of the crystallisation of complex of cpn 60/cpn 10 from *Thermus thermophilus*[119] should further our understanding of how these large multisubunit complexes assist protein folding. Eukaryotic cytosolic chaperones have been described in some detail[120-122] Many of the chaperonins are heat shock proteins (Hsp)[111-113]. Hsp 90 has been shown to chaperone the *in vitro* folding of citrate synthase[123]. Similarly the GroEL protein has been found to promote the in vitro renaturation of dodecameric glutamine synthetase[124] and other *Escherichia coli* proteins[125]. The TPC1 complex has been implicated in tubulin and actin assembly[120,122].

In the absence of a structure for a chaperone/protein complex only models can be proposed based on binding studies[126-130]. Many of the chaperones such as GroEL show a lack of specificity and will bind to 50% of the soluble proteins in *Escherichia coli*[125]. In the case of β-lactamase it has been shown that GroEL does not alter the folding pathway of the enzyme but prevents the accummulation of early folding intermediates[131]. Partially folded states of ribulosebisphosphate carboxylase have been found bound to chaperonin 60[129]. ATP participates in

Main chaperone		Cohorts (assistants)		Function
Prokaryotic	Eukaryotic	Prokaryotic	Eukaryotic	
HtpG	hsp90	?	hsp56 50 kDa (?) 23 kDa (?)	Bind unfolded polypeptides and either silence their function (e.g. steroid receptor protein), help them to fold properly or escort them to their proper cellular compartment (e.g. $pp60^{v\text{-}src}$).
DnaK	hsp70 BiP	DnaJ	Sec63 Scj1 Ydj1 (Mas5) Sis1	Bind to and maintain the unfolded state of polypeptides and release them following ATP hydrolysis; unfold partially folded polypeptides; disaggregate protein aggregates.
		GrpE	GrpE-like (?)	
GroEL	hsp60	GroES	cpn10	Bind to and maintain the unfolded state of polypeptides (molten globular structure?) and release them following ATP hydrolysis.
TF55	Tcp1	GroES-like (?)	GroES-like (?)	Role similar to GroEL/GroES (?)
SecB	SRP/docking protein; hsp70	SecA (PrlD) SecY (PrlA) SecE (PrlG)	? Sec61 ? Sec62 Sec63	Bind to nascent polypeptide chains destined to be translocated, localize them to the membrane and help them traverse the membrane.

Figure 6. Various classes of molecular chaperone machines (reproduced with permission from reference 114).

various stages of chaperone-assisted folding, although its exact role is not understood[132-134]. However, evidence is available for temperature dependent partial phosphorylation/dephosphorylation upon heat shock to regulate the protein binding properties of chaperones[135-139].

5 Protein Stability

A topic that is of increasing interest in the field of protein research in how proteins can remain stable at high temperatures. Studies with an enzyme found in all cells, the monomeric protein phosphoglycerate kinase has been used as a model enzyme to study protein thermostability[140]. Here we can see how nature has dealt with this problem since this enzyme can be isolated, its sequence and three-dimensional structure determined from a variety of sources e.g. yeast enzyme stable to 37°C, and two thermophilic bacteria, *Bacillus stearothermophilus* and *Thermus thermophilus* where the enzyme is stable to 69°C and 90°C respectively. The structure comparisons reveal that the main feature responsible for increased thermostability include additional ion pairs (some holding internal helices together and anchoring the N-terminus of the protein to the rest of the molecule), better α helix stability resulting from the removal of helix destabilizing residues and the addition of extra α helix dipole interactions. The extreme eubacteria such as *Thermus thermophilus* have a high G-C content in their DNA. This property is used at the codon level to increase the number of glycine and proline residues (codons GGG and CCC) and at the protein level these residues give rigidity (proline) and flexibility (glycine) to both stabilize and enable the protein chain to 'turn' in shorter surface loops. It also appears that the most easily deaminated amino acid asparagine is removed in all places at the active site where it is required for activity[142].

The small iron protein rubredoxin has been used to study thermostability by comparison of the structure of the enzyme from the

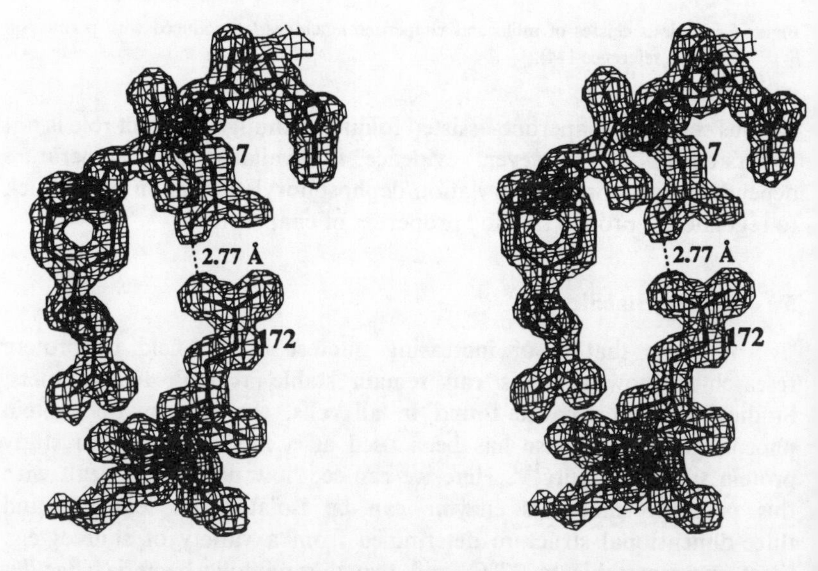

Figure 7. A typical ionic interaction that is present in the thermophilic *Bacillus stearothermophilus* phosphoglycerate kinase that is not present in mesophilic phosphoglycerate kinase. This interaction anchors the N terminus of the protein to the rest of the molecule[140,141].

hyperthermophilic Archaea *Pyrococcus furiosus* and from mesophilic sources[143].

A systematic study of understanding thermostability has been carried out over a number of years by the group of Matthews using the enzyme T4 lysozyme. Much of this work has been documented in a recent review[144] and involves site-specific mutagenesis of residues in combination with X-ray crystallographic studies of the mutant proteins to understand the principles of stability. Papers published in 1992 by this group include a study on mutation of residues at the N-terminus of helices in T4 lysozyme[145], substitution to alanine in an α helix of lysozyme to give a more stable protein[146], the ability of lysozyme to have proline substitutions in α helix[147]; a study of cavity-creating mutations and its relation to the hydrophobic effect[148]; a study of a cavity containing mutant which is stabilized by a benzene molecule[149], and a stabilizing disulphide bridge which closes the active site cleft of the enzyme[150].

Protein engineering has also been used to study the small protein barnase. A series of elegant experiments describe the analysis of mutant proteins and their stability. This work sets out to make a series of mutations in barnase and then to measure the changes in free energy of unfolding upon urea, guanidinium chloride or thermal induced denaturation[151]. A study was carried out to understand helix stability[152]. This study shows that the relative effect on helix stability of alanine and glycine residues is both position and context dependent. The relative free energies of alanine and glycine calculated for a model polyalanine helix show that alanine is more stable in the middle positions of an α helix but glycine is preferred at its ends[153]. The first and last residues of α helix are referred to as the carboxy-cap and the amino-cap.[154] Mutational experiments on the two carboxy-caps of barnase have generated a rank order of preference Gly > > Arg, His > Ala, Ser, Thr > Asp > Val. Whereas the rank order of preference at the amino-caps of barnase is Asp, Thr, Ser > Asn, Gly > Glu, Gln > His > Ala > Val.

A detailed analysis has been made of placing histidine residues at both caps of the major helices of barnase and has shown that the interaction energy between a single charge and the dipole is > 0.5 kcal mol^{-1} [155].

Site-directed mutagenesis has been used to examine amino acid substitutions at various positions on the second helix of barnase. Alanine is the most stabilizing residue whereas glycine destabilises the helix relative to alanine by 0.7-0.8 kcal mol^{-1} and proline by 3-4 kcal mol^{-1} [156]. The factors affecting the intrinsic effect of different amino acids at internal positions on helix stability could be burial of hydrophobic surface and van der Waals contacts, the entropy of folding[157] and different solvation effects[158].

Salt bridges that occur in proteins can be analysed as to their contribution to stability. Solvent exposed salt bridges are thought to contribute only 0.1→0.5 kcal mol^{-1} to stability[159]. This has been experimentally confirmed by mutagenesis studies with both T4 lysozyme and barnase[160,161]. Buried salt bridges can be contrasted with surface salt bridges in that removing the hydrogen-bonding partner to any buried charged group costs at least 3 kcal mol^{-1} and burying charges without a partner can lead to a major destabilising effect on protein stability which can range from 3-6 kcal mol^{-1} [162,163]. Specific hydrogen bonding in proteins has long been known to offer stability. Two recent papers published with ribonuclease T1 and barnase discuss these interactions[164,165]. In hydrogen bonds that have no access to water, removal of the uncharged partner/donor results in a decrease in stability of 1-2 kcal mol^{-1} per hydrogen bond. If the bond has access to water the decrease is smaller (0-0.5 kcal mol^{-1}).

The importance of the hydrophobic core to protein stability can be probed by using mutations that create detections in side chains without changing their geometry. Experiments on several proteins including barnase[165] and T4 lysozyme[148] show that a value of 1.5 ± 0.5 Kcal mol^{-1} per methylene group deleted can be found. The hydrophobic interactions appear to be cumulative and the largest decrease in stability is observed when several methylene groups are removed. Substitutions of hydrophobic side chains by others of different stereochemistry can often be tolerated[166,167].

Protein engineering studies have revealed the importance of histidine and aromatic amino acid interactions[168]. The interaction of this pair of residues in barnase is stronger when the histidine is protonated. This interaction contributes ~1 Kcal mol^{-1} to protein stability at low pH and also raises the pKa of the histidine by over half of a pH unit.

The active sites of enzymes are very often a source of instability in the protein. A recent paper has confirmed that some of the interactions which normally confer stability do not do so in the active site of barnase[169].

6 Proteins that Bind Nucleic Acid

An important group of proteins that have received considerable attention in the last two years are those proteins which interact with the genetic material of the cell, namely DNA. Proteins play important roles in binding to nucleic acids and in so doing control transcription and specifically repair, cut, or modify DNA. Other proteins specifically bind to RNA. These are the aminoacyl tRNA synthetases, ribosomal proteins, viral proteins and the protein component of small RNA protein particles, used in gene splicing.

Some of the first proteins studied in this group were found to have a fold in their polypeptide chain which produces a characteristic 'helix-turn-helix' motif that interacts with the bases and phosphate backbone of DNA (reviewed[170]). For a while it was thought that this was the way in which the protein chain would fold in all proteins in this group. The last two years has shown this is not the case and a variety of protein folds are now known to occur. Nature has provided us with a number of proteins which fold into different shapes in order that they can either grip or slide along the nucleic acid duplex.

6.1 *Transcription Factors.*

Transcription factors are proteins which bind to DNA and regulate gene expression by controlling mRNA transcription. They therefore play a vital role as to which genes will be turned on or off in response to other factors or during cellular development. Two reviews have been published during the past few years which discuss different classes of protein/DNA binding domains[171,172]. Many of the proteins have been crystallised with their target nucleic acid so specific details of their interactions are known.

In the past year the structures of another five new transcription factors have been solved by X-ray crystallography and they all belong to different classes. A recently reported structure that is related to the 'helix-turn-helix' motif is that of the yeast MATα2 homeodomain[173]. Homeodomains are 60 amino-acid DNA binding domains consisting of three α-helices and an amino-terminal arm. The protein binds DNA as a monomer inserting the third long α helix into the major groove of the DNA and the amino-terminal arm into the adjacent minor groove. The structure of a transcriptional activator of the bovine papilloma virus has recently been reported[174]. This a dimeric protein which is conserved among the papillomaviruses. It has a unique mode of binding to DNA by presenting an α helix from each monomer to contact DNA bases in the major groove which is connected to a eight stranded anti-parallel β-barrel, composed of four *met* repressor from *Escherichia coli* co-crystallised with the operator DNA reveals that DNA recognition is by β strands[175]. The *met* operator complex shows two dimeric repressor molecules bound to adjacent sites 8 base pairs apart on an 18 base-pair DNA fragment. Sequence specificity is achieved by insertion of double-stranded antiparallel protein β-ribbons into the major groove of the B-form DNA.

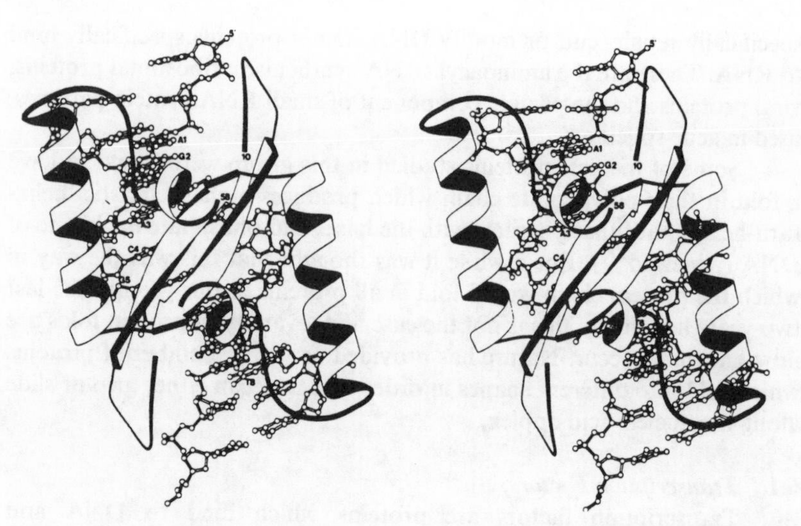

Figure 8. Stereo diagram of part of the *met* repressor-operator complex. The DNA is shown as a ball-and-stick model, with filled bonds for the sugar-phosphate backbone, and open bonds for the remainder of the ribose rings and the bases. The central region of the DNA shown corresponds to one *met* box and the bases have been labelled from one strand. The β-ribbon lies in the major groove. (Reproduced with permission from reference 175. Copyright 1992. Macmillan Magazines Limited.)

A second important protein structure is that of the TF11D TATA-box binding protein[176]. This protein is required to bind to a complex with other proteins to form a pre-initiation complex necessary for RNA polymerase mediated transcription. This protein has been described as 'saddle shaped' being a monomer with a 10 stranded β sheet lining the concave surface and two pairs of α-helices located at each end. Although this structure has not been solved bound to DNA it appears from mutagenesis studies that amino-acid substitutions affecting DNA binding are distributed across the concave surface of the 'saddle'.

Another class of DNA binding protein is that of the so called leucine zipper. This type of protein is found in a number of eukaryotic transcription factors and has recently been reviewed[177]. The leucine zipper is made up of two interacting α helices which contain a regular repeat of the amino-acid residue leucine. The helices interact to form a parallel coiled coil. Residues which are amino-terminal to the leucine zipper helices are important for DNA binding. These basic regions are usually disordered in the absence of DNA. A recently reported structure is that of a yeast transcriptional factor GCN4[178]. This protein grasps the DNA l

The final class of transcription factors to be discussed here is that of the zinc-binding proteins. A recently reported protein structure falling into this class is that of the GAL4 yeast transcriptional activator. This protein has a metal binding domain containing two Zn^{2+} ions co-ordinated by six cysteine residues which in the protein this forms a $Zn_2(Cys)_6$ binuclear cluster. The role of zinc is increasingly important in nucleic acid binding proteins. The structure of this protein has been solved by X-ray crystallography[179] and by NMR using Cd-substituted protein[180,181]. The protein is a dimer, each monomer consisting of a metal binding domain joined by a linker α helix which serves to form the dimer. The metal binding domains insert into the major grooves on opposite faces of the DNA. This type of protein is referred to as a 'zinc finger' binding protein because the metal ion helps to allow the protein to form 'finger like' projections which can bind to DNA. In the previous year, 1991, the structures of two other types of 'zinc twist' proteins had been reported. These were the oestrogen receptor protein[182] and the glucocorticoid receptor[183] which have two zinc atoms each co-ordinated to four cysteine residues. A central α helix in the 'zinc twist' forms the binding site to DNA and the second zinc binding site is important for dimer interaction[184]. This forms one of the Cys_2 His_2 group where two cysteines and two histidine residues are co-ordinated to the Zn^{2+} ion in each of the 9 fingers of this protein. Recently a paper has been published describing a detailed analysis of the TF111A-DNA complex predicted from footprinting data which shows fingers 1-3 and 7-9 wrap around the DNA whereas fingers 4-6 lie along one face of the DNA[185]. In addition, studies have been reported on a truncated version of TF111A[186]. Studies with adjacent zinc-finger motifs in multiple zinc-finger peptides from the yeast transcription factor SW15 have been shown to form structurally independent, flexibly linked domains[187] and solution studies have been carried out by 2D NMR to look at the different structures of the zinc fingers, 1, 2 and 3[188].

The structure of the human enhancer binding protein has been determined by NMR methods[189]. This is a two finger peptide which shows a different arrangement between domains than that reported in 1991 for a three finger peptide Zif 268[190]. Site-directed mutagenesis has been used to study the amino acid residues important for stabilizing the three dimensional structure of the zinc finger and for its recognition properties for DNA[191-195]. An important result that has now been proven to be correct is that retroviral nucleocapsid proteins such as HIV-1 contain zinc ions. These appear to be bound at Cys_3 His domains in the viral protein[196,197]. It is likely that these proteins play an important role in recognizing packaging signals within retroviral genomic RNA.

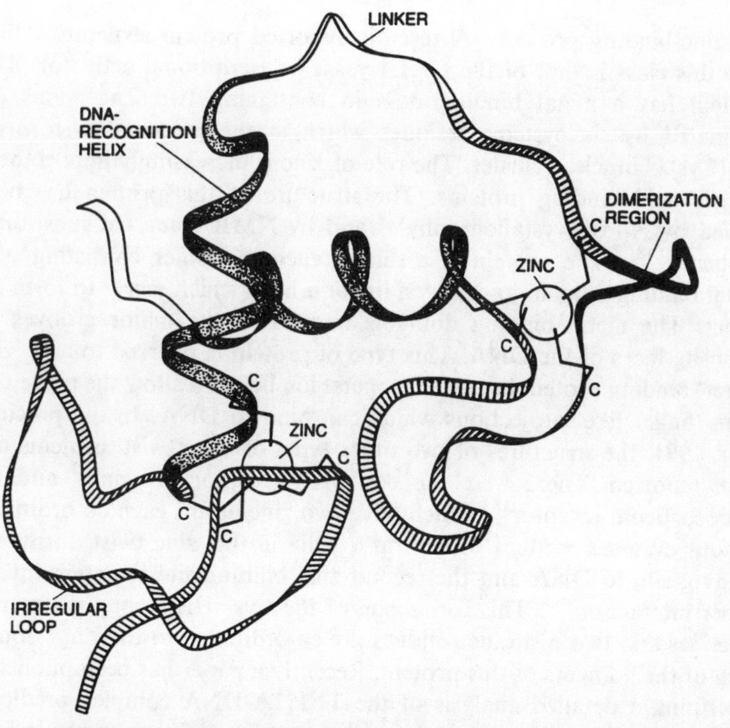

Figure 9. Three dimensional view of the DNA-Binding Domain of the Estrogen receptor. Showing the two zinc binding sites, one for interaction with DNA and the other for dimer formation (reproduced with permission from *Scientific American*, February 1993, 32–39. Copyright © 1993 by Scientific American, Inc. All rights reserved.)

6.2 *DNA Repair Enzymes.*

The enzymes that repair DNA are important molecules since they are responsible for repairing the accidental lesions in the genetic material that occur spontaneously in living cells. Three enzyme structures which carry out this role have been reported in the last year. These are of the bacteriophage T4 endonuclease V[198], the endonuclease III of *E. coli* [199] and the *Escherichia coli* recA protein[200].

When DNA is exposed to UV radiation pyrimidine dimers are produced which are mutagenic and lethal to cells. The T4 endonuclease V protein catalyses two reactions, the first of which is the incision of the N-glycosyl bond at the 5′-side of the pyrimidine dimer. The second reaction is the cleavage of the apyrimidinic phosphodiester bond. Despite these two independent activities the enzyme has a single compact domain composed of three α helices. One side of the protein presents a concave positively charged surface which is thought to contact DNA. Site-directed mutagenesis studies and structural studies have identified the catalytic centre of the N-glycosylase to be at Glu 23 and surrounding residues[198]. The recA protein[200] is a larger protein composed of a central major nucleotide binding domain and two flanking sub-domains one of which is involved in formation of a polymeric structure when the protein binds to DNA to form helical nucleoprotein filaments. To date the X-ray structural studies on these enzymes have been carried out without DNA.

6.3 *Restriction Enzymes.*

A group of proteins which have found unlimited use in modern day molecular genetics laboratories are the type II restriction nucleases. These enzymes cleave DNA at specific base recognition sites. The main role of these enzymes in nature is to protect bacteria against foreign DNA. Each restriction endonuclease has its accompanying methylase enzyme which protects the bacteria's own DNA from being cleaved. The methylase recognises the same DNA sequence specific site and transfers a methyl group from S-adenosyl methionine to a base in this sequence thereby protecting it from cleavage by the endonuclease. Over 2,000 different endonucleases have been discovered which recognise over 200 different DNA sequences[201]. Two structures already published of the restriction endonucleases show different modes of binding to the DNA. Both the E*co*RI[202] and E*co*RV[203] enzymes are dimers. The E*co*RI enzyme binds in the major groove of the DNA using a bundle of four helices whereas the E*co*RV enzyme contacts both major and minor grooves using four short loops. This difference may be attributed to the positions of the scissile bonds since E*co*RI cleaves DNA to leave single stranded 5′ overhangs, and E*co*RV cleaves DNA to leave 'blunt' ends with no overhang. Further studies on other endonucleases are needed to clarify these differences. The E*co*RV enzyme structure has been determined with and without DNA and it is clear that the enzyme undergoes large conformational changes when binding to DNA.

The catalytic sites of the two enzymes have been compared and many site-directed mutagenesis studies made. Several papers have been published[204-208] in 1992 which discuss this work. The catalytic sites appear to be quite different in E*co*RI and E*co*RV but they share a structural motif in which two acidic side chains and one basic side chain are in the vicinity of the scissile phosphodiester bond. As yet no structural information is available for the monomeric methylase enzymes.

6.4 *DNA Polymerases.*

These enzymes which are RNA or DNA dependent catalyse the reaction where the parent DNA strand is copied to make daughter strands. A lot of progress has recently been made in the study of these enzymes. An important member of this group is the AIDS HIV virus reverse transcriptase. The structure of this protein has proved difficult to solve due to the poor diffraction of the protein crystals obtained. However these problems have now been overcome and the X-ray structure of this protein was reported in 1992 by the group of Steitz[209]. The HIV reverse transcriptase is processed from a homodimer of 66 KDalton subunits which each contain a polymerase and an RNase H domain. In the mature protein the RNase H domain from one subunit is removed producing a heterodimer containing a 66 KDalton subunit and a 51 KDalton subunit. The heterodimer appears to have only one polymerase site, one RNase H site, one tRNA binding site and one binding[210] site for non-nucleotide inhibitors such as Nevirapine[211]. The 66 KDalton subunit is folded into 5 separate domains. The overall structure can be thought of as a right hand with domain being called 'fingers', and 'thumb' and a 'connection' subdomain. The conformation of the 51 KDalton subunit differs from the 66 KDalton subunit.

Also reported in 1992 was the structure of the β-subunit of DNA polymerase III[212]. Extensive mutagenesis studies have been reported on the polymerase region of DNA polymerase I[213,214]. The studies indicate that the polymerase domain in the different enzymes is related to a common evolutionary ancestor.

6.5 *Aminoacyl-tRNA Synthetases.*

These proteins are providing the first insights into protein-RNA interactions. The role of these enzymes is to specifically interact with the cognate tRNA and to charge that tRNA molecule with the correct amino acid *via* an aminoacyl-adenylate intermediate. These enzymes can be divided into two different classes based on their aminoacylation site on the terminal ribose of tRNA i.e. the 2′ OH of the terminal ribose for class 1 and 3′ OH for class 2 and different sequence motifs. For many years no structural information was available for these enzymes due to the lack of good quality co-crystals. A recent review of Moras[215] discusses these two structures and shows that they differ in their mode of tRNA recognition as revealed by two available crystal structures reported in 1989 and 1990[216,217].

In 1992 several advances in the determination of other crystal complexes were reported. These were the 6Å resolution structure of phenylalanyl-tRNA synthetase from the thermophilic bacteria, *Thermus thermophilus*[218], preliminary crystallographic results with aspartyl-tRNA synthetase-tRNA Asp complex[219] and a new crystal form complex between seryl-tRNA synthetase and tRNASer from *Thermus thermophilus*[220]. In the absence of crystallographic data, site-directed mutagenesis of the synthetase can be carried out to suggest sites of contact with the tRNA[221]. In mammalian cells the aminoacyl-tRNA synthetase enzymes are arranged in large complexes[222]. Similarly, engineering experiments have been reported to identify the regions of the protein involved in formation of the multi-synthetase complex. In the next few years it is hoped that further studies will resolve some confusion as to the function of some synthetase enzymes, particularly the phenylalanine aminoacyl synthetase enzyme where the sequence of this enzyme has characteristics of a class 2 synthetase but functions like a class 1 enzyme. In addition it is hoped that we will gain a further understanding of the multisynthetase complexes in mammalian systems.

7 Other Proteins of Special Interest

7.1 *Antibodies.*

Antibodies from the immune system in vertebrates provide a defence mechanism against foreign parasites, such as viruses and bacteria. The polypeptide chain of antibodies is divided into protein domains. The most common antibody IgG has two identical light chains and two identical heavy chains linked by disulphide bonds. The structure of these so called immunoglobulins is simply described by Branden and Tooze[223]. Antibody-antigen interactions are nature's way for proteins to specifically recognize and bind to other proteins and small molecules. Apart from their important involvement in the immune system, the study of these proteins is a topic of special interest. The work before 1990 has been documented in reviews by Davies *et al.*,[224] and Chothia[225]. Recent work has shown that the influenza virus protein neuraminidase interacts with a large area (900 Å^2) of the specific antibody which recognises it[226]. Antigen/antibody structures reported in 1992 are for peptides that produce an induced fit mechanism for interaction[227], a peptide hormone angiotensin[228], a cyclic peptide immunosuppressant[229] and the large neuraminidase protein discussed above[226-230]. The crystal structure of human immunoglobin fragment FAab New has now been refined at 2.0Å resolution[231]. The Fv structure from a human IgM immunoglobin has been reported by the group of Edmundson[232]. A human monoclonal-antibody Fab fragment against GP41 the HIV virus has been studied structurally by He *et al.*[233].

The mode of binding of antigens to the immunoglobulin molecule is via the six hypervariable loops (CDR's, see reference 223). It is a combination of these loops, minimum usually four , which determines

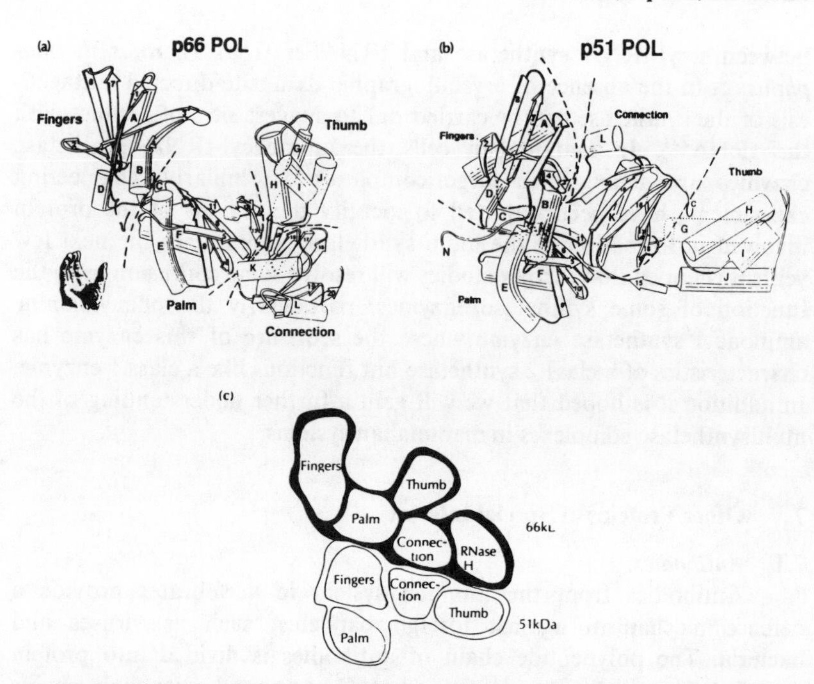

Figure 10. Structure of the HIV reverse transcriptase enzyme showing the 66 KDalton (a) and 51 KDalton (b) subunits. (c) shows how the two subunits come together in the mature protein (reproduced with permission from reference 209).

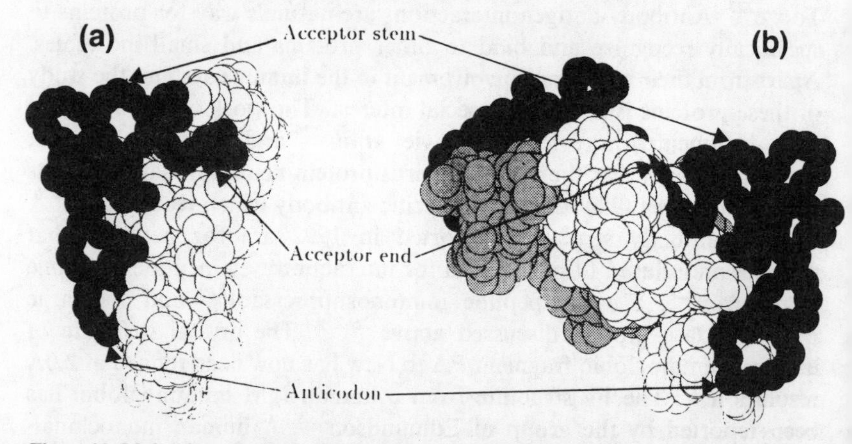

Figure 11. Molecular structure (Ca backbone) of *Escherichia coli* glutamine aminoacyl tRNA synthetase (a) and yeast aspartic acid aminoacyl tRNA synthetase complexes with their tRNA (phosphate backbone). The class 1 enzyme (a) binds to the minor groove of the tRNA acceptor stem whereas the class II enzyme (b) approaches the tRNA from the major groove side (reproduced with permission from reference 215).

the binding pocket. Common features of the antigen binding loops and applications to modelling loop conformations have been studied by Tramontana and Lesk[234]. In the large interaction of the antibody with the neuraminidase protein only five of the six CDR loops are used[226]. Usually a greater number of contacts involve the heavy chain[235,236] although both chains are needed for the specificity of the binding. The heavy chain CDR, H3, for which it is difficult to predict a common structure, seems to play a key role in recognition[237]. A controversial issue has been the question of whether conformation changes occur in antibody or antigen upon binding[238]. Structures published in 1992 show that substantial conformational changes can occur in the antibody[227,235].

Antibodies have been the subject of protein engineering experiments. Attempts have been made to humanize the mouse anti-lysozyme antibody[239]. It was found that substitutions also in the human framework region were necessary to obtain high efficiency binding to the antigen. Other engineered structures reported are of an immunoglobulin domain and a complementarity-determining region 1-grafted mutant[240] and a chimeric Fab fragment of an antibody binding tumour cells[241]. A functional antibody has been made with an insertion in the framework region -3 loop of the VH domain[242]. Calculations of antibody/antigen binding have been made by microscopic and semi-microscopic evaluation of free energies of binding of phosphorylcholine analogues to antibodies[243]. Two papers on anti-insulin antibody structure and conformation discuss molecular modelling and mechanics of an insulin antibody[244] and molecular dynamics with explicit solvent[245].

7.2 *Lipases.*

Much interest has been generated recently in the study of lipase enzymes. This has been triggered by an interest in their role in lipid digestion in mammalian systems and their potential use industrially. The crystal structures of three neutral triglyceride lipases have been reported: the fungal *Rhizomucor miehei* lipase[246] (RML), the human pancreatic lipase[247] and the *Geotrichum candidum* lipase[79]. In 1992 the high resolution structure of the fungal RML lipase was described[248]. All three enzymes have common features where a catalytic serine residue, which forms part of a catalytic triad similar to that found in the serine proteinases, is located at the top of a central β sheet. The 'catalytic triad' found in lipases (His, Glu or Asp and Ser) varies from that of the serine proteases (see section 3.2). Structural relationships between lipases and peptidases of the prolyl oligopeptidase family have recently been discussed by Polgar[249]. All the lipases have a 'lid' or α helices covering the active site which have to be moved when the substrate binds. The displacement of the lid in response to the enzymes absorption to an oil-water interface was postulated as the structural basis for the phenomenon of interfacial activation. The RML lipase has been co-crystallised with a covalent inhibitor, n-hexylphosphonate-ethylester[250] and the inhibitor diethyl p-nitrophenylphosphate[251]. These studies demonstrate the large movement of the 'lid' of RML on binding of the inhibitors.

A number of new lipase enzymes were isolated and sequenced and characterised in 1992. These include rabbit pancreatic triglyceride lipase[252], a lipase from *Penicillium simplicissimum*[253], lipases from *Geotrichum candidum* ATCC 66592[254], two lipases from *Candida cylindracea*[255], a lipase from *Pseudomonas glumae*[256], a lipase from *Staphylococcus hyicus*[257], a lipoprotein lipase from chicken[258], mono-acylglycerol, diacylglycerol lipase from *Penicillium camembertii*[259] a lipase from *Bacillus subtilis* 168[260], and a lipase from *Pseudomonas aeruginosa*[261]. Several of these lipases have been crystallised including a lipase from *Pseudomonas cepacia*[262], rabbit and human gastric lipases[263].

Substrate specificity studies have been carried out with lipase A and B from *Geotrichum candidum* CMICC 335426[264]. The effects of substitution of glycine and asparagine for serine on the activity and binding of human lipoprotein lipase to very low density lipoproteins[265] has been determined. This study showed that the substitution of glycine for the active site serine residue produced a protein which could still hydrolyse triolein and tributyrin. Site-directed mutagenesis has been used to investigate active site residues and the topology of the surface loop in lipoprotein lipase[266]. This study verified that the catalytic triad was serine, histidine, aspartic acid and that it is possible to replace the surface loop of lipoprotein lipase with that from hepatic lipase without loss of activity. Human pancreatic lipase has been studied by mutagenesis[267]. Again the active site residues serine, histidine and aspartic acid were found to be essential for activity. It was possible to substitute glutamic acid for aspartic acid without activity loss. Inactive serine mutants still bound to interfaces. The structural information available to date shows that although the amino acid sequences of these enzymes are very different they all show the α/β hydrolase fold (see section 3.1.2). A recent paper by Schrag *et al.*,[268] discusses the case of *Geotrichum candidum* and human pancreatic lipase where the acid residue of the catalytic triad is positioned differently. A comparison of these two lipases suggests that the pancreatic lipase is an evolutionary intermediate in the pathway of migration of the catalytic acid residue to a new position in the fold.

7.3 *Cholinesterases.*

The structure of *Torpedo* (electric eel) acetyl cholinesterase has recently been reported[269]. This structure has helped and revised the design of potential inhibitors of this enzyme. Cholinesterases have been implicated in diseased states such as Alzheimers disease, leukaemias and carcinomas. They are sensitive targets for natural and synthetic cholinergic toxins including toxic glyco-alkaloids, highly poisonous organophosphorus (OP) and carbamate insecticides, snake venom peptidases and diverse synthetic therapeutic agents. To obtain a deeper insight into the biochemical basis for choline ester catalysis and the complex mechanism of interaction of cholinesterases and their numerous ligands, Soreq *et al.*, have superimposed acetyl and butyrylcholinesterases into a single molecular mode[270]. Cholinesterases are another group of enzymes that have the α/β hydrolase fold (see section 3.1.2). Acetylcholine esterase has a similar structure to *Geotrichium candida* lipase. They have a catalytic triad (serine, glutamic acid and histidine) similar to that found in the lipases, which lies at the bottom of a 20Å deep gorge which is lined with aromatic residues. The previously predicted anionic binding site on which earlier inhibitors are based does not exist and binding mainly seems to be by hydrophobic interactions.

7.4 *Other X-ray Protein Structures.*

The X-ray structure of telokin, the C-terminal domain of the myosin light chain kinase, has been determined at 2.8Å resolution[271]. Telokin is an acidic protein identical to the C-terminal portion of smooth myosin light chain kinase from turkey gizzard. It contains 154 amino acid residues and the overall fold consists of 7 strands of anti-parallel β-pleated sheet forming a barrel with the β-strands adopting the same topological arrangement as observed for the constant domains of immunoglobulins.

The molecular structure of UDP-Galactose 4-Epimerase from *Escherichia coli* has been determined to 2.5Å resolution by Bauer *et al.*[272]. This enzyme catalyses the conversion of UDP galactose to UDP glucose during normal galactose metabolism. The epimerase is a dimer of 2 × 79 Daltons and each subunit is divided into two domains with a N-terminal domain that binds NAD^+. The substrate analogue, UDP-benzene, binds in a cleft located between the two domains. Each domain binds NAD^+ and a substrate analogue. The enzyme shows the unique feature of non-stereospecificity of hydride return from the B side of the nicotinamide ring. The structure of the quinoprotein methylamine dehydrogenase protein from *Paracoccus denitrificans* has been solved by molecular replacement[273]. The enzyme catalyses the oxidation of primary amines into the corresponding aldehyde and transfers electrons to C-type cytochromes through a mediating blue copper protein, amicyanin. The enzyme is a tetramer of different subunits H_2L_2 with molecular weights of 46.7 K Daltons for the heavy subunit and 15.5 K Daltons for the light subunit. The redox cofactor of this enzyme is tryptophan tryptophylquinone. This is unusual and is formed from two covalently linked tryptophan side chains of the L subunit, one of which contains orthoquinone.

The crystal structure of porcine pepsinogen has been reported by Hartsuck *et al.*[274]. This serves as a model for aspartic proteinase zymogens. The pepsinogen contains a single amino acid chain of 370 residues and upon activation, 44 residues are removed from the N-terminal end. It is a member of the family of aspartic proteinases (see section 9) which include pepsins A and C and chymosin from the stomach, cathepsins D and E from the lysosome, renin from the kidney, proteinase A from yeast granules, rhizopuspepsin, penicillopepsin and endothiapepsin from fungi and the retroviral proteinase from HIV-1 and Rous sarcoma virus. The activation peptide packs into the active site cleft and the N-terminus occupies the position of the mature N-terminus. It also makes specific interactions with the substrate binding subsites.

The crystal structure of subtilisin BL, an alkaline protease from *Bacillus lentus* with activity at pH 11 has been solved at 1.4Å resolution[275]. Almost all of the acidic side-chains of the enzyme are involved in some sort of electrostatic interaction either via ion-pairs or calcium binding. The overall structure of the enzyme is very similar to other subtilisins, but it has as a high isoelectric point, a high activity at pH 11 and is thermostable. It is therefore suited for industrial applications. X-ray structures of two crystal forms of a variant (asparagine 115 -> arginine) of the alkaline protease from *Bacillus alcalophilus* have been reported[276]. Preliminary X-ray studies have been carried out with the *Staphylococcal* enterotoxin type C[277]. *Bacillus stearothermophilus* mutants of lactate dehydrogenase have been studied crystallographically[278]. Analysis of these triple active site mutants should help our understanding of the catalytic mechanism of this enzyme. The structure of methanol dehydrogenase from two methylotrophic bacteria has been described at 2.6Å resolution[279]. The enzyme malate dehydrogenase with a complex of the apo enzyme and citrate has been studied at 1.8Å resolution[280]. Comparisons have been made of the native structure of thermitase and thermitase elgin-c-complexes[281]. Preliminary X-ray diffraction studies have been reported for the human major histocompatibility antigen HLA-B27[282]. The structure of rhombohedral R6 Insulin/Phenol complex has been reported[283]. The *Escherichia coli* biotin haloenzyme synthetase biorepressor crystal structure delineates the biotin-binding and DNA binding domains[284]. The crystal structure of a

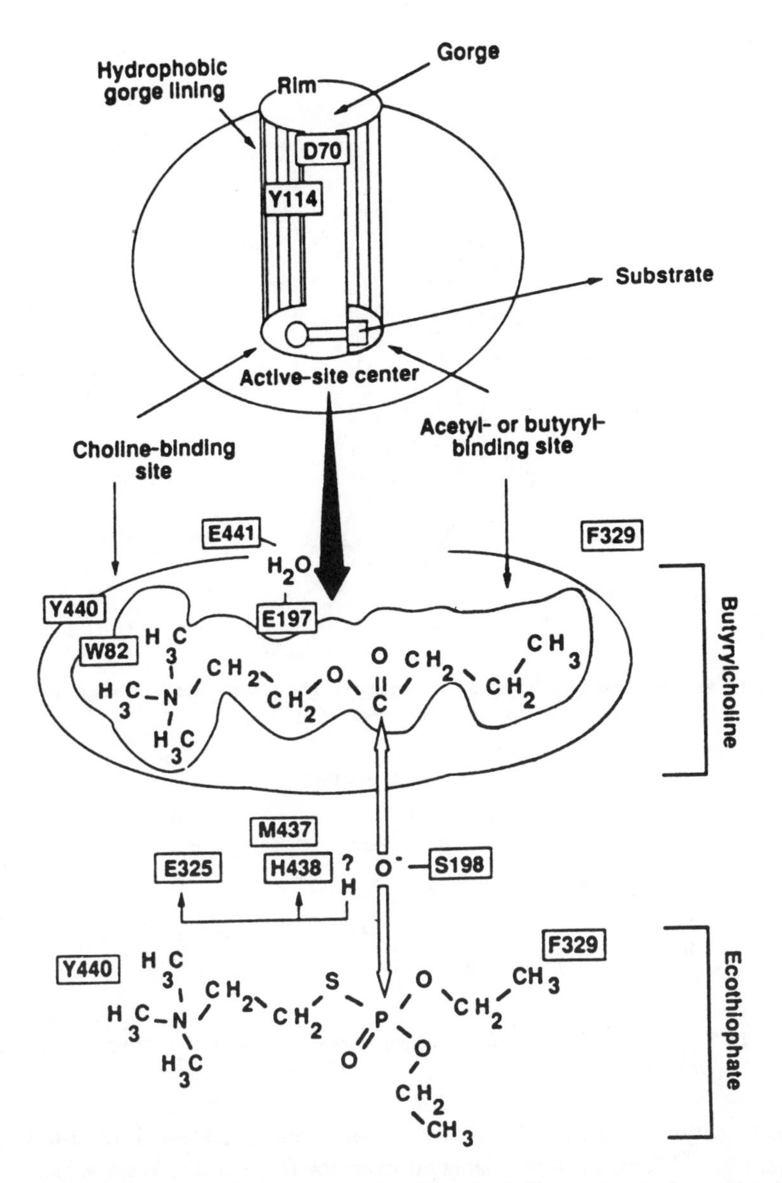

Figure 12. Positioning of key amino acid residues in cholinesterases. Amino acid residues that participate in catalysis and ligand binding are in the deep active site gorge. Regions rich in hydrophobic residues lining the gorge are striped. Butyrylcholine and ecothiophate are presented as representative substrate and inhibiting OP agents. Residue numbering according to human butyryl cholinesterase (reproduced with permission from reference 270).

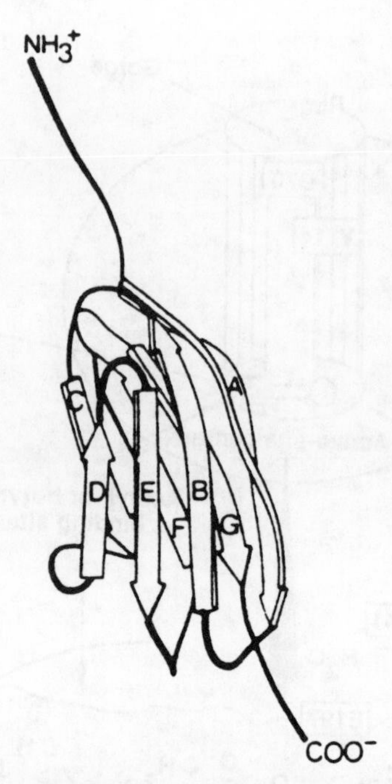

Figure 13. Ribbon drawing of the telokin molecule. The strands are labelled A to G (reproduced with permission from reference 271).

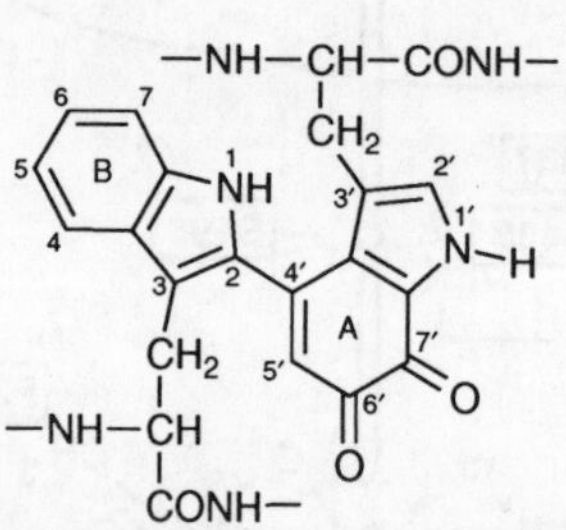

Figure 14. Tryptophan tryptophylquinone (reproduced with permission from reference 273).

sweet tasting protein, Thaumatin I has been determined at 1.65Å resolution[285]. This protein is isolated from the fruits of the West African shrub *Thaumatococcus danielli*. Studies on this protein are of interest to understand chemical senses and sensory perception of its sweet taste.

The crystal structure of an interesting membrane protein, Porin, has been reported in 1992[286]. Porins are integral membrane proteins that are found in the outer membrane of gram negative bacteria, mitochondria and chloroplasts. They act as molecular sieves to ensure the unhindered diffusion of nutrients into the periplasmic space while protecting the cell from hostile substances such as degrading enzymes or bile salts. This type of membrane protein is distinct from those that are composed of helices e.g. bacteriorhodopsin, in that it is made up entirely of β sheets. The porin structure from *Rhodobacter capsulatus* is a homotrimer with 301 amino acid residues and three calcium ions in each monomer.

The refined crystal structure of *Pseudomonas putida* lipoamide dehydrogenase complexed with the cofactor NAD^+ has been determined to 2.45Å[287]. The structure of native and apo carbonic anhydrase-II and the structure of some of its anion ligand complexes has been reported by the group of Liljas[288]. The crystal structure of the reduced form of p-hydroxybenzoate hydroxylase from *Pseudomonas fluorescens* has been refined to 2.3Å by the group of Drenth[289]. The structures of the oxidised and reduced forms of the enzyme-substrate complex were found to be very similar. This is a flavoprotein which also requires NADPH and molecular oxygen for activity. The flavin ring is planar in this structure despite models suggesting that bending of the ring may affect its properties. The enzyme catalyses the conversion of the aromatic substrate p-hydroxybenzoate into 3,4-dihydroxybenzoate. The crystal structure of the enzyme-substrate complex initially reported in 1979[290] consists of three domains: the FAD binding domain, the substrate binding domain and the interface domain. The active site is at the interface of all three domains. The structure of the small ribosomal protein, S5 was reported in 1992. This is the first protein structure determination for a ribosomal protein which interacts specifically with the 16S rRNA of the small ribosomal subunit[291].

The number of structures mentioned in this report is limited by space and concentrates on some of those reported in 1992. Atomic structures of protein molecules from both NMR and crystallographic studies reported in 1990, 1991 and 1992 are collected together in three new books edited by Hendrickson and Wüthrich[292]. These provide a wealth of information to those interested in protein structure and function.

8 Metal Containing Proteins

Metal ions can play different roles in proteins. In the past few years since detection methods for metal ion content have become more sensitive, many more metal containing proteins have been discovered. One of the most striking of these is the group of zinc containing

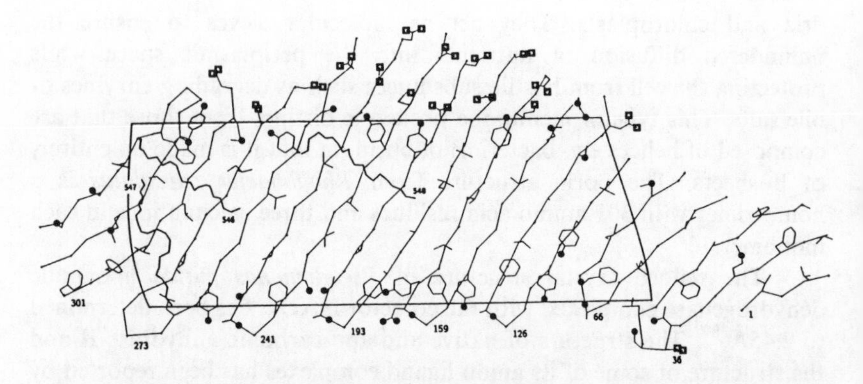

Figure 15. A projection of the outer surfaces of the β-barrel of porin on a cylinder. Polar atoms are marked by dots and ionogenic groups by quadrangles. The non-polar surface exposed to the membrane is boxed (reproduced with permission from reference 286).

eukaryotic transcription factors described in section 6.1. It is established that over 1% of the proteins found in the nucleus of human cells is made up of these zinc containing proteins. In this case the metal ion appears to play the role of holding the protein chain in a certain structure so that it can interact with DNA. Of course metal ions can play different roles in proteins; they are often part of the active site of an enzyme or complexed to a substrate.

8.1 *Zinc Proteins.*

The non-toxic nature of zinc and its chemical properties make it an ideal choice for nature. The co-ordination sphere of zinc is flexible and it can adopt co-ordination numbers from two to eight and it is the four, five and six co-ordination that is seen most frequently in biology. The zinc cluster of metallothionein shown by X-structure is unique to biology[293]. A recent review by Vallee and Auld[294] classify zinc binding proteins into three types of motifs: catalytic, structural or cocatalytic. This classification is based on the fifteen zinc containing enzymes for which a structure is known. In catalytic zinc sites three histidine, glutamic acid, aspartic acid or cysteine residues from the protein provide the zinc ligands, with the imidazolyl nitrogen of histidine being the most popular. The fourth ligand is H_2O which is a critical feature of this motif, which allows the activation of water so that it can be ionized, polarized or displaced. There is a regularity of amino acid spacing between the ligands of the catalytic site. It appears that there is a short spacing (1-3 amino acids) between the first two ligands and a long spacing (20-120

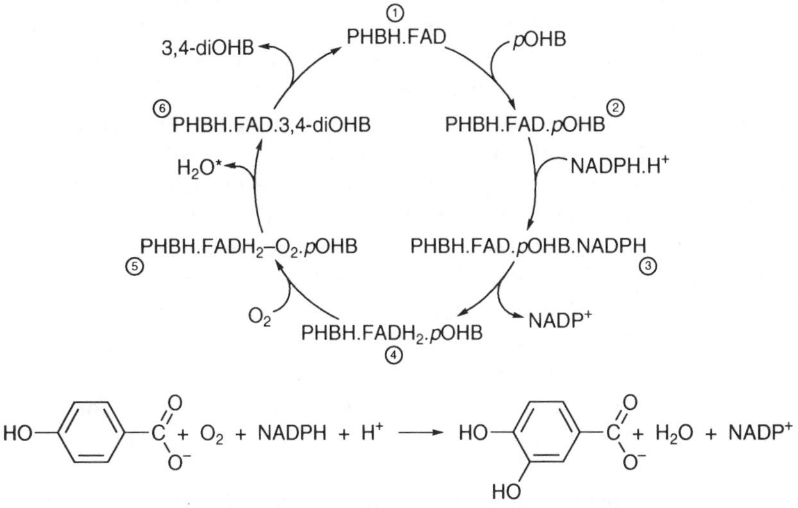

Figure 16. Catalytic cycle of p-hydroxybenzoate hydroxylase (reproduced from reference 289 with permission).

amino acids) between the third ligand and the first and second. In the three dimensional structure of the protein these are brought together to form the zinc binding site. The enzyme carbonic anhydrase is part of this group. A recent theoretical study of the catalytic mechanism of this enzyme has been described[295].

Coactive or cocatalytic zinc sites occur in enzymes which contain two or more metal atoms. The metal ions are close to each other and function as a unit for catalytic activity. Examples of this group are alkaline phosphatase[296], bovine lens aminopeptidase[297], and *Escherichia coli* DNA polymerase I, 3′-5′ exonuclease activity[298,299] with zinc and magnesium. In all of these enzymes an aspartic acid or a glutamic acid is the bridging amino acid. Aspartic acid and histidine are found most frequently as the zinc binding amino acids. Phospholipase C also falls into the class[300] as does superoxide dismutase[301] in which copper is the active site metal and zinc is the other metal (both metals being bound to histidine which is the bridging amino acid).

A group of proteins for which considerable medical and pharmaceutical interest has developed is the monozinc aminopeptidases. A recent paper has described the crystallographic structure of astacin[302]. Although the linear arrangement of the two histidines and one glutamic acid is similar to that of thermolysin[303] the overall homology of the aminopeptidases is different. The three dimensional structure of astacin exhibits a new type of zinc co-ordination, where three imidazole nitrogens and two oxygen atoms form a trigonal-bipyramidal co-ordination sphere. The five zinc ligands are three histidines, a tyrosine and a water molecule which is anchored to the carboxylate group of a glutamic acid. The active site cleft is wider in astacin than thermolysin which might explain the collagen-digesting activity of the former. Further studies with this group of enzymes will help to gain an understanding of substrate binding and will assist in the design of potential inhibitors which have obvious therapeutic importance. Families of metalloendopeptidases and their relationships have been discussed by Jiang and Bond[304]. A hydra matrix protease has been found to have homology with *Drosphilia* dorsal-ventral patterning protein and other members of the astacin family[305].

The eukaryotic transcription factors which rely on zinc for their structure and subsequent interaction with DNA are discussed in section 6.1.

Adenosine deaminase which is a key enzyme in purine metabolism is an example of a enzyme that previously had never been reported to be a metalloenzyme. A new crystal structure now indicates that zinc is present in the active site of the enzyme[306]. In this case zinc is co-ordinated by five atoms, 3 histidine nitrogens, and oxygen of aspartic acid and an oxygen of a hydroxyl group of an inhibitor bound at the active site. A recent report describes the fact that botulinum neurotoxins are also zinc proteins[307].

The histidine-glycine rich region of the light chain of high molecular weight kininogen is thought to be responsible for binding to negatively charged surfaces and initiation of the intrinsic coagulation, fibrinolytic and kinin-forming systems. This region is thought to act as a primary structure feature for zinc-dependent binding to an anionic surface[308].

An exciting experiment is the ability to engineer metal binding sites into proteins to carry out specific roles. This has been carried out by Regan and Clarke[309] who have engineered a tetrahedral zinc binding site into a model four-helix bundle protein by introducing a cysteine and histidine ligand into each of two adjacent helices. A recent review deals with the exciting prospect of engineering metalloregulation in enzymes[310]. A review of zinc proteins can be found in the Annual Review of Biochemistry 1992[311].

8.2 *Calcium Proteins.*

A common structural motif of 12 amino acids flanked by two α helices is found in calcium binding proteins[312]. Usually 2-8 copies of this motif occur and they are arranged in pairs. Residues 1, 3, 5, 9 and 12 ligate calcium through side-chain carboxyl or amide carbonyl oxygens, and residue 7 through the backbone carbonyl oxygen. An invariant glutamic acid provides bidentate co-ordination yielding a pentagonal bipyramid arrangement about the calcium ion[313]. Troponin C has two such domains. NMR studies with synthetic peptides corresponding to the calcium binding sites in this protein[314,315] and proteolytic fragments of troponin C[316] have been made to understand the calcium binding. In this case calcium appears to induce the peptides to fold into a troponin C like structure.

Site-directed mutagenesis has been used to investigate calcium binding to Calbindin D9K which has two calcium binding domains, one usual and one modified. Four surface carboxylate groups were modified which reduced the affinity for calcium by x2000 at low ionic strength and x25 at physiological ionic strength[317]. The infrared X-ray structure of calmodulin has been reported at high resolution318. Calmodulin which has two calcium binding domains has also been a target for mutagenesis[319,320]. The invariant glutamic acid was mutated in each calcium binding site to a glutamine and the conformational changes induced by calcium monitored by fluorescence measurements and ^{1}H-NMR[321]. Binding to both sites is important for the transition to occur. The activation of the Ca^{2+}-dependent Na^+ channel in *Paramecium* has been recently investigated using primary mutations in calmodulin[322].

Calcium binding sites have also been engineered into proteins. Toma *et al.*,[323] have grafted a calcium binding loop from thermolysin into a homologous region of a *Bacillus subtilis* neutral protease. The mutant binds one extra calcium ion which stabilizes the protein against thermal denaturation.

Calcium also appears to be important in another exciting and important area of protein recognition. Many calcium dependent animal lectins contain a common sequence motif with 14 invariant and 18 highly conserved residues as part of a specific carbohydrate recognition domain in the protein. The role of carbohydrates in protein function and activity is regarded as an increasingly important area of research. Important structural papers show the role that calcium plays in the formation of a structural motif in rat serum mannose binding protein[324,325]. The equatorial 3 and 4 hydroxyl groups of mannose are co-ordinated to one of the bound calcium ions and four of the protein side chains (two asparagines and two glutamic acids) are also co-ordinated to this calcium and hydrogen bonded to the same sugar hydroxyl groups. The remaining ligands for the calcium are contributed by the side chains of an aspartic acid residue.

The X-ray structure of an interesting enzyme which requires calcium for catalysis was reported in 1992. This was the first thiamine pyrophosphate dependent enzyme and is called transketolase[326]. The natural role of the enzyme is in the pentose phosphate pathway for the metabolism of glucose-6-phosphate. The enzyme catalyses ketol transfer from a ketose to an aldose amongst a variety of donor and acceptor sugar phosphates and has applications for biotransformation reactions in organic synthesis. The bakers yeast enzyme is a dimer with the active sites located at the interface between the two domains. The cofactor thiamine pyrophosphate is bound at the interface between the two subunits, with the diphosphate moiety interacting with the protein through a calcium ion, which has pentameric co-ordination with square pyramidal geometry. The calcium binding site is made up of three protein ligands (aspartic acid and asparagine side chains and isoleucine main chain oxygen) and two oxygen atoms of diphosphate.

8.3 *Tungsten Proteins.*
A new group of organisms which are neither prokaryotes or eukaryotes called the Archaea appear to have proteins which contain tungsten. A novel tungsten-iron-sulphur protein which is an aldehyde ferredoxin oxidoreductose has been isolated from an extremely hyperthermophilic member of this group, *Pyrococcus furiosus*.[327] The tungsten appears to be present as a pterin-type cofactor, analogous to the molybdopterin cofactor found in the majority of molybdenum-containing enzymes. It is suggested that tungsten may play a role in the proteins of other hyperthermophilic organisms such as *Thermatoga maritima*. In fact a novel iron-hydrogenase has been isolated from this extreme eubacteria whose cellular activity is dependent upon tungsten.[328]

8.4 *Iron Proteins.*
Rubredoxins are non-heme iron-proteins of low molecular weight which occur in several anaerobic bacteria. Little is known about the specific function of these proteins that are thought to participate in electron transfer reactions. Several high resolution X-ray structures of this protein had been previously reported. More recently the structure of the protein from the extremely thermophilic Archaea *Pyrococcus furiosus* has been determined, by both X-ray and NMR studies[329-332] and compared to other rubredoxins. The thermostable protein is similar in structure to other mesophilic rubredoxins but has a more extensive hydrogen bonding network in the β-sheet region and multiple electrostatic interactions. Further NMR studies have been carried out with zinc substituted rubredoxin[333], and for oxidised three-iron clusters in the

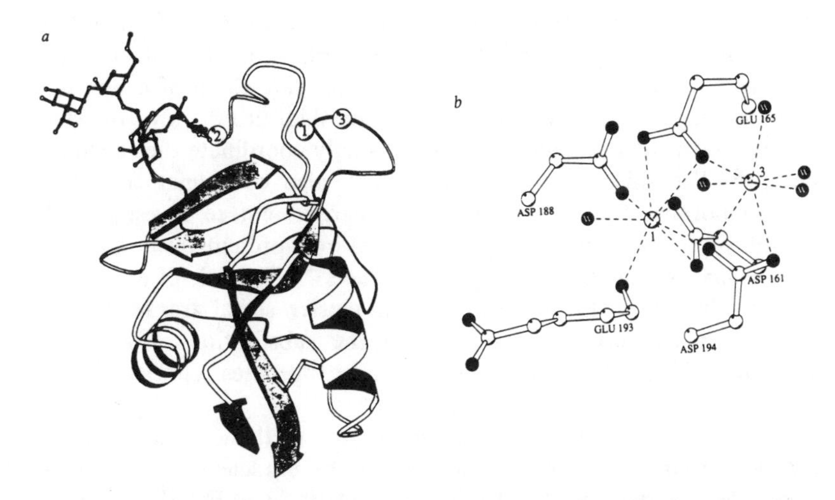

Figure 17. The figure shows the calcium and sugar binding sites of the C-type mannose-binding protein. a) A view of the oligosaccharide binding to one protomer. The protein is shown as Cα backbone with α helices and the β sheets as arrows. The three Ca^{2+} binding sites are shown. b) A closeup of the Ca^{2+} sites 1 and 3. White, black and light grey spheres represent carbon, oxygen and calcium respectively. Black spheres marked 'w' are water molecules. Dashed-line denotes calcium coordination bonds (reproduced with permission from reference 324).

ferredoxins from *Pyrococcus*[334]. It has been reported that conserved structural elements of the protein are detected in ferredoxins and cupredoxins by Raman-spectroscopy[335]. The three dimensional structure of the high-potential iron-sulphur protein (HiPIP) isolated from the purple phototrophic bacterium *Rhodocyclus tenius* has been determined at 1.5Å resolution[336]. As with other HiPIPs the iron-sulphur cluster is co-ordinated by four cysteinyl ligands and exhibits a cubane-like motif. High resolution crystal structures and comparisons of T-state deoxyhaemoglobin and two liganded T-state haemoglobins have been used to explain the structural basis of the reduced affinity of the T-state[337].

The crystallographic structure of a nitrogenase molybdenum iron protein from *Azotobacter vinelandii* has been reported[338,339] and structural models for the metal centres in this protein have been made[340,341]. Each centre consists of two bridged clusters as shown in figure 18. The FeMo-cofactor has 4Fe:3S and 1Mo:3Fe:3S clusters bridged by three non-protein ligands and the P clusters contain 4Fe-4S clusters bridged by two cysteine thiol ligands. Six of the seven Fe sites in the FeMo-cofactor appear to have trigonal coordination geometry, including one ligand provided by the bridging group. The remaining Fe site has tetrahedral geometry and is liganded to the side chain of a cysteine residue. The Mo site exhibits approximate octahedral coordination geometry and is liganded to three sulphurs in the cofactor, two oxygens from homocitrate and the imidazole side chain of a histidine. The P-clusters are liganded by six cysteine thiol groups, two of which bridge the two clusters and four which singly coordinate the remaining Fe sites. The side chain of a serine may also co-ordinate one iron. Electrophoretic studies have also been carried out to investigate the assembly of the nitrogenase molybdenum-iron protein from *Klebsiella pneumoniae* NIFD and NIFK gene products[342]. The role of an aspartic acid residue in nucleotide iron-sulphur cluster signal transduction in nitrogenase has been reported[343] and the substitution of a specific arginine residue 100[344]. Other papers on nitrogenases reported in 1992 are listed in references 345–350.

The active site structure of the enzyme methane monooxygenase has been found to be closely related to the binuclear iron centre of ribonucleotide reductase[351]. The Fe(II) oxidation and Fe(III) incorporation by the 66 KDalton microsomal iron protein that stimulates NADPH oxidation has been studied[352]. Identification of the iron ions of the high-potential iron protein from *Chromatium vinasum* has been made by 2D NMR experiments[353]. The techniques of multi-frequency EPR and high resolution Mossbauer-spectroscopy have been used to study a putative (6Fe-6S) Prismane-cluster containing protein from *Desulfovibrio vulgaris*[354]. Iron-protein interactions in ferritin have been studied by EXAFS and XRD[355]. Novel artificial non-heme iron protein has been used to investigate iron sulphur centres as endogenous blue-light sensitizers of cells[356]. The nomenclature recommendations of 1992 for electron-transfer proteins has been summarised by Palmer and Reedijk[357].

8.5 *Copper Proteins.*

Several reports of the copper containing oxidases have appeared in the literature. The evolution of these and related proteins has been discussed by Ryden and Hunt[358]. X-Ray structures have been reported of ascorbate oxidase[359,360], copper (II)-and substituted horse liver alcohol-dehydrogenase[361] and the refinement of polar plastocyanin[362]. The adsorbate oxidase from Zucchini is composed of subunits of 70 KDaltons which are arranged as tetramers. Two N-linked sugar moieties one attached to an asparagine amino acid and can be defined in the structure. Each subunit has four copper atoms bound as mononuclear and trinuclear species. The mononuclear copper has two histidine and a cysteine and a methionine ligand and represents the type-1 copper this is also found in plastocyanin and azurin. The trinuclear cluster has eight

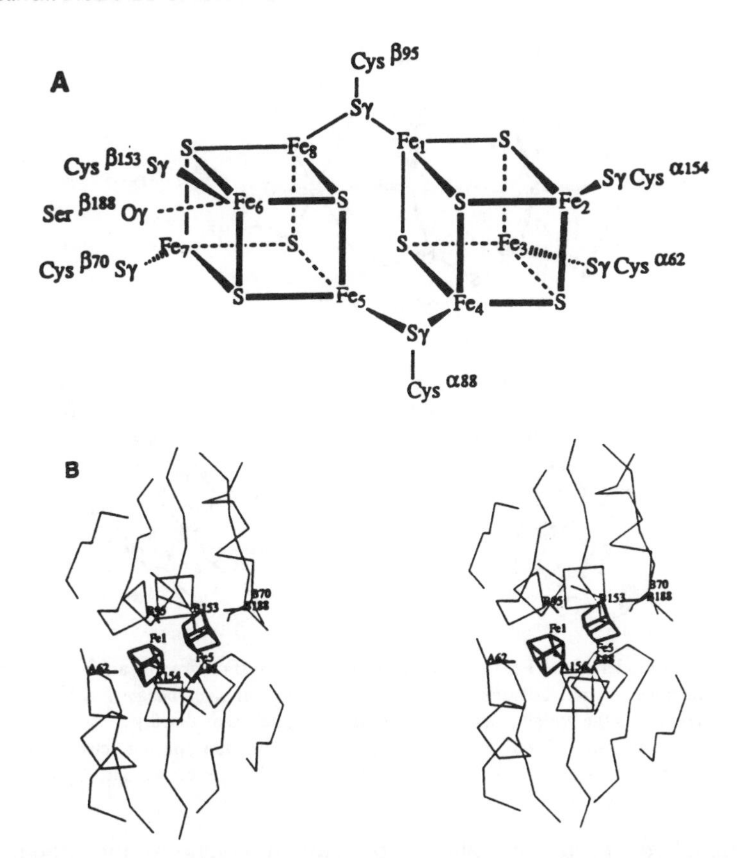

Figure 18. a) A model of the P-cluster based on the crystallographic analysis of the nitrogenase molybdenum-iron (MoFe) protein from *Azotobacter vinelandii*.
b) Stereoview of the P-cluster and surrounding protein model.

histidine ligands. It may be sub-divided into a pair of copper atoms with histidine ligands whose ligated N-atoms are arranged trigonal prismatic. The pair is the putative type-3 copper. The remaining copper has two histidine ligands and is the putative spectroscopic type-2 copper. Several studies with azurin have been carried out. These include site-directed mutagenesis studies to study the role of a specific asparagine residue[363], the study of long range intramolecular electron-transfer[364] protein-protein cross-reactions phosphorescence studies[365,366], and pH and temperature dependencies on reduction potentials[367]. The latter study also looked at plastocyanin and cytochrom-F. Photoinduced electron transfer has been studied in a ternary system involving zinc cytochrome c and plastocyanin[368]. The role of surface-exposed Tyr-83 of plastocyanin in electron transfer from cytochrome c has also been reported[369]. The

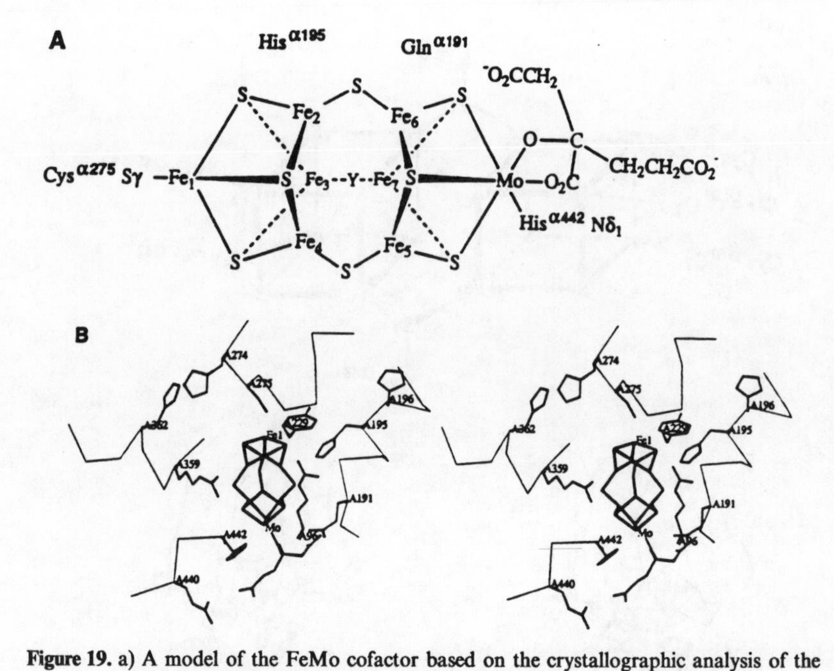

Figure 19. a) A model of the FeMo cofactor based on the crystallographic analysis of the nitrogenase molybdenum iron (MoFe) protein from *Azotobacter Vinelandii*. Y represents the bridging ligand with relatively light electron density.
b) Stereoview of the FeMo-cofactor and surrounding protein environment (Figures 18 and 19 reproduced with permission from reference 340. Copyright 1992 by the AAAS.)

importance of protein-protein electrostatic interactions and donor-acceptor coupling has also been the feature of a study of electron transfer reactions of cytochrome F with flavin semiquinones and with plastocyanin[370]. The reaction site charge on plastocyanin from *Cyanobacterium anabaena-variabilis* has been determined from electrostatic interactions with small inorganic complexes[371]. Spinach plastocyanin has also been used at a lipid bilayer-modified electrode as a probe of membrane protein interactions[372]. Auracyanins that are blue copper proteins have been isolated and characterised from the green photosynthetic bacterium *Chloroflexus aurantiacus*[373].

The T-transition to R-transition has been studied in the copper (II) substituted insulin hexamer[374]. Anion complexes of the R-state species exhibited type I and type 2 spectral characteristics. The reaction of cyanide with cytochrome-BA3 from the thermophilic bacterium *Thermus thermophilus* has been used to characterise the Fe (II) A3-CN-B-Cu (II)B-CN Complex and suggests that 4 N-14 atoms are coordinated to Cu (B)[375]. The c-type cytochromes of methylotrophic bacteria have been

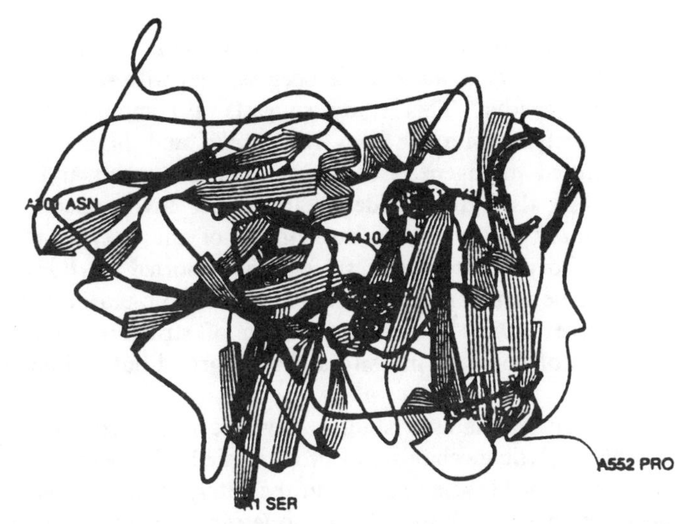

Figure 20. A schematic representation of the monomer structure of ascorbate oxidase showing the monomer plus the four copper ions (reproduced with permission from reference 359).

discussed in a recent review by Anthony[376]. Another recent review has examined the current knowledge concerning the electronic structures of active-sites in copper proteins and their contributions to reactivity[377]. Finally model complexes of blue copper proteins have been made that are able to catalyse the oxidation of alkenes with t-butyl hydroperoxide[378].

9 Catalytic Mechanism

The use of site-directed mutagenesis to investigate the catalytic mechanism of enzymes now makes it possible to test earlier proposals. A recent review covers 10 years of protein engineering with the past results and expectations for the future[379]. Individual amino acids in the active site can be changed specifically and their effect on the catalytic behaviour of the enzyme monitored. It is important when making these changes to be sure that only the structure of the enzyme at the point of mutation has changed and that the enzyme is folded correctly. This can be carried out by circular dichroism, NMR spectroscopy or X-ray crystallography. Experiments of this type have been mentioned in the appropriate sections of this report however, studies on many other enzymes have been reported in 1992. Papers in this area are increasing and only some of those reported in 1992 are listed and discussed.

The catalytic mechanism of N-hydroxyacrylamine O-acetyl-transferase from *Salmonella typhimurium* has been studied with respect to an essential cysteine residue. This study suggests a common catalytic mechanism of acetyltransferase from *Salmonella* and higher organisms[380]. Site-directed mutagenesis of a recombinant human arylamine N-acetyltransferase also gives evidence for direct involvement of a cysteine 68 residue[381]. The kinetic mechanism of the enzyme purine nucleoside phosphorylase from calf spleen was reported by Porter[382]. The crystallographic structure of the human erythocytic purine nucleoside phosphorylase enzyme[383] and application of structure modelling methods in design of potential inhibitors[384] was carried out in 1990 and 1991.

Two catalytic mechanism studies have been carried out with peptidases:- the catalytic mechanism of prokaryotic leader peptidase-I[385] and the importance of histidine residue in the catalytic mechanism of a *Streptomyces* peptidase[386]. A model for the tetrahedral intermediate in aspartic peptidase-mediated cleavage of amide bonds has been shown by direct observation of a potent renin tripeptide inhibitor complexed in crystals of the aspartic proteinase endothiapepsin[387]. This indicates a mechanism in which aspartic acid 32 is the proton donor and aspartic acid 215 carboxylate polarises a bound water for nucleophilic attack. The mechanism involves a carboxylate that is stabilized by extensive hydrogen bonding, rather than an oxyanion derivative of the peptide as in serine proteinase catalysis.

Computer-simulation of the initial proton-transfer step has been carried out for human carbonic anhydrase-I[388]. A molecular modelling study has also been carried out to study substrate enzyme interactions and the catalytic mechanism of phospholipase C[389]. Site-directed mutagenesis and X-ray crystallography of phospholipase -A2 mutants has been reported by Thunnissen *et al.*[390].Site-directed mutants have been characterised from human dihydrolipoamide dehydrogenases[391]. Studies on the catalytic mechanism of a glutamate dehydrogenase enzyme have shown an obligatory proton release step precedes hydride transfer for this enzyme[392,393].

The catalytic mechanism of *Streptomyces albus* serine beta-lactamase has been probed by molecular modelling of mutant enzymes[394]. Active-site residues of the transpeptidase domain of the penicillin-binding protein-2 from *Escherichia coli* has been shown to show similarity in catalytic mechanism to class A-beta lactamases[395]. Mutagenesis and modelling studies define a substrate binding role of serine in class A lactamases[396]. The catalytic mechanism of isoenzyme 3-3 of glutathione-S-transferase requires a contribution from a specific

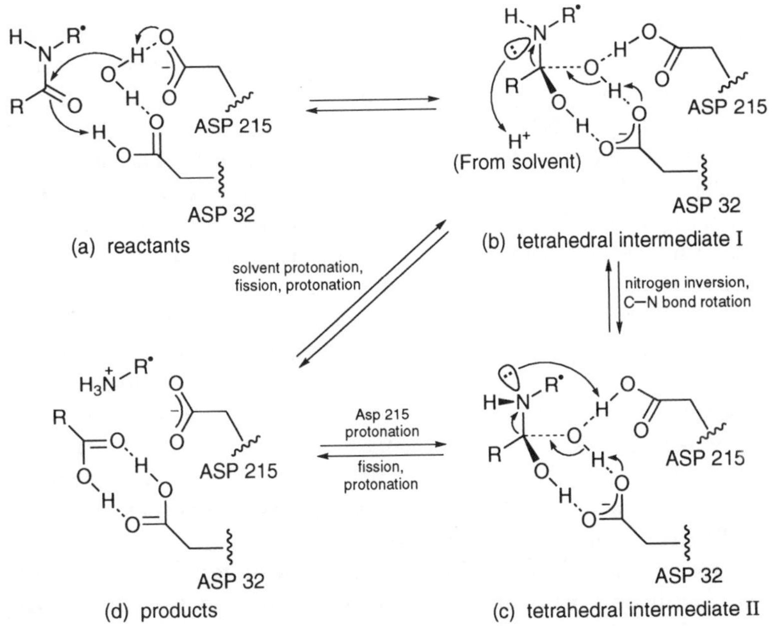

Figure 21. The proposed mechanism for proteolytic cleavage of the amide bond by an aspartic proteinase (reproduced with permission from reference 387, Cambridge University Press).

tyrosine residue[397,398]. Phosphotyrosyl protein phosphatase from bovine heart has been cloned and its catalytic mechanism studied[399].

Affinity labelling of a specific cysteine residue with N-(2,3-epoxypropyl)-N-amidinoglycine has been used to study the active-site of creatine kinase[400]. A study has been made to compare two mechanisms proposed for the hydration of carbon dioxide by the enzyme carbonic anhydrase[401,402]. The thermodynamic parameters of the interaction between Co(II) bovine carbonic anhydrase and anionic inhibitors has also been studied[403]. F-19 NMR has been used to study the catalytic mechanism of phosphoglycomutase with fluorinated substrates and inhibitors[404]. The role of a specific aspartic acid has important implications for the reverse reaction catalysed by phosphofructokinase I from *Escherichia coli*[405]. The enzyme phosphoglycerate kinase has been the subject of many site-directed mutagenesis studies to investigate its catalytic mechanism. This enzyme binds its two substrates ATP and 3-phosphoglycerate on two separate domains which come together in a proposed 'hinge-bending' mechanism during catalysis. This mechanism has been studied by emission and quenching fluorescence studies[406] on wild type and a mutant enzyme where a histidine in the hinge region has been mutated to a glutamine. Site-directed mutagenesis studies were reported on a histidine and arginine mutated at the 3-phosphoglycerate binding site. These mutants were studied by NMR experiments[407]. A combined study of NMR and X-ray crystallography has been used to study another active site arginine of this enzyme[408]. This arginine is buried just below the 3-phosphoglycerate binding site. Mutation of the residue to a methionine affects the Km for this substrate by changing the water structure in this area of the binding site. This is part of a long term study of mutagenesis of this enzyme to try to understand its catalytic mechanism. All of the crystallographic structures reported so far including that of a thermostable phosphoglycerate kinase[409] show the enzyme in an 'open' state. Although still in an 'open' conformation the substrate 3-phosphoglycerate has been co-crystallised with pig muscle phosphoglycerate kinase[410]. A schematic diagram showing interaction of the enzyme with this substrate is shown in Figure 22. A combined approach of solution NMR and X-ray crystallographic studies is therefore required to address this problem.

Key residues in the allosteric transition of *Bacillus stearothermophilus* pyruvate kinase have been identified by site-directed mutagenesis[411]. The structural consequences of exchanging tryptophan and tyrosine residues in *Bacillus stearothermophilus* lactate dehydrogenase have been studied by the group of Holbrook as part of a long term study of mutagenesis of the lactate dehydrogenase enzyme[412]. These changes have little effect of the structure of the mutant protein allowing the tryptophan of the mutant to act as a fluorescent probe to monitor protein folding.

Insights into the catalytic mechanism of orotidylate decarboxylase have been made from substrate specificity studies[413]. Analysis of the catalytic mechanism of juvenile-hormone esterase has been carried out by site-directed mutagenesis[414]. The catalytic mechanism of cytochrome c oxidase has been studied by Raman spectroscopy[415] and the catalytic mechanism of cytochrome -P-450 has been shown to involve a distal charge relay[416]. Studies with thymidylate synthase have found a specific asparagine residue to be a major determinant of pyrimidine specificity[417].

The group of Johnson has spent many years on the study of the allosteric enzyme phosphorylase. The control of phosphorylase B conformation has been studied crystallographically using a modified cofactor[418]. The interaction of catalytic site mutants of *Bacillus subtilis* α amylase has been investigated with substrates and acarbose[419]. The ionization of amino acid residues involved in the catalytic mechanism of aspartic transcarbamoylase has been studied[420].

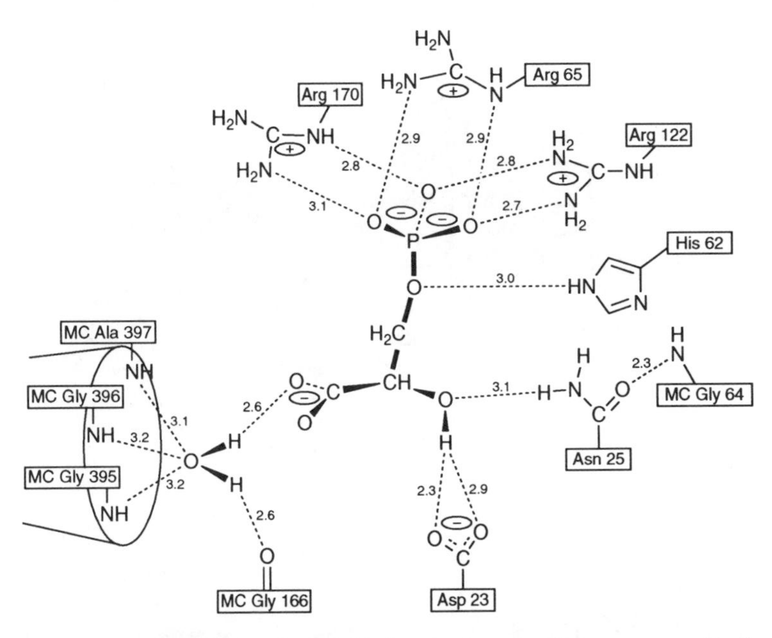

Figure 22. Interaction of the pig muscle phosphoglycerate kinase enzyme with the 3-phosphoglycerate substrate (reproduced with permission from reference 410).

A specific aspartic acid residue has been found to enhance the function to the enzyme-bound coenzyme pyridoxal 5′-phosphate in the catalytic mechanism of aspartic aminotransferase[421]. Mutations made in a loop region of the α subunit of tryptophan synthase affect the transition from an open to a closed conformation in the bienzymecomplex[422]. The role of histidine in the catalytic mechanism of *Bordetella pertussis* adenylate cyclase has been studied by Munier *et al.*[423]. The substrate polarisation by residues in Δ^5-3-ketosteroid isomerase has been probed by site-directed mutagenesis and UV resonance Raman spectroscopy[424]. This enzyme from *Pseudomonas testosteroni* catalyses the isomerization of Δ^5-3-ketosteroids to Δ^4-3-ketosteroids by the specific transfer of the steroid 4β proton to the 6β-position, using a tyrosine residue as a general acid and aspartic acid as a base. Analysis of several key active-site residues of the ricin A chain have been studied by mutagenesis[425]. The role of threonine and serine residues of *Escherichia coli* asparaginase II has been investigated by site-specific mutagenesis[426]. In contrast 6β lactamase structure and function has been studied by a method of random mutagenesis. This approach involves randomizing the DNA sequence of a short stretch of a gene (3-6 codons) and determining the percentage of all possible random sequences that produce a functional protein.

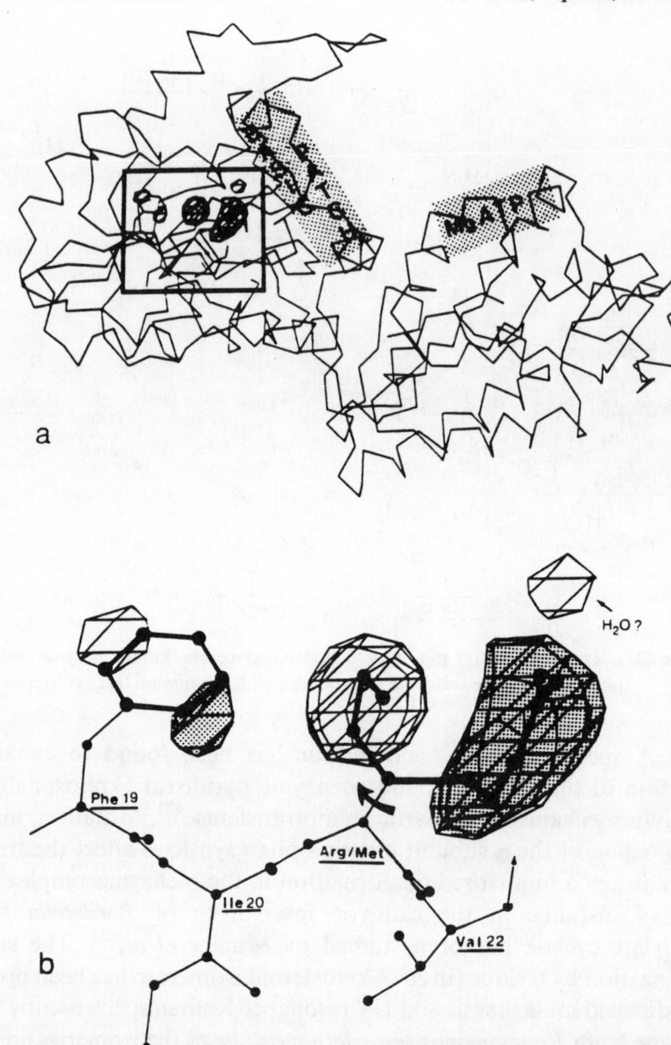

Figure 23. The major peaks close to residue 21, one negative the other positive, have maxima close to ten times the value of the root-mean-square deviation calculated for the complete map. (a) The difference map with a-carbon skeleton overlaid showing the positions of the major electron-density peaks relative to the basic-patch and ATP-binding regions of the molecule. The N-terminal domain is to the left and the C-terminal domain to the right of the figure. (b) An enlarged section of the difference map showing the peaks associated with the replacement of the Arg21 side chain with that of Met. The negative peaks have been shaded to distinguish them from positive peaks. The two minor peaks, one positive the other negative, close to the phenylalanine side chain are indicative of a small conformational change (reproduced with permission from reference 408).

Repeating the mutagenesis over many regions throughout a protein gives a global perspective of which amino acid sequences in a protein are critical[427].

9.1 *Substrate Specificity.*

It is now possible to change the substrate specificity of an enzyme by site-directed mutagenesis of the active site residues. This has been well documented for the enzyme lactate dehydrogenase which has been changed to malate dehydrogenase[428,429]. A report in 1992 demonstrated how adenylate cylase could be changed to guanylate cylase[430].

It was assumed that nature had evolved a particular enzyme to have the optimum catalytic activity for the reaction it carries out. It appears however that the catalytic activity of several enzymes can be increased by specific site-directed mutagenesis of the active site residues. For example a mutant of the enzyme yeast phosphoglycerate kinase where an active site histidine residue has been changed to a glutamine residue results in the specific activity of the enzyme being increased two fold[431]. This is interpreted as the result of changing the hydrogen-bonding of the 1,3 diphosphoglycerate product to the enzyme thereby reducing the rate limiting step of the reaction. Recently the structure of a specific mutant of *Escherichia coli* alkaline phosphatase has been determined which has a higher catalytic activity than the wild-type enzyme[432]. The changes in the mutant enzyme are primarily the side chain of arginine 166 which by losing the hydrogen bond interaction with the carboxyl side chain of aspartic acid 101, becomes more flexible thereby allowing the enzyme to achieve a quicker catalytic turnover by again allowing a faster release of product. A journal *Protein Engineering*, IRL Press has emerged which specifically reports these type of experiments.

9.1.1 Chimeric Proteins. - By using genetic manipulation procedures, so called ‘fusion’ proteins can be made. This can be used to join one protein domain to another (see section 7.1) or to investigate or change substrate specificity. This has been carried out with human aldolase A and B to understand the difference in specificity between the two isoenzymes for the substrates fructose-1-phosphate and fructose 1,6 bisphosphate. Aldolase BAB chimeras, aldolase A fragments inserted into aldolase B, showed activity of aldolase B type[433].

The functional domain of the endonucleases from *Clostridium cellulovorans* have been studied by chimeric protein construction[434]. The construction of several other chimeric proteins of interest have been recently reported. The *Pseudomonas* exotoxin A-epidermal growth factor (EFG) mutant chimeric protein has been used as an indicator for identifying amino acid residues important in EFG receptor interaction[435]. Chimeric protein A, protein G and protein G alkaline phosphatase have been used as reporter molecules[436]. Functional structures of the recA protein have been found by chimera analysis[437]. A functional protein hybrid has been made between the glucose transporter and the N-acetylglucosamine transporter of *Escherichia coli*[438]. Some of the chimeric proteins have direct medical applications such as SCD4[178] - PE40 for aids therapy. Many other chimeric proteins will be studied and characterised in the near future.

10 Protein-Protein Interactions

Site-directed mutagenesis can be used to study protein-protein interactions. An example of this approach has been used to study the nature of interaction of insulin monomers into the hexamer structure commonly found complexed with zinc. The aggregation of insulin is thought to be related to the slow effect of injected insulin in diabetics. A paper in 1992 discusses the role of a glutamic acid of the insulin B chain in the structure of a recombinant mutant where glutamic acid 13 has been mutated to glutamine[439]. Another study examines the NMR solution structure of the B9 aspartic acid mutant of insulin which forms a dimer[440]. Previous studies to produce 'superactive' insulin molecules were reported in 1988 and 1991[441-443]. Some of these engineered protein molecules are already undergoing clinical trials.

I have tried to cover some of the current trends in protein research reported in 1992. Improved technologies to sequence proteins from both protein sequencing and DNA sequencing methods have been developed. Our understanding of protein structure in relation to protein primary amino acid sequence has increased. The rapidly increasing number of proteins for which three dimensional structural information is available will add to this knowledge. The ability to specifically manipulate protein structures by site-directed mutagenesis will help us to understand how nature has evolved protein molecules with such a diverse range of functions found in biological systems. The construction of new protein molecules to carry out specific functions can be undertaken. This knowledge will have far reaching applications in the use of proteins for both medical and industrial purposes.

References

1. K. Grandlundmoyer. American Biotechnology Laboratory, 1992, **10**, 34.
2. B.Wittmann-Liebold, *Pure Applied Chemistry*, 1992, **64**, 537-543.

3. Short Communications. Ninth International Conference on Methods in Protein Sequence Analysis (Sept. 20-24, 1992, Ostu, Japan), *J. of Protein Chem.*, 1992, **11**.
4 B.L. Karger, A.S. Cohen, and A. Guttman, *J. Chromatogr.*, 1989, **492**, 585-614.
5. T.Bergman, *J. Protein Chem.*, 1992, **11**, 356-357.
6. R.L. Moritz and R.J. Simpson, *J. Chromatogr.*, 1992, **599**, 119-130.
7. M. Kamo, M. Sano, C.S. Jone and A. Tsugita, *J. Protein Chem.*, 1992, 11, 358-359.
8. A. Tsugita, K. Takamoto, H. Iwadata, M. Kamo, and K. Satake, *J. Protein Chem.*, 1992, **11**, 362-363.
9. H. Bradaczek and B. Wittman-Liebold, *J. Prot. Chem.*, 1992, **11**, 365.
10. K. Nokihara, J. Kondo, R. Yamamoto, A. Ueda, and M. Hazama, *J. Prot. Chem.*, 1992, **11**, 365.
11. J.M. Bailey, N.R. Shenoy, M. Ronk and J.E. Shively, *Protein Sci.*, 1992, **1**, 68-81.
12. N.R. Shenoy, J.M. Bailey and J.E. Shively, *Protein Sci.*, 1992, **1**, 58-68.
13. S. Tsunasawa and H. Hirano, *J. Prot. Chem.*, 1992, **11**, 382.
14. G.C. Harbour, R.L. Garlick, S.B. Lyle, F.W. Crow, R.H. Robins and J.G. Hoogerheide, 1992, in Techniques in Protein Chemistry III (R.H. Angeletti, ed.) 487-495. The Protein Society, Academic Press.
15. B.B. Reinhold, E.L. Reinherz and V.N. Reinhold, 1992, in Techniques in Protein Chemistry III (R.H. Angeletti, ed.) 287-294. The Protein Society, Academic Press.
16. G.K. Meng, M. Mann and J.B. Fenn, *Z. Phys. D.*, 1988, 10, 361-368.
17. M. Karas and F. Hillenkamp, *Anal. Chem.*, 1988, **60**, 2299-2301.
18. R.C. Beavis, T. Chaudhary and B.T. Chait, *Org. Mass Spectrom*, 1992, **27**, 156.
19. B. Spengler and R. Kaufman, *Analysis*, 1992, **20**, 91.
20. M.M. Vestling, C.M. Murphy, C. Fenselau, J. Dedinas, M.S. Doleman, P.B. Harrsch, R. Kutny, D.L. Ladd and M.A. Olsen, 1992, in Techniques in Protein Chemistry III (R.H. Angeletti, ed.) 477-485. The Protein Society, Academic Press.
21. P.C. Andrews, M.H. Allen, M.L. Vestal and R.W. Nelson, 1992, in Techniques in Protein Chemistry III (R.H. Angeletti, ed.) 515-523. The Protein Society, Academic Press.
22. S.E. Brockerhoff, C.G. Edmonds and T.N. Davis, *Protein Sci.* 1992, **1**, 504-516.
23. S.C. Hall, F.R. Schindler, F.R.Masiarz and A.L. Burlingame, 1992, in Techniques in Protein Chemistry III (R.H. Angeletti, ed.) 525-532. The Protein Society, Academic Press.
24. K. Biemann, *Fresenius Journal of Analytical Chemistry*, 1992, **343**, 25-26.
25. R.S. Johnson and K.A. Walsh, *Protein Sci.*, 1992, **1**, 1083-1091.
26. R. Aebersold, E.J. Bures, M. Namchuk, M.H. Goghari, B. Shushan, and T.C. Covey, *Protein Sci.*, 1992, **1**, 494-503.
27. B. Ganem, Y.T. Li, and J.D. Henion, *J. Am. Chem. Soc.*, 1991, **113**, 6294-6296.
28. M. Baca and S.B.H. Kent, *J. Am. Chem. Soc.*, 1992, **114**, 3992-3993.
29. B.T. Chait and S.B.H. Kent, *Science*, 1992, **257**, 1885-1894.
30. M.J. Geisow, *Tibtech*, 1992, **10**, 432-441.
31. S.A. Benner, M.A. Cohen and G. Gonnet. *Abstracts of Papers of the American Chemical Society* ,1992, **203**, 61.
32. G.H. Gonnet, M.A. Cohen and S.A. Benner, *Science*, 1992, **256**, 1443-1445.
33. A. Bairoch and B. Boeckmann, *Nucl. Acids Res.*, 1992, **20**, 2019-2022.
34. W.C. Barker, D.G. George, H.W. Mewes and A. Tsugita, *Nucl. Acids Res.*, 1992, **20**, 2023-2026.
35. W.C. Barker, D.G. George, G.Y. Srinivasarao and L.S. Yeh, *FASEB Journal,* 1992, **6**, 1.

36. SEQNET facility, Daresbury Laboratory, User Interface Group (UIG), Daresbury Laboratory, Warrington, U.K.
37. F.C. Bernstein, T.F. Koetzle, G.J. Williams, E.F. Meyer, M.D. Brice, J.R. Rodgers, O. Kennard, T. Shimanouchi, and M. Tasumi, *J. Mol. Biol.*, 1977, **112**, 535-542.
38. C. Chothia, *Nature*, 1992, **357**, 543-544.
39. Biosym Technologies, 9685 Scraton Road, San Diego, C.A. USA.
40. Polygen Corporation, 200 Fifth Avenue, Waltham, M.A. USA.
41. J. Overington, D. Donnelly, M.S. Johnson, A. Sali, and T.L. Blundell, *Protein Sci.*, 1992, **1**, 216-226.
42. J.U. Bowie, R. Lüthy, and D. Eisenberg, *Science*, 1991, **253**, 164-170.
43. R. Lüthy, J.U. Bowie, and D. Eisenberg, *Nature*, 1992, **356**, 83-85.
44. J.W. Ponder and F.M. Richards, *J. Mol. Biol.*, 1987, **193**, 775-791.
45. M.J. Sippl and S. Weitckus, *Proteins*, 1992, **13**, 258-271.
46. D.T. Jones, W.R. Taylor, and J.M. Thornton, *Nature*, 1992, **358**, 86-89.
47. A. Godzik and J. Skolnick, *Proc. Nat. Acad. Sci. USA*, 1992, **89**, 12098-12102.
48. R.A. Goldstein, Z.A. Luthey-Schulten, and P.G. Wolynes, *Proc. Natl. Acad. Sci. USA*, 1992, **89**, 9029-9033.
49. R.A. Goldstein, Z.A. Luthey-Schulten, and P.G. Wolynes, *Proc. Natl. Acad. Sci. USA*, 1992, **89**, 4918-4922.
50. J.M. Rondeau, F. Tête-Favier, A. Podjarny, J.M. Reymann, P. Barth, J.F. Biellmann, and D. Moras, *Nature*, 1992, **355**, **217**, 737-750.
52. R.L. Brady, A.M. Brzozowski, Z.S. Derewenda, E.J. Dobson, and G.G. Dobson, *Acta Crystallogr.*, 1991, **47**, 527-535.
53. B. Mikami, M. Sato, T. Shibata, M. Hirose, S. Aibara, Y. Katsube, and Y. Morita, *J. Biochem.*, 1992, **112**, 541-546.
54. D.K. Wilson, F.B. Rudolph and F.A. Quiocho, *Science*, 1992, **252**, 1278-1284.
55. P.M.G. Curmi, D. Casico, R.M. Sweet, D. Eisenberg, and H. Schreuder, *J. Biol., Chem.*, 1992, **267**, 16980-16989.
56. M. Wilmanns, J.P. Priestle, T. Niermann, and J. N. Jansonius, *J. Mol. Biol.*, 1992, **223**, 477-507.
57. S.J. Gamblin, G.J. Davies, J.M. Grimes, R.M. Jackson, J.A. Littlechild, and H.C. Watson, *J. Mol. Biol.*, 1991, **219**, 573-576.
58. G. Hester, O. Brenner-Holzach, F.A. Rossi, M. Struck-Donatz, K. H. Winterhalter, J.D.G. Smit, and K. Piontek, *FEBS Lett.*, 1992, **292**, 237-242.
59. D.J. Neidhart, P.L. Howell, G.A. Petsko, V.M. Powers, R. Li, G.L. Kenyon, and J.A. Gerlt, *Biochemistry*, 1991, **30**, 9264-9273.
60. M. Hennig. B. Schlesier, Z. Dauter, S. Pfeffer, C. Betzel, W.E. Höhne, and K.S. Wilson, *FEBS Lett.*, 1992, **306**, 80-84.
61. J. Piatigorsky, *J. Biol. Chem.*, 1992, **267**, 4277-4280.
62. M. Wilmanns, C.C. Hyde, D.R. Davies, K. Kirschner, and J.N. Jansonius, *Biochemistry*, 1991, **30**, 9161-9169.
63. H.M. Ke, *J. Mol. Biol.*, 1992, **228**, 539-550.
64. Z.X. Xia and F.S. Mathews, *J. Mol. Biol.*, 1990, **212**, 837-863.
65. Y. Lindqvist, *J. Mol. Biol.*, 1989, **209**, 151-166.
66. S.D. Pickett, M.A.S. Saqi, and M.J.E. Sternberg, *J. Mol. Biol.*, 1992, **228**, 170-187.
67. A. Godzik, J. Skolnick and A. Kolinski, *Proc. Natl. Acad. Sci. USA*, 1992, **89**, 2629-2633.
68. R. Rudolph, R. Siebendritt and T. Kiefhaber, *Protein Sci.*, 1992, **1**, 654-666.
69. J. Eder and K. Kirschner, *Biochemistry*, 1992, **31**, 3617-3625.
70. J. Eder and W. Wilmanns, *Biochemistry*, 1992, **31**, 4437-4444.

71. J.P.Y. Scheerllinck, I. Lasters, M. Claessend, M. DeMaeyer, F. Pio, P. Delhaise, and S.J. Wodak, *Proteins*, 1992, **12**, 299-313.
72. R. Urfer and K. Kirschner, *Protein Sci.*, 1992, **1**, 31-45.
73. D.L. Ollis, E. Cheah, M. Cygler, B. Dijkstra, F. Frolow, S.M. Franken, M. Harel, S.J. Remington, I. Silman, J. Schrag, J.L. Sussman, K.H.G. Verschueren, and A. Goldman, *Protein Engineering*, 1992, **5**, 199-211.
74. J.L. Sussman, M. Harel, F. Frolow, C. Oefner, A. Goldman, L. Toker, and I. Silman, *Science*, 1991, **253**, 872-879.
75. D-I Liao and S.J. Remington, *J. Biol. Chem.*, 1990, **265**, 6528-6531.
76. T. Thomas, A. Cooper, H. Bussey, and G.J. Thomas, *J. Biol. Chem.*, 1990, **265**, 10821-10824.
77. D. Pathak, and D. Ollis, *J. Mol. Biol.*, 1990, **214**, 497-525.
78. S.M. Franken, H.J. Rozeboom, K.H. Kalk, and B.W. Dijkstra, *EMBO J.*, 1991, **10**, 1297-1302.
79. J.D. Schrag, Y. Li, S. Wu and M. Cygler, *Nature*, 1991, **351**, 761-764.
80. B.L. Sibanda, T.L. Blundell and J.M. Thornton, *J. Mol. Biol.*, 1989, **206**, 759-777.
81. A.V. Efimov, Molecular Conformation and Biological Interactions, P. Balaram and S. Ramaeshan ed., *Indian Acad. Sciences*, 1991, 19-29.
82. C.S. Ring, D.G. Kneller, R. Langridge and F.E. Cohen, *J. Mol. Biol.*, 1992, **224**, 685-699.
83. A.V. Efimov, *FEBS Lett.*, 1992, **298**, 261-265.
84. R.F. Doolittle, *Protein Sci.*, 1992, **1**, 191-200.
85. D. Wen, E. Peles, R. Cupples, S.V. Suggs, S.S. Bacus, Y. Luo, G. Trail, S.M. Silbiger, R.B. Levy, R.A. Koski, H.S. Lu, and Y. Yarden, *Cell*, 1992, **69**, 559-572.
86. J. Partanen, E. Armstrong, T.P. Makela, J. Korhonen, M. Sandberg, R. Renkonen, S. Knuutila, K. Huebner, and K. Alitalo, *Mol. Cell. Biol.*, 1992, **12**, 1698-1707.
87. G. Montelione, K. Wüthrich, A.W. Burgess, E.C. Nice, G. Wagner, K.D. Gibson, and H.A. Sheraga, *Biochemistry*, 1992, **227**, 271-282.
88. U. Hommel, T.S. Harvey, P.S. Driscoll, and I.D. Campbell, *J. Mol. Biol.*, 1992, **227**, 271-282.
89. M. Baron, D.G. Norman, T.S. Harvey, P.A. Handford, M. Mayhew, G.G. Brownlee, and I.D. Campbell, *Protein Sci.*, 1992, **1**, 81-90.
90. M. Ullner, M. Selander, E. Persson, J. Stenflo, T. Dakenberg and O. Telleman, *Biochemistry*, 1992, **31**, 5974-5983.
91. M. Selander-Sunnerhagen, M. Ullner, E. Persson, O. Telleman, J. Stenflo, and T. Drakenberg, *J. Biol. Chem.*, 1992, **267**, 19642-19649.
92. I.D. Campbell and P. Bork, *Curr. Opin. Struct. Biol.*, 1993, **3**, 385-392.
93. R.B. Freedman, in Protein Folding, edited by T.E. Creighton, N.Y., W.H. Freeman, 1992, 457-541.
94. R. Noiva and W.J. Lennarz, *J. Biol. Chem.*, 1992, **267**, 3553-3556.
95. K. Vuori, R. Myllyla, T. Pihlajaniemi, and K.I. Kivirikko, *J. Biol. Chem.*, 1992, **267**, 7211-7214.
96. C.T. Walsh, L.D. Zydowsky, and F.D. McKeon, *J. Biol. Chem.*, 1992, **267**, 13115-13118.
97. S.J. O'Keefe, J. Tamura, R.L. Kincaid, M.J. Tocci, and E.A. O'Neill, *Nature*, 1992, **357**, 692-694.
98. L.D. Zydowsky, F.A. Etzkorn, H.Y. Chang, S.B. Ferguson, L.A. Stolz, S.I. Ho, and C.T. Walsh, *Protein Sci.*, 1992, **1**, 1092-1099.
99. P.K.K. Tai, M.W. Albers, H. Chang, L.E. Faber, and S.L. Schreiber, *Science* 1992, **256**, 1315-1318.

100. J. Liu, M.W. Albers, T.J. Wandless, S. Luan, D.G. Alberg, P.J. Belshaw, P. Cohen, C. Mackintosh, C.B. Klee, and S.L. Schreiber, *Biochemistry*, 1992, **31**, 3896-3901.
101. A.W. Yen, A.G. Tomasselli, R.L. Heinrikson, H. Zurcher-Neely, V.A. Ruff, R.A. Johnson, and M.R. Deibel, Jr., *J. Biol. Chem.*, 1992, **267**, 2868-2871.
102. S. Base and R.B. Freedman, *Biochem. Soc. Trans.*, 1992, **20**, 256S.
103. T. Kiefhaber, H.H. Kohler, and F.X. Schmid, *J. Mol. Biol.*, 1992, **224**, 217-229.
104. T. Kiefhaber and F.X. Schmid, *J. Mol. Biol.*, 1992, **224**, 231-240.
105. R. Schonbrunner and F.X. Schmid, *Proc. Natl. Acad. Sci. USA*, 1992, **89**, 4510-4513.
106. R.K. Harrison and R.L. Stein, *J. Am. Chem. Soc.*, 1992, **114**, 3464-3471.
107. S.T. Park, R.A. Aldape, O. Futer, M.T. De Cenzi, and D.J. Livingston, *J. Biol. Chem.*, 1992, **267**, 3316-3324.
108. C. Spitzfaden, H.-P. Weber, W. Braun, J. Kallen, G. Wider, H. Widmer, M.D. Walkinshaw, and K. Wuthrich, *FEBS Lett.*, 1992, **300**, 291-300.
109. S.W. Fesik, P. Neri, R. Meadows, E.T. Olejniczak, and G. Gemmecker, *J. Am. Chem. Soc.*, 1992, **114**, 3165-3166.
110. D. Altschun, O. Vix, B. Rees, and J.-C. Thierry, *Science*, 1992, **256**, 92-94.
111. M.J. Gething and J. Sambrock, *Nature*, 1992, **355**, 33-45.
112. F.U. Hartl, J. Martin and W. Neupert, *Annu. Rev. Biophys. Biomol. Structure*, 1992, **21**, 293-322.
113. C. Georgopoulos, *Trends Biochem. Sci.*, 1992, **17**, 295-299.
114. R.J. Ellis, *Nature*, 1992, **358**, 191-192.
115. E.J. Hutchinson, W. Tichelaar, G. Hofhaus, H. Weiss, and K. Leonard, *EMBO J.*, 1989, **8**, 1485-1490.
116. P.V. Vitanen, G.H. Lorimer, R. Seetharam, R.S. Gupta, J. Oppenheim, J.O. Thomas, and N.J. Cowan, *J. Biol. Chem.*, 1992, **267**, 695-698.
117. T. Langer, G. Pfeifer, J. Martin, W. Baumeister, and F.U. Hartl, *EMBO J.*, 1992, **11**, 4757-4765.
118. N. Ishii, H. Taguchi, M. Sumi, and M. Yoshida, *FEBS Lett.*, 1992, **299**, 169-174.
119. N.M. Lissin, S.E. Sedelnikova, and S.N. Ryazantsev, *FEBS Lett.*, 1992, **311**, 22-24.
120. M.B. Yaffe, G.W. Farr, D. Miklos, A.L. Horvich, M.L. Sternlicht, and H. Sternlicht, *Nature*, 1992, **358**, 245-248.
121. V.A. Lewis, G.M. Hynes, D. Zheng, H. Saibil, and K. Willison, *Nature*, 1992, **358**, 249-252.
122. Y. Gao, J.O. Thomas, R.L. Chow, G.-H. Lee, and N.J. Cowan, *Cell.*, 1992, **69**, 1043-1050.
123. H. Wiech, J. Buchner, R. Zimmerman, and U. Jakob, *Nature*, 1992, **358**, 169-170.
124. M. Fischer, *Biochemistry*, 1992, **31**, 3955-3963.
125. P.V. Vitanen, A.A. Gatenby, and G.H. Lorimer, *Protein Sci.*, 1992, **1**, 361-369.
126. S.J. Landry and L. M. Gierasch, *Biochemistry*, 1991, **30**, 7359-7362.
127. M. Schmidt and J. Bucher, *J. Biol. Chem.*, 1992, **267**, 16829-16833.
128. S.J. Landry, R. Jordon, R. McMackea, and L.M. Gierasch, *Nature*, 1992, **355**, 455-457.
129. S.M. Van der Vies, P.V. Vijtanen, A.A. Gatenby, G.H. Lorimer, and R. Jaenicke, *Biochemistry*, 1992, **31**, 3635-3644.
130. H. Roy, M. Kupferschmid, and J.A. Bell, *Protein Sci.*, 1992, **1**, 925-934.
131. R. Zahn and A. Plueckthun, *Biochemistry*, 1992, **31**, 3246-3255.
132. E.S. Bochkareva, N.M. Lissin, G.C. Flynn, J.E. Rothman, and A.S. Girshovich, *J. Biol. Chem.*, 1992, **267**, 6796-6800.
133. T.E. Gray and A.R. Fersht,*FEBS Lett.*, 1991,**292**, 254-258.
134. F. Baneyx and A.A. Gatenby, *J. Biol. Chem.*, 1992, **267**, 11637-11644.

135. J. Landry, H. Lambert, M. Zhou, J.N. Lavole, E. Hickey, L.A. Weber and C.W. Anderson, *J. Biol. Chem.*, 1992, **267**, 794-803.
136. M.Y. Sherman and A.L. Goldberg, *Nature*, 1992, **357**, 167-169.
137. Y. Miyata and I. Yahara, *J. Biol. Chem.*, 1992, **267**, 7042-7047.
138. P.J. Freiden, J.R. Gaut, and L.M. Hendershot, *EMBO J.*, 1992, **11**, 63-70.
139. D.R. Palleros, K.L. Reid, M.S. McCarty, G.C. Walker, and A.L. Fink, *J. Biol. Chem.*, 1992, **267**, 5279-5285.
140. G.J. Davies, S.J. Gamblin, J.A. Littlechild, and H.C. Watson, *Proteins*, 1993, **15**, 283-289.
141. J.A. Littlechild, S.D. Davies, S.G. Gamblin and H.C. Watson. International Conference. Thermophiles: Science and Technology, Reykjavik, Iceland, 1992,132.
142. G.Davies, J.A. Littlechild, H.C. Watson, and L. Hall, *Gene*, 1991, **109**, 39-45.
143. J.E. Wampler, E.A. Bradley, M.W.W. Adams, and D.E. Steward, In *Biocatalysis at Extreme Temperatures: Enzyme Systems Near and Above 100°*C. M.W.W. Adams and R.M. Kelly eds. p.153-174, Americal Chemical Society, Washington D.C.
144. B.W. Matthews, *Annual Reviews of Biochemistry*, 1993, **62**, 139-160.
145. J.A. Bell, W.J. Becktel, U. Sauer, W.A. Baase, and B.W. Matthews, *Biochemistry*, 1992,**31**, 3590-3596
146. X.J. Zhang, W.A. Baase, and B.W. Matthews, *Protein Sci.*, 1992, **1**, 761-776.
147. U.H. Sauer, D.P. Sun, and B.W. Matthews, *J. Biol. Chem.*, 1992, **267**, 2393-2399.
148. A.E. Eriksson, W.A. Baase, X.J. Zhang, D.W. Heinz, M. Blaber, E.P. Baldwin, and B.W. Matthews, *Science*, 1992, **255**, 178-183.
149. A.E. Eriksson, W.A. Baase, J.A. Wozniak, and B.W. Matthews, *Nature*, 1992, **355**, 371-373.
150. R.H. Jacobson, M. Matsumura, H.R. Faber and B.W. Matthews, *Protein Sci.*, 1992, **1**, 46-58.
151. A.R. Fersht, A. Matouschek, and L. Serrano, *J. Mol. Biol.*, 1992, **224**, 771-782.
152. L. Serrano, J.L. Neira, J. Sancho and, A.R. Fersht, *Nature*, 1992, **356**, 453-455.
153. L. Serrano, J. Sancho, M. Hirshberg, and A.R. Fersht, *J. Mol. Biol.*, 1992, **227**, 544-549.
154. W.G.J. Hol, *Prog. Biophys. Mol. Biol.* 1987, **45**, 149-195.
155. J. Sancho, L. Serrano, and A.R. Fersht, *Biochemistry*, 1992, **31**, 2253-2258.
156. A. Horovitz, J.M. Matthews, and A.R. Fersht, *J. Mol. Biol.*, 1992, **227**, 560-568.
157. T.P. Creamer and G.D. Rose, *Proc. Natl. Acad. Sci. USA*, 1992, **89**, 5937-5941.
158. D.J. Tobias and C.L. Brooks, *Biochemistry*, 1991, **30**, 6059-6070.
159. P.C.C. Lyu and N.R. Kallenbach, *J. Mol. Biol.*, 1992, **223**, 343-350.
160. D. Sali, M. Bycroft, and A.R. Fersht, *J. Mol. Biol.*, 1991, **220**,779-788.
161. S. Dao-Pin, H. Nicholson, and B.W. Matthews, *Biochemistry*, 1991, **30**, 7142-7153.
162. A.R. Fersht, *J. Mol. Biol.*, 1972, **64**, 497-509.
163. D.E. Anderson, W.J. Becktel and F.W. Dahlquist, *Biochemistry*, 1990, **29**, 2403-2408.
164. B.A. Shirley, P. Stanssens, U. Hahn, and C.N. Pace, *Biochemistry*, 1992, **31**, 725-732.
165. L. Serrano, J.T. Kellis, P. Cann, A. Matouschek, and A.R. Fersht, *J. Mol. Biol.*, 1992, **224**, 783-804.
166. W.A. Lim, D.C. Farruggio, and R.T. Sauer, *Biochemistry*, 1992, **31**, 4324-4333.
167. J.H. Hurley, W.A. Baase, and B.W. Matthews, *J. Mol. Biol.*, 1992, **224**, 1143-1159
168. R. Lowenthal, J. Sancho, and A.R. Fersht, *J. Mol. Biol.*, 1992, 224, 759-770.
169. E.M. Meiering, L. Serrano, and A.R. Fersht, *J. Mol. Biol.*, 1992, **225**, 585-589.
170. R.G. Brennan, *Curr. Opin. Struct. Biol.*, 1992, **2**, 100-108.
171. S.C. Harrison, *Nature*, 1991, **353**, 715-719.

172. C.O. Paco and R.T. Sauer, *Annu. Rev. Biochem.*, 1992, **61**, 1053-1095.
173. C. Wolberger, A.K. Vershon, B. Liu, A.D. Johnson, and C.O. Pabo, *Cell.*, 1991, **67**, 517-528.
174. R.S. Hegde, S.R. Grossman, L.A. Laimins, and P.B. Sigler, *Nature*, 1992, **359**, 505-512
175. W.S. Somers and S.E.V. Phillips, *Nature*, 1992, **359**, 387-392.
176. D.B. Niklow, S.H. Hu, J.P. Lin, A. Gasch, A. Hoffmann, M. Horikoshi, N.-H. Chua, R.G. Roeder, and S.K. Burley, *Nature*, 1992, **360**, 40-46.
177. D. Pathak and P.D. Sigler, *Curr. Opin. Struct. Biol.*, 1992, **2**, 116-123.
178. T.E. Ellenberger, C.J. Brandi, K. Struhl, and S.C. Harrison, *Cell.*, 1992, **71**, 1223-1237.
179. R. Marmorstein, M. Carey, M. Ptashne, and S.C. Harrison, *Nature*, 1992, **356**, 408-414.
180. P.J. Kraulis, A.R.C. Raine, P.L. Gadhavi, and E.D. Laue, *Nature*, 1992, **356**, 448-450.
181. J.D. Baleja, R. Marmorstein, S.C. Harrison, and G. Wagner, *Nature*, 1992, **356**, 450-453.
182. J.W.R. Schwabe and D. Rhodes, *Trends in Biochem. Sci.*, 1991, **16**, 291-296.
183. B.F. Luisi, W.X. Xu, Z. Otwinowski, L.P. Freedman, K.R. Yamanoto, and P.B. Sigler, *Nature*, 1991, **352**, 497-505.
184. J. Miller, A.D. McLachlan, and A. Klug, *EMBO Journal*, 1985, **4**, 1609-1614.
185. J.J. Hayes and T.D. Tullis, *J. Mol. Biol.*, 1992, **227**, 407-417.
186. X. Liao, K.R. Clements, L. Tennant, P.E. Wright, and J.M. Gottesfeld, *J. Mol. Biol.*, 1992, **223**, 857-871.
187. Y. Nakaseko, D. Neuhaus, A. Klug and D. Rhodes, *J. Mol. Biol.*, 1992, **288**, 619-636.
188. D. Neuhaus, Y. Nakaseko, J.W.R. Schwabe and A. Klug, *J. Mol. Biol.*, 1992, **288**, 637-651.
189. J.G. Omichinski, G.M. Clore, M. Robien, K. Sakaguchi, E. Appella, and A.M. Gronenborn, *Biochemistry*, 1992, **31**, 3907-3917.
190. N.P. Pavletich and C.O. Pabo, *Science*, 1991, **252**, 809-817.191.
X. Qian and M.A. Weiss, *Biochemistry*, 1992, **31**, 7463-7476.
192. R.J. Mortshire-Smith, M.S. Lee, L. Bolinger, and P.E. Wright, *FEBS Lett.*, 1992, **296**, 11-15.
193. S.F.Michael, V.J. Kilfoil, M.H. Schmidt, B.T. Amann, and J.M. Berg, *Proc. Natl. Acad. Sci. U.S.A*, 1992, **89**, 4796-4800.194.
J.R. Desjarlais and J.M. Berg, *Proteins*, 1992, **12**, 101-104.
195. S.K. Thukral, M.L. Morrison, and E.T. Young, *Mol. Cell. Biol.*, 1992, **12**, 2784-2792.
196. J.W. Bess, Jr., P. Powell, H.J. Issaq, L.J. Schumack, M.K. Grimes, L.E. Henderson, and L.O. Arthur, *J. Virol.*, 1992, **66**, 840-847.
197. M.F. Summers, L.E. Henderson, M.R. Chance, J.W. Bess, Jr., T.L. South, P.R. Blake, I. Sagi, G. Peret Alvarado, R.C. Sowder, D.R. Hare, and L.O. Arthur, *Protein Sci.*, 1992, **1**, 563-574.
198. K. Morikawa, O. Matsumoto, M. Tsujimoto, K. Katayanagi, M. Ariyoshi, T. Doi, M. Ikehara, T. Inaoka and E. Ohtsuka, *Science*, 1992, **256**, 523-526.
199. C.-F. Kuo, D.E. Mcree, C.L. Fisher, S.F. O'Handley, J.A. Cunningham, and J.A. Tainer, *Science*, 1992, **258**, 434-440.
200. R.M. Story, I.T. Weber, and T.A. Steitz, *Nature*, 1992, **355**, 318-325.
201. D. Macelis and R.J. Roberts, *Nucleic Acid Res.*, 1992, **20**, 2167-2180.

202. J.M. Rosenberg, *Curr. Opin. Struct. Biol.*, 1991, **1**, 104-113.
203. F.K. Winkler, *Curr. Opin. Struct. Biol.*, 1992, **2**, 93-99.
204. V. Thielking, U. Selent, E. Kohler, A. Landgraf, H. Wolfes, J. Alves, and A. Pingoud, *Biochemistry*, 1992, **31**, 3727-3732.
205. C.L.M. Vermote and S.E. Halford, *Biochemistry*, 1992, **31**, 6082-6089.
206. C.L.M. Vermote, I.B. Vipond, and S.E. Halford, *Biochemistry*, 1992, **31**, 6089-6097.
207. U. Selent, T. Rüter, E. Köhler, M. Liedtke, V. Thielking. J. Alves, T. Oelgeschläger, H. Wolfes, F. Peters, and A. Pingoud, *Biochemistry*, 1992, **31**, 4808-4815.
208. A. Jeltsch, J. Alves, G. Maass, and A. Pingoud, *FEBS Lett.*, 1992, **304**, 4-8.
209. L.A. Kohlstaedt, J. Wang, J.M. Friedman, P.A. Rice, and T.A. Steitz, *Science*, 1992, **256**, 1783-1790.
210. Z. Hostomsky, Z. Hostomska, T.-B. Fu, and J. Taylor, *J. Virol.*, 1992, **66**, 3179-3192.
211. J.C. Wu, T.C. Warren, J. Adams, J. Proudfoot, J. Skiles, P. Raghaven, C. Perry, I. Potocki, P.R.Farina, and P.M. Grob, *Biochemistry*, 1991, **30**, 2022-2026.
212. X.-P. Kong, R. Onrust, M. O'Donnell, and J. Kuriyan, *Cell*, 1992, **69**, 425-437.
213. A.H. Polesky, M.E. Dahlberg, S.J. Benkovic, N.D.F. Grindley, and C.M. Joyce, *J. Biol. Chem.*, 1992, **267**, 8417-8428.
214. V. Derbyshire, N.D.F. Grindley, and C.M. Joyce, *EMBO J.*, 1991, **10**, 17-24.
215. D. Moras, *Trends Biochem. Sci.*, 1992, **17**, 159-164.
216. M.A. Rould, J.J. Perona, D. Soll, and T.A. Steitz, *Science*, 1989, **246**, 1135-1142.
217. S. Cusack, C. Berthet-Colominas, M. Hartlein, N. Nassar, and R. Leberman, *Nature*, 1990, **347**, 249-255.
218. L. Reshetnikova, M. Chernaya, V. Ankilova, O. Lavrik, M. Delanie, J.C. Thierry, D. Moras and M. Safro, *Eur. J. Biochem.*, 1992, **208**, 411-417.
219. S. Eiler, M. Boeglin, F. Martin, G. Eriani, J. Gangloff, J.C. Thierry, and D. Moras, *J. Mol. Biol.*, 1992,**224**, 1171-1173.
220. A. Yaremchuk, M.A. Tukalo, I. Kriklivty, N. Malchenko, V. Biou, C. Berthet-Colominas, and S. Cusack, *FEBS Lett.*, 1992, **310**, 157-161.
221. L. Despons, B. Senger, F. Fasiolo, and P. Walter, *J. Mol. Biol.*, 1992, **225**, 1-11.
222. M. Mirande, M. Lazard, B. Martinez, and M.T. Latreille, *Eur. J. Biochem.*, 1992, **203**, 459-466.
223. C. Branden, and J. Tooze, *Introduction to Protein Structure*, Chapter 12, 179-199.
224. D.R. Davies, E.A. Padlan, and S. Sheriff, *Ann. Rev. Biochem.*, 1990, **59**, 439-473.
225. C. Chothia, *Curr. Opin. Struct. Biol.*, 1991, **1**, 43-59.
226. W.R. Tulip, J.N. Varghese, W.G. Laver, R.G. Webster, and P.M. Colman, *J. Mol. Biol.*, 1992, **227**, 122-148.
227. J.M. Rini, U. Schulze-Gahmen, and I.A. Wilson, *Science* 1992, **255**, 959-965.
228. K.C. Garcia, P.M. Ronco, P.J. Verroust, A.T. Brunger, and L.M. Amzel, *Science*, 1992, **257**, 502-507.
229. D. Altschuh, O. Vix, B. Rees and J.C. Thierry, *Science*, 1992, **256**, 92-94.
230. W.R. Tulip, J.N. Varghese, R.G. Webster, W.G. Laver, and P.M. Colman, *J. Mol. Biol.*, 1992, **227**, 149-159.
231. F.A. Saul and R.J. Poljak, *Proteins*, 1992, **14**, 363-371.
232. Z.-C. Fan, L. Shan, L.W. Guddat, X.-M. He, W.R. Gray, R.L. Raison and A.B. Edmundson, *J. Mol. Biol.*, 1992, **228**, 188-207.
233. X.M. He, F. Ruker, E. Casale and D.C. Carter, *Proc. Natl. Acad. Sci., USA*, 1992, **89**, 7154-7158.
234. A. Tramontana and A.M. Lesk, *Proteins*, 1992, **13**, 231-245.
235. T.A. Collet, P. Roben, R. O'Kennedy, C.F. Barbas, D.R. Burton, and R.A. Lerner, *Proc. Natl. Acad. Sci. USA*, 1992, **89**, 10026-10030.

236. S.L. Zebedee, C.F. Barbas, Y.L. Hom, R.H. Caothein, R. Graff, J. DeGraw, J. Pyati, R. LaPolla, D.R. Burton, and R.A. Lerner, *Proc. Natl. Acad. Sci. USA*, 1992, **89**, 3175-3179.
237. C.F. Barbas, J.D. Bain, D.M. Hoekstra, and R.A. Lerner, *Proc. Natl. Acad. Sci. USA*, 1992, **89**, 4457-4461.
238. D.R. Davies and E.A. Padlan, *Curr. Biol.*, 1992, **2**, 254-256.
239. J. Foote and G. Winter, *J. Mol. Biol.*, 1992, **224**, 487-499.
240. B. Steipe, A. Pluckthun, and R. Huber, *J. Mol. Biol.*, 1992, **225**, 739-753.
241. R.L. Brady, D.J. Edwards, R.E. Hubbard, J.S. Jiang, G. Lange, S.M. Roberts, R.J. Todd, J.R. Adair, J.S. Emtage, and D.J. King, *J. Mol. Biol.*, 1992, **227**, 253-264.
242. T. Simon and K. Rajewsky, *Protein Engineering*, 1992, **5**, 229-234.
243. F.S. Lee, Z.T. Chu, M.B. Bolger, and A. Warshel, *Prot. Eng.*, 1992, **5**, 215-228.
244. L.J. Nell, J.A. McCammon and S. Subramaniam, *Biopolymers*, 1992, **32**, 11-21.
245. J.J. Tanner, L.J. Nell and J.A. McCammon, *Biopolymers*, 1992, **32**, 23-31.
246. L. Brady, A.M. Brzozowski, Z.S. Derewenda, E. Dodson, G. Dodson, S. Tolley, J.P. Turkenburg, L. Christiansen, B. Huge-Jensen, L. Norskov, L. Thim, and U. Menge, *Nature*, 1990, **343**, 767-770.
247. F.K. Winkler, A.D'Arcy, and W. Hunziker, *Nature*, 1990, **343**, 771-774.
248. Z.S. Derewenda, U. Derewenda, and G.G. Dobson, *J. Mol. Biol.*, 1992, **227**, 818-839.
249. L. Polgar, *FEBS Lett.*, 1992, **311**, 281-284.
250. A.M. Brzozowski, U. Derewenda, Z.S. Derewenda, G.G. Dobson, D.M. Lawson, J.P. Turkenburg, F. Bjorkling, B. Huge-Jensen, S.A. Patkar, and L. Thim, *Nature*, 1991, **351**, 491-494.
251. U. Derewenda, A.M. Brzozowski, D.M. Lawson and Z.S. Derewenda, *Biochemistry*, 1992, **31**, 1532-1541.
252. J.A. Aleman-Gomez, N.S. Colwell, T. Sasser, and V.B. Kumar, *Biochem. and Biophys. Res. Com.*, 1992, **188**, 964-971.
253. H. Sztajer, H. Lünsfdorf, H. Erdmann, U. Menge, and R. Schmid, *Biochim et Biophys. Acta.*, 1992, **1124**, 253-261.
254. T. Jacobsen and O.M. Poulsen, *Can. J. Microbiol/*, 1992, **38**, 75-80.
255. S. Longhi, F. Fusetti, R. Grandori, M. Lotti, M. Vanoni, and Lilia Alberghina, *Biochim. et Biophys. Acta.*, 1992, **1131**, 227-232.
256. L.G.J. Frenken, M.R. Egmond, A.M. Batenburg, J.W. Bos, C. Visser, and C.T. Verrips, *App. and Environ. Microbiology*, 1992, **58**, 3787-3791.
257. S. Jäger, G. Demleitner, and F. Götz, *FEMS Micro. Lett.*, 1992, **100**, 249-254.
258. D.A. Cooper, S-C. Lu, R. Viswanath, R.N. Freiman, and A. Bensadoun, *Biochim. et Biophys. Acta.*, 1992, **1129**, 166-171.
259. K. Isobe, K. Nokihara, S. Yamaguchi, T. Mase, and R.D. Schmid, *Eur. J. Biochem.*, 1992, **203**, 233-237.
260. V. Dartois, A. Baulard, K. Schank, and C. Colson, *Biochim and Biophys. Acta.*, 1992, **1131**, 253-260.
261. M. Chihara-Siomi, K. Yoshikawa, N. Oshima-Hirayama, K. Yamamoto, Y. Sogabe, T. Nakatani, T. Nishioka, and J. Oda, *Arch. Biochem. and Biophys.*, 1992, **296**, 505-513.
262. K.K. Kim, K.Y. Hwang, H.S. Joen, S. Kim, R.M. Sweet, C.H. Yang, and S.W. Suh, *J. Mol. Biol.*, 1992, **227**, 1258-1262.
263. H. Moreau, C. Abergel, F. Carriere, F. Ferrato, J.C. Fontecilla-Camps, C. Cambillau, and R. Verger, *J. Mol. Biol.*, 1992, **225**, 147-153.

264. E. Charton, and A.R. Macrae, *Biochim. et Biophys. Acta.*, 1992, **1123**, 59-64.
265. J. Tashiro, J. Kobayoshi, K. Shirai, Y. Saito, I. Fukamachi, H. Hashimoto, T. Nishida, T. Shibui, Y. Morimoto, and S. Yoshida, *FEBS Lett.*, 1992, **298**, 36-38.
266. F. Faustinella, L.C. Smith, and L. Chan, *Biochemistry*, 1992, **31**, 7219-7223.
267. M.E. Lowe, *J. Biol. Chem.*, 1992, **267**, 17069-17073.
268. J.D. Schrag, F.K. Winkler, and M. Cygler, *J. Biol. Chem.*, 1992, **267**, 4300-4303.
269. J.L. Sussman, *Science*, 1991, 253, 872-879.
270. H. Soreq, A. Gnatt, Y. Loewenstein, and L.F. Neville, *Trends in Biochem. Sci.*, 1992, **17**, 353-358.
271. H.M. Holden, M. Ito, D.J. Hartshorn and I. Rayment, *J. Mol. Biol.*, 1992, **227**, 840-851.
272. A.J. Bauer, I. Rayment, P.A. Frey and H.M. Holden, *Proteins*, 1992, **12**, 372-381.
273. L. Chen, F.S. Mathews, V.L. Davidson, E.G. Huizinga, F.M.D. Vellieux and W.G.J. Hol, *Proteins*, 1992,14, 288-299.
274. J.A. Hartsuck, G. Koelsch and S.J. Remington, *Proteins*, 1992, **13**, 1-26.
275. D.W. Godette, C. Paech, S.S. Yang, J.R. Mielenz, C. Bystroff, M.E. Wilke and R.J. Fletterick, *J. Mol. Biol.*, 1992, 228, 580-595.
276. H. Sobek, H.-J. Hecht, W. Aehle and D. Schomburg, *J. Mol. Biol.*, 1992, **228**, 108-117.
277. G.A. Bohch, Y.-In Chi and C.V. Stauffacher, *Proteins*, 1992, **13**, 152-157.
278. K. Huang, R. Kodandapani, H. Kallwass, J.K. Hogan, W.Parris, J.D. Friesen, M. Gold, J.B. Jones and M.N.G. James, *Proteins*, 1992, **13**, 158-161.
279. Z.X. Xia, W.W. Dai, J.P. Xiong, Z.P. Hao, V.L. Davidson, S.White and F.S. Mathews, *J. Biol. Chem.*, 1992, **267**, 22289-22297.
280. M.D. Hall, D.G. Levitt and L.J. Banaszak, *J. Mol. Biol.*, 1992, **226**, 867-882.
281. P. Gros, A.V. Teplyakov, and W.G.J. Hol, *Proteins*, 1992, **12**, 63-74.
282. J.C. Gorga, D.R. Madden, J.K. Prendergast, D.C. Wiley and J.L. Strominger, *Proteins*, 1992, 12, 87-90.
283. G.D. Smith and G.G. Dodson, *Proteins*, 1992, **14**, 401-408.
284. K.P. Wilson, L.M. Shewchuk, R.G. Brennan, A.J. Otsuka and B.W. Matthews, *Proc. Natl. Acad. Sci. USA*, 1992, **89**, 9257-9261.
285. C.M. Ogata, P.F. Gordon, A.M. de Vos and S.-H. Kim, *J. Mol. Biol.*, 1992, 893-908.
286. M.S. Weiss and G.E. Schulz, *J. Mol. Biol.*, 1992, **227**, 493-509.
287. A. Mattevi, G. Obmolova, J.R. Sokatch, C. Betzel and W.G.J. Hol, *Proteins*, 1992, **13**, 336-351.
288. K. Hakansson, M. Carlsson, L.A. Svensson and A. Liljas, *J. Mol. Biol.*, 1992, **227**, 1192-1204.
289. H.A. Schreuder, J.M. van der Laan, M.B.A. Swarte, K.H. Kalk, W.G.J. Hol and J. Drenth, *Proteins*, 1992, **14**, 178-190.
290. R.K. Wierenga, R.J. de Jong, K.H. Kalk, W.G.J. Hol and J. Drenth, *J. Mol Biol.*, 1979, **131**, 55-73.
291. V. Ramakrishnan and S.W. White, *Nature*, 1992, **358**, 768-771.
292. Macromolecular Structures, 1990, 1991, 1992. Ed. W.A. Hendrikson and K. Wüthrich. Current Biology Ltd., London, U.K.
293. A.H. Robbins, D.E. McRee, S.A. Williamson, N.H. Collett, W. Xuong, W.F. Furey, B.C. Wang and C.D. Stout, *J. Mol. Biol.*, 1991, **221**, 1269-1293.
294. B.L. Vallee and D.S. Auld, *Faraday Discuss.*, 1992, **93**, 1-19.
295. K.M. Merz, *J. Mol. Biol.*, 1990, **214**, 799-802.
296. E.E. Kim and H.W. Wyckoff, *J. Mol. Biol.*, 1991, **218**, 449-464.

297. S.K. Burley, P.R. David and W.N. Lipscomb, *Proc. Natl. Acad. Sci. USA*, 1991, **88**, 6916-6920.
298. L.S. Beese and T.A. Steitz, *EMBO, J.*, 1991, **10**, 25-33.
299. V. Derbyshire, N.D.F. Grindley and C.M. Joyce, *EMBO, J.*, 1991, **10**, 17-24.
300. E. Hough, L.K. Hansen, B. Birknes, K. Jynge, S. Hansen, A. Horvik, C. Little, E. Dobson and Z. Derewenda, *Nature*, 1989, **338**, 357-360.
301. J.A. Tainer, D. Getzoff, K.M. Beem, J.S. Richardson and D.C. Richardson, *J. Mol. Biol.*, 1982, **160**, 181-217.
302. W. Bode, F.X. Gomis-R̃th, R. Huber, R. Zwilling and W. Stõcker, *Nature*, 1992, **358**, 164-167.
303. B.M. Matthews, J.N. Jansonius, P.M. Colman, B.P. Schoenborn and D. Dupourque, *Nature*, 1972, **238**, 37-41.
304. W.P. Jiang and J.S. Bond, *FEBS Lett.*, 1992, **312**, 110-114.
305. G. Pollock, H. Nagase, J.K. Huff, L. Sanderson and M.P. Sarras, *Mol. Biol. of the Cell*, 1992, **3**, 319.
306. D.K. Wilson, F.B. Rudolph and F.A. Quiocho, *Science*, 1991, **252**, 1278-1284.
307. R.A.D. Cadena and R.W. Colman, *Protein Sci.*, 1992, **1**, 151-160.
308. G. Schiavo, O. Rossetto, A. Santucci, B.R. Dasguta and C. Montecucco, *J. Biol. Chem.*, 1992, **267**, 23479-23483.
309. L. Regan and N.D. Clarke, *Biochemistry*, 1990, **29**, 10878-10883.
310. J.N. Higaki, R.J. Fletterick and C.S. Craik, *Trends in Biochem Sci.*, 1992, 100-104.
311. J.E. Coleman, *Ann. Review of Biochemistry*, 1992, **61**, 897-946.
312. C.W. Heizmann and W. Hunziker, *Trends Biochem. Sci.* 1991, **16**, 98-103.
313. N.C.J. Strynadka and M.N.G. James, *Ann.Rev. Biochem.*, 1989, **58**, 951-998.
314. G.S. Shaw, R.S. Hodges and B.D. Sykes, *Biochemistry*, 1991, **30**, 8339-8347.
315. G.S. Shaw, L.F. Golden, R.S. Hodges and B.D. Sykes, *J. Am. Chem. Soc.*, 1991, **113**, 5557-5563.
316. L.E. Kay, J.D. Forman-Kay, W.D. McCubbin and C.M. Kay, *Biochemistry*, 1991, **30**, 4323-4333.
317. S. Linse, C. Johansson, P. Brodin, T. Grundstrom, T. Drakenberg and S. Forsen, *Biochemistry*, 1991, **30**, 154-162.
318. R. Chattopadhyaya, W.E. Meador, A.R. Means, and F.A. Quiocho, *J. Mol. Biol.*, 1992, **228**, 1177-1192.
319. K. Beckingham, *J. Biol. Chem.*, 1991, **266**, 6027-6030.
320. S. Linse, A. Helmersson and S. Forsen, *J. Biol. Chem.*, 1991, **266**, 8050-8054.
321. M.A. Starovasnik, D.-R. Su, K. Beckingham and R. Klevit, *Protein Sci.*, 1992, **1**, 245-253.
322. K.-Y. Ling, R.R. Preston, R. Burns, J.A. Kink, Y. Saimi and C. Kung, *Proteins*, 1992, **12**, 365-372.
323. S. Toma, S. Campagnoli, I. Margarit, R. Gianna, G. Bolognesi, V. De Filippis and A. Fontana, *Biochemistry*, 1991, **30**, 97-106.
324. W.I. Weis, R. Kahn, R. Fourme, K. Drickamer and W.A. Hendrickson, *Science*, 1991, **254**, 1608-1615.
325. W.I. Weis, K. Drickamer and W.A. Hendrickson, *Nature*, 1992, **360**, 127-134.
326. Y. Lindqvist, G. Schneider, U. Ermler and M. Sunström, *EMBO Journal*, 1992, **11**, 2373-2379.
327. G. George, R.C. Prince, S. Mukund, M.W.W. Adams, *J. Amer. Chem. Soc.*, 1992, **114**, 3521-3523.
328. A. Juszczak, A. Shigotoshi, and M.W.W. Adams, *J. Biol. Chem.*, 1991, **266**, 13834-13841.

329. M.W. Day, B.T. Hsu, L. Joshua-Tor, J-B. Park, Z.H. Zhou, M.W.W. Adams, and D.C. Rees, *Protein Sci.*, 1992, **1**, 1494-1507.
330. S.J. George, J. von Elp, J. Chen. C.T. Chen, Y. Ma, J-B. Park, M.W.W. Adams, F. de Groot, J.C. Fuggle, B.G. Searle, S.P. Cramer, *J. Amer. Chem. Soc.*, 1922, **114**, 4426-4427.
331. P.R. Blake, J-B. Park, Z.H. Zhou, D.R. Hare, M.W.W. Adams, and M.F. Summers, *Protein Sci.*, 1992, **1**, 1508-1521.
332. P.R. Blake, J-B. Park, M.W.W. Adams, and M.F. Summers, *J. Amer. Chem. Soc.*, 1922, **114**, 4931-4933.
333. P.R. Blake, M.F. Summers, M.W.W. Adams, J-B. Park, Z.H. Zhou, and A. Boux, *J. Biomol. NMR*, 1992, **2**, 527-533.
334. S.A. Busse, G.N. La Mar, L.P. Yu, J.B. Howard, E.T. Smith, Z.H. Zhou, and M.W.W. Adams, *Biochemistry* , 1992, **31**, 11952-11962.
335. T.M. Loehr, *J. of Raman Spectroscopy,* 1992, **23**, 531-537.
336. I. Rayment, G. Wesenberg, T.E. Meyer, M.A. Cusanovich, and H.M. Holden, *J. Mol. Biol.*, 1992, **228**, 672-686.
337. R. Liddington, Z. Derewenda, E. Dodson, R. Hubbard, and G. Dodson, *J. Mol. Biol.*, 1992, **228**, 551-579.
338. J.S. Kim and D.C. Rees, *Nature*, 1992, **360**, 553-560.
339. M.M. Georgiadis, H. Komiya, P. Chakrabarti, D. Woo, J.J. Kornuc and D.C. Rees, *Science*, 1992, **257**, 1653-1659.
340. J.S. Kim and D.C. Rees, *Science*, 1992, **257**, 1677-1682.
341. L.C. Seefeldt, T.V. Morgan, D.R. Dean and L.E. Mortenson, *J. Biol. Chem.*, 1992, **267**, 6680-6688..
342. T.C. White, G.S. Harris, and W.H. Ormejohnson, *J. Biol. Chem.* 1992, 24007-24016.
343. D. Wolle, D.R. Dean, and J.B. Howard, *Science*, 1992,**258**, 992-995.
344. D. Wolle, C.H. Kim, D. Dean, and J.B. Howard, *J. Biol. Chem.*, 1992, **267**, 3667-3673.
345. K.K. Surerus, M.P. Hendrich, P.D. Christie, D. Rottgardt, W.H. Ormejohnson and E. Munck, *J. Am. Chem. Soc.*, 1992, **114**, 8579-8590.
346. I. Bohm, A. Halberr, S. Smaglinski, A. Ernst, and P. Bodger, *J. Bact.*, 1992, **174**, 6179-6183.
347. F.E. Caballero, M.I. Igeno, R. Quiles, and F. Castillo, *Arch. Microbiol.*, 1992, **158**, 14-18.
348. A. Soliman and S. Nordlund, *Arch. Microbiol.*, 1992, **157**, 431-435.
349. R.N.F. Thorneley, *Phil. Trans. of Royal Society of London.* Series B-Biological Sciences, 1992, **336**, 73-82.
350. P.B. Hallenbeck, *Biochim. Biophys. Acta.FEBS. Lett.*, 1992, **307**, 257-262.
352. G. Minoltti and M. Ikedasaito, *J. Biol Chem.*, 1992, **267**, 7611-7614.
353. I. Bertini, F. Capozzi, S. Ciurli, C. Lunchinat, I. Messori, and M. Piccioli, *J. Amer. Chem. Soc.*, 1992, **114**, 3332-3340.
354. A.J. Pierik, W.R. Hagen, W.R. Dunham, and R.H. Sands, *Eur. J. Biochem.* 1992, **206**, 705-719.
355. E.C. Theil, D.E. Sayers, Q. Wang, A.M. Edwards, G.S. Waldo, and P.C. Weber, *FASEB Journal*, 1992, **6**, 1.
356. C.S. Kim and J. Jung, *Photochemistry and Photobiology,* 1992, **56**, 63-68.
357. G. Palmer and J. Reedijk, *J. Biol. Chem* ., 1992, **267**, 665-677.
358. L.G. Ryden and L.T. Hunt, *J.Mol. Evol.*, 1993, **36**, 41-66.
359. A. Messerschmidt, R. Ladenstein, R. Haber, M. Bolognesi, L. Avigliano, R. Petruzzeli, A. Rossi, and A. Finazziagro, *J. Mol. Biol.*, 1992, **224**, 179-205.

360. A. Messerschmidt, W. Steigemann, R.Huber, G. Lang, and P.M.H. Kroneck, *Eur. J. Biochem*., 1992,**209**, 597-602.
361. G. Formicka, M. Zeppezauer, F. Fey, and J. Huttermann, *FEBS. Lett*, 1992, **309**, 92-96.
362. J.M. Guss, H.D. Bartunik, and H.C. Freeman, *Acta Cryst. B. Structural Science*, 1992,**48**, 790-811.
363. C.W.G. Hoitink, and G.W. Canters, *J. Biol. Chem*., 1992, **267**, 13836-13842.
364. O. Farver, and I. Pecht, *J. Amer. Chem. Soc*., 1992, **114**, 5764-5767.
365. D.G.A.H. Desilva, D. Beokubetts, P. Kyritsis, K. Govindaraju, R. Powls, N.P. Tomkinson, and A.G. Sykes, *J. Chem. Soc. Dalton Trans*., 1992, 2145-2151.
366. F.K. Klemens, and D.R. McMillin, *Photochemistry and Photobiology*, 1992, **55**, 671-676.
367. C.S. Stclair, W.R. Ellis, and H.B. Gray, *Inorganica Chimica Acta* 1992, **191**, 149-155.
368. J.S. Zhou, and N.M. Kostic, *Biochemistry*, 1992, **31**, 7543-7550.
369. S. Modi, H.E. Sp. J.C. Gray, and D.S. Bendall, *Biochim. et Biophys. Acta*., 1992, **1101**, 64-68.
370. L. Qin, and N.M. Kostic, *Biochemistry*, 1992, **31**, 5145-5150.
371. H.E.M. Christensen, L.S. Conrad, and J. Ulstrup, *Acta. Chemica Scandinavica*, 1992, **46**, 508-514.
372. Z. Salamon and G. Tollin, *Arch. Biochem. and Biophys*., 1992, **294**, 382–387.
373. J.D. McManus, D.C. Brune, J. Han, J. Sandersloehr, T.E. Meyer, M.A. Cusanovich, G. Tollin, and R.E. Blankenship, *J. Biol. Chem*., 1992, **267**, 6531-6540.
374. M.L. Brader, D. Borchardt, and M.F. Dunn, *Biochemistry*, 1992, **31**, 4691-4696.
375. K.K. Surerus, W.A. Oertling, C.L. Fan, R.J. Gurbiel, O. Einarsdottir, W.E. Antholine, R.B. Dyer, B.M. Hoffman, W.H. Woodruff, and J.A. Fee, *Proc. Natl. Acad. Sci.USA*, 1992, **89**, 3195-3199.
376. C. Anthony, *Biochim. et Biophys. Acta*., 1992, **1099**, 1-15.
377. E.I. Soloman, M.J. Baldwin, and M.D. Lowery, *Chemical Reviews*, 1992, **92**, 521-542.
378. Z.R. Lu, Y.Q. Yin, and D.S. Jin, *J.Mol. Catalysis* 1991, **70**, 391-397.
379. A. Fersht, and G. Winter, *Trends in Biochem. Sci*., 1992, **17**, 292-294.
380. M. Watanabe, I. Sofuni, and I. Nohmi, *J. Biol. Chem*., 1992, 267, 8429-8436.
381. J.M. Dupret, and D.M. Grant, *J. Biol. Chem*., 1992, **267**, 7381-7385.
382. D.J.T. Porter, *J. Biol. Chem*., 1992, **267**, 7342-7351.
383. S.E. Ealick, S.A. Rule, D.C. Charter, T.J. Greenhough, Y.S. Babu, W.J. Cook, J. Habash, J.R. Helliwell, J.D. Stoeckler, R.E. Packs, and C.E. Bugg, *J. Biol. Chem*., 1990, **265**, 1812-1820.
384. S.E. Ealick, Y.S. Babu, C.E. Bugg, M.D. Erion, W.C. Guida, and J.A. Montgomery, *Proc. Natl. Acad. Sci. USA*, 1991, **88**, 11540-11544.
385. M.T. Black, J.G.K. Munn, and A.E. Allsop, *Biochem. J*., 1992, **282**, 539-543.
386. A.M. Hadonou, M. Jamin, M. Adam, B. Joris, J. Dusart, J.M. Ghuysen, and J.M. Frere, *Biochem. J*., 1992, **282**, 495-500.
387. B. Veerapandian, J.B. Cooper, A. Sali, T.L. Blundell, R.L. Rosati, B.W. Dominy, D.B. Damon, and D.J. Hoover, *Protein Sci*., 1992, **1**, 322-328.
388. J. Aqvist, and A. Warshel, *J. Mol. Biol*., 1992, **224**, 7-14.
389. J.R. Byberg, F.S. Jorgensen, S.Hansen, and E. Hough, *Proteins*, 1992, **12**, 331-339.
390. M.M.G.M. Thunnissen, P.A. Franken, G.H. Dehaag, J. Drenth, K.H. Kalk, H.M. Verheij, and B.W. Dijkstra, *Protein Engineering*, 1992, **5**, 597-603.
391. H. Kim, and M.S. Patel, *J. Biol. Chem*., 1992, **267**, 5128-5123.
392. H.F. Fisher, S. Maniscalco, N. Singh, R.N. Mehrotra, and R. Srinivasan, *Biochim. et Biophys. Acta*., 1992, **1119**, 52-56.

393. H.F. Fisher, N. Singh, and S.J. Maniscalco, *FASEB Journal* 1991, **6**, No. 1.
394. J. Lamottebrasseur, F. Jacobdubuisson, G. Dive, J.M. Frere, and J.M. Ghuysen, *Biochem. J.*, 1992, **282**, 189-195.
395. H. Adachi, M. Ishiguro, S. Imajoh, T. Ohta, and H. Matsuzawa, *Biochemistry*, 1992, **31**, 430-437.
396. J.M. Juteau, E. Billings, J.R. Knox and R.S. Levesque, *Protein Engineering*, 1992, **5**, 693-701.
397. S.X. Liu, P.H. Zhang, X.H. Ji, W.W. Johnson, G.L. Gilliland, and R.N. Armstrong, *J. Biol. Chem*, 1992, **267**, 4296-4299.
398. R.H. Kolm, G.E. Sroga and B. Mannervik, *Biochem J.*, 1992, **285**, 537-540.
399. Y.Y.P. Wo, M.M. Zhou, P. Stevis, J.P. Davies, Z.Y. Zhang, and R.L. Vanmetten, *Biochemistry*, 1992, **31**, 1712-1721.
400. D.D. Buechter, K.F. Medzihradszky, A.L. Burlingame, and G.L. Kenyon, *J. Biol. Chem*, 1992, **267**, 2173-2178.
401. M. Sola, A. Lledos, M. Duran, and J. Bertran, *J. Amer. Chem. Soc.* 1992, **114**, 869-877.
402. C. Engstrand, C. Forsman, Z.W. Liang and S. Lindskog, *Biochem. et Biophys. Acta.*, 1992, **1122**, 321-326.
403. J.M. Moratal, M.J. Martinezferter, A. Donaire, and L. Aznar, *J. Inorg. Biochem.*, 1992, **45**, 65-71.
404. M.D. Percival and S.G. Withers, *Biochemistry*, 1992, **31**, 505-512.
405. I. Auzat and J.R. Garel, *Protein Sci.*, 1992, **1**, 254-258.
406. D.T.F. Dryden, P.G. Varley, and R.H. Pain, *Eur. J. Biochem.*, 1992, **208**, 115-123.
407. M.A. Sherman, W.J. Fairbrother, and M.T. Mas, *Protein Sci.*, 1992, **1**, 752-760.
408. P.A. Walker, H.C. Joao, J.A. Littlechild, R.J.P. Williams, and H.C. Watson, *Eur.J. Biochem.*, 1992, **207**, 29-37.
409. G.J. Davies, S.J. Gamblin, J.A. Littlechild, and H.C. Watson, *J. Mol. Biol.*, 1992, **227**, 1263-1264.
410 K. Harlos, M. Vas, and C.F. Blake, *Proteins*, 1992, **12**, 133-144.
411. D. Walker, W.N. Chia and H. Muirhead, *J. Mol. Biol.*, 1992, **228**, 265-276.
412. D.I. Roper, K.M. Moreton, D.B. Wigley and J.J. Holbrook, *Protein Engineering*, 1992, **5**, 611-615.
413. K. Shostak and M.E. Jones, *Biochemistry*, 1992, **31**, 12155-12161.
414. U.K. Ward, B.C. Bonning, T. Huang, T. Shiotsuki, V.N. Griffeth, and B.D. Hammock, *Int. J. Biochem.*, 1992, **24**, 12, 1993-1941.
415. D.L. Rousseau, S.H. Han, S.H. Song and J.C. Ching, *J. of Raman Spect.*, 1992, **23**, 551-556.
416. N.C. Gerber and S.G. Sligar, *J. Am. Chem. Soc.*, 1992, **114**, 8742-8743.
417. L.W. Hardy and E. Nalivaika, *Proc. Nat. Acad. Sci., USA*, 1992, **89**, 9725-9729.
418. D.D. Leonidas, N.G. Oikonomakos, A.C. Papageorgiou, K.R. Acharya, D. Barford and L.N. Johnson, *Protein Sci.*, 1992, **1**, 1112-1122.
419. K. Takase, *Biochim. et Biophys. Acta.*, 1992, **1122**, 278-282.
420. J.L. Turnbull, G.L. Waldrop and H.K. Schachman, *Biochemistry*, 1992, **31**, 6390-6395.
421. T. Yano, S. Kuramitsu, S. Tanase, Y. Morino and H. Kogamiyama, *Biochemistry*, 1992, **31**, 5878-5887.
422. P.S. Brzovic, Y. Sawa, C.C. Hyde, E.W. Miles and M.F. Dunn, *J. Biol. Chem.*, 1992, **267**, 13028-13038.
423. H. Munier, A. Bouhss, E. Krin, A. Danchin, A.M. Gilles, P. Glaser and O. Barzu, *J. Biol. Chem*, 1992, **267**, 9816-9820.

424. J.C. Austin, A. Kuliopulos, A.S. Mildvan and T.G. Spiro, *Protein Sci.*, 1992, **1**, 259-271.
425. Y. Kim and J.D. Robertus, *Protein Engineering*, 1992, **5**, 775-780.
426. C. Derst, J. Henseling and K.H. Rõhm, *Protein Engineering*, 1992, 5, 785-790.
427. T. Palzkill and D. Botstein, *Proteins*, 1992, **14**, 29-44.
428. C.R. Dunn, H.M. Wilks, D.J. Halsall, T. Atkinson, A.R. Clarke, H. Muirhead, and J.J. Holbrook, *Phil. Trans. R. Soc. London B*, 1991, **332**, 117-184.
429. G. Casy, T.V. Lee, H. Lovell, B.J. Nichols, R.B. Sessions and J.J. Holbrook, *J. Chem. Soc. - Chem. Comm.*, 1992, **13**, 924-926.
430. A. Beuve and A. Danchin, *J. Mol. Biol.*, 1992, **225**, 933-938.
431. W.J. Fairbrother, L. Hall, J.A. Littlechild, P.A. Walker, H.C. Watson and R.J.P. Williams, *FEBS Lett.*, 1989, **258**, 247-250.
432. L. Chen, D. Neidhard, W.M. Kohlbrenner, W. Mandecki, S.Bell, J. Sowadski and C. Abad-Zapatero, *Protein Engineering*, 1992, **5**, 605-610.
433. Y. Takasakai and K. Hori, *Protein Engineering*, 1992, **5**, 101-104.
434. T. Hamamoto, F. Foong, O. Shoseyov and R.H. Doi, *Mol. Gen. Genetics*, 1992, **231**, 472-479.
435. H.S. Shiah, T.Y. Chen, C.M. Chang, J.T. Chow, H.J. Kung and J.L. Hwang, *J. Biol. Chem.*, 1992, **267**, 24034-24040.
436. S.H. Sun and A.M. Lew, *J. Immun. Methods*, 1992, **152**, 43-48.
437. T. Ogawa, A. Shinohara, H. Ogawa and J. Tomizawa, *J. Mol. Biol.*, 1992, **226**, 651-660.
438. U. Hummel, C. Nuofter, B. Zanolari and B. Erni, *Protein Sci.*, 1992, **1**, 356-362.
439. G.A. Bentley, J. Brange, Z. Derewenda, E.J. Dodson, G.G. Dodson, J. Markussen, A.J. Wilkinson, A. Wollmer and B. Xiao, *J. Mol. Biol.*, 1992, **228**, 1163-1176.
440. A.M.M. Jorgensen, S.M. Kristensen, J.J. Led and P. Balschmidt, *J.Mol.Biol.*, 1992, **227**, 1146-1163.
441. J. Brange, U. Ribel, J.F. Hansen, G. Dodson, M.T. Hansen, S. Havelund, G. Melberg, F. Norris, K. Norris, L. Snel, A.R. Sorensen and H.O. Voight, *Nature*, 1988, **333**, 679-682.
442. G.P. Burke, S.Q. Hu, N. Ohnta, S.P. Schwatz, L. Zong, P.G. Katsoyannis, *Biochem. Biophys. Res. Com.*, 1991, **173**, 982-987.
443. R.A. Mirmira, S.H. Nakagawa, H. Tager, *J. Biol. Chem.*, 1991, **266**, 1428-1436.

6
Metal Complexes of Amino Acids and Peptides

By K.B. NOLAN, A.A. SOUDI and R.W. HAY

1 Introduction

This chapter deals with synthesis, structures and reactions of metal-amino acid and metal-peptide complexes and covers material published in 1991 and 1992. A number of reviews in this area have been published. Hydroxamic acids, aminohydroxamic acids and their complexes with metal ions are the subject of one review.[1] Complexes of cobalt(II), nickel(II), copper(II), zinc(II), palladium(II) and cadmium(II) with N-protected amino acids are covered in a second review.[2] Ternary complexes of platinum(II) and other metal ions with amino acids or peptides and nucleosides or nucleotides are described.[3] The binding sites of these biologically important ligands in the ternary complexes and the interactions between the side chains of the amino acids or peptides and the aromatic rings or phosphate groups in the nucleosides or nucleotides are discussed. Crystal structure data which confirms the existence of inter- and intra-molecular interactions in the ternary complexes are also presented.

The mechanism of action of platinum antitumour drugs of general formula cis-$PtX_2(NHR_2)_2$, in which R is an organic fragment and X is a leaving group such as chloride, sulfate or (chelating bis) carboxylate as well as the key role of hydrogen bonding is the topic of another review.[4] Complexes of the platinum group metals with amino acids and peptides containing non-coordinating and coordinating side chains,[5] and the complexing properties of peptides towards alkali, alkaline earth and transition metal ions,[6] are also reviewed. Other reviews deal with metal catalysed ester, amide and peptide hydrolysis,[7] metal to metal intramolecular electron transfer across peptide and protein bridges,[8] electron transfer reactions in metalloproteins,[9] cadmium binding peptides from plants,[10] chiral complexes of nickel(II), copper(II) and copper(I) as reagents, catalysts and receptors for asymmetric synthesis and chiral recognition of amino acids,[11] metal complexes of amino acids and peptides,[12] synthesis, structure and acid base properties of cobalt(III) complexes with aminoalcohols,[13] iron carriers and iron proteins,[14] interrelations between metal ions, enzymes and gene expression,[15] X-ray diffraction studies on carboxypeptidase A complexes the zinc stereochemistry,[16] coordination compounds and life processes,[17] the comparative intestinal absorption of metal-amino acid complexes and inorganic metal salts.[18]

2 Amino Acid Complexes

2.1 *Crystal and Molecular Structures.*

The crystal and molecular structures of several amino acid complexes of copper(II) have been reported. Many of these contain N-substituted amino acid ligands. In the complex Cu(N-Bu^t-N-Me-$Gly)_2H_2O$ the metal is in a distorted square pyramidal environment with trans 2N, trans 2O coordination in the equatorial plane and an axial water ligand.[19] The Cu-O (1.992 Å), Cu-N (2.081 Å) and Cu-O_{H_2O} (2.464 Å) bond distances are comparable to those found in other amino acid complexes of copper(II). The structure is composed of discrete molecules linked by C=O···HOH hydrogen bonds. The synthesis and crystal structures of the N,N-bis(2-hydroxyethyl)glycinate complexes Cu[$(OHEt)_2$-Gly]Br **(1)** which contains copper(II) in a trigonal bipyramidal geometry and Cu[$(OHEt)_2$-Gly]Br(H_2O) **(2)** which contains copper(II) in a distorted octahedral geometry have been reported.[20] In both cases the amino acid acts as a tetradentate N, 3O donor.

The crystal and molecular structures of monoclinic **(3)** and orthorhombic forms of the complex Cu(L-N,N-Et_2-$Ala)_2H_2O$ have been determined.[21,22] In both forms the coordination around the metal is distorted square pyramidal with trans 2N, trans 2O equatorial coordination and an apical water ligand. The Cu-N and Cu-O bond distances are close to normal for bis(amino-acidato)copper(II) complexes while the Cu-O_{H_2O} distances of 2.263 Å (monoclinic) and 2.260 Å (orthorhombic) are extremely short. The complex Cu(DL-N,N-Et_2-$Ala)_2$ contains discrete molecules disordered over two sites in a 72:28 ratio with the metal atom on an inversion centre.[23] The coordination around the copper is irregular square planar and O and N atoms trans. In the complex Cu(L-N,N-Me_2-$Ile)_2$ **(4)** the metal lies in a tetrahedrally distorted square planar environment involving trans 2N, trans 2O coordination of the ligand.[24] The conformations of the above complexes have all been analysed by molecular mechanics calculation.

Crystal structures, electronic absorption spectra and crystal field parameters are reported for the complexes Cu(N-Ac-$Gly)_2(H_2O)_2.2H_2O$,[25] and Cu(L-$Leu)_2$.[26] The crystal structure of the N-benzenesulphonylglycine (Bs-Gly) complex Cu(bipy)(Bs-Gly)H_2O **(5)** has been determined and this shows that the metal is in a tetrahedrally

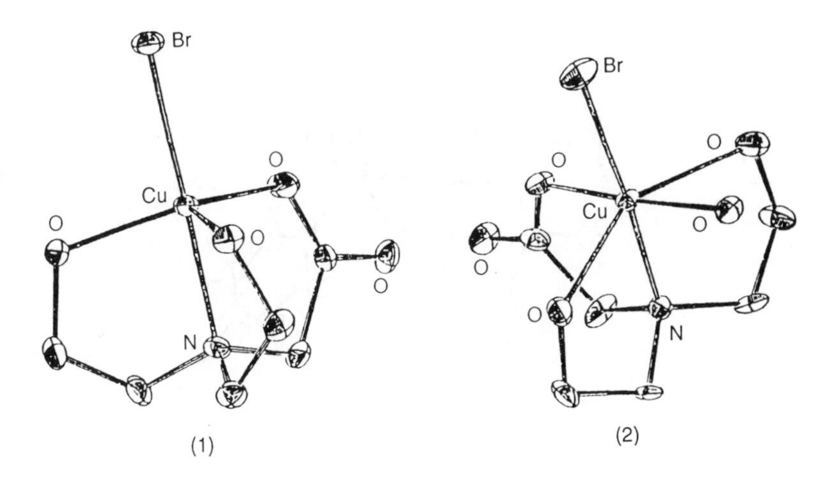

(1) (2)

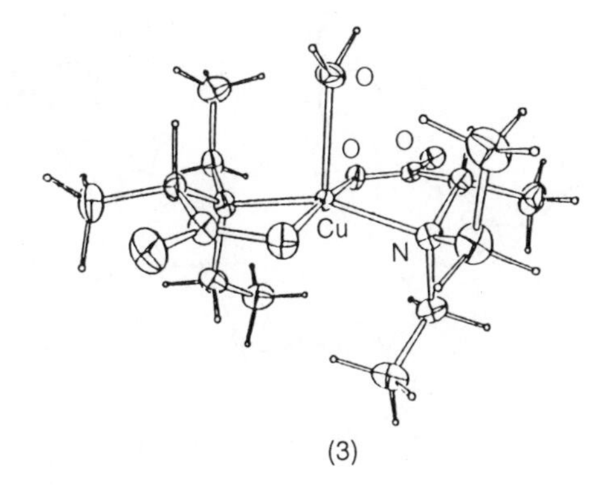

(3)

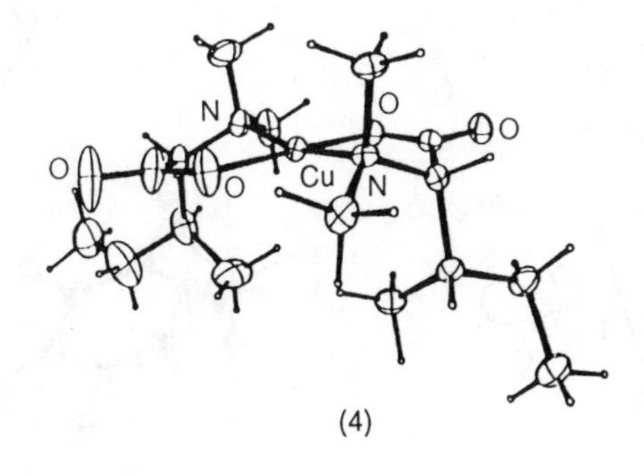

(4)

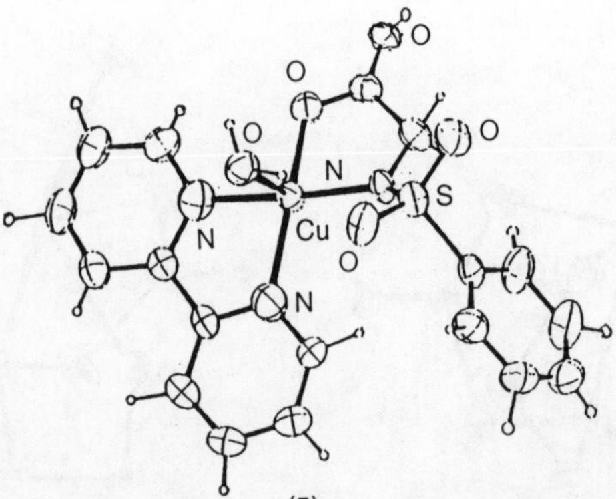

(5)

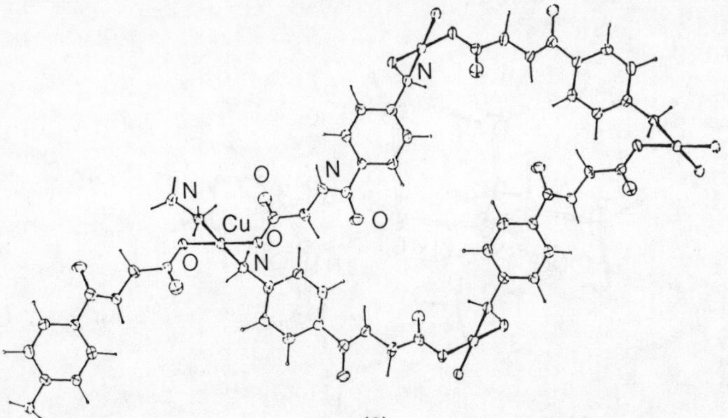

(6)

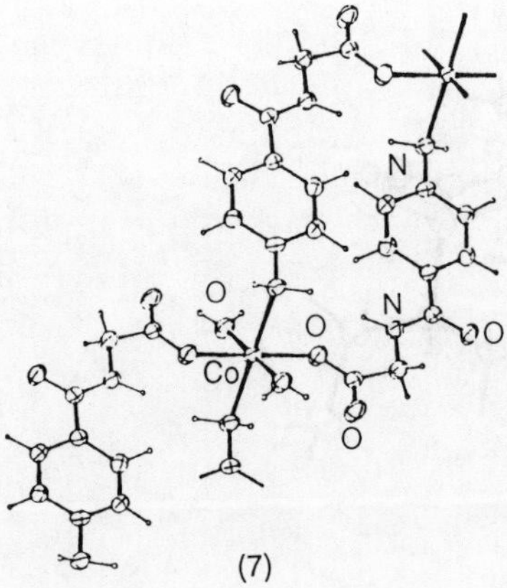

(7)

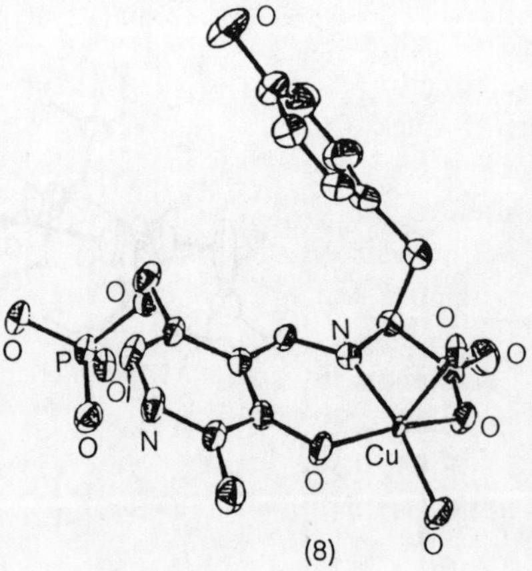

(8)

distorted square pyramidal environment involving N_3O_2 donor atoms.[27] The structural and spectroscopic properties of the N-tosyl-DL-asparagine complex, Cu(bipy)(Tos-Asn)$_2$.2H_2O have been reported.[28] In this complex the metal is in a square planar environment of two bipy nitrogen atoms and two carboxylate oxygen atoms. The structural and spectroscopic properties of cobalt(II), nickel(II) and copper(II) complexes of N-4-aminobenzoylglycine, HL, have been investigated.[29] In the complex $(CuL_2)_n$.4nH_2O **(6)** the metal is in a tetragonally distorted octahedral geometry with two oxygens and two nitrogens of different amino acid ligands in the basal plane and two oxygens in the axial sites. This gives a polymeric complex made up of two dimensional layers. The complexes $[CoL_2(H_2O)_2]$n.2nH_2O **(7)** and its nickel(II) analogue are isomorphous and have two carboxylate and two water oxygens in the basal plane with two amino nigrogens coordinated axially.

Two complexes of formula $[CuL(H_2O)]$.nH_2O containing the Schiff base ligands (H_2L) 5′-phosphopyridoxylidene-DL-tyrosine (n = 3.5) **(8)** and 5′-phosphopyridoxylidene-DL-phenylalanine (n = 2.5) **(9)** have been obtained in the crystalline form and the structures of these show parallel orientations of the aromatic side chains of the amino acid residues with the pyridoxal π systems.[30] Such interactions are postulated to account for activation of the α-CH groups and unusually rapid racemisation in solution in the corresponding complexes containing the optically pure L-amino acid residues. In the 5′-phosphopyridoxylidene-DL-tyrosine complex, $[CuL(H_2O)]$.4H_2O the coordination geometry around the copper is distorted square pyramidal with phenolic O, imino N, carboxylate O and the aquo ligand occupying the basal plane and a phosphate O from a neighbouring molecule positioned at the axial site.[31] This complex is weakly antiferromagnetic and shows π-π stacking interactions similar to that in the other Schiff base complexes. Aromatic ring stacking is also implied from the crystal structure of [Cu(bipy)(L-Tyr)ClO_4].2H_2O **(10)**.[32] This complex has a tetrahedrally distorted square pyramidal structure with the 2N atoms of bipy and the N and O (carboxylate) atoms of tyrosine in the basal plane and an O atom of the perchlorate ligand in the axial position. The average spacing between the tyrosine and bipyridyl rings is 3.35 Å. The implications of this for conformational changes resulting from tyrosine phosphorylation in proteins is discussed.

The crystal and molecular structures of the pyridoxylidenehistamine (HL′) and 5′-phosphopyridoxylidenehistamine (HL) copper (II) complexes [CuL′(H_2O)Br]NO_3 and $[Cu_2L_2(NO_3)_2]$.6H_2O have been determined and four other complexes of these ligands have been synthesised.[33] These are [Zn(L′)CH_3COO].2H_2O, [Cd(L′)Cl].2H_2O,

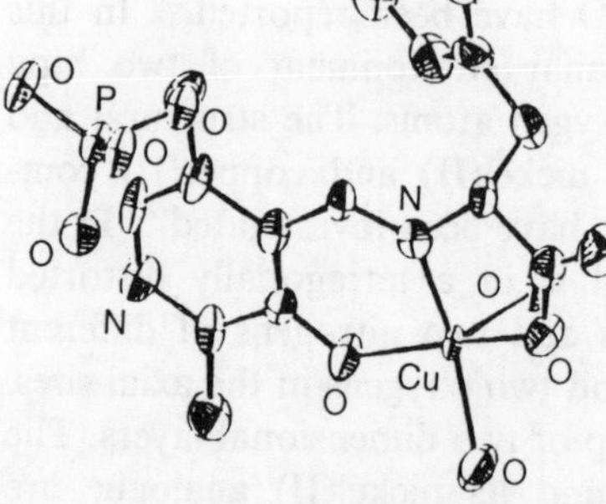

(9)

(10)

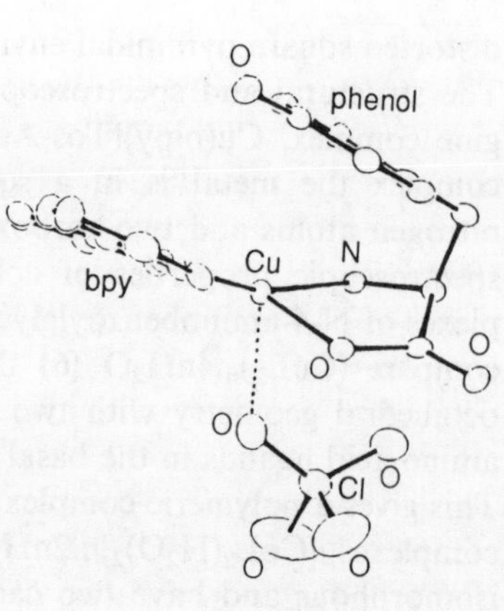

(11)

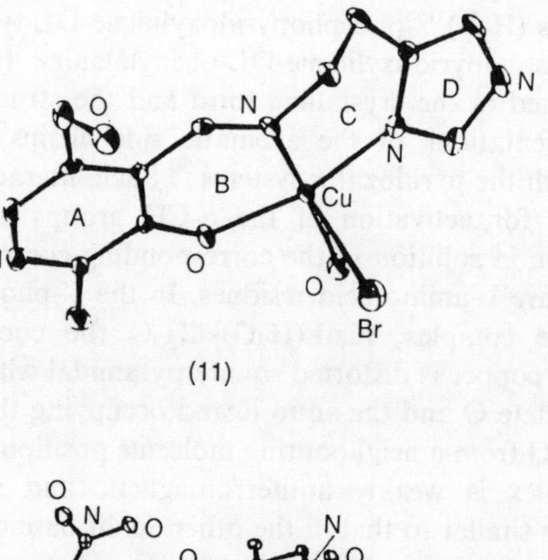

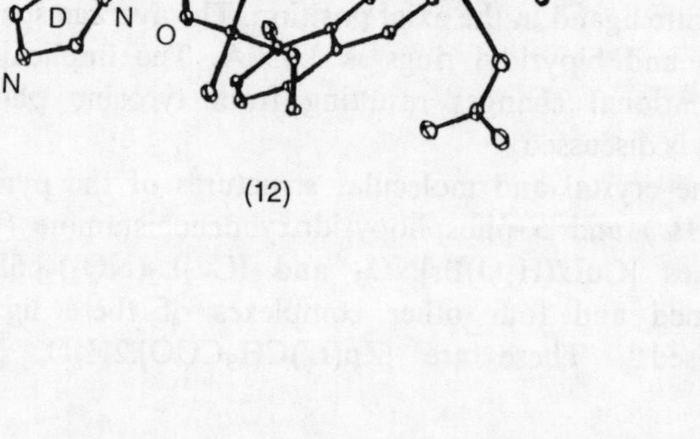

(12)

$[ZnL(CH_3COO)].3H_2O$ and $[Cd(L)Cl].3H_2O$. In $[CuL'(H_2O)Br]NO_3$ **(11)** the metal lies in a square pyramidal ligand field with the tridentate, zwitterionic Schiff base ligand (imino and imidazolyl nitrogen atoms and phenolate oxygen donors) and a water ligand in the basal plane and Br^- situated axially. The metal ion in $[Cu_2L_2(NO_3)_2].6H_2O$ **(12)** has the same coordination geometry but in this case the Schiff base acts as a tetradentate monoanionic ligand. The crystal and molecular structure of the pyridoxylideneglycine (H_2L'') complex $Me_4N[Co(L'')_2].4\frac{1}{2}H_2O$ **(13)** has been reported.[34] N.m.r. studies showed that for the diastereotopic protons of the methylene group of the Gly residue the fastest exchange occurs for the proton of the CH bond which is closer to being perpendicular to the plane of the π system of the aromatic rings. The molecular structure of the complex reveals features which may account for the different labilities of the C-H bonds.

The structures of three complexes containing Schiff base ligands derived from 1-aminocyclopropanecarboxylic acid and pyridoxal (HL) or pyridoxal-5′-phosphate (H_2L') have been determined.[35] These complexes are $[CuL'(H_2O)]NO_3.H_2O$ **(14)**, which contains square pyramidal copper(II), $[NiL(H_2O)_2]NO_3.H_2O$ which contains square bipyramidal nickel(II) and $[CuL'(H_2O)_2]_2.4H_2O$ which contains independent $CuL'(H_2O)$ structural units linked by phosphate bridges. In the first two complexes the basal plane is defined by the imino N, the phenolic and carboxylate O atoms of the Schiff base ligand with water ligand(s) in the apical position(s). These complexes may serve as models for the Schiff base derived from pyridoxal and 1-aminocyclopropanecarboxylate, a postulated reaction intermediate in the biosynthesis of ethene.

The crystal and molecular structures of two copper(II) complexes of alanine have been reported. The complex Cu(DL-Ala)$2H_2O$ **(15)** obtained from an aqueous solution of racemic alanine and basic copper carbonate contains an elongated octahedral arrangement of donor atoms around copper(II) with amino nitrogen atoms and carboxylate oxygens of two symmetry related (D and L) alanine ligands occupying trans equatorial sites and oxygen atoms of two water ligands at the apical positions.[36] These apical positions are common to neighbouring octahedra so that the water molecules act as bridging ligands **(16)**. There is strong H-bonding between the water ligands and the carboxylate groups of two different neighbouring molecules. This provides a short superexchange pathway and accounts for the high degree of magnetic coupling inferred from magnetic susceptibility studies and e.p.r. data. This superexchange pathway is different from those observed in other bis(amino acidato)copper(II) complexes. The crystal and molecular structure of Cu(L-Ala)$SO_4.2H_2O$ **(17)** obtained from equimolar aqueous

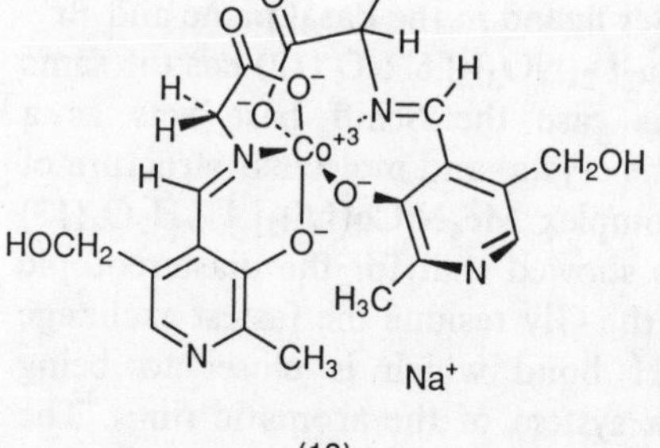

(13)

(14)

(15)

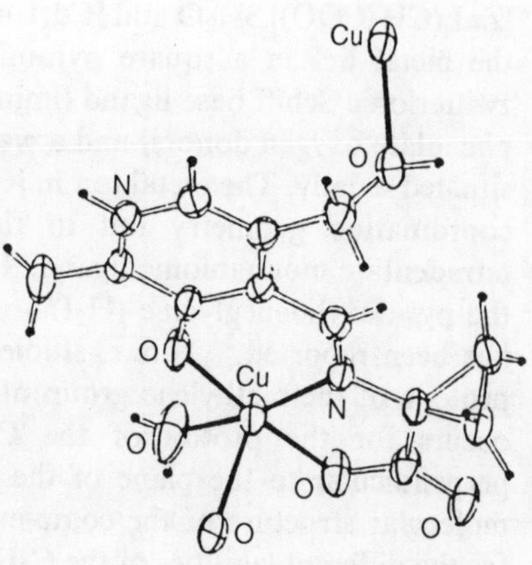

(16)

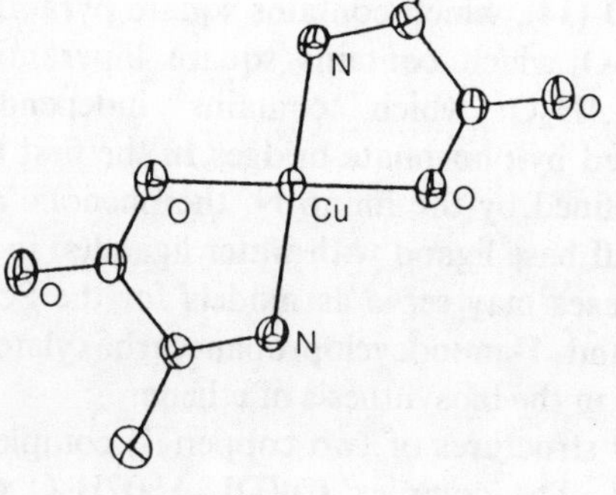

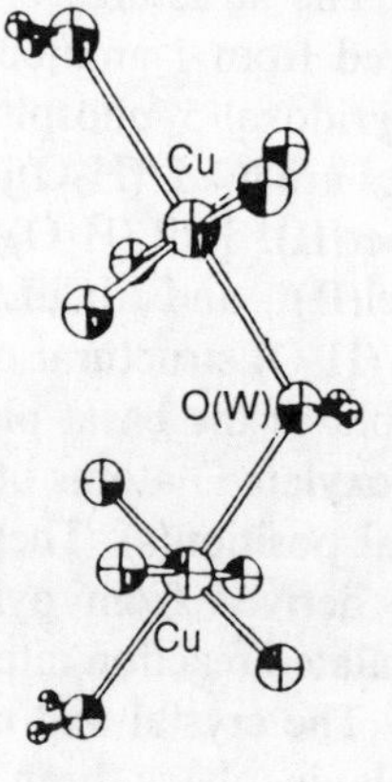

(17)

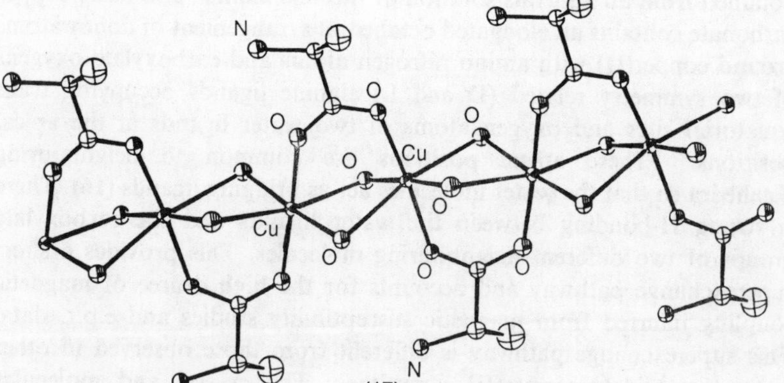

solutions of $CuSO_4.5H_2O$ and L-Ala (~0.5 mol dm^{-3}) is also described.[37] This consists of a polymeric chain made up of $Cu_2(\mu\text{-}H_2O)_2(\mu\text{-L-AlaH}^+)$ units in which pairs of copper ions are bridged by two aquo ligands and the carboxylate group of the zwitterionic amino acids. The metal lies in an environment of four 'in plane' bridging water ligands with two axial carboxylate oxygens. Similar bridging amino acid complexes of nickel(II) are obtained by reacting $[Ni_2L(MeOH)_2(ClO_4)_2].2Et_3N.HClO_4$ containing the macrocyclic ligand derived from **(18)** with Gly, β-Ala and Gly-Gly (L′) to give the complexes $[Ni_2L(\mu\text{-}L')(H_2O)_2](ClO_4)_2.nH_2O$.[38] These contain μ-carboxylato bonded amino acid and peptide ligands. The crystal and molecular structure of the glycine derivative $[Ni_2L(\mu\text{-}O_2CCH_2NH_3)(H_2O)_2](ClO_4)_2.2H_2O$ has been determined.

The crystal and molecular structure of a copper(II) complex of β-alaninehydroxamic acid (A) has been reported.[39] This complex has formula $[Cu_5A_4H_{-4}]^{2+}$ **(19)** and is a major species in solution in the pH range 4.5-9. In this complex the hydroxamic acid ligand is doubly deprotonated and the structure consists of an almost planar $Cu_4A_4H_{-4}$ tetrameric core made up of CuN_2O_2 units with a central cation bonded to four bridging hydroxamate oxygens.

Copper(II) complexes of the diuretic furosemide, FuH, **(20)** have been synthesised and the crystal structure of $CuFu_2(MeOH)_2$ **(21)** determined.[40] This contains elongated CuO_6 bipyramidal units linked by carboxylate bridges. The equatorial plane contains two MeOH ligands with Cu–O = 1.951 Å, and two bridging carboxylate oxygens with Cu′-O = 1.924 Å. The remaining oxygens on each carboxylate ligand occupy the apical positions with Cu-O = 2.720 Å. The Cu-Cu distance is 4.749 Å. The aminobenzoate (Amb) complex $[Cu(4\text{-Amb})(1,10\text{-phen})(H_2O)]_2(NO_3)_2.2(4\text{-Amb}).2H_2O$ and the 3,4-dimethoxyhydrocinnate (DPP) complex $[Cu(DPP)(4,7\text{-phen})(H_2O)]_2(NO_3)_2.2H_2O$ have been synthesised and characterised by magnetic measurements e.s.r. and diffuse reflectance spectra.[41] The e.p.r data are consistent with antiferro-magnetic coupling and the crystal structure of the first complex shows carboxylato bridging with a Cu-Cu distance of 3.060 Å and with the phenanthroline ligands stacked and separated by a minimum distance of 3.39 Å. Single crystals of copper(II) complexes of L- and DL-Met and 2-aminobutyric acid have been studied by e.s.r. methods and exchange coupling mechanisms are discussed.[42]

The copper(II) directed condensation of amino acids with H_2CO and NO_2 in basic MeOH produces open chain tetradentate ligands with a pendant NO_2 substituent.[43] The crystal structures of copper(II) complexes, CuL, of three of these ligands, H_2L, containing 5-nitro-3,7-

(18)

(19)

(20)

(21)

(22)

(23)

diazanonanedioic acid with 5-methyl, **(22)** (2S, 8S)-2,5,8-trimethyl **(23)** and (2R, 8S)-2,5,8-trimethyl **(24)** substituents have been determined. The reaction of Cu(β-Ala)$_2$ with H_2CO and NH_3 gives 3N, 7N-[1,3,5,7-tetraazabicyclo[3.3.1]nonyl-di-3-propionato]copper(II) trihydrate at pH 6.6-8.8 and the crystal structure of which has been reported.[44]

A range of 2,6-pyridinedicarboxylate (PDC) complexes of formula Cu(PDC)L(S) containing substituted imidazole (L) and solvent (S = H_2O or MeOH) ligands have been investigated by X-ray diffraction and by electronic and e.s.r. spectroscopy.[45] These complexes have distorted square pyramidal structures with the tridentate ligand and the substituted imidazole in the basal plane and the solvent at the axial position **(25)**. The structure of the complex Cu_2(TCC)(4-MeIm)$_2$.2H_2O where H_4TCC is N,N,N′,N′-tetrakis(carboxylmethyl)cystamine, -[$SCH_2CH_2CH_2N(CH_2COOH)_2$]$_2$, and 4-MeIm is 4-methylimidazole has also been reported and is compared with those of [Cu_2(TCC)(H_2O)$_2$.4H_2O]$_n$ and Cu_2TCC(Im)$_2$.2H_2O which had been previously determined.[46]

The preparation isolation and characterisation of the complexes cis-β-[CoL(AA)]PF_6 where HL = N-(2-pyridyl)methyl-2-2-amino-ethylthioacetamide and HAA = Gly, L-Ala, L-Ile, L-Leu, Sar, Aib, β-Ala, are described.[47] The crystal and molecular structure of β-[CoL(Gly)]PF_6.2H_2O **(26)**, has been determined. The complex β-[CoLCl_2]PF_6 was found to promote hydrolysis of amide and ester bonds in peptides and amino acid esters giving β-[CoL(AA)]PF_6 products. The synthesis and crystal structure of the complex [Co(phen)$_2$Gly].4H_2O have been reported.[48] The crystal and molecular structure of the cobalt(III) complex, [Co(HL)Cl](ClO_4)$_2$ containing the pendant arm macrocyclic ligand 13-amino-13-methyl-1,4,8,11-tetradecane-6-carboxylic acid (HL) **(27)** has been determined.[49] The complex **(28)** which was obtained by reacting HL.5HCl with Co(II) and air contains the chloride ligand cis to the pendant amine, the four secondary amines coordinated in a folded geometry with RRRR stereochemistry and the carboxylic acid group uncoordinated. When the reaction mixture was left at pH 8 over charcoal the complex [CoL](ClO_4)$_2$.2H_2O **(29)** in which the pendant amine and carboxylate groups are also coordinated (trans) was obtained. The crystal structure of this complex is also reported.

The complexes Co(1,4-diaminobutane)(NO_2)$_2$(AA), where AA = aminoacidato can exist as three geometric isomers cis-(NO_2)$_2$ cis-(NH_2)$_2$, cis-(NO_2)$_2$ trans-(NH_2)$_2$ and trans -(NO_2)$_2$ as shown in structures **(30)**-**(32)**. Additionally each of the cis-(NO_2)$_2$ isomers can exist as two optical isomers. The synthesis of the trans-(NO_2)$_2$ and cis-(NO_2)$_2$trans-(NH_2)$_2$ isomers of Co(1,4-diambut)(Gly)(NO_2)$_2$ as well as the trans-(NO_2)$_2$

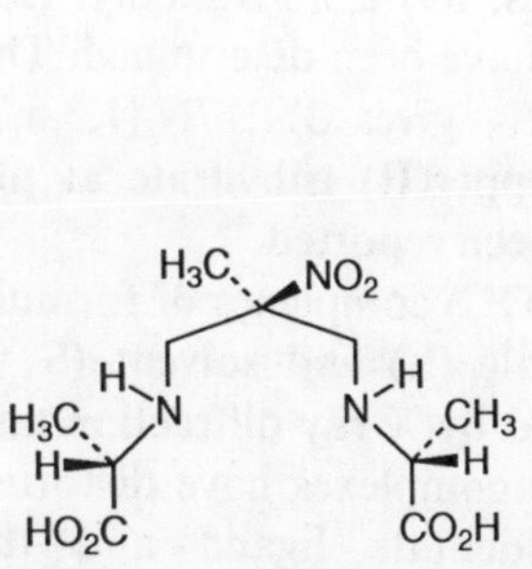

(24)

(25)

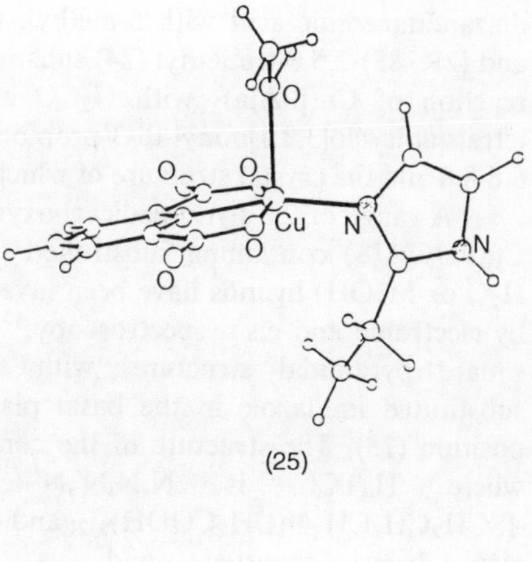

(26)

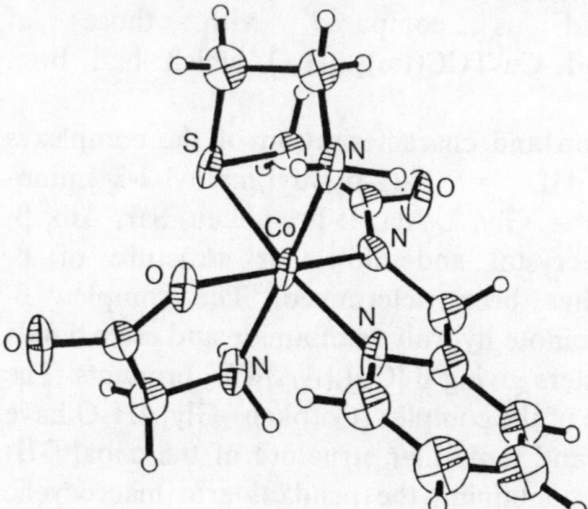

(27)

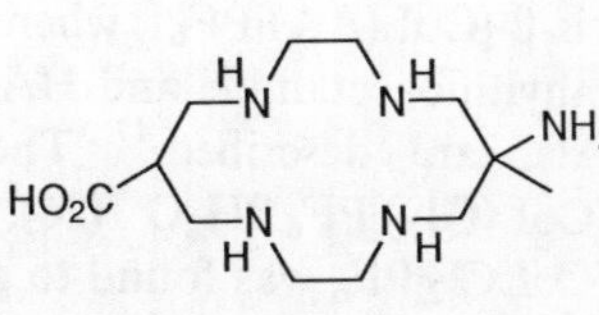

(28)

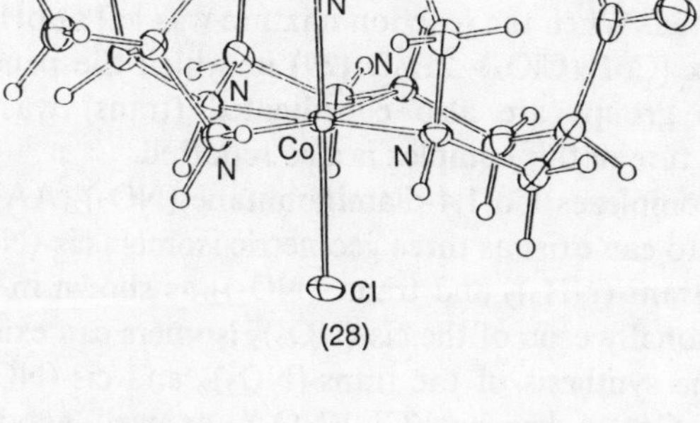

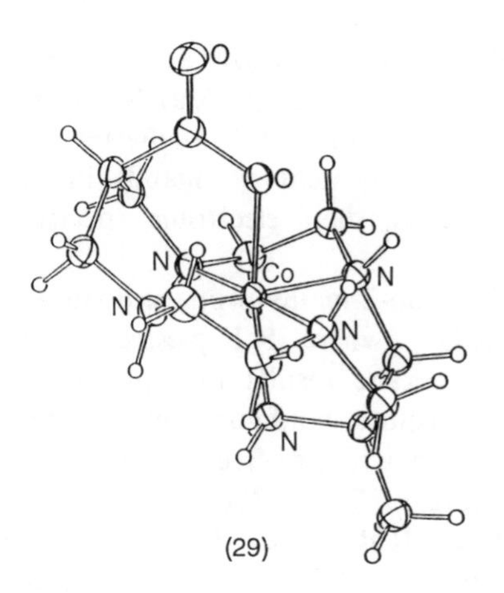

(29)

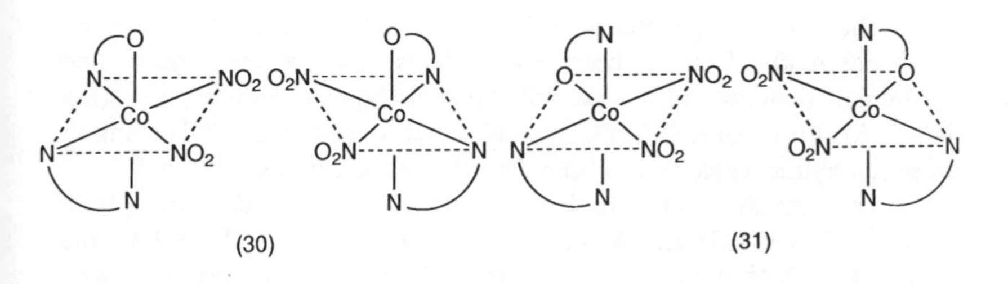

(30)

(31)

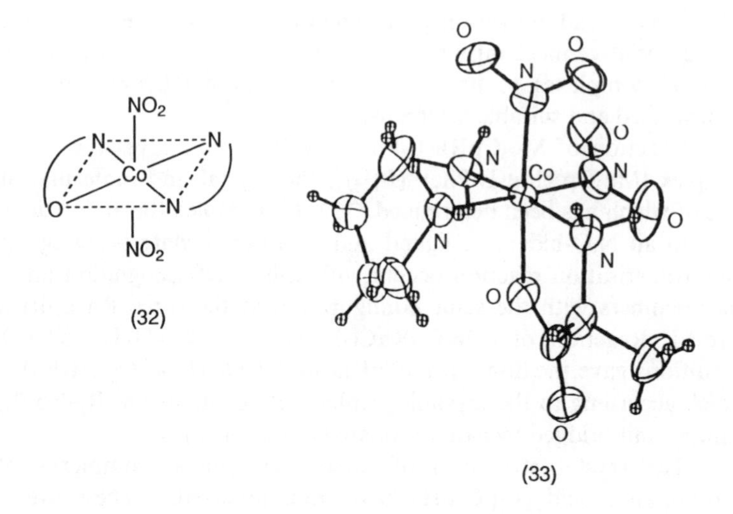

(32)

(33)

and Δ-(−)$_{589}$-cis$(NO_2)_2$-trans-$(NH_2)_2$ and Λ-(+)$_{589}$-cis$(NO_2)_2$-trans-$(NH_2)_2$ isomers of Co(1,4-diambut)(R-Ala)$(NO_2)_2$ have been reported.[50] The crystal structure of the Δ-(−)$_{589}$-diastereomer **(33)** has been determined and the arrangement of ligands in the other complexes established by comparing their electronic spectra with that of this diastereomer.

Reaction of 2-amino-2-methylpropanedioate ($AMMA^{2-}$) with Λ-β-[Co(R,R-picstein)$Cl_2]^+$ where R,R-picstein is 3R,4R-diphenyl-1,6-di(2-pyridyl)-2,5-diazahexane results in replacement of the Cl^- ions with retention of absolute configuration of the metal atom and AMMA adopting the β and pro-S configurations.[51] Decarboxylation of this gave a mixture of 93 ± 3% Λ-$β_1$-[Co(R,R-picstein)R-Ala$]^{2+}$ and 7 ± 3% Λ-β-[Co(R,R-picstein)S-Ala$]^{2+}$. The crystal and molecular structure of Λ-β-[Co(R,R-picstein)AMMA]$ClO_4.2H_2O$ at −135°C has been determined.

The synthesis, geometric and absolute configuration of the isomers of $[Co(S\text{-}Arg)_3]^{3+}$ have been reported as well as the crystal and molecular structure of (−)$_{589}$-anti(N)-Δ-cis(N)-cis-(O)-Λ-cis(N)-cis(O)-$[Co(S\text{-}Arg)_2\mu\text{-}OH]_2$ **(34)**.[52] The crystal structure of a cobalt(III) complex of pyridoxylideneglycine has been described earlier in this section.[34]

The crystal and molecular structures of the complexes $K[Ru^{III}(HL)Cl]$ **(35)** and $K[Ru^{III}(HL')Cl]$ where HL = EDTA-H and HL′= PDTA-H have been determined.[53] These complexes react with NaOCl to give the oxoruthenium(V) species $K[Ru^VO(L)]$ and $K[Ru^VO(L')]$ and activation parameters for the reactions are consistent with associative mechanisms. Oxygen atom transfer from the oxoruthenium(V) complexes to unsaturated and saturated hydrocarbons have been studied and suitable mechanisms proposed.

Reaction of Me_5CpRh(L-Phe)Cl with $AgBF_4$ gives the trinuclear complex $[Me_5CpRh(\mu\text{-L-Phe})]_3(BF_4)_3$, the crystal and molecular structure of which has been determined.[54] In this complex the amino acid acts as both an N,O-bidentate ligand and as a carboxylate bridging ligand. This trimerisation reaction occurs with chiral self recognition and only diastereomers with the same configuration at the three Rh centres are formed. Reaction of $[Me_5CpRhCl_2]_2$ with Cbz-GlyNH_2 under basic conditions gave the dimer $[Me_5CpRh(\mu\text{-}NHCOCH_2NCO_2CH_2Ph)]_2$ **(36)** which according to the crystallographic data contains an (R)-Rh,(S)-Rh dimeric unit bridged by a deprotonated amide nitrogen.

The crystal structures of three alkali metal complexes of L-pyroglutamic acid (L-pGluH) have been reported.[55] These are M(L-pGlu)L-pGluH where M = Li,Na **(37)** and K **(38)**. Aqueous solutions of the complexes have also been studied by ^1H and ^{13}C n.m.r. spectroscopy.

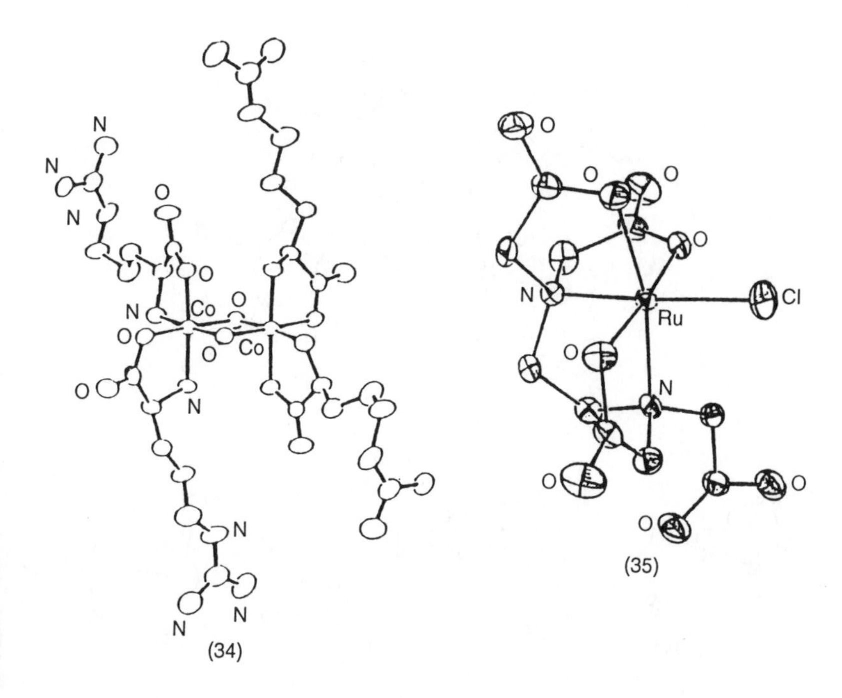

(34)

(35)

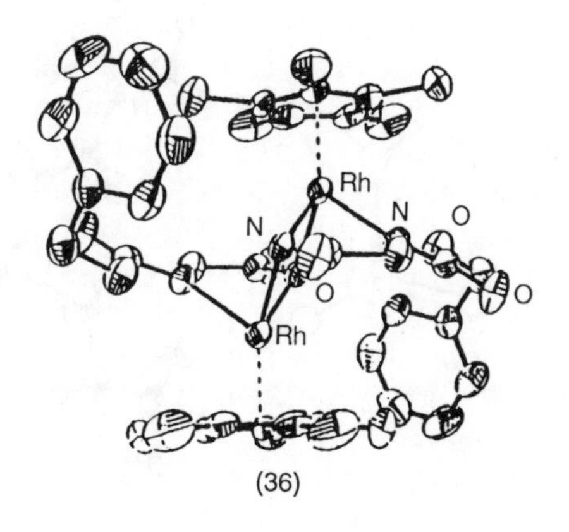

(36)

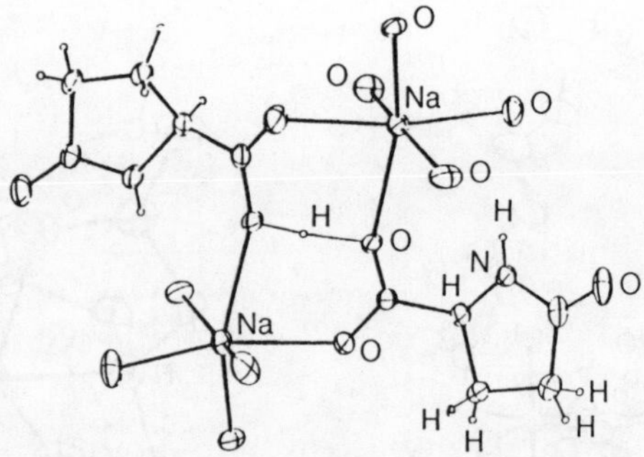

(37)

(38)

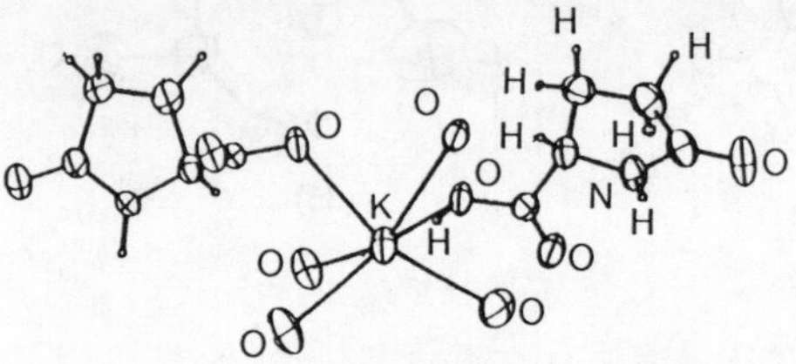

(39)

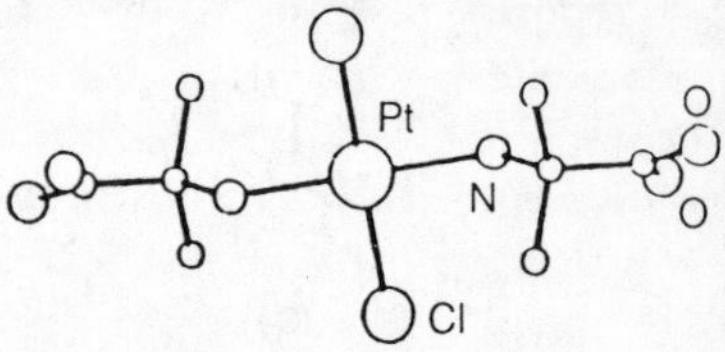

(40)

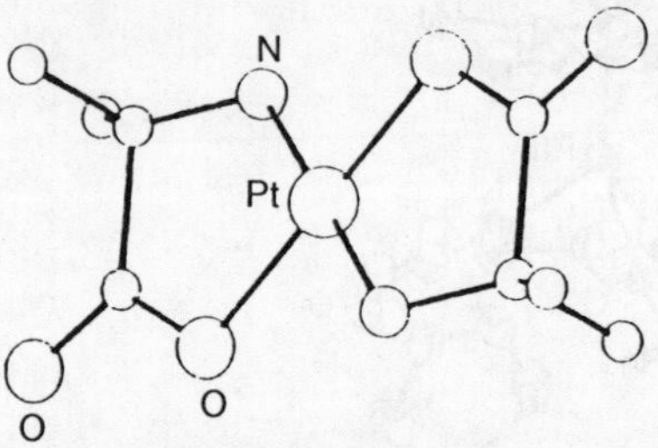

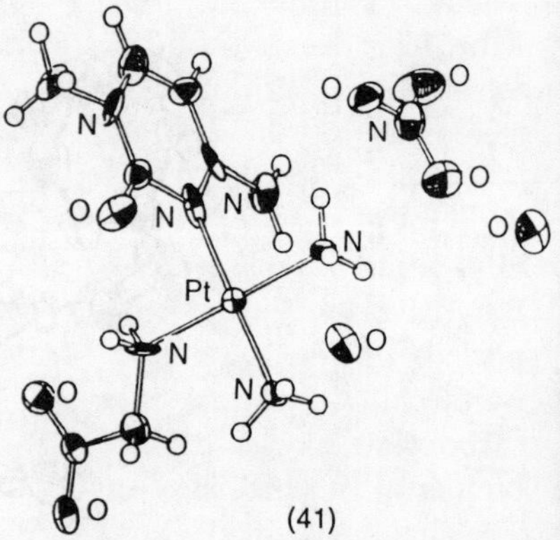

(41)

The preparation and crystal structure of bis(pyroglutamato)calcium (II) and an investigation of its solution behaviour by n.m.r. (^{1}H, ^{13}C and ^{17}O) has also been described.[56]

The crystal and molecular structures of the distorted square-planar platinum(II) complexes trans-$PtCl_2$(Aib-OH)$_2$ **(39)** and trans Pt(Aib-O^-)$_2$ **(40)** containing α-aminoisobutyric acid ligands, Aib-OH, have been determined.[57] Several platinum(II) complexes of the type cis-$Pt(NH_3)_2$AA(L) where HAA = Gly, L-Ala, L-2-aminobutyric acid, L-Val, L-norvaline and HL = 1-methylcytosine or 9-methylguanine were prepared by reacting cis-[$Pt(NH_3)_2$AA]NO_3 with HL or cis-[$Pt(NH_3)_2$L(Cl)]NO_3 with HAA.[58] In these complexes the amino-acids act as monodentate N-bonded ligands with their carboxylic acid groups ionized while the nucleobases are coordinated through N3 (1-MeC) or N7 (9-MeG). The crystal structure of cis-[$Pt(NH_3)_2$(1-MeC)Gly]NO_3 **(41)** is reported.

Stereoselectivity in the coordination of prochiral olefinic alcohols in complexes of the type Pt(AA)(olefin)Cl where AA = Gly, α-Aib, Sar and S-Pro and olefin = allyl alcohol, 3-buten-2-ol, or 2-methyl-3-buten-2-ol (2-Mb) has been studied by n.m.r. (^{1}H, ^{13}C and ^{195}Pt) spectroscopy and by X-ray crystallography.[59] Whereas for trans-N,olefin complexes stereoselectivity of olefin coordination is insignificant, for cis-N, olefin isomers of N-chiral sarcosine and S-Pro stereoselectivity is substantial. The X-ray structures of the preferred isomers of cis(N, olefin)-Pt(Sar)(2-Mb)Cl has been determined in order to establish the stereochemistry and to identify intermolecular interactions responsible for the preferred stereoselectivies.

The crystal and molecular structures of the cadmium(II) complexes [Cd(bipy)(Bs-Gly)]$_4$.8H_2O **(42)** prepared at pH 11 and [Cd(bip)(Ts-Gly)$_2$]$_2$ prepared at pH 7 where Bs-Gly and Ts-Gly are N-(benzenesulfonyl)glycine and N(toluene-p-sulfonyl)glycine respectively are reported.[60] In the tetrameric Bs-Gly complex each Cd is hexacoordinated by the bipy nitrogens, the deprotonated sulfonamide nitrogen and the carboxylate oxygens. In the Ts-Gly complex the metal is bonded to the bipy nitrogens and the 4 carboxylate oxygens of the amino acid anions. The analogous zinc(II) complexes have also been prepared and these too show evidence for zinc(II)-promoted sulphonamide group deprotonated. The crystal structure of the betaine complex [Cd(Et_3NCH_2COO)(μ-Cl)$_2$]$_n$ has been reported.[61] Each cadmium(II) is in a distorted octahedral environment composed of 4 bridging Cl^- ligands and a bidentate carboxylato group. The crystal structures of the polymeric betaine complexes [$Ag_2(Me_3NCH_2COO)_2(H_2O)_2(NO_3)_2$]$_n$ and [$Ag_2(C_5H_5NCH_2COO)_2(ClO_4)_2$]$_n$ have also been reported.[62] The

(42)

(43)

(44)

(45) R = CH_2
R = $(CH_2)_2$
R = o-C_6H_4

structures are similar and consist of stairs-like polymer based on bis(carboxylato-O,O′)-bridged centrosymmetric dimers.

Reaction of Ph_3PAuCl with the silver salt of PhCO-Ala gave the O-bonded amino acid complex $Ph_3PAu(OOCCH(CH_3)NHCOPh)$ **(43)** the structure of which has been determined.[63] The asymmetric unit contains two independent molecules both with linear coordination around gold but having different amino acid conformations.

The crystal structure of the amino acid analogue piperidinomethylphosphinic acid hydrate **(44)** has been reported. In the solid state this exists as a zwitterion and forms dimers linked by H-bonds. The dimers are connected by water molecules to form endless chains.[64]

2.2 *Synthesis.*

The oxovanadium(IV) complex $VO(Pen\text{-}OMe)_2$ where HPen-OMe is penicillamine methyl ester has been isolated and a square pyramidal structure containing trans S and trans N atoms in the basal plane proposed.[65]

The reduction of L-CysOEt at pH 7-8 in the presence of AMP, Asp aspartame (L-Asp-L-Phe-OMe) and 2,5-pyridinedicarboxylic acid (2,5-PDCA) has been studied.[66] These ligands are relevant to the formation of DNA-Cr(III)-protein crosslinks and in the course of the study the complexes $[Cr(aspartame)_2]X$, $[Cr(aspartame)]X_2$ ($X = Cl^-$, NO_3) and $Cr(2,5\text{-}PDCA)_3$ were synthesised. The pink complex $Cr(S\text{-}Ala)_3$ was prepared and characterised and studied by circular dichroism.[67] The reaction of $Mn(OAc)_2$ or $CrCl_3$ with salicylideneamino acids $H_2Sal\text{-}AA$, AA = Gly, Ala, Phe, in $EtOH/H_2O$ gave the products $Cr(Sal\text{-}AA)Cl.2H_2O$ and $Mn(Sal\text{-}AA).2H_2O$ which were fully characterised by spectroscopic, conductimetric and magnetic susceptibility measurements.[68] The coordination geometry around Cr(III) is distorted octahedral while the manganese complexes are dimeric, low symmetry octahedral species containing carboxylate bridges. Several octahedral complexes of chromium(III) and nickel(II) with N-acylaminoacids have been synthesised and characterised.[69] These complexes have formulae $[CrL_2(OH)]_2.nH_2O$ where HL = Phe, Bz-Phe, Bz-Gly and Bz-Leu, $[CrL_2]_2.nH_2O$, $NiL'_2.2H_2O$ where HL′ = HCO-L-Phe, PhCO-DL-Ala, PhCO-Gly, PhCO-L-Leu, Ac-L-Phe, ClAc-L-Phe, F_3Ac-L-Phe, ClAc-DL-Ala, Cl_3Ac-D-Ala, F_3Ac-D,L-Ala and $NiL'_2\,L''_2.nH_2O$ where L″ = imidazole, L-methylimidazole and 1,2-dimethylimidazole.

Condensation of $BocNH(CH_2)_2CON(OCH_2Ph)(CH_2)nCO_2R$ where R = succinimido and n=2-5 with $N(CH_2CH_2NH_2)_3$ gave the tripodal oligoamidehydroxamic acids $[BocNH(CH_2)_2CON(OH)(CH_2)n\,CONHCH_2CH_2]_3N$.[70] Iron(III) complexes of these hydroxamic acids were synthesised in DMF/H_2O. The ligands show substantial growth promoting activity in *Aureobacterium Clavescens* although weaker relative to the natural siderophore desferrioxamine B.

Iron(III) complexes of the dihydroxamic acids $(CH_2)_n[CON(R)OH]_2$, H_2L, where n = 2-4, 6, 8; R = H, Ph, o-tolyl, p-tolyl and the amino hydroxamic acids $NH_2CH(R)CONHOH$, HL′, where R = H, Me and $XC_6H_4CONHOH$,HL″, where X = o-NH_2, p-NH_2 have been synthesised and characterised by microanalysis, molecular weights, magnetic susceptibilities and spectroscopy (electronic, i.r. and Mossbauer).[71] Based on these studies the formulae $Fe_2(LH)_2L_2$, Fe_2L_2O and Fe_2L_3 are proposed for the dihydroxamate complexes and $Fe(OH)_4L'_2(H_2O)_2$ and FeL''_3 for the aminohydroxamate complexes.

Iron(III) and cobalt(II) complexes with a series of tridentate Schiff bases H_2L, derived from 2-hydroxy-1-naphthaldehyde and amino acids have been prepared and investigated by microanalysis, conductance measurements and spectroscopy (i.r. and electronic).[72] These complexes have formulae $[Fe(L)HL].xH_2O$ and $CoL(H_2O)$, x = 1-4 and are postulated to have octahedral and square planar geometries respectively. Iron(III) complexes of general formula FeL_2Cl where HL represents Schiff bases derived from isatin **(45)** with Gly, β-Ala, anthranilic acid and various hydrazides and carbazides have been synthesised.[73] Nickel(II) complexes of the same ligands having formula $NiL_2.xH_2O$ have also been synthesised.[74]

Iron(II) and iron(III) complexes of formulae $[FeL_2(H_2O)_2]X$ where $X = SO_4^{2-}$ or $3Cl^-$ and L = o, m- and p-aminobenzoic acid hydrazides have been synthesised.[75] Ternary complexes of iron(III) of formulae $K[Fe(L)H_2O]$, $K[FeA_2B_2]$, $K[FeX_2(Y)H_2O]$ and $K[FeX_4Z]$ where HA = succinimide; H_2L = o-, m-, p-aminobenzoic acid; HB = L-Pro, Gly, L-Leu; HY = L-Cys, Z = L-Tyr, Ac-D-L-Met, were prepared and characterised.[76]

Five copper(II) complexes CuL_2 with N,N-dialkylaminoacid ligands where HL = Me_2-Ile, Me_2-alloIle, Me_2-Val, Et_2-Ala and Et_2-Val have been synthesised and studied by electronic and e.s.r. spectroscopy.[77] The e.s.r. spectra were found to be sensitive to solvent and ligand substituents and the effect of ligand stereochemistry on the spectra are explained. The structures of several copper(II) complexes with N-N-dialkylamino acid ligands have been determined by X-ray diffraction and these are described in section 2.1 of this chapter. The synthesis of bipyridylalanine and related bipyridyl peptide derivatives **(46)** and their copper(II) complexes are described.[78] These complexes and others such as $[Cu(bipy)_2]BF_4$ have been characterised by electrospray mass spectrometry.

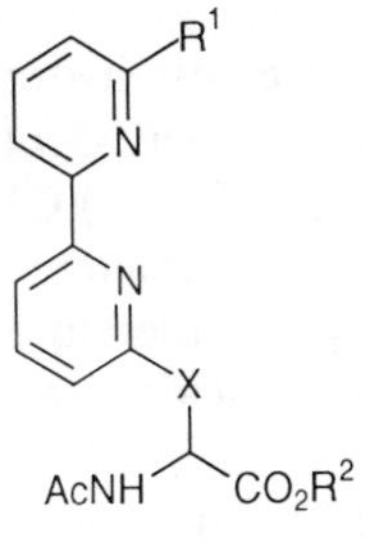

R^1 N N X AcNH CO_2R^2

(46)

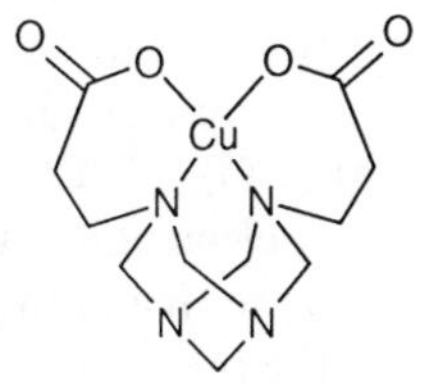

O O O O Cu N N N N

(47)

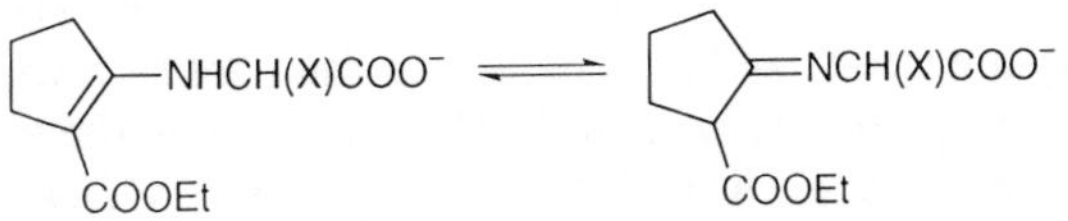

$-NHCH(X)COO^-$ ⇌ $=NCH(X)COO^-$

COOEt COOEt

(48)

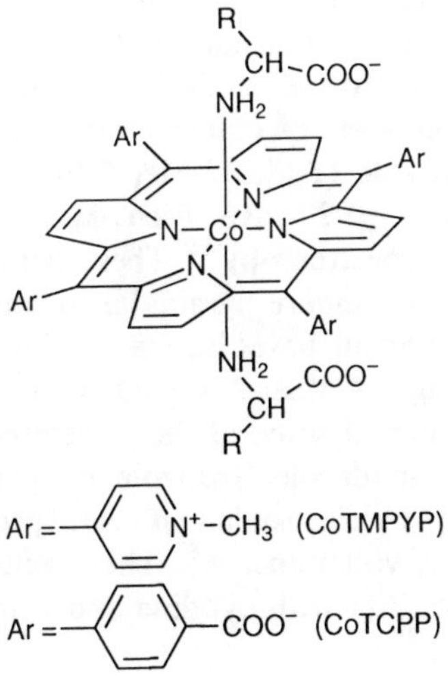

R CH COO^- NH_2 Ar Ar N N–Co–N N Ar Ar NH_2 COO^- CH R

(49) Ar = N^+-CH_3 (CoTMPYP)

Ar = $-COO^-$ (CoTCPP)

The reaction of mixtures of elemental Cu and Zn powder with Gly or Asp in 1:1:2 ratios in the presence of excess hydrogen peroxide at 50°C results in the formation of $Cu(Zn)_2(O_2)(CO_3)_2(H_2O)_4$ while the same reaction with Cu alone gives $Cu(O_2)(Gly)_2H_2O$ and $Cu(O_2)(Asp)(H_2O)_2$.[79] These complexes were characterised by microanalysis, e.s.r., electronic and i.r. spectroscopy. It therefore appears that under the above reaction conditions amino acids are converted to carbonate in the presence of zinc.

The condensation of Cu(β-Ala)$_2$.6H_2O with formaldehyde and ammonia gives the tetraaza ligand **(47)** which has been characterised by electronic and i.r. spectroscopy and by cyclic voltammetry.[80] Condensation products of several other copper(II)-amino acid complexes with formaldehyde or acetaldehyde have also been prepared and studied.[81,82] Several copper(II) complexes of Schiff bases derived from the amino acids DL-Phe,[83] DL-Leu, L-Ile and L-His,[84] with ethyl-α-ketocyclopentylcarboxylate, H_2L **(48)**, were synthesised and characterised. These complexes have formulae CuL.H_2O and contain the doubly deprotonated Schiff base as a tridentate (N-, COO-, COOEt), ligand. The ν(C=O) band for the ester appears at 1630-1650 cm^{-1} indicative of its involvement in complexation. Copper(II) and zinc(II) complexes of Schiff base ligands, H_2L, obtained from amino acids (Gly, D,L-Ala, D,L-Met, L-Leu and L-Phe) and 2,4-dihydroxybenzaldehyde have also been prepared and characterised.[85] These complexes have formula ML(H_2O) and contain the ligand coordinated through the carboxylate and phenolate oxygen atoms as well as the imino nitrogen. N-Salicylideneglycinatocopper(II) complexes of composition Cu(SalGly)L(H_2O)n where L=substituted imidazole (1-Me, 2-Me, 2-Et, 2-Et-4-Me, 2-phenyl) or benzimidazole and n=0-2 have been synthesised and studied by electronic and e.s.r. spectroscopy.[86] These complexes appear to have structures ranging from square pyramidal to square planar. Copper(II) complexes of similar Schiff bases but containing a range of amino acid residues of composition Cu(Sal-AA)H_2O where AA = Gly, L-Ala, L-Ser, L-Thr and β-Ala and some of the N-heterocyclic adducts. Cu(Sal-AA)L where L = imidazole, pyrazole or pyridine have also been prepared and studied by electronic and e.s.r. spectroscopy and by cyclic and differential pulse voltammetry.[87] The reaction of Cu(Sal-AA)H_2O where AA = Gly or Ala with cytidine and cytosine, L, gives adducts Cu(Sal-AA)L(H_2O)n.[88]

Cobalt(II), nickel(II) and zinc(II) complexes of 4-(butylamino)benzoic acid, HL,[89,90] and of m- and p- aminobenzoic and hydrazide, HL′,[91] were prepared and characterised by spectroscopic and magnetic studies. The complexes have formulae $M(HL)_2Cl_2$ and

$[M(HL')_2(H_2O)_2]Cl_2$. Cobalt(II), nickel(II), copper(II), zinc(II), cadmium(II) and palladium(II) complexes of acetic acid and phenylacetic acid hydrazones of 2-aminonicotinaldehyde have been reported.[92] Several metal complexes of the type $M(AA)TSA.nH_2O$ where M = Co, Ni, Cu, Cd, HAA = Phe, Ser, Tyr and HTSA = thiosalicylic acid have been synthesised. In these complexes the ligands act as bidentate N,O and O,S donors forming five- and six-membered chelate rings.[93]

Complexes of general formula trans-$[Pt(guan)_2(HAA)_2]Cl_2$ where guan = guanosine and HAA = Gly, L-Ala, L-Val, L-Ile were isolated by the reaction of trans $Pt(HAA)_2Cl_2$ with guanosine in aqueous solution.[94] The complexes have been characterised by microanalysis, i.r. and n.m.r. spectroscopy and conductivity measurements. Guanosine coordinates through N(7) and the amino acids through their amino groups. Evidence for participation of amino acid protons in non-equivalent H-bonds and for inter-ligand hydrophobic interactions is presented and the various possible conformations of guanosine are discussed in relation to DNA-Pt-protein cross links involving cisplatin and its trans isomer. The synthesis of some cis-$Pt(NH_3)_2AA(L)$ complexes where HL = 1-methylcytosine or 9-methylguanine has been referred to in the section on crystal structures.

The reactions of trans$[Pt(NH_3)_2(L)Cl]X$ where L = 9-methylguanine or 1-methylcytosine, X = Cl or NO_3 with $AgNO_3$ followed by amino acids HAA(HAA = Gly, L-Ala, L-aminobutyric acid, L-Val, L-norVal) in aqueous solution produced trans-$[Pt(NH_3)_2L(HAA)]NO_3.mH_2O$ which when protonated with HNO_3 or $HClO_4$ gave trans-$[Pt(NH_3)_2L(HAA)]X_2.nH_2O$, X = NO_3 or ClO_4, n = 0,1.[95] Interligand hydrophobic interactions in these complexes is evident in their ^1H n.m.r. spectra. Mixed ligand complexes of Pd(II) and Pt(II) with Met and 2,4-disubstituted pyrimidines (2-amino-4-hydroxy; 2-hydroxy-4-hydroxy; 2,4-dione; 2-thio-4-amino) have been synthesised.[96] In these complexes Met acts as an N,S bidentate ligand while the pyrimidines are monodentate N-donors except thiocytosine which acts as a bidentate SH,N donor. The DL-selenomethionine complex $PdCl_2(SeMet)$ in which the SeMet ligand is bidentate has been isolated and studied by ^1H and ^{13}C n.m.r. and i.r. spectroscopy.[97] Binding of the selenoether group to the metal generates a new chiral centre which results in two sets of ^1H and ^{13}C methyl signals.

Complexes of the type $M(daba)X_2$ where M = Pt(II), Pd(II), X = Cl, Br, I and daba = 3,4-diaminobenzoic acid acting as a 2N, bidentate ligand have been synthesised and characterised and their sodium salts tested for antitumour activity.[98] Four binuclear complexes of formula

$[M_2(bipy)_2(DAA)]Cl_2$, M = Pd(II), Pt(II), DAA = meso-α,α′-diaminoadipate or meso-α,α′-diaminosuberate have been synthesised, characterised and tested for antitumour activity. Binding studies to calf thymus DNA have also been carried out.[99]

Nine complexes of general formula $[Pd(phen)AA]^+$, HAA = Gly, L-Ala, L-Leu, L-Phe, L-Tyr, L-Trp, L-Val, L-Pro or L-Ser have been synthesised and characterised and the binding of these and the analogous platinum(II) complexes to calf thymus DNA studied by u.v. difference absorption and fluorescence spectroscopy.[100] Some of these complexes have also been tested for cytotoxicity but only $[Pd(phen)Gly]^+$ and $[Pd(phen)Val]^+$ show comparable activity to cisplatin.

Several platinum(II) complexes containing trans-R,R- or cis-1,2-diaminocyclohexane (DACH) and amino acid HAA ligands (HAA = Gly, L-Pro, L-Ser, L-Ala, Sar and Me_2Gly) ligands have been synthesised and characterised by microanalysis, i.r., ^1H, ^{13}C and ^{195}Pt n.m.r. spectroscopy.[101] Complexes of palladium(II) and platinum(II) with the thioamides $R'NHCHRC(S)NH_2$, R = H, R′ = H or $PhCH_2OCO$-; R = CH_2Ph, R′ = H or $PhCH_2OCO$- have been synthesised and characterised.[102] The synthesis, structure and properties of the complex $K_2[Ni(L\text{-}Cys)_2]1\frac{1}{2}H_2O$ containing N,S coordinated cysteinate have been described.[103] The complexes $Ni(His)_2.H_2O$, $Ni(Gly)_2.2H_2O$, $Pd(His)Cl_2.H_2O$ and $Pd(Gly)_2.2H_2O$ have been prepared and studied.[104] Mixed ligand complexes of nickel(II) with S-methyl-N-arylidenehydrazinecarbodithioate and amino acid ligands have been synthesised.[105] The synthesis of C-allylglycine and C-vinylglycine complexes of Pd(II), Pt(II), Rh(III) and Ir(III) and their reactions with nucleophiles are described.[106]

Twelve metal complexes of general formula MX_2L where M = Zn, Cd, Pd, Pt, Cu; X = Cl, Br, I and L = 4-amino-2-methyl-5-pyrimidinylmethylthioacetic acid have been prepared and characterised.[107] The ternary complexes $[M(H\text{-}terpy)A(X)_2]H_2O$ where AA = amino acids and M = Co, Ni, Cu, Zn and X = Cl and NO_3 have also been synthesised.[108] In these complexes terpyridine appears to act as a bidentate, protonated ligand.

The reaction of ammonium tetrathiomolybdate with the barium salts of dithiocarbamates(dtc) from Gly, Ala and 2-aminobutyric acid. $^-OOCCH(R)NHC(=S)S^-$, yielded $Ba[MO_2S_4(dtc)_2]$.[109] The low magnetic moment (0.4 BM), electronic and i.r. spectra and CVM peak potentials are indicative of the dimeric structure for this product and the i.r. spectra suggest bidentate dithiocarbamate groups and no involvement of carboxylate in coordination. Complexes of Au(III) with dithiocarbamates derived from D,L-Ala, D,L-Val and D,L-Leu have also been synthesised.[110] These complexes appear to contain hexacoordinated gold in a distorted trigonal prismatic ligand field.

In an attempt to model the active site of isopenicillin N-synthetase complexes of cysteinate derivatives with high oxidation state metal species, in this case oxoruthenium(VI), have been synthesised.[111] The complexes are of the type $Ru(O)_2L_2(SCH_2CH(R)COO)$ where L_2 = 2 py or bipy, R = H, NHCHO, NHCOMe. Dioxouranium(VI) complexes of the type $UO_2(L)L'(H_2O)_n$ where HL = 2-hydroxy-, 2-amino- or 2-mercaptobenzoic acid, HL′ = Gly, L-Ala or L-Val and n = 0 or 2 have been synthesised and studied.[112] In these complexes the metal is either 6- or 8-coordinate and the ligands are both bidentate COO^- and O, N or S donors.

Other amino acid complexes which have been reported include $K[Ru(L)Cl_2H_2O]$ where H_2L are Schiff bases from amino acids and salicylaldehyde,[113] copper(II) complexes of amino acid derivatives of 6-methoxy-3-methylcoumarilic acid,[114] zinc(II), cadmium(II) and mercury(II) complexes of two galactose α-amino acid derivatives,[115] $NEt_4[PtCl_3(PBu^n{}_3)]$,[116] lanthanide(III) complexes of Schiff bases derived from 2-hydroxybenzaldehyde and 3-methoxy-2-hydroxybenzaldehyde with the amino acids Ala, Val, Tyr and Glu,[117] and the mixed valent complexes $[Rh_2(Hap)_4X]X_2$ where Hap = 2-aminopyridine, X = Cl, SCN BF_4, F, Br, I.[118]

2.3 *Solution Studies: Structures and Reactions.*

Binding of Ca^{2+} to Gly, DL-Ala and β-Ala has been studied by ^{14}N, ^{15}N and ^{17}O n.m.r. spectroscopy.[119] The addition of the metal ion to the amino acid solutions causes broadening of the ^{14}N and ^{17}O signals indicative of involvement of both the amino and carboxylate groups in complexation.

Anation reactions involving $[Cr(H_2O)_6]^{3+}/[Cr(H_2O)_5OH]^{2+}$ and the amino acids L-Arg, Met, L-Ile have been investigated as a function of amino acid concentration, $[H^+]$, solvent composition and ionic strength.[120-122]

The effects of the presence of the amino sugar 2-amino-2-deoxy-D-gluconic acid and additional metal ions such as Cu(II), $UO_2{}^{2+}$, Pb(II), Ni(II) and Cd(II) on the reduction of iron(III) to iron(II) by D-galacturonic acid has been studied.[123] Stoichiometries and stability constants for complexes of the metal ions with the amino sugar and for ternary complexes which also involve D-galacturonic acid are reported. The results are discussed in relation to the mobilisation and bioavailability of iron.

In anaerobic acid solutions L-dopa, H_3L, reacts with iron(III),[124] to form the complexes $Fe(LH)^+$ which has λ_{max} at 442 and 700 nm with ε_{700} = 1380 $M^{-1}cm^{-1}$ and $Fe(H_2L)^{2+}$ which has λ_{max} 435, 660 nm with ε_{660} = 520 $M^{-1}cm^{-1}$. These decompose by intramolecular electron transfer in $Fe(H_2L)^{2+}$ to iron(II) and dopasemiquinone which is then oxidised by iron(III) to dopaquinone λ_{max} 385 nm, ε_{385} 1650 $M^{-1}cm^{-1}$. This cyclises by an intramolecular Michael addition to give leucodopachrome. Rate constants for the above reactions have been determined by stopped flow spectrophotometry. The formation and γ-radiolysis of complexes of iron(III) by 1,2-bis(β-aminoethoxy)ethane-N,N,N',N'-sodium sulfonate triacetic acid has been studied.[125]

The effects of L-amino acids on the binding of iron(III) to ovatransferrin was investigated by electronic spectroscopy.[126] While Gly and Glu were found to form the most stable iron(III)-ovatransferrin-amino acid ternary complexes, amino acids with OH, amide or S-containing side chains form less stable complexes and Leu, Ile, Val, Lys, Arg, Tyr and Trp do not form complexes of this type. Binding of Zn(II), Co(II) and Cd(II) to ovatransferrin in the presence of Gly has also been studied and results of ^{113}Cd n.m.r. spectroscopy shows that the amino group of Gly binds directly to the metal ion.

The coordination of amino acid esters to $[Fe_4S_4]^{2+}$ clusters has been studied by ^1H n.m.r. and Mossbauer spectroscopy.[127] The results indicate deprotonation of the amino acid esters and shed new light on the binding of such clusters in biological systems. Activation parameters for DMF exchange in the iron(III) complex of o-phenylenediamine-N,N,N',N'-tetraacetate in DMF were determined by high pressure n.m.r. methods.[128]

The structures of complexes formed between amino acids, HAA, and water soluble cobalt(III) porphyrins **(49)** have been investigated by n.m.r. spectroscopy.[129] The porphyrins, H_2Q, used in this study were tetrakis(4-N-methylpyridinimyl)porphyrin chloride, H_2TMPyP, and sodium tetrakis(4-carboxylatophenyl)porphyrin, $Na_4(H_2TCPP)$. At pH > 7 the predominant species in solution is $[CoQ(AA)_2]^-$ while at lower pH $CoQ(AA)H_2O$ is also present. In these the AA is N-coordinated to the metal ion. Three different complexes of His with Co(TMPyP) were detected. At pH 7 His is bound to cobalt(III) through the imidazole N(3) and at pH > 10 through the NH_2 group. Similar behaviour was observed for the binding of Met and Lys. Conformational analysis shows that in these complexes the amino acid side chains may interact with the porphyrins by stacking (aromatic amino acids), hydrophobic (e.g. Leu) or electrostatic (e.g Asp) interactions.

The reactions of several complexes of the type

$[Co(en)_2(NH_2CHRCOO)]^{2+}$ where R = CH_2OH, CH_2SMe, CMe_2SMe, CHMeOH **(50)** with thionyl chloride in DMF have been studied.[130] These complexes undergo β-elimination to chelated enamines **(51)** which rearrange to the imine complexes or react further with $SOCl_2$ to give isothiazole-3-carboxylate chelates **(52)**, Scheme 1. The mechanisms of these reactions are discussed. The oxidation of $[Co(en)_2Sar]^{2+}$ by SO_2Cl_2 in DMF was found to give an N-methylthiooxamato complex.[131] This occurs via an acid chloride chelate **(53)** which deprotonates at the α-carbon and then captures $SOCl^+$ to give a sulfine **(54)** following loss of HCl, Scheme 2. Addition of $SOCl_2$ across the S=O bond followed by decomposition gives SO_2Cl_2 and on hydrolysis the N-methylthioxamates complex **(55)**.

The reactions of α-aminomalonate, AM^{2-}, with trans-$[CoN_4Cl_2]^+$ complexes where N_4 = $(en)_2$; 2,3,2-tet or 3,2,3-tet- were investigated by spin trapping and e.s.r. spectroscopy.[132] These reactions involve oxidation of AM^{2-} to an iminomalonate intermediate which in the en and 2,3,2-tet complexes proceeds via radical formation at the α-carbon. The role of cobalt(II) intermediates and their spin states as well as the effects of the polyamine ligands on reactivity are fully discussed.

The kinetics and mechanism of oxidation of the cobalt(II) complexes $[Co(EDTA)]^{2-}$ and $[Co(HEDTA)]^-$ by peroxomonophosphoric acid in acetate buffer have been investigated spectrophotometrically.[133] These reactions follow second order kinetics and show a marked inverse dependence on $[H^+]$. The Complexes $[Co_2(\mu\text{-}OH)(\mu\text{-}O_2)L]$- where H_2L = $RNH(CH_2)_2NHR$ and R = CH_2COOH, CH(Me)COOH, CH(Et)COOH, CH(Ph)COOH, $CH(CH_2OH)COOH$ undergo metal centred oxidation to the corresponding cobalt(III) complex and H_2O_2 in basic solution.[134] The kinetics and mechanism of this base catalysed reaction have been studied. Enthalpies and entropies of activation have been determined for the aerial oxygenation of cobalt(II) complexes containing the linear tetradentate amino acid ligands ethylenebis(amino acidato) and a heterocyclic N base.[135] The rates of acid hydrolysis of the complexes $[Co(BigH)_2AA]^{2+}$ where BigH = biguanide, HAA = Gly, Ala, β-Ala, vary linearly with $[H^+]$, while for the cysteinate complex $1/k_{obs}$ vs. $1/[H^+]$ is linear.[136] The reactivity order is β-Ala < Gly < Ala < Cys.

The quantitative oxidation by O_2 of the copper(II) complex of the Schiff base from pyridoxal phosphate, PLP and p-sulfophenylglycine to the keto acid p-sulphophenylglyoxylic acid and the copper(II) complex of PLP-oxime has been investigated.[137] In the presence of excess amino acid the copper(II)-oxime complex is an intermediate in the oxidative deamination of the amino acid with Cu(II) and PLP as catalysts.

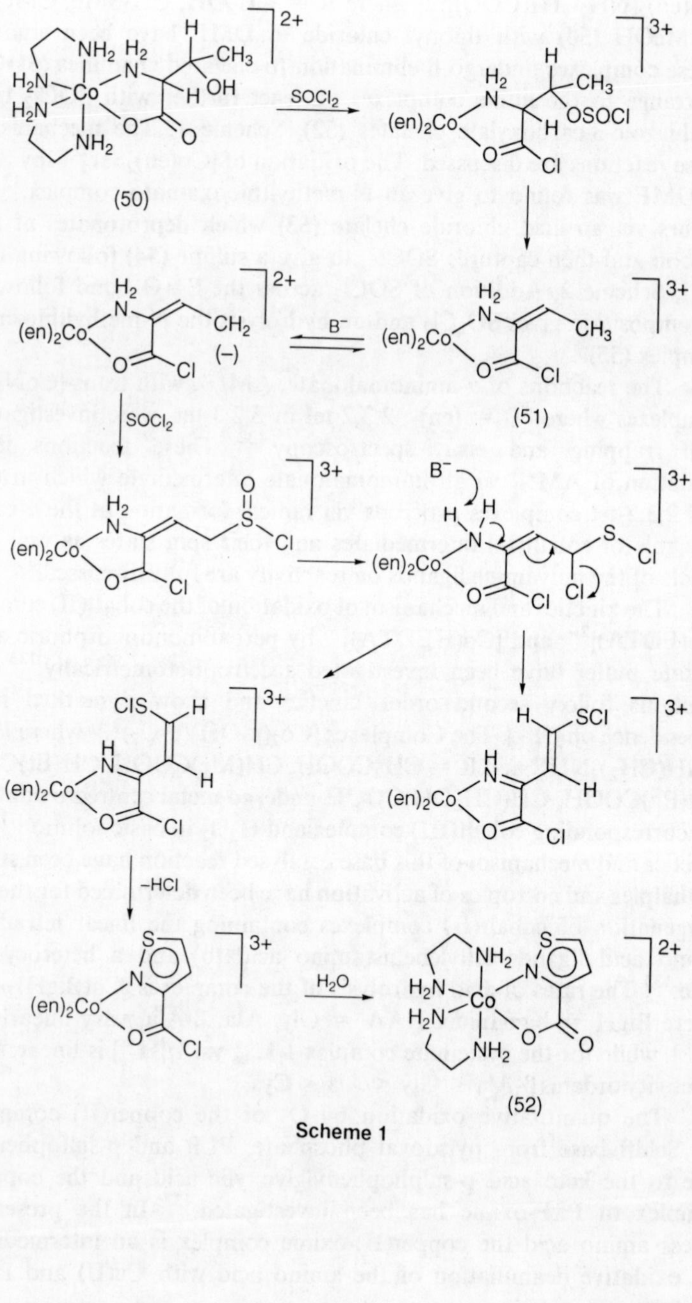

Scheme 1

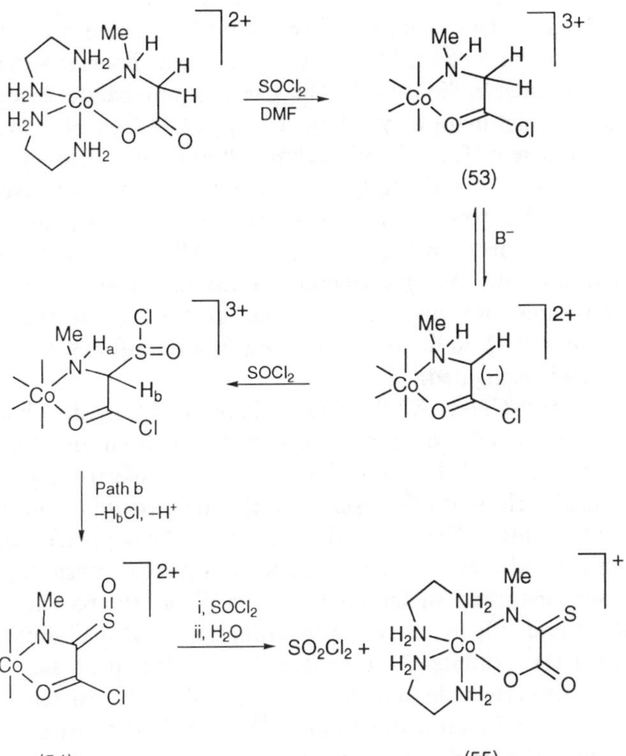

(53)

(54)

(55)

Scheme 2

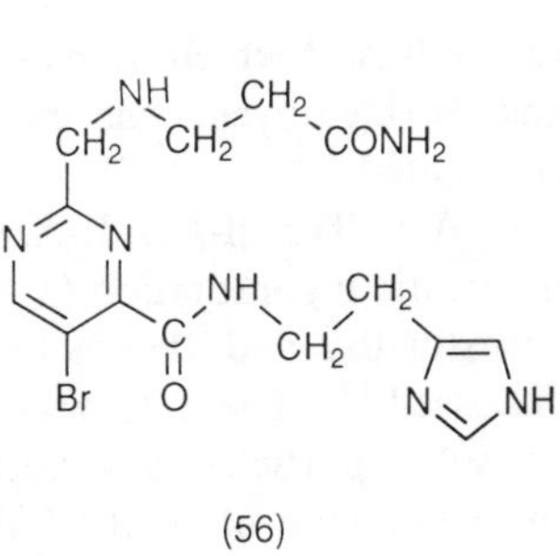

(56)

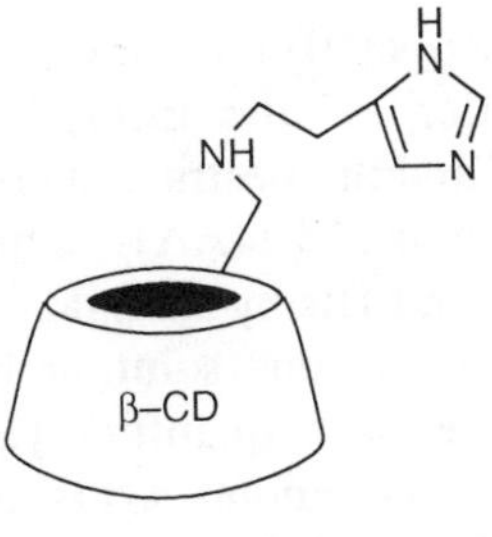

(57)

Rate constants for the reactions of the radical $^{+}NH_3C(CH_2CH_3)COO^-$ derived from 2-methylalanine with Cu^{2+}_{aq} and Cu^{+}_{aq} to form unstable Cu-C σ bonded intermediates were determined by pulse radiolysis.[138] The Cu(II) intermediate decomposes by β-carbonyl elimination yielding Cu^{+}_{aq}, CH_3COCH_3 and CO_2 whereas in the case of Cu^+ β-amine elimination occurs yielding Cu^{2+}_{aq}, $CH_2=C(CH_3)COO^-$ and NH_3. Plausible pathways for biological damage caused by free radicals are discussed. The kinetics of the exchange reaction between [Y(APTA)] where APTA is o-aminophenol-N,N,O-triacetate with Cu(II) have been studied and possible mechanisms and intermediates are discussed.[139] The kinetics of interaction of copper(II), nickel(II) and cobalt(II) with α-amino-β-methyl-n-valeric acid have been investigated.[140]

The complex Cu(sal-AA)L$(H_2O)_n$ where sal-AA is salicylideneglycine or alanine and L = cytidine or cytosine have been investigated by electronic, i.r., e.s.r. and ^{1}H n.m.r. (line broadening) spectroscopy and by electrochemical techniques.[88] Formation of a deoxycytidine adduct has also been investigated. The interaction of L with the copper(II) complex depends on several factors which include coordinative unsaturation of the metal and distortion of the coordination plane, the basicity, 'ortho effect' and steric effects of L. The complex ethylenediamine-N,N′-diacetocopper(II) exhibits 2 reversible 1e$^-$ redox processes at the hanging Hg drop electrode and this is attributed to the existence of the complex as a dimer in aqueous solution.[141] Several bis(alicyclic-α-amino acidato)copper(II) complexes have been investigated by spectrophotometry and voltammetry.[142] The electrical conductivity of Cu(II), Co(II) and Ni(II) complexes of 2-aminobenzoic acid hydrazide was measured at 300-500K.[143]

The copper(II) complex of **(56)** a synthetic ligand which mimics a major portion of the metal binding loans of bleomycin has been synthesised and its spectra and properties investigated.[144]

Photolysis of $Cu(AA)_2$ complexes where AA = Tris, β-Ala, L-Asp, L-Glu, Gly and His by sunlight and monochromatic u.v. radiation (λ = 310 mm) in aqueous solution has been investigated and the factors affecting rates and quantum yields are discussed.[145] The interaction between Cu(II), amino acids and marine solid particles has been investigated by FT i.r. spectroscopy.[146] The behaviour of copper(II)-amino acid complexes on reversed phase C_{18}-silica for analysis at the picomole level in marine matrixes has been studied.[147] The simultaneous spectrophotometric determination of cystine and cysteine in amino acid mixtures using copper(II)-neocuproin is described.[148]

The c.d. spectrum of the complex anion $[Mo_2(\mu\text{-}O)_2(O)_2(R\text{-}Cys)_2]^{2-}$

undergoes major changes on addition of cationic surfactants such as cetyltrimethylammonium bromide, CTAB, but no changes on addition of anionic surfactants or simple salts.[149] Similar spectral changes occur for $[Mo_2(\mu\text{-}O)_2(O)_2(S\text{-}Pen)_2]^{2-}$, $[Mo_2(\mu\text{-}O)(\mu\text{-}S)(O)_2(R\text{-}Cys)_2]^{2-}$ and $[Mo_2(\mu\text{-}S)_2(O)_2(R\text{-}Cys)_2]^{2-}$ on addition of CTAB but in contrast the complex $[Mo_2(\mu\text{-}O)_2(O)_2R\text{-}PDTA]^{2-}$ is unaffected. The change in spectrum is accounted for by partial dissociation of the complex with loss of R-Cys or by partial dissociation of a COO^- arm.

The kinetics of aqua ligand substitution in cis-$[Ru(bipy)_2(H_2O)_2]^{2+}$ by L-Cys in aqueous solution has been studied spectrophotometrically.[150] The effect of pH on the reaction of $[Ru(bipy)_3]^{3+}$ with amino acids and its implications for their detection by chemiluminescence has been described.[151]

The reaction of $Pd(Me_4en)Cl_2$ where Me_4en is N,N,N′,N′-tetramethylethylenediamine with inosine and 5′-inosine monophosphate was investigated as a function of $[Cl^-]$ and pH.[152] Evidence for the formation of 1:1 complexes was observed with $Pd(Me_4en)Cl(H_2O)^+$ being the only reactive species. These systems serve as useful models for the interactions of antitumour active platinum complexes with DNA.

The interaction of Pd(II) with DL-selenomethionine, $^+NH_3CH(CH_2CH_2SeMe)COO^-$, in aqueous acid solution has been studied by ^1H and ^{13}C n.m.r. spectroscopy.[97] Binary and ternary complexes of Pd(II) with L-2-amino-3-(4-hydroxyphenyl)propanoic acid and bipy or 1,10-phen has been studied by polarography.[153] Complex formation between K_2PdCl_4 and phosphonate analogues (α- and β-) of aspartic acid has been studied by ^{31}P and ^{13}C n.m.r. spectroscopy.[154]

The helix coil transition of poly (α-L-Glu) induced by Cd^{2+} and/or Zn^{2+} was investigated by direct current polarography and c.d. spectroscopy.[155]

2.4 *Formation Constants.*

The synthesis of the functionalised cyclodextrin 6-deoxy-6-(N-histamino)-β-cyclodextrin **(57)** has been described and its protonation and copper(II) complexation studied by n.m.r., e.s.r., pH-metric and calorimetric methods.[156] Inclusion of the imidazole ring in the cavity is promoted by protonation while the coordination of copper(II) is markedly affected by the presence of the cavity. Ternary complexes of copper(II) with this functionalised cyclodextrin and amino acids have also been investigated and it has been established that the D-enantiomer of aromatic amino acids forms more stable ternary complexes than the L-isomer.[157,158]

Copper(II) complexes of various bidentate aliphatic, alicyclic and aromatic aminophosphonous and aminophosphinic acids and the α- and β-phosphinic acid derivatives of Asp have been studied by pH-metric and spectroscopic (visible and e.s.r.) methods.[159] The binding abilities of the bidentate ligands are weaker than the corresponding amino carboxylates due to the difference in basicities of the PO_2R^- and COO^- groups. The Asp derivatives which are potentially tridentate bind only via their aminocarboxylate groups when an α-phosphinate group is present which also coordinates. Stability constants of divalent metal ion (Mg, Ca, Pb, Co, Ni, Cu, Cd, Zn) complexes with piperidinomethylphosphinic acid and piperazine-1,4-diylbismethylene-bisphosphinic acid were found to be lower than their carboxylic or phosphonic acid analogues. The ^1H, ^{33}P and ^{13}C n.m.r. spectra of these ligands in D_2O as a function of pH have also been reported.[64]

Copper(II) complexes of β-alaninehydroxamic acid (A) have been studied by potentiometric methods and by electronic and e.s.r. spectroscopy.[39] In the pH range 4-9 the major species in solution is the e.s.r. silent (at liquid N_2 temperatures and above) pentanuclear $[Cu_5A_4H_{-4}]^{2+}$ for which a stability constant log $\beta = 46.66$ has been obtained and for which a crystal structure has been reported. Formation constants have also been reported for copper(II) and iron(III) complexes of L-leucinehydroxamic acid, HL.[160] In the Cu(II)-HL system ligand coordination is via the NH_2 and NHO^- groups whereas in the iron(III) complex coordination occurs through the carbonyl and hydroxamate (O^-) oxygen atoms. Co(II), Ni(II), Cu(II) and Zn(II) complexes of hydroxamic acids of Phe, Tyr and Dopa have also been studied and stabilities, constants and ligand binding modes reported.[161] The binding of Zn^{2+} to amino hydroxamic acid analogues of Ala, Sar, Leu and His has been followed by potentiometric and ^1H n.m.r. methods.[162] Binding of the metal ion occurs initially at the hydroxamate O atoms but then deprotonation of the $^+NH_3$ group occurs and results in the formation of Zn_2L_3 species in the case of the first three ligands but tridentate coordination in the case of His. Potentiometric, polarographic and n.m.r. methods have been used to study Cd(II) coordination to β-alaninehydroxamic acid, $NH_2(CH_2)_2CONHOH$.[163] Very stable oligonuclear complexes are formed. The stability constants of ternary complexes of cadmium(II) with D-penicillamine and L-histidine and with these ligands in the presence of oxalate have also been determined.[164]

Copper(II) complexes of $^-O_2CCH_2(H_2NOCCH_2)NCH_2CHRN(CH_2CONH_2)CH_2CO_2^-$ where R = H or Me have been investigated by potentiometric and spectrophotometric methods.[165] These ligands are hydrolytic metabolites of the anticancer bis(3,5-dioxopiperazin-l-yl)alkanes and by their chelating properties may be responsible for the activity of the parent drugs. At low pH the complexes contain 2N, 2O carboxylate in-plane donor groups but at higher pH deprotonation of the amide groups occur with concomitant displacement of carboxylates from the in-plane positions by the deprotonated amide groups. The formation of dinuclear copper(II) complexes of aminopolycarboxylic acids with OH groups attached to the alkyl linkage e.g. $HOOCH_2CNCH(OH)CH_2CH_2N(CH_2COOH)_2$, have been studied by potentiometry, e.s.r. and other spectroscopies.[166] Two or three dinuclear species are formed depending on the ligand and pH. At low pH the copper(II) are coordinated independently by the aminocarboxyl groups and show only weak dipolar magnetic interactions. As the pH is raised the alcoholic OH groups deprotonate to give μ-alkoxo-bridged species and in some cases μ-alkoxo-μ-hydroxo species which show strong spin exchange interactions between the paramagnetic centres.

The formation of copper(II) complexes of EDTA, EDDA, IDA, NTA and $(HOCH_2CH_2)(HOOCCH_2)NCH_2CH_2N(CH_2COOH)_2$ has also been studied by e.s.r. and electronic absorption spectroscopy.[167]

Complex formation between N-(phenylsulfonyl)glycine and N-(tolylsulfonyl)glycine and the metal ions Zn(II) and Cd(II) in the absence and in the presence of bipy has been studied by dc polarography, pH-metric titration and ^1H n.m.r. spectroscopy.[168] The presence of bipy in the Cd(II) complex makes the sulfonamide NH group more acidic with a lowering in pK_a from 8 in the binary system to 7.6 in the ternary. A similar observation was previously made in the case of Cu(II). The presence of the bipy ligand in the Zn(II) complex causes the sulfonamide NH group to be acidic whereas no deprotonation is observed in the binary system. Formation constants have also been determined for complexes of Cd(II) and Pb(II) with dithiocarbamates of Gly, DL-Ala, DL-Val, DL-Leu, DL-Ile and L-Pro.[169] Complex formation between Hg(II) and Gly or β-Ala (HL) has been investigated under physiological conditions (37°C, 0.15M aqueous NaCl solution) by glass electrode potentiometry.[170] In both cases the complexes HgL^+ and HgL_2 are formed with the Gly complexes being the more stable. The stability constants of the Gly complexes are much lower than those observed in other ionic media e.g. KNO_3 due to the high affinity of Cl^- for Hg(II). Stability constants are reported for the molybdenum(VI) complexes $[MoO_3(Asp)]^{2-}$, $[Mo_2O_5(Asp)_2]^{2-}$, $[Mo_2O_4(OH)(Asp)_2]^-$ and $Mo_2O_4(Asp)_2$.[171]

Formation constants have been reported for the following systems, Cu(II)-Leu,[172] Fe(III)-Gly, Fe(III)-OH-Gly,[173] Cu(II)-Zn(II)-IDA, Ni(II)-Zn(II)-IDA where IDA is iminodiacetate,[174] Th^{4+}, UO_2^{2+}, VO^{2+} with NTA, HEDTA and EDTA in the presence and absence of ohydroxynaphthalidene benzoyl (or salicoyl)hydrazones,[175] Pb(II)-His or Orn,[176] Ln(III)-penicillamine,[177] Cu(II)-catechol-AA where A = Trp, Tyr, Phe,[178] M(II)-salicylic acid-amino acids,[179] Cu(II) and Ni(II)-amino acids (standard molar enthalpies of formation),[180,181] M-AA-imidazole where M = Cu(II) or Zn(II) and AA = Gly, DL-Ala or DL-Val,[182] Cu(II)-AA and Cu(II)-A-L where AA = Ala, Phe, Trp, Lys, Arg, Ser, Thr, Asp, His and L = N-1-(naphthyl)ethylenediamine or ethylenediamine.[183] Cu(II)/6-aminopenicillamine acid, bipy, Gly or adenine,[184] M(II)/6-aminopenicillanic acid where M = Co, Ni, Cu and Zn,[185] Zn(II) and Cd(II)-His, Gly-His, Ala-His, Gly_2 and Gly_3,[186] Cu(II)-amino acids-tartrate,[187] Cu(II)-poly(N-methacryloyl-AA) where AA = L-Ala, L-Asp, L-Glu or L-Asn,[188] Cu(II)-cimetidine-AA where AA = L-Ala, β-Ala, L-Phe or L-Tyr,[189] Cu(II)-terpy-AA,[190] M(II)/2-benzylamino-2-deoxy-D-Glycero-D-guloheptonic acid where M = Co, Ni, Cu, Zn, Cd, Hg,[191] Cu(II)-AA-succinate where AA = Asp, Glu or Lys,[192] M(II)-L-AA where M = Co, Ni, Cu or Zn; L = iminodiacetic acid or nitrilotriacetic acid and AA = Gly, Ala, Val, Leu, norLeu, Phe, Trp, Ser, Thr, Met, Asp,[193] M(II)/L-2-amino-3-(4-hydroxyphenyl)propanoic acid/bipy or phen where M = Co, Ni, Cu or Zn,[194] Cu(II)-4-aminopicolinic acid N oxide,[195] Cu(II) or Cd(II)/oxalic acid/2-amino-3-hydroxypyridine,[196] M(II)/Asp or Glu/2-aminophenol where M = Co, Ni, Cu, Zn,[197] Zn(II)-AA-acetic acid,[198] M(II)/inosine or xanthosine/AA where M = Cu, Ni, Zn or Co and AA = Ala, α-aminoisobutyric acid, norvaline or norleucine,[199] Pb(II)-nitrilotrimethylenephosphonic acid.[200]

2.5 *Catalysis.*

Several synthesis of amino acid derivatives are reported to be catalysed by metal complexes. These include the formation of cyclobutane-β-amino acid derivatives **(58)** in high yield in the presence of Ru(cod)cot where cod = cycloocta-1-5-diene, cot = cycloocta-1,3,5-triene,[201] the synthesis of optically pure pyridylalanines **(59)** from the corresponding enamides in the presence of a rhodium complex catalysts,[202] asymmetric synthesis of furylalanine **(60)** derivatives by the hydrogenation of dehydroamino acids in the presence of aminophosphine-phosphinite-rhodium complexes,[203] asymmetric synthesis of fluorine containing phenylalanines $RCH_2CH(NHBz)CO_2R'$, by hydrogenation of the corresponding (fluorophenyl)acrylic acids using rhodium catalysts,[204] the enantioselective synthesis of β-amino acids by hydrogenation of the corresponding acrylic acids using ruthenium(II) catalysts,[205] the asymmetric hydroformylation of N-acylaminoacrylic acid derivatives to give aldehydes regioselectively in high yields in the

COR^2
R^3 R
NR^1

(58)

CH_2CHCO_2H
N
NH_2

(59)

R^1 O CH_2CHCO_2R
NHBz

(60)

H
Me N CO_2Me
O CH_2
Fe
$(CO)_3$

(61)

H
N
NH
H
CO_2Me
NH
N N
MeO_2C
H NH Zn
S S O
NHZ
H
O
ZHN H

(62)

R^3 O
O N O N R^2
Tc
R^4 S N O
R^1 X

(63) MAG_3; $X = CO_2H$; $R^1 = R^2 = R^3 = H$

O
HO
CH_3
O O
N CH_3
H CH_3 N
H
CH_3 $N-O^-$ O
N
O Fe^{III} O^-
NH O
O^- O HN
O N H
N N CH_3
CH_3 H
O O
CH_3

(64)

presence of rhodium-chiral diphosphine catalysts,[206] the synthesis of protected hydroxyamino acids by selective hydrogenation of dehydro-amino acids in the presence of chiral, homogenous rhodium catalysts,[207] asymmetric synthesis of 2S, 3R- and 2S, 3S-2-(aminomethyl)-3-hydroxy-3-phenylpropanoic acids via a Ni(II) complex of a Schiff base from β-Ala and S-2-[N-(N′-benzyl)prolylamino]benzophenone,[208] asymmetric synthesis of phosphorus analogues of α-amino dicarboxylic acids, (S)-$NH_2CHRCOOH$ where R = $CH_2CH_2PO(Me)(OH)$ or $-(CH_2)_nPO_3H_2$, n = 1-3, in the presence of a Ni(II)-Schiff base derived from $PhCH_2$-L-Pro-NHC_6H_4 and Gly,[209] the synthesis of β,β,β-trialkyl-α-amino acids by the addition of MeLi to **(61)**,[210] in the presence of iridium catalysts, the kinetic resolution of racemic diphosphines and their application in catalytic asymmetric hydrogenation,[211] influence of β-substituents in chiral seven membered rhodium diphosphine rings on asymmetric hydrogenation of amino acid precursors.[212]

3 Peptide Complexes

The chemistry of metal peptide complexes is the subject of an increasing number of articles with many of the complexes described having direct or indirect important biological applications.

3.1 *Synthesis, crystal structures.*

The design of peptides that mimic the structural and/or functional features of naturally occurring proteins is an area of much current interest. Initial results with designed metallopeptides have been reported,[213,214] although the ability of such peptides to bind or activate ligands has not been demonstrated. Berg and coworkers,[215] have recently developed a strategy to convert a peptide with a structural site in which all of the donor atoms are provided by the peptide to one that has an open coordination site for substrate binding and potential activation. The approach involves truncation of a metal binding peptide to remove one of the ligands with the hope that such a truncated peptide will still bind metal ions. The zinc finger peptide CP1 has been truncated by removal of the last four amino acids to produce a peptide CP1-C4 which binds cobalt(II) in a tetrahedral site very strongly via two cysteinate and two histidine donors.

A novel metal ion-assisted self-organising molecular process has been described in which a small peptide has been assembled into a large and topologically predetermined protein tertiary structure.[216] A 15-residue amphiphilic peptide with a 2,2′-bipyridine functionality at the N-terminus was designed and shown to undergo spontaneous self-assembly, in the presence of transition metal ions to form a 45-residue triple-helical coiled-coil metalloprotein. The basic zinc finger template **(62)** (Z = $PhCH_2O_2C$) is formed in a single step from Z-Cys(Acm)-His-OMe (Acm = $ACNHCH_2$) and $ZnCl_2$ by a tandem complexation deprotection sequence.[217]

Metal complexes of cadmium(II) with thiol-organic and thiol-peptide ligands related to metallothioneins have been the subject of a recent thesis.[218]

Several complexes of 2,2′-bipyridine, 1,10-phenanthrozine and 2,9-dimethyl-1,10-phenanthroline with copper(II) dipeptides have been synthesised.[219] These complexes are neutral and display a broad absorption band (d-d) in the 630-640 nm range. The e.s.r. data suggest that these complexes have distorted square pyramidal geometry about Cu(II). The superoxide dismutase activity of several of these complexes has been studied. They display higher activity than the corresponding Cu(II)-dipeptide complexes because of a strong axial bond of one of the nitrogen atoms of the α-diimine.

Various synthetic routes to peptide ligand precursors bearing the S-benzoyl protected mercaptoacetyl group at the N-terminus have been evaluated.[220] The S-benzoyl protected mercaptoacetylglycylglycylglycine(MAG_3) is a ligand precursor for the synthesis of ^{99m}Tc-MAG_3 **(63)** which has been successfully used as a radiopharmaceutical imaging agent for renal studies. The inorganic carboxyterminus protecting group, $(NH_3)_5Co^{III}$, was evaluated in the preparation of MAG_3 analogues containing bulky amino acids at the N-terminus. Use of this inorganic protecting group appears to be a useful new synthetic route to such analogues.

The complexes $[Ru(glyglyH_{-1})(PPh_3)(CH_3OH)]$ and $[Ru(glyglyglyH_{-1})(PPh_3)_2]_2$ can be prepared by reaction of $[RuCl_2(PPh_3)_3]$ with diglycine (glyglyOH) and triglycine (glyglyglyOH) respectively in methanol at reflux in the presence of base.[221] Their molecular structures havebeen determined by X-ray crystallography. The diglycine complex has N(amino), N(peptide), O(carboxyl)-coordination to Ru with the PPh_3 ligands positioned trans to a methanol oxygen and to a peptide nitrogen, respectively. The triglycinate ion is tetradentate with N(amino), N(peptide), O(peptide), (O-carboxyl) coordination leading to the formation of a dimeric complex containing a 14-membered central ring system.

Three cobalt(III) complexes containing N-β-alanyl-(S)-aspartic acid (H_3L), [CoL(en)] (en = ethylenediamine), [CoL(dien)] (dien = diethylenetriamine) and $[Co(HL)(dien)]^+$ have been prepared and characterised by ^{13}C n.m.r., u.v.-vis-absorption and c.d. spectra.[222] The dipeptide L functions as a quadridentate ligand in [CoL(en)] and as a terdentate ligand in [Co(dien)] and $[Co(HL)(dien)]^+$. A monapeptide with an Ala-(HO)Gly-Ala sequence has been prepared by condensation of appropriately protected tripeptide units.[223] ^{1}H NMR and CD spectra indicate that the monapeptide has disordered structure in DMSO and water. The peptide forms a 1:3 complex with iron(III) at neutral pH which is believed to have the structure **(64)**. The complex has λ_{max} 410nm (ε = 2160 $M^{-1}cm^{-1}$) at pH 7. Alanine residues influence the hydroxamate groups to produce a chiral complex.

A number of complexes of the type $[Co(R,R\text{-benzet})(dipeptidato)]^+$ where R,R-benzet is N-benzyl-N'-(2-picolyl)-1R,2R-diaminocyclohexane and dipeptidato is the dianion of GlyGly, Gly-S-Val, Gly-R-Val, Gly-S-Leu, Gly-R-Leu, S-LeuGly, R-LeuGly, S-ValGly, S-Leu-S-Ala and S-Leu-S-Phe have been prepared and characterised.[224] All of the ternary complexes contain the dipeptide coordinated meridionally and are formed stereospecifically in every case except for S-Val-Gly. The stereochemistry has been conformed by X-ray crystallography and the Gly-R-Val and Gly-Gly derivatives (perchlorate salts).

Small peptides interact with alkaline earth metal ions under conditions of fast atom bombardment to fom abundant metal-bis(peptide) complexes having the composition $[2pept + M^{2+}\text{-}3H^+]$.[225] The structure of these gas-phase bis(peptide) complexes has been determined by collisionally activated decompositions (CAD) and they are shown to involve metal ion binding to the deprotonated C-terminal carboxylate groups of both peptides and to a deprotonated amide nitrogen of one of the constituent peptides. The location of the alkali metal ion in gas-phase peptide complexes has also been studied.[226]

The use of metal ions to organise peptides into specific structures is now attracting considerable attention. An interesting example has been recently published,[227] in which iron(II) organised a synthetic peptide into a three helix bundle, Scheme 3.

A stable complex of rhenium(V) with mercaptoacetylglycylglycylglycine (MAG_3) has been prepared by the reaction of $ReO_2(en)Cl$ (en = ethylenediamine) or Re(V) citrate with MAG_3 at pH 10.[228] The complex can be isolated as salts of $X[ReO(MAG_3)]$ ($X = Bu_4N^+$ or Ph_4As^+). Crystallography establishes that the rhenium atom in $[ReO(MAG_3)]^-$ is bound to three nitrogens (amide) one sulphur (thiolate) and one oxygen (yl) atom in a distorted square pyramidal geometry. The potential radiotherapeutic applications of ^{186}Re and ^{188}Re have stimulated interest in the chemistry of such rhenium complexes.

Model complexes of reduced rubredoxin $[Fe^{II}(Z\text{-Cys-Pro-Leu-Cys-Gly-NH-}C_6H_4\text{-}p\text{-}X)_2]^{2-}$ (X = OMe, H, F, CN) have been prepared by

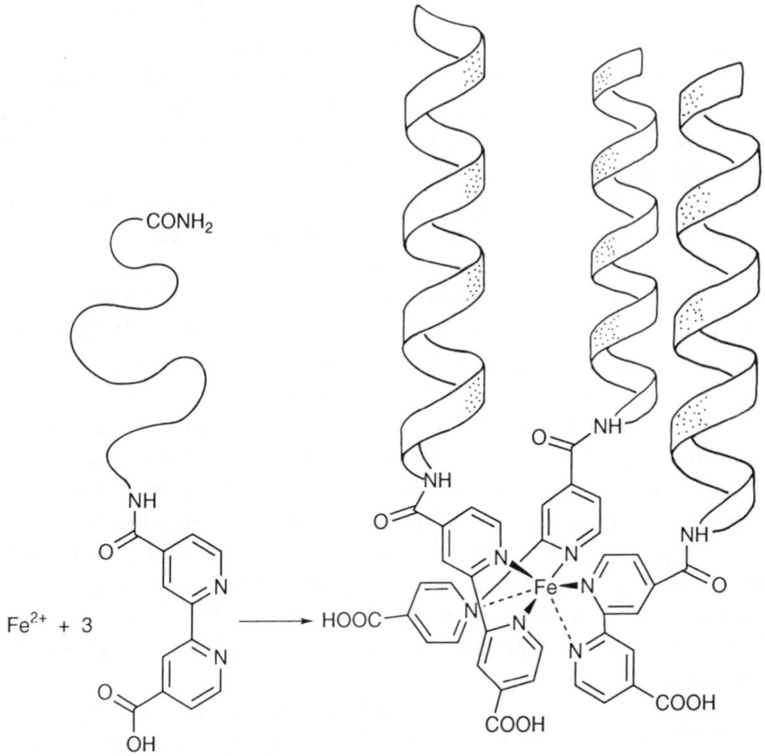

Scheme 3

ligand exchange reactions of $[Fe^{II}(S\text{-}t\text{-}Bu)_4]^{2-}$ with Z-Cys(SH)-Pro-Leu-Cys(SH)-Gly-NH-C_6H_4-p-X.[229] These complexes give positively shifted redox potentials compared with other peptide model complexes. Evidence for intramolecular NH—S hydrogen binds has been obtained from ^{2}H n.m.r. spectroscopy of reduced-rubredoxin Fe(II) model complexes with bidentate peptide ligands.[230]

The $[Fe_2S_2]^{2+}$ complex of an artificial 20-peptide Ac-Pro-Tyr-Ser-Cys-Arg-Ala-Gly-Ala-Cys-Ser-Thr-Cys-Ala-Gly-Pro-Leu-Leu-Thr-Cys-Val-NH_2 containing an invariant Cys-A-B-C-D-Cys-X-Y-Cys(A, B, C, D, X, Y = amino acid residues) fragment of plant type ferredoxins has been prepared,[231] by a ligand exchange method with $[Fe_2S_2(SCMe_3)_4]^{2-}$. ^{1}H NMR and electrochemical data indicates the presence of two isomers. One of the isomers has a Cys-X-Y-Cys bridging coordination to the two iron centres with a $[Fe_2S_2]^{2+}$ core environment similar to those of the denatured plant-type ferredoxins and a positively shifted redox potential at -0.64 V *versus* SCE in DMF. The other isomer has Cys-A-B-C-D-Cys bridging coordination and a more negative redox potential (-0.96 V).

Complexes of iron(III) with amino sugars and small glycopeptides related to peptidoglycan monomer (PGM) have been prepared and characterised.[232] Separation of the reaction products on a molecular sieve, allowed the characteristation of low molecular weight monomeric complexes.

The preparation and crystal structure of $[ML(H_2O)_2].4H_2O$ (M = Co(II), Cu(II) or Zn(II); HL = 2,2-diacetoamidopropionic acid) has been described.[233] An uncommon carbonyl-O(amide) bonded structure is observed in a two-dimensional layered polymeric structure.

3.2 *Reactivity.*

The half life for hydrolysis of the amide bond in neutral aqueous solution is ca. 7 years.[234] Since many proteolytic enzymes require metal ions for activity the role of metal ions in these processes has attracted considerable attention. Certain platinum(II) and palladium(II) complexes attached to the sulphur atoms of cysteine, S-methylcysteine and methionine in peptides and in other amino-acid derivtives promote, under relatively mild conditions, selective hydrolysis of the unactivated amide bond involving the carboxylic group of the amino acid anchoring the metal complex.[235,236] Detailed kinetic studies of these reactions are reported.

Iron(II)-cysteine-containing peptide complexes have been found to exhibit catalytic activity for air oxidation of benzoin to benzil and methyl DL-mandelate to methyl benzoylformate.[237] In the presence of $[Fe(2\text{-}Cys\text{-}Gly\text{-}Val\text{-}OMe)_4]^{2-}$, the rates of catalytic air oxidation of p-substituted benzoin follows the reaction sequence Br > H > CH_3 > MeO and the isotope effect k_H/k_D = 3.4. These results indicate that the methine hydrogen of benzoin is released as a protein in the rate-determining step. Catalytic oxidation of benzoin to benzil by p-benzoquinone in the presence of the complex $[Fe_4S_4(Z\text{-CysS-Pro-Val-OMe})_4]^{2-}$ (CysS = S-deprotonated cysteinate) has an activity five times greater than that of $[Fe_4S_4(Z\text{-CysS-Pro-Gly-OMe})_4]$, so that bulky ligands appear to exert a specific effect.[238]

The kinetics and mechanism of the oxidation of $[Co(EDTA)]^{2-}$ by four bis(dipeptide)nickel(III) complexes has been investigated at 25°C and 0.10 mol dm^{-3} perchlorate.[239] The reactions are first order in each reagent and have a complex dependence on pH. The dominant pathway over the pH range 4-10 involves the acid-catalysed formation of a precursor complex through which electron transfer can take place.

3.3 *Solution Chemistry.*

The electrochemistry of Cu(II)-peptide complexes containing histidine residues has attracted attention. Anomalous current fluctuation in the reduction of Cu(II)-GlyGlyHis has been observed,[240] and appears to be related to the absorption and desorption behaviour of the reduction product. Copper(II)-peptide complexes (peptide = Gly-L-His-L-Lys, Gly-L-His-Gly and Gly-L-His) have also been studied in aqueous solution by various electrochemical methods at different pH values.[241]

Complexation of copper(II) wih three peptides containing a tyrosine residue (Gly-Gly-L-Tyr), L-Tyr-Gly-Gly and Leu-enkephalin (H_2A) have been studied by pH measurements, visible spectra at 25°C and 0.2 mol dm^{-3} KNO_3 also by e.s.r. and calorimetry at 25°C and I = 0.1 mol dm^{-3} KNO_3. Four species are reported for the Cu(II)-tripeptide systems, while five species occur with Cu-H_2A.[242] The phenolic OH group in the side chain does not play a direct role in coordination to copper(II). Eight pentapeptides have been synthesised as models for the N-terminal pentapeptide fragment of Substance P a peptide containing 11 amino acid residues in the sequence Arg-Pro-Lys-Pro-Gln-Gln-Phe-Phe-Gly-Leu-MetNH_2 which is a member of the group of tachykinins.[243] A potentiometric and spectrophotometric study of the complexes formed with H^+ and Cu(II) and a potentiometric study of the complexes with substance P have been carried out. The results demonstrate the profound effect which the prolyl residue can have on the formation of copper(II)-peptide complexes. The prolyl residue acts as a break-point to coordination and encourages the formation of folded peptide chains, through β-turns, resulting in large, but very stable, chelate rings. The coordination behaviour of Substance P is almost identical to that of the N-terminal fragment, Substance P_{1-5}, with chelation through the N-terminal amino N and the ε-amino N of the Lys residue to form a large chelate ring of high stability, the peptide being forced into a bent conformation by the prolyl residue.

Recent studies of biologically active peptides such as thymopoietin (Arg-Lys-Asp-Val-Tyr), the N-terminal tetrapeptide fragment of fibrinopeptide A (Ala-Asp-Ser-Gly) and angiotensin (Asp-Art-Val-Tyr-Ile-His-Pro-Phe) have shown that the β-carboxylate of the aspartic acid residue may dramatically influence the binding ability of oligopeptide ligands to metal ions. A recent paper,[244] reports a study of the influence of Asp and Glu residues in a peptide chain on the ability of the peptide to coordinate to copper(II). The results establish that while the Glu residue has little influence on coordination equilibria, an N-terminal Asp residue significantly stabilises the copper(II) complex with only one nitrogen atom coordinated due to chelation through the β-carboxylate group rather than the peptide C=O oxygen. This effect is much more pronounced than with aspartic acid itself. Asp residues in the second or third position of the peptide sequence stabilise 2N and 3N complexes respectively significantly delaying, and in some cases preventing completely the formation of 4N complexes. An Asp residue in the fourth position has a much smaller effect.

Complexes formed in aqueous solution by copper(II) with Gly-L-HisGly(L-) have been studied by potentiometry, calorimetry, u.v.-vis and c.d. techniques.[245] Only monomeric species are formed in the acidic range of pH, ($[CuL]^+$, $[CuLH_{-1}]$, $[CuL_2]$ and $[CuL_2H_{-1}]^-$) while at higher pH (>8) polynuclear $[Cu_4L_4H_{-8}]^{4-}$ is formed in addition to $[CuLH_{-2}]^-$. In the tetrameric complexes there is probably dissociation of the N(1)-pyrrole hydrogen with formation of imidazolate bridges. The thermodynamic stereoselectivity and spectroscopic characteristics of ternary complexes of copper(II) with cyclo-L-histidyl-L-histidine and L- or D- amino acids has been studied.[246] Ternary complex formation with L-amino acids is more enthalpy and less entropy favoured compared with the analogous D-amino acids.

The interaction between mercury(II) complexes of penicillamine and glutathione and some transition metal ions has been studied potentiometrically.[247] Mixed metal complexes of the type $Hg(Pen)_2M$ and $Hg(GS)_2M$ (M = Co(II) or Ni(II)) were detected. Complexes of glutathione disulphide with Zn(II), Ni(II), Co(II) and Cd(II) were also investigated. Zinc(II) and nickel(II) form the complexes M(GSSG)H and M(GSSG) while the other metal ions form only the fully deprotonated complex M(GSSG).

Reaction of potassium dichromate with the tripeptide glutathione gives a red 1:1 complex of Cr(VI) with glutathione which is stable for ca. 60 minutes at 4°C and I = 1.5 mol dm^{-3}, ^{1}H, ^{13}C and ^{17}O n.m.r. studies establish that glutathione acts as a monodentate ligand and binds to Cr(VI) via the cysteinyl thiolate group, forming a $GSCrO_3^-$ complex.[248]

In recent years a large class of proteins has been discovered that is characterised by the presence of one or more sequences that closely approximate to the type (Tyr, Phe)-X-Cys-$X_{2,4}$-Cys-X_3Phe-X_5-Leu-X_2-His-X_3,X_4-His where X represents relatively variable amino acids.[249] Each of these sequences appear to form a small domain (often termed a "zinc finger" domain) organised around a zinc ion tetrahedrally coordinated by the cysteine and histidine residues. A single zinc finger peptide, ProTyrLysCysProGluCysGlyLysSerPheSerGlnLysSerAsp LeuValLysHisGlnArgThrHisThrGly has now been designed,[250] with the use of a data base of 131 zinc finger sequences. This peptide binds metal ions such as Zn(II) and Co(II) and folds in their presence. The affinity of this peptide for metal ions is greater than for any other zinc finger peptide characterised to date. Histidine residues can be protonated and dissociated from the metal centre with only local loss of structure. In addition, a histidine residue can also be replaced by a cysteine to give a peptide which has a $(Cys)_3(His)$ rather than a $(Cys)_2(His)_2$ metal binding site.

Structural and kinetic features of the Mn(II)-Leu-enkephalin binding equilibria have been defined by measureing ^{13}C and ^{1}H n.m.r. spin-lattice relaxation rates.[251] The temperature dependence of such rates showed that some carbons were experiencing slow exchange regimes such that kinetic parameters at room temperature could be calculated (k = 1400 s^{-1}, $\Delta H‡$ = 12.0 kcal mol^{-1}; $\Delta S‡$ = -9.9 e.u.).Carbon-Mn(II) distances were calculated allowing a model of the 1:1 complex to be developed.

Formation constants for mixed ligand complexes of the type [CuAL] and $[CuA_{-H}L]^-$ (A = Gly-Gly, Gly-S-Ala and Gly-S-Leu; L = indole acetic acid, indole propionic acid and indole butyric acid) have been determined using 1:1 (v/v) dioxane-water as solvent at 30°C and I = 0.2 mol dm^{-3} ($NaClO_4$).[252] The stabilisation of the ternary complexes was attributed to intramolecular hydrogen bond formation between the two ligands.

A number of papers have appeared dealing with the interaction of dioxygen with Co(II)-dipeptides. A set of computer programs have been published,[253] for data processing in potentiometric and gasometric titrations in the cobalt-dipeptide-oxygen system. Spectrophotometric characterisation of complex equilibria of Co(II) with dipeptides under oxygen free conditions has been studied,[254] as has the effect of non-coordinating aliphatic side chains on complex formation and dioxygen uptake.[255] A further paper deals with the effect of dipeptide structure on the stability of cobalt(II)-dipeptide complexes and their ability to take up oxygen.[256] Absorption and e.s.r. spectra of the cobalt-dipeptide-oxygen system have also been reported.[257] The oxidation of various cobalt(II)-dipeptide complexes by dioxygen and hydrogen peroxide has been examined.[258] The nature of the oxidant is of importance as the oxo-intermediate species apparently possess different structures depending on the oxidant. When O_2 is employed the resulting cobalt(III)-dipeptide complexes are more inert suggesting a different degree of protonation.

3.4 *Miscellaneous.*

The dynamical behaviour of the 25- residue 'zinc finger' peptide xfin31 has been modelled through molecular dynamics simulations in vacuum and in water, and by normal mode and by Langevin mode analysis.[259] Protein n.m.r. and optical spectral studies on a heme undecapeptide in aqueous SDS micelles and different pH values have been reported.[260] In a micellar solution the monodispersed heme peptide is found to be encapsulated inside the hydrophobic micellar cavity. Raman spectral evidence has been obtained for an anhydride intermediate in the catalysis of ester hydrolysis by carboxypeptidase A.[261] Coordination modes of histidine in copper(II) dipeptide complexes has been studied by multifrequency e.s.r.[262]

The biuret reaction makes peptides electrochemically active. A recent paper discusses the influence of tyrosine, an electroactive amino acid in the detection of copper(II)-peptide complexes containing tyrosine.[263]

The role of the carboxy-terminus of the polypeptide D1 in the assembly of a functional water-oxidising manganese cluster in photosystem II of the cyanobacterium *Synechocystis* sp. PCC 6803 has been studied.[264] Assembly requires a free carboxyl group at C-terminal position 344.

Copper(II) complexes are formed by dipeptides (Gly-Gly, Gly-Sar, Gly-Glu and Gly-Asp) on amorphous Al hydrous oxide surfaces during the simultaneous absorption of Cu(II) and dipeptides from aqueous solution.[265] E.s.r. studies indicate that these complexes are effectively anchored to the surface by carboxylate group binding to surface Al atoms.

References

1. B. Kurzak, H. Kozlowski and E. Farkas, *Coord. Chem. Rev.*, 1992, **114**, 169.
2. C.A. Bonamartini, *Coord. Chem. Rev.*, 1992, **117**, 45.

3. A. Garoufis, S. Kasselouri, M. Lamera-Hadjiliadis and N. Hadjiliadis, *Chem. Chron.*, 1992, **22**, 11.
4. J. Reedijk, *Inorg. Chim. Acta*, 1992, **198-200**, 873.
5. H. Kozlowski and L.D. Pettit, *Stud. Inorg. Chem.*, 1991, **11**, 530.
6. L.D. Pettit, J.E. Gregor, H. Kozlowski, 'Perspectives on Bioinorganic Chemistry', Vol. 1, ed. R.W. Hay, J.R. Dilworth and K.B. Nolan, JAI Press, Connecticut, 1991, pp. 1-14.
7. T.H. Fife, , in 'Perspectives on Bioinorganic Chemistry', Vol. 1, ed. R.W. Hay, J.R. Dilworth and K.B. Nolan, JAI Press, Connecticut, 1991, pp. 43-93.
8. S.S. Isied, *Top. Mol. Organ Eng.*, 1991, **7**, 63.
9. Metal Ions in Biological Systems, Vol. 27, ed. H. Sigel and A. Sigel, Marcel Dekker, New York, N.Y. 1991.
10. W.E. Rauser, *Methods Enzymol.*, 1991, **205**, 319-333.
11. Y.N. Belokon, *Pure Appl. Chem.*, 1992, **64**, 1917.
12. R.W. Hay and K.B. Nolan in Amino Acids and Peptides, Royal Society of Chemistry, Specialist Periodical Reports, 1991, **22**, Ch.6.
13. O.N. Stepanenko and L.G. Reiter, *Ukr. Khim. Sh. (Russ. Ed.)*, 1992, **58**, 1047.
14. 'Iron Carriers and Iron Proteins', Physical Bioinorganic Chemistry, Vol. 5, ed. T.M. Loehr, VCH, New York, N.Y., 1989.
15. Metal Ions in Biological Systems, Vol. 25, ed. H. Sigel and A. Sigel, Marcel Dekker, New York, 1989.
16. S. Mangani, P. Carloni and P. Orioli, *Coord. Chem. Rev.*, 1992, **120**, 309.
17. I. Bertini, L. Messori and M.S. Viezzoli, *Coord. Chem. Rev.*, 1992, **120**, 163.
18. H. DeWayne Ashmead, *ACS Symp. Ser.*, 1991, **445**, 306-319.
19. B. Kaitner, G. Ferguson, N. Paulic and N. Raos, *J. Coord. Chem.*, 1992, **26**, 95.
20. H. Yamaguchi, Y. Inomata and T. Takeuchi, *Inorg. Chim. Acta*, 1991, **181**, 31.
21. B. Kaitner, *J. Coord. Chem.*, 1992, **25**, 337.
22. B. Kaitner, N. Paulic and N. Raos, *J. Coord. Chem.*, 1991, **24**, 291.
23. B. Kaitner, G. Ferguson, N. Paulic and N. Raos, *J. Coord. Chem.*, 1992, **26**, 105.
24. B. Kaitner, N. Paulic and N. Raos, *J. Coord. Chem.*, 1991, **22**, 269.
25. J. Li, M. Xu. and C. Ge, *Cryst. Res. Technol.*, 1991, **26**, K82.
26. J. Li. and M. Xu, *Cryst. Res. Technol.*, 1991, **26**, K87.
27. G.B. Gavioli, M. Borsari, L. Menabue, M. Saladini and M. Sola, *J. Chem. Soc., Dalton Trans.*, 1991, **11**, 2961.
28. L. Forti, M. Saladini and M. Sola, *Inorg. Chim. Acta*, 1991, **187**, 197.
29. L. Forti, L. Menabue and M. Saladini, *J. Chem. Soc., Dalton Trans*, 1991, **11**, 2955.
30. I.I. Matthews and H. Manohar, *Polyhedron*, 1991, **10**, 2163.
31. I.I. Matthews, P.A. Joy, S. Vasudevan and H. Manohar, *Inorg. Chem.*, 1991, **30**, 2181.
32. O. Yamauchi, A. Odani and H. Masuda, *Inorg. Chim. Acta*, 1992, 198.
33. I.I. Mathews and H. Manohar, *J. Chem. Soc., Dalton Trans.*, 1991, 2289.
34. A.G. Sykes, R. D. Larsen, J.R. Fischer and E.H. Abbott, *Inorg. Chem.*, 1991, **30**, 2911.
35. K. Aoki, N. Hu. and H. Yamazaki, *Inorg. Chim. Acta*, 1991, **186**, 253.
36. R. Calvo, P.R. Levstein, E.E. Castellano, S.M. Fabiane, O.E. Piro and S.B. Oseroff, *Inorg. Chem.*, 1991, **30**, 216.
37. H.O. Davies, R.D. Gillard, M.B. Hursthouse and A. Lehmann, *J. Chem. Soc., Chem. Commun.*, 1993, 1137.
38. R. Das, K.K. Nanda, K. Venkatsubramanian, P. Paul and K. Nag, *J. Chem. Soc., Dalton Trans.*, 1992, **7**, 1253.

39. B. Kurzak, E. Farkas, T. Glowiak and H. Kozlowski, *J. Chem. Soc., Dalton Trans.*, 1991, 163.
40. P. Bonchev, K. Kadum, G. Gochev, B. Evtimova, J. Macicek and C. Nachev, *Polyhedron*, 1992, **11**, 1973.
41. L.P. Battaglia, A.B. Corradi, S. Ianelli, M.A. Zoroddu and G. Sanna, *J. Chem. Soc., Faraday Trans.*, 1991, **87**, 3863.
42. P.R. Levstein, H.M. Pastawski and R. Calvo, *J. Phys. Condens. Matter*, 1991, 3, 1877.
43. P. Comba, T.W. Hambley, G.A. Lawrance, L.L. Martin, P. Renold and K. Varnagy, *J. Chem. Soc., Dalton Trans.*, 1991, **2**, 277.
44. S-B. Teo, C-H. Ng, S-G. Teoh and H-K. Fun, *J. Coord. Chem.*, 1991, **24**, 151.
45. N. Choi, H.G. Ang, J.A. Crowther, G.R. Hansen, W.L. Kwik and M. McPartlin, *J. Inorg. Biochem.*, 1991, **43**, 202.
46. E. Abarca Garcia, N-H. Dung, B. Viossat, A. Busnot, J.M. Gonzalez Perez and J. N. Gutierrez, *J. Inorg. Biochem.*, 1991, **43**, 226.
47. P.J. Toscano, K.A. Belsky, T.C. Hsieh, T. Nicholson and J. Zubieta, *Polyhedron*, 1991, **10**, 977.
48. B. Ye, T. Zeng, H. Zhuang, and L. Ji, *J. Inorg. Biochem*, 1991, **43**, 467.
49. T.W. Hambley, G.A. Lawrance, M. Maeder and E.N. Wilkes, *J. Chem. Soc., Dalton Trans.*, 1992, **7**, 1283.
50. M.J. Malinar, M.B. Celap, R. Herak and B. Prelesnik, *Polyhedron*, 1992, **11**, 1169.
51. R.R. Fenton, F.S. Stephens, R.S. Vagg and P.A. Williams, *Inorg. Chim. Acta*, 1991, **182**, 59.
52. P.N. Radivojsa, N. Juranic, M.B. Celap, K. Toriumi and K. Saito, *Polyhedron*, 1991, **10**, 271.
53. M.M.T. Khan, D. Chatterjee, R.R. Merchant, P. Paul, S.H.R. Abdi, D. Srinivas, M.R.H. Siddiqui, M.A. Moiz, M.M. Bhadbhade and K. Venkatasubramanian, *Inorg. Chem.*, 1992, **31**, 2711.
54. R. Kraemer, K. Polborn, C. Robl and W. Beck, *Inorg. Chim. Acta*, 1992, 198.
55. O. Kumberger, J. Riede and H. Schmidbaur, *Chem. Ber.*, 1992, **125**, 1829.
56. H. Schmidbaur, P. Kiprof, O. Kumberger and J. Riede, *Chem. Ber.*, 1991, **124**, 1083.
57. A. Lombardi, O. Maglio, E. Benedetti, B. Di Blasio, M. Saviano, F. Nastri, C. Pedone and V. Pavone, *Inorg. Chim. Acta*, 1992, **196**, 241.
58. A. Iakovidis, N. Hadjiliadi, J.F. Britten, I.S. Butler, F. Schwarz, B. Lippert, *Inorg. Chim. Acta*, 1991, **184**, 209.
59. L.E. Erickson, G.S. Jones, J.L. Blanchard and K.J. Ahmed, *Inorg. Chem.*, 1991, **30**, 3147.
60. A.B. Corradi, L. Menabue, M. Saladini, M. Sola and L.P. Battaglia, *J. Chem. Soc., Dalton Trans.*, 1992, **17**, 2623.
61. X.M. Chen, T.C.W. Mak, W. Huang and L. Lu, *Acta Crystallogr., Sect. C, Cryst. Struct. Commun.*, 1992, **C48**, 57.
62. X.M. Chen and T.C.W. Mak, *J. Chem. Soc., Dalton Trans.*, 1991, **5**, 1219.
63. P.G. Jones and R. Schelbach, *Inorg. Chim. Acta*, 1991, **182**, 239.
64. I. Lukes, K. Bazakas, P. Hermann and P. Vojtisek, *J. Chem. Soc., Dalton Trans.*, 1992, **5**, 939.
65. U.A. Bagal, C.B. Cook and T.L. Riechel, *Inorg. Chim. Acta*, 1991, **180**, 57.
66. W. Weng, L. Tian, G. Nowels and N. Rowan Gordon, *Inorg. Chim. Acta*, 1991, **188**, 85.
67. J. P. De Jesus, T.M. Dos Santos and P. O'Brien, *Polyhedron*, 1991, 10, 575.
68. D. Sattari, E. Alipour, S. Shirani and J. Amighian, *J. Inorg. Biochem.*, 1992, **45**, 115.

69. S.S. Sandhu and N.S. Aulak, *J. Indian Chem. Soc.*, 1991, **68**, 29.
70. A. Katoh and M. Akiyama, *J. Chem. Soc., Perkin Trans.*, 1991, **8**, 1839.
71. M.K. Das, K. Chaudhury, N. Roy and P. Sarkar, *Trans. Met. Chem*, 1990, **15**, 468.
72. A.M. Abdel-Mawgoud, S.A. El-Gyar and M.M.A. Hamed, *Synth. React. Inorg. Met. -Org. Chem.*, 1991, **21**, 1061.
73. A.M.A. Hassaan, *J. Chem. Soc. Pak.*, 1992, **14**, 108.
74. A.M.A. Hassaan, *Trans. Met. Chem.*, 1990, **15**, 283.
75. H.A. Hammad and S.H. Salah, *Spectrosc. Lett.*, 1992, **25**, 1067.
76. M.S. Islam, R.K. Roy and M.A. Ali, *Bangladesh J. Sci. Ind. Res.*, 1992, **27**, 29.
77. N. Paulic, V. Nothig-Laslo and V. Simeon, *Z. Anorg. Allg. Chem.*, 1992, **613**, 132.
78. S.R. Wilson, A. Yasmin and Y. Wu, *J. Org. Chem.*, 1992, **57**, 6941.
79 M.S. Sastry, S.S. Gupta, V. Natarajan and A.J. Singh, *J. Inorg. Biochem.*, 1992, **45**, 159.
80. N. Arulsamy and P.S. Zacharias, *Trans. Met. Chem.*, 1991, **16**, 645.
81. N. Arulsamy, C. R.K. Rao and P. S. Zacharias, *Trans. Met. Chem.*, 1991, **16**, 606.
82. N. Arulsamy, B. Srinivas and P.S. Zacharias, *Trans. Met. Chem.*, 1990, **15**, 309.
83. J. L. Manzano, P. Marquez, E. Rodriguez and D. Sanchez, *J. Inorg. Biochem.*, 1991, **43**, 228.
84. A. Angoso, J.M. Martin-Llorente, J.L. Manzano, M. Martin, R. Martin, E. Rodriguez and J. Soria, *Inorg. Chim. Acta*, 1992, **195**, 45.
85. G. Wang and J.C. Chang, *Synth. React. Inorg. Met. -Org. Chem.*, 1991, **21**, 897.
86. G. Plesch, C. Friebel, O. Svajlenova and J. Kratsmar-Smogrovic, *Polyhedron*, 1991, **10**, 893.
87. N. Arulsamy, P.S. Zacharias, *Trans. Met. Chem.*, 1991, **16**, 255.
88. I. Samasundaram, M.K. Kommiya and M. Palaniandavar, *J. Chem. Soc., Dalton Trans.*, 1991, 2083.
89. J.R. Allan, G.H.W. Milburn, F. Richmond, D.L. Gerrard, J. Birnie and A.S. Wilson, *Thermochim. Acta*, 1991, **177**, 213.
90. J.R. Allan, G.H.W. Milburn, F. Richmond, D.L. Gerrard, J. Birnie and A.S. Wilson, *Eur. Polym. J.*, 1991, **27**, 419.
91. H.A. Hammad, A.M. Hassen, S.M. Yossif, M.M. Hussien and M.M. El-Okr, *Bol. Soc. Quim. Peru*, 1992, **58**, 7.
92. B. Swamy and J.R. Swamy, *Asian J. Chem.*, 1991, **3**, 152.
93. S.A. Ibrahim, *Phosphorus, Sulfur Silicon Relat. Elem.*, 1991, **60**, 139.
94. A. Garoufis, J. Hatiris and N. Hadjiliadis, *J. Inorg. Biochem.*, 1991, **41**, 195.
95. V. Aletras, N. Hadjiliadis and B. Lippert, *Polyhedron*, 1992, **11**, 1359.
96. B.T. Khan, K.M. Mohan and G. N. Goud, *Trans. Met. Chem.*, 1990, **15**, 407.
97. A.A. Isab and A.R.A. Al-Arfaj, *Trans. Met. Chem.*, 1991, **16**, 304.
98. N. Jain and T.S. Srivastava, *Indian J. Chem. Sect. A, 1992,* **31A**, 102.
99. H. Mansuri-Torshizi, T.S. Srivastava, H.K. Parekh and M.P. Chitnis, *J. Inorg. Biochem.*, 1992, **45**, 135.
100. R. Mital, T.S. Srivastava, H.K. Parekh and M.P. Chitnis, *J. Inorg. Biochem.*, 1991, **41**, 93.
101. A.R. Khokhar and G.J. Lumetta, *J. Coord. Chem.*, 1992, **26**, 251.
102. Z. Velkov, R. Pelova, S. Stoev and E. Golovinsky, *J. Coord. Chem.*, 1992, **26**, 75.
103. N. Baidya, D. Ndreu, M.M. Olmstead and P.K. Mascharak, *Inorg. Chem.*, 1991, **30**, 2448.
104. Y. Zhang, Q. Su, G. Zhao and J. Li, *Spectrosc. Lett.*, 1992, **25**, 521.
105. A.M.A. Hassaan, Egypt. *J. Pharm. Sci*, 1991, **32**, 871.
106. I. Zahn, K. Polborn, B. Wagner and W. Beck, *Chem. Ber.*, 1991, **124**, 1065.

107. G.A. Kolawole and A.O. Adeyemo, *Synth. React. Inorg. Met. - Org. Chem.*, 1992, **22**, 1073.
108. A.M. Abdel-Mawgoud, S.A. El-Gyar, S.A. Ibrahim and L.N. Abdel-Rahman, *Synth. React. Inorg. Met. - Org. Chem.*, 1992, **22**, 815.
109. V. Lakshmanan, M.R. Udupa and K.S. Nagaraja, *Polyhedron*, 1992, **11**, 1387.
110. J.J. Criado, J. A. Lopez-Arias, B. Macias, L.R. Fernandez-Lago and J.M. Salas, *Inorg. Chim. Acta*, 1992, **193**, 229.
111. W.S. Bigham and P.A. Shapley, *Inorg. Chem.*, 1991, **30**, 4093.
112. M.K. Hassan, *Trans. Met. Chem.*, 1991, **16**, 618.
113. M.M.T. Khan, R.I. Kureshy and N.H. Khan, *Tetrahedron: Asymmetry*, 1991, **2**, 1015.
114. T.M. Ibrahim, A.A. Shabana and H.A. Hammad, *Arch. Pharmacol. Res.*, 1992, **15**, 130.
115. M.A. Diaz Diez, F.J. Garcia Barros, E. Sabio Ray and C. Valenzuela Calahorro, *J. Inorg. Biochem.*, 1991, **43**, 499.
116. I. Michaud-Soret and J.C. Chottard, *Biochem. Biophys. Res. Commun.*, 1992, **182**, 779.
117. R. Roy, M.C. Saha and P.S. Roy, *Trans. Met. Chem.*, 1990, **15**, 51.
118. A. Massaferro, M. Queirolo and B. Sienra, *An. Quim.*, 1992, **88**, 230.
119. M. Maeda, K. Okada and K. Ito, *J. Inorg. Biochem.*, 1991, **41**, 143.
120. K. Ud-Din and G.J. Khan, *J. Coord. Chem.*, 1992, **26**, 351.
121. I.A. Khan, M. Shad and K. Ud-Din, *Trans. Met. Chem.*, 1991, **16**, 18.
122. K. Ud-Din and G.J. Khan, *Trans. Met. Chem.*, 1990, **15**, 39.
123. S. Deiana, C. Gessa, P. Piu and R. Seeber, *J. Chem. Soc., Dalton Trans.*, 1991, **5**, 1237.
124. W. Linert, R.F. Jameson and E. Herlinger, *Inorg. Chim. Acta*, 1991, **187**, 239.
125. M.M. El-Dessouky, *Isotopenpraxis*, 1991, **27**, 399.
126. G. Battistuzzi and M. Sola, *Biochim. Biophys. Acta*, 1992, **1118**, 313.
127. D.J. Evans and G.J. Leigh, *J. Inorg. Biochem.*, 1991, **42**, 25.
128. M. Mizuno, S. Funahashi, N. Nakasuka and M. Tanaka, *Bull Chem. Soc. Jpn.*, 1991, **64**, 1988.
129. E. Mikros, F. Gaudemer and A. Gaudemer, *Inorg. Chem.*, 1991, **30**, 1806.
130. R.M. Hartshorn and A.M. Sargeson, *Aust. J. Chem.* 1992, **45**, 5.
131. L. Grondahl, A. Hammershoi, R.M. Hartshorn and A.M. Sargeson, *Inorg. Chem.*, 1991, **30**, 1800.
132. T. Kojima, J. Tsuchiya, S. Nakashima, H. Ohya-Nishiguchi, S. Yano and M. Hidai, *Inorg. Chem.*, 1992, **31**, 2333.
133. P.S. Swaroop, K.A. Kumar, P.V.K. Rao, *Trans. Met. Chem.* (London), 1991, **16**, 416.
134. M. Strasak and A. Kubeczkova, *J. Mol. Catal.*, 1992, **71**, 383.
135. M. Strasak and J. Kavalek, *J. Mol. Catal.*, 1991, **66**, 239.
136. R. Bhattacharya and B. Chakravarty, *Indian J. Chem., Sect. A*, 1992, **31A**, 187.
137. V.M. Shanbhag and A.E. Martell, *Proc. Int. Symp. Vitam. B_6 Carbonyl Catal., 8th*, 1990, 365.
138. S. Goldstein, G. Czapski, H. Cohen and D. Meyerstein, *Inorg. Chem.*, 1992, **31**, 2439.
139. H. Zhang and P. Li, *Trans. Met. Chem.* (London), 1991, **16**, 352.
140. H.C. Malhotra, Y. Sharma and J.A. Yogeshwar, *Proc. Indian Natl. Sci. Acad., Part A*, 1992, **58**, 133.
141. N. Arulsamy and P.S. Zacharias and K. Rajeshwar, *Polyhedron*, 1992, **11**, 1837.

142. N. Arulsamy and P.S. Zacharias, Proc. - *Indian Acad. Sci., Chem. Sci.*, 1992, **104**, 361.
143. M.G. Abd El Wahed, A.M. Hassen, H.A. Hammad and M.M. El-Desoky, *Bull. Korean Chem. Soc.*, 1992, **13** 113.
144. L.A. Scheich, P. gosling, S.J. Brown, M.M. Olmstead and P.K. Mascharak, *Inorg. Chem.*, 1991, **30**, 1677.
145. K. Hayase and R.G. Zepp, *Environ. Sci. Technol.*, 1991, **25**, 1273.
146. X. Wang, *Analyst (London)*, 1992, **117**, 165.
147. F. Baffi, M.C. Ianni, A.M. Cardinale, E. Magi, R. Frache and M. Ravera, *Anal. Chim. Acta*, 1992, **260**, 99.
148. E. Tutem and R. Apak, *Anal. Chem. Acta*, 1991, **255**, 121.
149. Y. Sasaki and M. Miyashita, *Inorg. Chim. Acta*, 1991, **183**, 15 .
150. D. Mallick and G.S. De, *Trans. Met. Chem.*, 1992, **17**, 34.
151. S.N. Brune and D.R. Bobbit, *Talanta*, 1991, **38**, 419.
152. M. Shoukry, H. Hohmann and R. van Eldik, *Inorg. Chim. Acta*, 1992, **198-200**, 187.
153. R.K. Bapna, K.D. Gupta and K.K. Saxena, *Asian J. Chem.*, 1992, **4**, 239.
154. E. Matczak-Jon, W. Wojciechowski, *Pol. J. Chem.*, 1992, **66**, 617.
155. T. Kurotu, *Inorg. Chim. Acta*, 1992, **191**, 141.
156. R.P. Bonomo, V. Cucinotta, F. D'Alessandro, G. Impellizzeri, G. Maccarrone, G. Vecchio and E. Rizzarelli, *Inorg. Chem.*, 1991, **30**, 2708.
157. R. Corradini, G. Impellizzeri, G. Maccarrone, R. Marchelli, E. Rizzarelli and G. Vecchio, *J. Inorg. Biochem.*, 1991, **43**, 227.
158. R. Corradini, G. Impellizzeri, G. Maccarrone, R. Marchelli, E. Rizzarelli and G. Vecchio, *Top. Mol. Organ. Eng.*, 1991, **8**, 209.
159. T. Kiss, M. Jezowska-Bojczuk, H. Kozlowski, P. Kafarski and K. Antczak, *J. Chem. Soc., Dalton Trans.*, 1991, **9**, 2275.
160. E. Leporati and G. Nardi, *Bull Chem. Soc., Jpn*, 1991, **64**, 2488.
161. E. Farkas and A. Kovacs, *J. Coord. Chem.*, 1991, **24**, 325.
162. B. Kurzak, H. Kozlowski and P. Decock, *J. Inorg. Biochem.*, 1991, **41**, 71.
163. B. Kurzak, J. Spychala and J. Swiatek, *J. Coord. Chem.*, 1992, **25**, 95.
164. J. Urbanska, H. Kozlowski and B. Kurzak, *J. Coord. Chem.*, 1992, **25**, 149.
165. N. Nic Daeid, K.B. Nolan and L.P. Ryan, *J. Chem. Soc., Dalton Trans.*, 1991, 2301.
166. S. Kawata, H. Kosugi, H. Uda and M. Iwaizumi, *Bull.Chem. Soc. Jpn.*, 1992, **65**, 2910.
167. G. Micera, D. Sanna, R. Dallocchio and A. Dessi, *J. Coord. Chem.*, 1992, **25**, 265.
168. G.G. Battistuzzi, M. Bosari, L. Menabue, M. Saladini and M. Sola, *Inorg. Chem.*, 1991, **30**, 498.
169. B. Macias, J.J. Criado, M.V. Vaquero, M.V. Villa and M. Castillo, *J. Inorg. Biochem.*, 1991, **42**, 17.
170. M. Maeda, M. Tsunoda and Y. Kinjo, *J. Inorg. Biochem.*, 1992, **48**, 227.
171. A. Domenech, E. Llopis, E. Garcia-Espana and A. Cervilla, *Trans. Met. Chem.*, 1990, **15**, 525.
172. F. U. Khan, G.M. Khan and Z. Iqbal, *Proc. Pak. Acad. Sci.*, 1991, **28**, 401.
173. P. Djurdjevic, *Trans. Met. Chem.*, 1990, **15**, 345.
174. P. Rajathirumoni, P.T. Arasu, M.S. Nair, *Indian J. Chem., Sect. A*, 1992, **31A**, 760.
175. A. Maleque and A.K. Chaudhury, *Indian J. Chem., Sect. A*, 1992, **31A**, 764.
176. E. Bottari, G. Carletti and M.R. Festa, *Ann. Chim. (Rome)*, 1992, **82**, 357.
177. A.K. Roul, R.K. Patnaik, *J. Inst. Chem. (India)*, 1991, **63**, 108.
178. V. Manjula and P.K. Bhattacharya, *J. Inorg. Biochem.*, 1991, **41**, 63.
179. A.A. Abd El-Gaber, M.B. Saleh and I.T. Ahmed, *J. Indian Chem. Soc.*, 1992, **69**, 17.

180. M.A.V. Ribeiro da Silva, M.D.M.C. Ribeiro da Silva, M.M.C.Bernardo and L.M.N.B.F. Santos, *Thermochim. Acta*, 1992, **205**, 99.
181. M.A.V. Ribeiro da Silva, M.D.M.C. Ribeiro da Silva, J.A.B.A. Tuna, L.M.N.B.F. Santos, *Thermochim. Acta*, 1992, **205**, 115.
182. Z. Khatoon and K. Ud-Din, *Trans. Met. Chem.*, 1990, **15**, 217.
183. P. Venkataiah, M.S. Mohan and Y.L. Kumari, Proc. - *Indian Acad. Sci., Chem. Sci.*, 1992, **104**, 453.
184. G.N. Mukherjee and T.K. Ghosh, *Indian J. Chem., Sect. A*, 1992, **31A**, 478.
185. G.N. Mukherjee and T.K. Ghosh, *J. Indian Chem. Soc.*, 1991, **68**, 194.
186. A. Lu, L.D. Pettit and J.E. Gregor, *Gaodeng Xuexiao Huaxue Xuebao*, 1992, **13**, 322.
187. H.M. Killa, E.S. Mabrouk and M.M. Ghoneim, *Trans. Met. Chem.* (London), 1992, **17**, 59.
188. M. Morcellet, *Thermochim. Acta*, 1992, **195**, 335.
189. P. Jorda, R. Ortiz and J. Borras, *J. Inorg. Biochem.*, 1991, **41**, 149.
190. M.A. Abdel-Mawgoud, S.A. El-Gyar and L.H. Abdel-Rahman, *Afinidad*, 1992, **49**, 241.
191. C. Valenzuele-Calahorro, M.A. Diaz-Diez, E. Sabio-rey, F.J. Garcia-Barros and E. Roman-Galan, *Polyhedron*, 1992, **11**, 563.
192. H.K. Killa, M.E. Mabrouk and M.M. Moustafa, *Croat. Chem. Acta*, 1992, **64**, 585.
193. A.K. Rao, G.N. Kumar, M.S. Mohan and Y.L. Kumari, *Indian J. Chem., Sect. A*, 1992, **31A**, 256.
194. R.K. Bapna and K.K. Saxena, *Acta Cienc. Indica, Chem.*, 1991, **17C**, 149.
195. S. Raj, C.S. Devi, G. Seshikala, M.G. Reddy, Proc.- *Indian Acad. Sci., Chem. Sci.*, 1992, **104**, 15.
196. R.S. Sindhu, S. Tikku, S.K. Bansal, *J. Indian Chem. Soc.*, 1991, **68**, 289.
197. S.Das and M.N. Srivastava, *Natl. Acad. Sci. Lett. (India)*, 1991, **14**, 133.
198. F. Khan and K. Nema, *J. Indian Chem. Soc.*, 1991, **68**, 290.
199. P.R. Reddy and M.R.P. Reddy, *Indian J. Chem., Sect. A*, 1991, **30A**, 1028.
200. V.P. Vasil'ev, A.V. Katrovtseva, R. Perez Matos, *Zh. Neorg. Khim.*, 1991, **36**, 2306.
201. T. Mitsudo, S. Zhang, N. Satake, T. Kondo and Y. Watanabe, *Tetrahedron Lett.*, 1992, **33**, 5533.
202. J.J. Bozell, C.E. Vogt and J. Gozum, J. *Org. Chem.*, 1991, **56**, 2584.
203. H.W. Krause, F.W. Wilcke, H.J. Kreuzfeld and C. Doebler, *Chirality*, 1992, **4**, 110.
204. H.W. Krause, H.J. Kreuzfeld and C. Doebler, *Tetrahedron: Asymmetry*, 1992, **3**, 555.
205. W.D. Lubell, M. Kitamura and R. Noyori, *Tetrahedron: Asymmetry*, 1991, **2**, 543.
206. S. Gladiali and L. Pinna, *Tetrahedron: Asymmetry*, 1991, **2**, 623.
207. U. Schmidt, A. Lieberknecht, U. Kazmaier, H. Griesser, G. June and J. Metzger, *Synthesis*, 1991, **1**, 49.
208. Y.N. Belokon, S. Mociskite, V.I. Maleev, S. Orlova, N.S. Ikonnikov, E.B. Shamuratov, A.S. Batsanov, Y.T. Struchkov, *Mendeleev Commun.*, 1992, **3**, 89.
209. V.A. Soloshonok, Y.N. Belokon, N.A. Kuzmina, V.I. Maleev, N.Y. Svistunova, V. A. Solodenko and V.P. Kukhar, *J. Chem. Soc., Perkin Trans.*, 1992, **12**, 1525.
210. J. Barker, S.L. Cook, M.E. Lasterra-Sanchez and S.E. Thomas, *J. Chem. Soc., Chem. Commun.*, 1991, **11**, 830.
211. J.M. Brown and P.J. Maddox, *Chirality*, 1991, **3**, 345.
212. H. Krause and C. Sailer, *J. Organomet. Chem.*, 1992, **423**, 271.
213. T. Handel and W.F.J. DeGrado, *J. Am. Chem. Soc.*, 1990, **112**, 6710.
214. L. Regan and N.D. Clarke, *Biochemistry*, 1990, **29**, 10878.
215. D.L. Merkle, M.H. Schmidt and J.M. Berg, *J. Am. Chem. Soc.*, 1991, **113**, 5450.

216. M. Reza Ghadiri, C. Soares and C. Choi, *J. Am. Chem. Soc.*, 1992, **114**, 825.
217. S. Ranganathan and N. Javaraman, *Tetrahedron Lett.*, 1992, **33**, 6681.
218. M. Mallouris, *Diss. Abstr. Int. B.*, 1992, **52**, 6397.
219. R.G. Bhirud and T.S. Srivastava, *Inorg. Chim. Acta*, 1991, **179**, 125.
220. R.M. Mobashar, A. Taylor and L.G. Marzilli, *Inorg. Chim. Acta*, 1991, **186**, 139.
221. W.S. Sheldrick and R. Exner, *Inorg. Chim. Acta*, 1991, **184**, 119.
222. T. Ama, R. Maki, H. Kawaguchi and T. Yasui, *Bull. Chem. Soc. Jpn.*, 1991, **64**, 459.
223. M. Akiyama, A. Katoh, M. Iijima, T. Takagi, K. Natori and T. Kojima, *Bull. Chem. Soc. Jpn.*, 1992, **65**, 1356.
224. P.D. Newman, P.A. Williams, F.S. Stephens and R.S. Vagg, *Inorg. Chim. Acta*, 1991, **183**, 145.
225. P. Hu. and M.L. Gross, *J. Am. Chem. Soc.*, 1992, **114**, 9161.
226. L.M. Teesch, R.C. Orlando and J. Adams, *J. Am. Chem. Soc.*, 1991, **113**, 3668.
227. M. Lieberman and T. Sasaki, *J. Am. Chem. Soc.*, 1991, **113**, 1470.
228. T.N. Rao, D. Adhikesavalu, A. Camerman and A.R. Fritzbert, *Inorg. Chim. Acta*, 1991, **180**, 63.
229. W-Y. Sun, N. Ueyama and A. Nakamura, *Inorg. Chem.*, 1991, **30**, 4026.
230. N. Ueyama, W-Y. Sun and A. Nakamura, *Inorg. Chem.*, 1992, **31**, 4053.
231. N. Ueyama, S. Ueno, A. Nakamura, K. Liada, H. Matsubara, S. Kumagi, S. Sakakibara and T.Tsukihara, *Biopolymers*, 1992, **32**, 1535.
232. B. Ladesic, D. Kantoci, H. Meider and O. Hadzija, *J. Inorg. Biochem.*, 1992, **48**, 56.
233. E.M. Zarate, J. Gomez-Lara, R.A. Toscano, G. Negron and A. Campero, *J. Crystallogr. Spectrosc. Res.*, 1992, **22**, 281.
234. K. Kahne and W.C. Still, *J. Am. Chem. Soc.*, 1988, **110** 7529.
235. I.E. Burgeson and N.M. Kostic, *Inorg. Chem.*, 1991, **30**, 4299.
236. L. Zhu and N.M. Kostic, *Inorg. Chem.*, 1992, **31**, 3994.
237. W-Y. Sun, N. Ueyama and A. Nakamura, *Tetrahedron*, 1992, **48**, 1557.
238. T. Sugawara, N. Ueyama and A. Nakamura, *J. Chem. Soc., Dalton Trans.*, 1991, 249.
239. S.E. Schadler, C. Sharp and A.G. Lappin, *Inorg. Chem.*, 1992, **31**, 51.
240. K. Takehara and Y. Ide, *Inorg. Chim. Acta*, 1991, **186**, 73.
241. K. Takehara and Y. Ide, *Inorg. Chim. Acta*, 1991, **183**, 195.
242. B. Mrabet, M. Jouini, J. Huet and G. Lapluye, *J. Chim. Phys. Phys.-Chim. Biol.*, 1992, **89**, 2187.
243. L.D. Pettit, W. Bal, M. Bataille, C. Cardin, H. Kozlowski, M. Leseine-Delstanche, S. Pyburn and A. Scozzafara, *J. Chem. Soc., Dalton Trans.*, 1991, 1651.
244. J-G. Galey, B. Decock-Le Reverene, A. Lebkiri, L.D. Pettit, S.I. Pyburn and H. Kozlowski, *J. Chem. Soc., Dalton Trans.*, 1991, 2281.
245. P.G. Daniele, O. Zerbinati, V. Zelano and G. Ostacoli, *J. Chem. Soc., Dalton Trans.*, 1991, 2711.
246. G. Arena, R.P. Bonomo, L. Casella, M. Gullotti, G. Impellizzeri, G. Maccarrone and E. Rizzarelli, *J. Chem. Soc., Dalton Trans.*, 1991, 3203.
247. M.M. Shoukry, *Transition Met. Chem.*, 1990, **15**, 1.
248. S.L. Brauer and K.E. Wetterhahn, *J. Am. Chem. Soc.*, 1991, **113**, 3001.
249. J.M. Berg, *Ann. Rev. Biophys. Biophys. Chem.*, 1990, **19**, 405 and references therein.
250. B.A. Krizek, B.T. Amann, V.J. Kilfoil, D.L. Merkle and J.M. Berg, *J. Am. Chem. Soc.*, 1991, **113**, 4518.
251. E. Gaggelli, A. Macotta and G. Valensin, *J. Inorg. Biochem.*, 1992, **38**, 173.
252. D. Chakraborty and P.K. Bhattacharya, *J. Inorg. Biochem.*, 1991, **42**, 57.
253. A. Kufelnicki, *Chem. Anal. (Warsaw)*, 1992, **37**, 177.

254. A. Kufelnicki, *Pol. J. Chem.*, 1991, **65**, 1879.
255. A. Kufelnicki, *Pol. J. Chem.*, **65**, 1887.
256. A. Kufelnicki, *Pol. J. Chem.*, 1991, **65**, 17.
257. A. Kufelnicki, *Pol. J. Chem.*, **65**, 269.
258. S. Canepari, V. Carunchio and A. Messina, *Transition Met. Chem.*, 1991, **16**, 296.
259. A.G. Palmer and D.A. Case, *J. Am. Chem. Soc.*, 1992, **114**, 9059.
260. S. Mazumbar, O.K. Medhi and S. Mitra, *Inorg. Chem.*, 1991, **30**, 700.
261. B.M. Britt and W.L. Peticolas, *J. Am. Chem. Soc.*, 1992, **114**, 5295.
262. R. Basosi, R. Pogni and G. Della Lunga, *Bull. Magn. Reson.*, 1992, **14**, 224.
263. S.G. Weber, *Anal. Chem.*, 1992, **64**, 2897.
264. P.J. Nixon, J.T. Trost and B.A. Diner, *Biochemistry*, 1992, **31**, 10859.
265. G. Micera, L.S. Ewe and R. Dallochio, *J. Chem. Res., Synop.*, 1992, 234.